AF182238

WALTER GREINER

Memorial Volume

WALTER GREINER
Memorial Volume

Edited by

Peter O Hess

Universidad Nacional Autónoma de México, Mexico
& Goethe-Universität, Frankfurt, Germany

Horst Stöcker

Goethe-Universität, Frankfurt, Germany

World Scientific

NEW JERSEY • LONDON • SINGAPORE • BEIJING • SHANGHAI • HONG KONG • TAIPEI • CHENNAI • TOKYO

Published by

World Scientific Publishing Co. Pte. Ltd.
5 Toh Tuck Link, Singapore 596224
USA office: 27 Warren Street, Suite 401-402, Hackensack, NJ 07601
UK office: 57 Shelton Street, Covent Garden, London WC2H 9HE

Library of Congress Cataloging-in-Publication Data
Names: Hess, Peter O., 1953– editor. | Stöcker, Horst, editor. | Greiner, Walter, 1935–2016, honouree.
Title: Walter Greiner memorial volume / edited by Peter O. Hess (Universidad Nacional Autónoma de México, Mexico & Goethe-Universität, Frankfurt, Germany),
Horst Stöcker (Goethe-Universität, Frankfurt, Germany)
Description: Singapore ; Hackensack, NJ : World Scientific, [2018] | Includes bibliographical references.
Identifiers: LCCN 2018001202| ISBN 9789813234277 (hardcover) | ISBN 981323427X (hardcover)
Subjects: LCSH: Nuclear physics. | Particles (Nuclear physics) | Quantum electrodynamics. |
Quantum theory. | Greiner, Walter, 1935–2016.
Classification: LCC QC782 .W35 2018 | DDC 539.7092 [B] --dc23
LC record available at https://lccn.loc.gov/2018001202

British Library Cataloguing-in-Publication Data
A catalogue record for this book is available from the British Library.

Copyright © 2018 by World Scientific Publishing Co. Pte. Ltd.

All rights reserved. This book, or parts thereof, may not be reproduced in any form or by any means, electronic or mechanical, including photocopying, recording or any information storage and retrieval system now known or to be invented, without written permission from the publisher.

For photocopying of material in this volume, please pay a copying fee through the Copyright Clearance Center, Inc., 222 Rosewood Drive, Danvers, MA 01923, USA. In this case permission to photocopy is not required from the publisher.

For any available supplementary material, please visit
http://www.worldscientific.com/worldscibooks/10.1142/10828#t=suppl

Printed in Singapore

Preface

Professor Walter Greiner was born on October 29, 1935, in Judenbach-Neuenbau, Sonneberg, Thuringen, Germany. He was just a boy when his family flew to West Germany after World War II to escape the communist regime in East Germany. Few people knew that his first professional activity was a locksmith apprentice at Farbwerke Hoechst, a large pharmaceutical and chemical plant in Frankfurt.

Walter studied physics at the Goethe-Universität, Frankfurt, where he obtained his Vordiplom, and received his Physics Diploma in 1960 at the Technische Hochschule Darmstadt, under the guidance of Prof. Otto Scherzer. He received his doctorate in 1961, after working for just a year at the University of Freiburg with his teacher, Prof. Hans Marschal, on the theory of nuclear polarization in muonic atoms.

Walter married his wife, Barbara, and moved with her to the University of Maryland as a young assistant professor during years 1962–1964. He returned to the University of Freiburg as a research associate and swiftly received several offers of professorship. Professor Greiner and his family moved to Frankfurt in 1965, where he was appointed Ordinarius and Director of the Institute for Theoretical Physics at Johann Wolfgang Goethe-Universität. He founded the Frankfurt School of Theoretical Physics and his first priority was to educate and nurture the best students, guiding them to pioneer exciting forefront scientific topics, as witnessed by one hundred and fifty students and coworkers, of whom fifty became professors at prestigious universities and research labs worldwide. Many of his students went on to successful careers in industry. Professor Greiner's renowned series of textbooks on theoretical physics, specially tailored to help the students grasp the physics ideas with mathematics and intuition, have appeared in many different languages and nurtured students all over the world.

Professor Greiner's research covered numerous exciting topics, to mention just a few: atomic physics, quantum electrodynamics of strong fields, nuclear physics — from nuclear structure physics (the geometric model of the nucleus and nuclear molecules), super-asymmetric fission to super-heavy elements, from relativistic hydrodynamics and shock waves to the equation of state and phase structure of QCD and to strange matter and antimatter, elementary particle physics and general relativity, from neutron stars to black holes — his feeling for and impact on fascinating physics issues is legendary, and so is his international network of scientific collaborators and friends, which includes more than one hundred prestigious international fellowships and awards of the Alexander von Humboldt Foundation, as well as hundreds of international postdocs and visitors. These collaborations lead to numerous (more than 1000) outstanding scientific papers and discoveries, and Walter's Hirsch Index is greater than 100.

Professor Greiner was co-founder of GSI, the world renowned national laboratory for heavy ion physics in Darmstadt, Germany. He founded in 2004 as Director the international interdisciplinary Frankfurt Institute for Advanced Studies (FIAS). Among the many honors Prof. Greiner received are the Doctor Honoris Causae of the Universities of Tel Aviv, Witwatersrand, Strasbourg, Bucharest, Nantes, Dubna, and others, the Otto Hahn Prize of the City of Frankfurt and the Max Born Prize of, jointly, the Institute of Physics, London and the German Physical Society. He had been appointed honorary and visiting professor by numerous prestigious institutes, universities and national laboratories.

Professor Greiner passed away on October 6, 2016, in his home at Kelkheim am Taunus. His wife, Barbara, and their two sons, Martin and Carsten, their families and five grandchildren miss him dearly. Walter is remembered by the science community as a great teacher, mentor and friend, an excellent scientist and a true scientific leader.

This memorial volume is a special tribute to Prof. Greiner honoring his contributions and passion in science. The volume begins with an interesting writing by Walter about his early days in science, in which he described his childhood activities, his studies in science and professional activities with his collaborators. The subsequent articles in this volume, comprising personal and scientific reminiscences of Walter Greiner by his former colleagues, students, and friends, which also serve as timely reviews on various topics of current interest.

Peter O. Hess and Horst Stöcker
Editors

Walter Greiner with Edward Teller.

Horst Stöcker presented to Walter Greiner the "academic genealogical tree" of all Greiner's students of the first generation and some of the second generation.

Walter Greiner with participants of the conference at Makutsi, South Africa, November 18, 2011.

Walter Greiner with M. Itkis (right) and a friend during a coffee break at the conference "Nuclear Physics: Present and Future", Boppard, Germany, 2013.

Contents

Preface v

1. Reflections on My Youth and Early Years in Science 1
Walter Greiner

2. The Early Work of Walter Greiner (1960–1968) 9
Şerban Mişicu

3. Photon Scattering off Nuclei 23
Hartmuth Arenhövel

4. The QCD Phase Diagram from Statistical Model Analysis 41
Marcus Bleicher, Jan Steinheimer and Reinhard Stock

5. Why May Hydrodynamics Work for Classical Radiation Field? 65
Tamás S. Biró

6. Chiral Symmetry Restoration and Deconfinement in Heavy-Ion Collisions 71
E. L. Bratkovskaya, W. Cassing, P. Moreau and A. Palmese

7. The Physics Case for the $\sqrt{s_{NN}} \approx 10$ GeV Energy Region 91
J. Cleymans

8. Untangling Simple Patterns in Intricate Atomic Nuclei 97
Jerry P. Draayer, Kristina D. Launey, Tomáš Dytrych, Alison C. Dreyfuss, Grigor H. Sargsyan and Robert B. Baker

9. Can One Determine the Neutrino Mass by Electron Capture? 123

Amand Faessler

10. Open and Hidden Charm in My Collaboration with Walter Greiner 139

Mark I. Gorenstein

11. Dark Matter Compact Stars in Pseudo-Complex General Relativity 149

D. Hadjimichef, G. L. Volkmer, R. O. Gomes and C. A. Zen Vasconcellos

12. Some Aspects of Nuclear Structure 169

J. H. Hamilton, A. V. Ramayya and E. H. Wang

13. The Power of the Geometrical Model of the Nucleus 183

Peter O. Hess

14. Pseudo-complex General Relativity: Theory 199

Peter O. Hess and Thomas Boller

15. Observational Tests of the Pseudo-complex Theory with Black Hole Imaging 209

Thomas Boller and Peter O. Hess

16. From Strangeness Enhancement to Quark–Gluon Plasma Discovery 221

Peter Koch, Berndt Müller and Johann Rafelski

17. Time-Dependent Perturbation Theory as a Basis for Combined Many-Body-Perturbation and QED 249

Ingvar Lindgren

18. The Fullerene-like Structure of Superheavy Element $Z = 120$ (Greinerium) — A Tribute to Walter Greiner 263

Ş. Mişicu and I. N. Mishustin

19. Cluster Radioactivity — Past and Future 275

D. N. Poenaru and R. A. Gherghescu

20. Nuclear Mean-Field Models and Super-Heavy Elements 283

P.-G. Reinhard

21. The Octupole Collective Hamiltonian.
Does It Follow the Example of the Quadrupole Case? 309

Stanisław G. Rohoziński and Leszek Próchniak

22. Modeling Hybrid Stars 327

S. Schramm

23. Elliptic Flow and the Nuclear Equation of State 339

W. Trautmann and H. H. Wolter

24. Black Holes and High Energy Physics:
from Astrophysics to Large Extra Dimensions 359

Michael Florian Wondrak, Marcus Bleicher and Piero Nicolini

Reflections on My Youth and Early Years in Science*

Walter Greiner

Frankfurt Institute for Advanced Studies, Frankfurt am Main, Germany

Received 6 April 2006

At the end of this remarkably interesting, scientifically broad and absolutely topical conference I would like to thank you all: the colleagues and friends who came from near and far away, who gave exciting presentations. Nuclear and Heavy Ion Science is explosively developing like a big thunderstorm. When I say this I am thinking about the *Superheavy Element Research*, the determination of the *Nuclear Equation of State*, the *Mach-Cone Shock Wave* type of physics repeating itself in the plasma phase during high energy nucleus-nucleus encounters, and the implications of the *Overcritical Vacuum* in various areas of physics. I would like, also, to thank the many people who helped in setting the organisation in action and keeping it well going. Here I mention in particular Dr. Joachim Reinhardt — as long as I know him he does everything he is responsible for with care, reliability and style. I also mention Mrs. Gabriela Meyer, Professor Stöcker's secretary, who was of greatest assistance during the course of the conference, and my secretary Mrs. Laura Quist, who helped wherever she could. I am exceptionally grateful, of course, to my colleague Professor Horst Stöcker. ISHIP was his idea; he initiated it and everything that goes with it. With his enthusiasm, wisdom and charm he overcame all stumbling blocks and difficulties. The conference dinner in the historical Römer–Hallen — even though quite noisy at some time — was a wonderful evening, bringing many friends and former students together, thus creating great joy. This again was Horst's idea.

Horst is one of my former students I am particularly proud of. Sometimes it seems as if we have similar genes. His interest in physics, striving for the understanding of topical problems, always alert and closely working with experimentalists, is admirable. I like this type of theoreticians. At the same time Horst cares about the problems in our FIAS institute, in our faculty, in our university. We are not always

*The Final Remarks presented by Walter Greiner at the International Symposium on Heavy Ion Physics, Frankfurt, Germany, 3–6 April 2006, reprinted from *Int. J. Mod. Phys. E*, Volume 16, 1285–1291 (2007).

of the same opinion, sometimes we collide violently, but an hour later our friendship is restored. He knows that I can trust him and that he can trust me. Thank you, Horst, for arranging this high-level conference; thank you for your friendship!

I am a very fortunate man: I grew up in a small village close to the Rennsteig in the Thüringer Wald, close to nature. Catching birds with a glue-rod I learned from my grand father; he also showed me how to catch foxes with traps. As a consequence of the bird catching I can even nowadays identify the birds by their songs, whether a Stieglitz, a Grasmücke, a Hänfling, a Zeisig, a Zaunkönig, a Schnäpper, a Nachtigal, an Amsel, a Drossel, a Fichtenkreuzschnabel, a ..., a ... I came out of a workers family. For generations my ancestors were glass-blowers, more recently also tailors and shoe makers. In fact, my ancestors brought the glass-blowing to Thuringia during the 14th century, coming originally from the Kraina (Croatia) via Bohemia (Grein at the Danube river) and the Schurwald, a part of the Schwäbische Alb (Baierseck near Stuttgart).

Our small farm bordered a river, the border between Thuringia and Bavaria, which in 1946 became the iron curtain. As a boy I fished in that river with bare hands for trouts. Thus I am one of the few people who can say that they fished in the iron curtain. With my boy friends I built water mills and was wondering how out of disordered (equilibrum) water of a pond came ordered flow which converted energy into work. Even more I was struck by the steam engine my father bought and presented to me at one of the early Christmas holidays I can remember: When I figured out how it functioned, I was fascinated. I think this was an important experience capturing me for and guiding me into physics. Observing the — occasionally very severe — thunderstorms in the Thüringer Wald was another intriguing wonder. I dreamed of building sometimes a power station on top of one of the mountains converting the electrical energy in lightenings into "ordered" electrical energy in power lines.

Our family fled the communists in East Germany (Thuringia) to the west in 1947/1948. School was not easy for me. The change from a 7th grade elementary school to a Gymnasium Quarta caused problems: while I was top in mathematics, physics, chemistry, biology and geography I had difficulties with languages. I left school and learned practical engineering, became a "Schlosser". During the first two years of training for practical engineering in the Hoechst Chemical Plant (Farbwerke Hoechst) I went daily to evening classes in Frankfurt (Abendschule Adam). It was an exhausting, but fascinating time: Leaving home in the morning at 6:30 hours for the Hoechst company, where I spent the day until 17:00 hours, left for downtown Frankfurt evening school, spent time there from 18:00 to 21:00 hours and came home at 22:00 hours. After 2 years I took three months off the engineering training at Hoechst and prepared myself for the Externes Abitur at Wiesbaden. Luckily, I was one of the very few who passed, after having been examined in all 10 disciplines: Mathematics, Physics, Chemistry, Biology, Geography, German, English and French languages, History and Social/Political Structures (Sozialkunde). While I passed the first five disciplines with highest marks, I degraded step by step into my final examination in French which showed that I could not understand what the French teacher wanted from me. I simply laughed, because I knew that I passed and nothing could change this anymore.

Right away I began studying Mathematics and Physics at the JohannWolfgang-Goethe-University in Frankfurt am Main. My teachers Hermann Dänzer (Physics), Wolfgang Franz and Ruth Moufang (both mathematicians) recommended me for the Studienstiftung des Deutschen Volkes (German National Merit Foundation), which helped me financially. During the break after the first semester I went back to the Hoechst company and completed my "Facharbeiterprüfung" as a "Schlosser". I remember that the piece I had to construct was quite complicated. After completing it, I checked finally all the measures once more, cleaned my tools, put them into the tool-box, closed it, gave the key and the "Gesellenstück" to Meister Dietrich and said "no more". This I kept all my life! Nevertheless I was highly graded and could choose a present. I wished for the book by Groebner on Algebraic Geometry. I still remember when Prof. Karl Winnacker (General Director of Hoechst, father of the present chief of the Deutsche Forschungsgemeinschaft) presented to me this book in front of several hundred people: He looked at the book, looked at me, looked again at the book, looked again at me, shook his head and gave me the book. I understood what was going on in his mind: "What in the world does this Schlosser want to do with this high-level math book, which even I do not understand". Prof. Winnacker was a chemist!

I did my Vordiplom directly after the 4th semester in Frankfurt, went to the Technische Hochschule in Darmstadt to Prof. Otto Scherzer for studying Theoretical Physics (the chair in Frankfurt was vacant due to Prof. Friedrich Hund's move to Göttingen). My diploma thesis was on the feasibility of obtaining directly electrical energy out of a fusion reactor. I became Diplomphysiker at the middle/end of my seventh semester and went right away to Freiburg/Breisgau University, because I had heard great praise about Prof. Hans Marshall. I liked him right away. He had a special human touch and used to invite me to his home for lunch, where his wife Elisabeth Marshall had prepared pork with Sauerkraut, which I loved to eat!

Hans Marshall suggested working on nuclear polarization in myonic atoms. It didn't take me long. A year later I had calculated this effect, which was quite important for interpreting the measurements of the transition energies of myons in heavy atoms. I utilized the Jensen-Steinwedel model for proton-neutron vibrations in nuclei (giant resonances). My thesis stood, I got my PhD (Dr. rer. nat.) in my 9th semester.

As an Assistent I had to take care of younger students: Amand Faessler, Florian Scheck, Max Georg Huber, ... who worked for their diploma thesis. I got Amand Faessler interested in the Bohr-Mottelson-Model. Soon we wrote our first joint paper on the Rotation-Vibration-Interaction in Deformed Nuclei. It was the first solution — even analytically — of the Bohr Hamiltonian (Davydov had solved the triaxial static rotor but we had solved the axially symmetric rotor with β- and γ-vibrations) and it made us known in many places world wide.

I had sent our work to Aage Bohr, but he replied in a long letter essentially that he knew all that. I doubt it even today. When I showed this letter to my teachers Hans Marshall and Theodor Schmidt (the "Schmidt-Line"-Schmidt) they shook their heads and said "forget Bohr — go on!"

Meanwhile I had married my wonderful wife Bärbel, daughter of a Lutheran Minister in Frankfurt. I had known her for several years. This was one of the luckiest highlights of my life.

In late summer 1962 I went as a post-doc to the University Maryland in College Park/MD, a suburb of Washington D.C. My wife Barbara accompanied me. She was pregnant. William Mac Donald, who had invited me (I met him at a Varenna school with V. Weisskopf, A. de Shalit, G.E. Brown, ...) but went on a sabbatical. R.A. Ferrell had just left nuclear physics and worked in solid state physics, superconductivity theory. Peter Fulde was at that time his PhD-student. I didn't like to enter solid state physics: Nuclear Physics fascinated me. Thus I was in Maryland more or less on my own. I had to give a colloquium on the "Rotation-Vibration-Model". There I met Michael Danos, who frequently came over from the National Bureau of Standards to Maryland University. People were quite excited: John Toll, the chairman of the Md. Physics Department, promoted me right away to Assistant Professor and I had to teach the graduate course on Nuclear Theory.

With Mike Danos began an extremely fruitful, productive scientific collaboration and friendship. In his thesis under Hans D.J. Jensen Mike had studied the splitting of the giant resonance in a statically deformed nucleus. Utilizing the rotation vibration model, we could extend this to rotating-vibrating nuclei and, what was a tremendous step forward, to surface vibrating spherical nuclei, leading to a tremendous structure in the giant resonances. It was very soon experimentally verified and lead to a new boom in photo-nuclear structure physics. I summarized this all in the first volume of my books with Judah Eisenberg: Nuclear Theory — Collective and Single Particle Aspects. By the way, I met Judah at a Washington Spring Meeting of the APS. He invited me down to the University of Virginia in Charlottesville, and he, Dexter Whitehead and Morris E. Rose, tried two years later to bring me to Virginia with an offer for a full professorship. Also Larry Biedenharn and Henry Newson tried to get me to Duke University at Durham, NC. After two years in Maryland and some months spent with R. Sheline at Florida State University in Tallahassee, I wanted to be back in Freiburg for habilitation. Equally important, our son Martin was born in Washington and the grandparents wanted to see their first grandchild and, in particular, grandfather Fritz, the lutheran minister, wanted to baptize his grandson.

Having returned in May 1964 I was invited to various places for colloquia, in particular to Darmstadt and to Frankfurt. Soon I heard that both places wanted me as a professor. In July 1964 I received the offer from Frankfurt, which I accepted and I became full professor. It was the historical chair in Theoretical Physics which before was held by Max v. Laue, Max Born, Erwin Madelung, Friedrich Hund and Helmut Steinwedel. I took on this position and became director at the Institut für Theoretische Physik on January 1, 1965. Heinz Bilz and Georg Süßmann were my colleagues. When I attended the first faculty meeting, my former teachers welcomed me. Reinhold Baer, the famous mathematician, asked me: "Don't we know each other?" I answered: "You were examining me at my Vordiplom in Mathematics. You were lying on your sofa and I was standing with chalk in my hand at the blackboard". He replied: "What would have happened if I hadn't let you pass?" My prompt answer was: "Most probably I would also be here". He got tremendously happy, put his two hands on my shoulders — just as if he wanted to bless me — and said: "This is the answer I expect from an ordinarius!"

Heinz Bilz, Georg Süßmann, Hermann Dänzer, Werner Martienssen and Erwin Schopper were my close colleagues. Erwin Schopper became a particularly dear colleague and friend, teaching me how to run an institute and how to handle all problems coming up with the chair in physics.

The principal message was: If you don't want to be occupied by the central administration, stealing your time from science, build up a counter administration which is loyal to you.

Some of my former student colleagues came to me for examination in their PhD and some (still!) in their diploma exams. I began with Max G. Huber (now professor at and former rector of the University of Bonn, Vice-President of the DAAD), Dieter Drechsel (now professor at Mainz), and H.G. Wahsweiler to set up a research group, having H.J. Weber (now professor at the University of Virginia), Hartmut Arenhövel (now professor at Mainz), Christian Toepffer (now professor at Erlangen), Wolfgang Donner, R. Ligensa, U. Mosel (now professor at Giessen), Bernd Fink (now professor at the Bayer company, Leverkusen), W. Pieper (now professor at the Fachhochschule Giessen), Carsten Pruess (leading scientist at the Lawrence Berkely Laboratory), and many others as early students. Frankfurt offered a great number of outstanding students. This is the best aspect of Frankfurt.

In the course of years I had over 140 PhD students and habilitants; out of them nearly 40 later became professors. Some of them I already mentioned, but let me not forget Berndt Müller, Johann Rafelski, Gerhard Soff, Michael Soffel, Horst Stöcker, Werner Scheid, Burkhard Fricke, Ulrich Heinz, Paul-Gerhard Reinhard, Joachim Maruhn, Volker Oberacker, Peter Hess, Dietmar Kolb, Werner Scheid, Adolf Grauel, Udo Müller-Nehler, Andreas Schäfer, Eliah Mshelia, Wolfram Weise, Steffen Bass, Dirk Rischke, and ..., and ...

Most of them I am very proud of. A few, however, disappointed me later on scientifically and by outing their true character in the course of time. That's life!

I do not want to reflect here on the various outstanding problems in nuclear physics and field theory we addressed in the course of 40 years. Let me just quote some keywords: the Dynamic Collective Model of Giant dipole and quadrupole resonances, the Generalized Collective Model (also known as the Gneuß-Greiner Model), QED of Strong Fields, the Overcritical Vacuum, Superheavy Elements, Nuclear Molecules, Electronic Molecules in Heavy Ion Collision, Nuclear Shock Waves, the Nuclear Equation of State, Phase Transitions in Nuclear Matter, Precritical Scattering in Heavy Ion Collisions, the Two-Center Shell Model, Potential Energy Surfaces of Nuclei, Cluster Radioactivities, Channeling in Bent Chrystals, the idea of a Gamma Laser, Extension of the Periodic System into the Directions of Strangeness and Antimatter, Clusters of Matter and Antimatter, Cold Compression of Nuclear Matter, Fullerene Clusters, Orientation of Migrating Birds in the Magnetic Field of the Earth, etc., etc.

But I would like to close with a few words on the initiation and foundation of GSI, which began in late summer 1965: My colleague Erwin Schopper talked with me again and again on various topical physics problems and also on the question of how to ensure future top-level nuclear physics research in Germany. In a telephone conversation with Peter Brix (Darmstadt) this also came up and I invited him for a discussion together with Schopper and me to Frankfurt. We quickly agreed that something special and novel had to be done and that an extended experimental facility could no more be run by one university alone: The tandem accelerator age was gone; the Rhein-Main area with the universities of Darmstadt, Frankfurt, Mainz, Giessen, Marburg and Heidelberg close-by was ideally suited for a joint "Ausbildungs- und Forschungszentrum für Kernphysik". We founded the "Arbeitsgemeinschaft Hessischer Kernphysiker" and met practically every Saturday in Erwin Schopper's Institute. Joined by Rudolf Bock, Wilhelm Walcher and Egbert Kankeleit we had intensive discussions. I proposed heavy ion physics as the main topic for future research (superheavy elements, nuclear molecules, strong-field QED and atomic processes connected with that) and Friedrich Beck (Darmstadt) pleaded for a proton-synchrotron (similar to the Indiana facility) to investigate nucleon-nucleon correlations and shell structure. Overwhelmingly we decided for heavy ions, wrote a proposal, in which even the research on tumour therapy with heavy ions was shortly mentioned (this idea came from Berkeley).

My colleague and friend, Prof. D. Allan Bromley (Yale University) encouraged me in pursuing this field of research. It was the idea of nuclear molecules which bound us together, but later the scope of our collaboration widened tremendously. Through him I met my friends Friedel Sellshop (Johannesburg) and André Gallmann (Strasbourg). Both played a very important role in my life, as did Joe Hamilton (Vanderbilt University). But this is another story.

Parallel to the political struggle for the establishment of the grand heavy ion laboratory with the colleagues from Heidelberg, who woke up and wanted to drag the emerging new laboratory to Karlsruhe where they imagined to be more influential, we organized weekly seminars on heavy ion physics: Every Tuesday afternoon we had a seminar either in Frankfurt or Darmstadt or Marburg. With buses we brought our collaborators and students together and created a stimulating atmosphere for heavy ion research. This became later the GSI-colloquium. Rudolf Bock was very essential; he was the only experimentalist with experience in heavy ion physics (at that time the ions were, in fact, light ions). In his contribution he tells more about the various difficulties, especially which type of accelerator should be built, and about Christoph Schmelzer's maneuvres. Even though Christoph Schmelzer played a strange role in those early years, we became good friends later on. Anyhow, in 1969 due to the strong interference of Georg August Zinn, the prime minister of the State of Hessen, the Federal Minister for Research and Sciences (Forschungsminister) — Gerhard Stoltenberg — decided that the Gesellschaft für Schwerionenforschung (GSI) was to be founded, funded and put into reality in the woods at Wixhausen between Frankfurt and Darmstadt. Nowadays GSI is the leading laboratory for heavy ion physics in the world, with outstanding, remarkable successes.

The Early Work of Walter Greiner (1960–1968)

Şerban Mişicu
Department of Theoretical Physics,
National Institute for Nuclear Physics and Engineering,
P.O. Box MG-6, RO-077125 Bucharest-Magurele, Romania
misicu@theory.nipne.ro

The first decade of Walter's Greiner scientific career is analyzed. It is a timespan of outstanding achievements in various subjects of theoretical nuclear physics: nuclear collective motion, and theory of nuclear reactions.

1. Scientific Genealogy

Walter Albin Erhard Greiner steems from one of the most prodigious scientific lineages. According to the Mathematical Genealogy Project[1] his genealogic tree plounges its roots deep into the fertile soil of German culture and science. No doubt it can be compared with other German scientific dynasties such as the one starting with Gottfried Leibniz, continuing with Fr. Gauss and Felix Klein, which subsequently evolves in the modern physics branch (Sommerfeld – Pauli – Heisenberg – Bethe), continuum mechanics branch (Prandtl – Timoshenko – von Kármán), and mathematical physics branch (Hilbert – Courant – von Mises – Weyl). The genealogic tree of Walter Greiner, displayed in Fig. 1, starts with Friedrich Leibniz (the father of Gottfried Leibniz) and includes names that are forever engraved in the Pantheon of German culture (Georg Lichtenberg), mathematics (Karl Weiserstrass, Carl Runge), and modern physics (Max Born). The "grandfather", Siegfried Flügge (1912–1997), was also a remarkable theoretical physicist and teacher. He published in the thirties influential articles on the exploitation of nuclear energy and during the war, in collaboration with C. F. Weizsäcker, he worked on the theoretical basis of uranium reactors and extended the Bohr–Wheeler theory on nuclear fission to the transuranic elements.[2] The famous collection of problems in Quantum Mechanics was first issued in 1947 and as Walter Greiner recalled in 1983,[3] this book, known also as the "Flügge–Marschall", was an ubiquitous didactic material for the German post-war generations of theoretical physicists.

Scientific Genealogy of Walter Greiner
(according to the Mathematics Genealogy Project)

Friedrich Leibniz — Universität Leipzig (1622)
Jakob Thomasius — Universität Leipzig (1643)
Otto Mencke — Universität Leipzig (1665)
Johann Wichmannshausen — Universität Leipzig (1685)
Christian Hausen — Martin-Luther-Universität Halle-Wittenberg (1713)
Kästner Abraham — Universität Leipzig (1739)
Georg Lichtenberg — Georg-August-Universität Göttingen (1765)
Bernhard Thibaut — Georg-August-Universität Göttingen (1796)
Christoph Gudermann — Georg-August-Universität Göttingen (1823)
Karl Weierstraß — Westfälische Wilhelms-Universität Münster (1841)
Carl Runge — Universität Berlin (1880)
Max Born — Georg-August-Universität Göttingen (1906)
Siegfried Flügge — Georg-August-Universität Göttingen (1933)
Hans Marschall — Humboldt-Universität zu Berlin (1944)
Walter Greiner — Albert-Ludwigs-Universität Freiburg im Breisgau (1961)

Fig. 1. Scientific lineage of Walter Greiner. The succesion from the top to the bottom of the list follows the order "adviser"–"student". For each member of the list the place where and year when the dissertation was defended is provided.

2. Doktorarbeit in Freiburg im Breisgau

In February 1960 Walter Greiner passed the Diplomphysiker-Hauptprüfung (Master of Science — Main Examination) at Technische Hochschulle (TH) Darmstadt with a thesis entitled "Plasmareaktoren", a work that to the date seems to be lost. His advisor was Otto Scherzer (1909–1982), an outstanding German physicist, student of Arnold Sommerfeld and pioneer of Electron Optics, who authored in collaboration with E. Brüche the first book (*Geometrische Elektronenoptik*, 1934) in this field, and fathered the so-called *Scherzer theorem* which, according to the Nobel Laureate Dennis Gabor, grounded the foundations of Holography.[4]

Stimulated by Hans Marschall (1913–1986), who was at the time a Diätendozent (sessional lecturer) at TH Darmstadt and Albert-Ludwigs Universität Freiburg, Greiner moves in May 1960 to Freiburg with a fellowship offered by the "Studienstiftung des Deutschen Volkes" (Foundation for Education of the German People). Three months later he married Bärbel Chun.

Starting several years before, Hans Marschall has been dedicating his efforts to the theory of muonic atoms, especially the description of μ-capture reactions $(pp)\mu$, $(dp)\mu$, $(dd)\mu$.[3] During the fifties he systematically worked on problems related to charge distributions in nuclei and how they are influenced by muons, excitation of nuclei induced by the capture process and during the atom excitation via Auger and radiative transitions.

2.1. *Nuclear polarization in muonic atoms. Role of nuclear structure*

In less than a year (April 1961) Walter Greiner succeeded to complete and submit his Dissertation entitled "Die Kernpolarisation in μ-Mesoatomen" (Nuclear Polarization in Muonic Atoms) for the "Erlangung der Doktorwürde" (Attainment of the Doctorate).

Muonic atoms consists essentially of a simple negative muon μ^- bound to a nucleus due to the electromagnetic force. In contrast to the ordinary atoms, in muonic atoms the nuclear degrees of freedom determine a series of interesting properties due to the large mass difference between muon and electron, as well because of the corresponding difference in scale of sizes and energies. One of the effects arising from the internal degrees of freedom of the nucleus is the nuclear polarization, which consists in energy shifts of the muon levels. In his first paper[5] Walter Greiner applied the formal theory of nuclear polarization to the muonic ^{208}Pb. This was previously developed by Cooper and Henley,[6] with the difference that Greiner is giving-up the single-particle feature of nuclei and instead adopts the macroscopic (hydrodynamical) approach of Steinwedel and Jensen.[7,a] In this picture the muon induces a relative motion between the neutron and proton fluids such that a modification of the potential produced by the nucleus takes place. This variation of the potential leads to the muon level energy shift. Numerical estimations, produced with the help of his younger collegue Amand Faessler, also student of Hans Marschall, show that the inclusion of proton fluid small-amplitude oscillations raise the bottom of the polarization potential by $\approx 5\%$ and by using the exact meson wave-functions instead of the hydrogen, one lowers drastically the sojourn probability of the muon inside the nucleus. For the $1s$ state in μ^{208}Pb atom he obtained a shift $\Delta E_{1s} \approx 1$ keV. At the end of the paper Greiner noted that extrapolating the conclusions drawn from a previous study on nuclear polarization in deuterium, the single-particle contribution in the studied case should not be expected to exceed 100 eV. It is also interesting to note that the Hydrodynamical Model used by him in calculating the nuclear polarization in muonic atoms played an important role in the model of giant resonances developed together with Danos a few years later.

In a subsequent paper with Hans Marschall he extended this framework in order to account for the quadrupole deformation of the nuclear surface induced by the muon.[9] They estimated that the energy shift of the $1p$ state in μ^{208}Pb due to the quadrupole distortion should be ≈ 4 keV. The contribution brought by Refs. 5 and 9 was recognized in reviews and textbooks on muonic atoms[10,11] especially concerning the step forward undergone by Greiner's framework that goes beyond the closure approximation of Cooper and Henley, by using matrix elements of the perturbation from nuclear wave-functions. In these early publications Walter Greiner proved

[a]Actually it was Nuding who had first the idea to apply the concept of low and high proton tide effect on the muon orbit.[8] However he failed to couple the proton density fluctuations to the muon quantum dynamics in the wave equation.

that he was already on the route of becoming a real master in handling quantum problems involving strongly coupled modes.

Other contributions to this domain concern: (a) a study on the reconciliation between the experimental value for the Bi-Pb energy difference and the value calculated from electron scattering data by using the single-particle shell-model for the $(Z+1)$-th proton in the muonic atom ${}_{Z+1}X^{A'}$ and the Ford–Willis wave-functions for muon;[12] (2) the effect of nuclear collective motion (rotations, surface oscillations, giant resonances) in the X-ray spectrum of a muonic atom.[13] These two studies demonstrate the prevailance of rotational excitations on the muon state compared to β and γ vibrations. In Ref. 14 Greiner is for the first time concerned with the issue of vacuum polarization, and he provides an estimate of the vacuum polarization potential for sharp nuclear surfaces. In the following years and up to the end of his life he will devote continuous efforts to this subject.

3. The Rotation–Vibration (RV) Model

In 1962 together with Amand Faessler, Greiner published a landmark paper,[15] which gives a detailed quantum mechanical treatment of collective nuclear motion derived from the classical Hamiltonian of a symmetric rotator with a vibrating surface. More precisely, the Rotation–Vibration (RV) framework assumes quadrupole distortions of the nuclear shape with amplitude small compared to that of the equilibrium deformation. An essential assumption of this model, where rotations and vibrations are fully coupled, concerns the hydrodynamical nature of surface vibrations and of the inertia moments. The same problem was treated a few years earlier in the Soviet Union by Davydov and collaborators in the lowest-order adiabatic approximation and only in a few cases (see Ref. 16 and references therein). The merit of the Faessler–Greiner approach was the exact treatment of (β, γ) vibrations by splitting the total Hamiltonian in a purely rotational, a purely vibrational and a coupled term. In the uncoupled case (zeroth-order) the energy eigenvalue is expressed as the sum of vibrational and rotational energies, a fact that provides a convenient way to classify various rotational bands that can be built on the β and γ vibrations. The effect of the remaining terms, describing the coupling between rotations and vibrations, is estimated in the first-order perturbation theory. For larger spins $(J \gtrsim 4)$, the ratio between the rotational and vibrational quantum energies ceases to be very small and an exact diagonalization of the total Hamiltonian is necessary, taking the wave-functions of the uncoupled Hamiltonian as basis states. In the mass range $156 \leq A \leq 184$ the diagonalization procedure applied by Faessler and Greiner resulted in a fitting error generally less than 0.4%! Of particular importance in the Faessler–Greiner model was the treatment of γ vibrations. They showed that for vibrations around $\gamma \neq 0$, the term $\hbar J_3^2/(2I_3)$ in the Hamiltonian for γ vibrations takes the form of a centrifugal potential, viz. $\sim 1/\gamma^2$, such that a qualitative distinction between $K = 0$ and $K \neq 0$ states occurs. In other words, the centrifugal repulsion for the second class of states shifts the average value of γ^2 to non-vanishing values, thus making the gamma vibrational bands effectively more triaxial. The power of the Faessler–Greiner framework compared to the predictions of the Bohr–Mottelson model became transparent from their first paper (see Figs. 4 and 5 of Ref. 15). A similar comparative study was made a year and a half later[17] and demonstrated that the $E2$-transition probabilities from the states of the γ-band to the states of the ground state band are better described in the RV model than in the Davydov theory. A comprehensive summary of the RV model in deformed nuclei was given in Ref. 18. Here, the Hamiltonian is diagonalized up to $I = 20$ and the energies and wave-functions for various states in the ground state, β and γ bands are tabulated.

In a short time Refs. 15, 17–20 became standard references in the literature dedicated to nuclear collective motion, e.g. Refs. 16, 21–25, and the framework developed by Faessler and Greiner is sometimes evoked in the literature under the name of Bohr–Mottelson–Frankfurt model. Even in the rare occasions when criticism was raised against this model, mainly related to its hydrodynamical foundations, it was stressed the progress achieved in understanding the collective feature of the atomic nucleus.[23] As Walter Greiner remembered almost four decades later about the inovative role played by the RV model developed together with A. Faessler:[26] "It was the first dynamical solution of the Bohr–Hamiltonian and up to now also the only analytical solution (Bohr and Mottelson had never solved it; they only talked qualitatively about it)."[b]

Other outstanding contributions of Walter Greiner using the RV framework concern the magnetic properties of even nuclei, Coriolis anti-pairing and blocking effects, and the octupole vibrations of deformed nuclei. Thus, in Refs. 27–29 he established a remarkable relation between the lowering of the g_R-factors and the magnetic dipole transitions, using the idea that the proton moment of inertia is less than the neutron moment of inertia on account of unequal pairing forces. Therefore this disbalance should be reflected in energies and magnetic transitions. By 2016, Ref. 29 received 152 citations (see for example the discussion from Ref. 30). The ideas developed by Greiner and collaborators in these papers were instrumental in the elaboration of the two liquid drops model used to describe the M1 properties of the rotational bands.[31]

In a very simple picture consisting of a core and a j-shell on top of it, which contains sufficiently enough nucleons such that core deformations are engendered, Greiner calculated the collective potential energy up to the fourth order in the quadrupole moment using the quasi-spin formalism and showed that in order to describe nuclear deformations one needs to take into account anharmonic terms.[32] Analytical formulas for the strengths of the harmonic, cubic, and quartic potential terms are given in terms of the strengths of short-range (pairing) and long-range (quadrupole–quadrupole) residual interactions, the number of nucleons in the j-shell, the quasi-spin and the reduced matrix element of the quadrupole operator.

[b] A thorough discussion on the question of the complete solution of the Bohr–Mottelson Hamiltonian is given in the monography.[25]

This study served for a later analysis (in collaboration with Faessler and Sheline) on the effects induced by Coriolis anti-pairing and blocking effects in the second-order rotational–vibrational Hamiltonian.[33] This paper is still cited in the literature today.[25]

In Ref. 34 he described octupole states as octupole phonons strongly coupled to a quadrupole deformed nucleus in a manner analogous to the Nilsson model. The Hamiltonian splits in rotational, quadrupole and octupole vibrational terms, and a quadrupole–octupole interaction term. Decades later the interest on octupole states was revived especially in connection with pear-shaped nuclei[35] and magnetic collective modes.[31] The $E1$ mechanism to the low-lying negative parity states suggested by Donner and Greiner for rare-earth nuclei was confirmed experimentally by Guhr *et al.* for light actinides.[36]

4. Washington D.C. and Collaboration with Michael Danos

In 1962 Walter Greiner crossed the Atlantic and begun one of the most fruitfull periods in his career, first as a postdoc and soon after as an assistant professor at the University of Maryland at College Park, a suburb of Washington D.C. The first published paper, while on American soil, concerns a theme on the electromagnetic radiation of nuclei, i.e. the braking radiation produced in proton–neutron scattering.[37] This work has two important results: (1) Greiner and Green pointed out that the model dependence of the bremmstrahlung at low energies ($E_{\text{lab}} = 10$ MeV) should be small; (2) the Bloch–Nordsieck transformation was proved for the first time for wave-packets and stationary bound states. This paper provided a simple tool for the calculation of probability amplitudes of single-particle $E1$ transitions between continuum states. It is regrettably that this nice paper was overlooked for so long. The simple but efficient framework proposed by Greiner and Green can be readily applied to radiative capture reactions involving halo nuclei with a single valence neutron or proton that are relevant for modern studies of nuclear astrophysical reactions of the type ${}^{A}Z(n,\gamma)^{A+1}Z$, ${}^{A}Z(p,\gamma)^{A+1}Z+1$. This model could be also applied to the soft dipole mode in halo nuclei. However in 1962 the existence of a considerable dipole strength in neutron-rich nuclei was not yet known.

According to his recollections, published in 2001,[26] while at College Park, and following a seminar at the Physics Colloquia, Greiner made acquaintance to Michael (Mike) Danos who was at that time a member of the Photonuclear Data Center of the US National Bureau of Standards (present NIST).[c] This center of excellence was carrying out a project of significant importance that consisted in the systematic collection of bibliography and data regarding the interaction of photons with atomic nuclei. Danos, a former student of Hans Jensen in Heidelberg, had already acquired a worldwide recognition especially for his earlier contributions to the

[c] According to Amand Faessler, Michael Danos visited previously Hans Marschall in Freiburg at the suggestion of Hans Jensen, and it was there he got to know Walter Greiner (communication to the author, May 26, 2017).

hydrodynamical model of dipole proton–neutron oscillations (Giant Dipole Resonance, GDR) where he extended the Steinwedel–Jensen model. In 1958 he established a remarkable relation between the ratio of the two components of the GDR in deformed nuclei and the ratio of the long and short axes lengths of the nuclear ellipsoid. The theoretical description of the GDR splitting in deformed nuclei is frequently referred to in the literature as the Danos–Okamoto model. Their meeting resulted not only in what became a life-lasting friendship but also in an outburst of publications "one better than the other" as Walter Greiner recalled long after.[26] Returning to Germany, Walter Greiner will send many of his students to NBS to work with Michael Danos: M. G. Huber, H. Arenhövel, D. Drechsel, B. Fink, and J. Rafelski.

4.1. *Dynamical theory of nuclear collective model*

The first pillar of the Greiner–Danos collaboration is the unification of the collective surface degrees of freedom, as encoded in the earlier developed rotation–vibration model, with giant resonance motion. For the first time in the literature it was revealed that the strong interaction between dipole or quadrupole density fluctuations and surface degrees of freedom leads to the experimentally observed substructures in the giant resonances. They laid down the foundations of the Dynamical Collective Model (DCM) theory in Refs. 38–40. In this framework the giant resonances are described within the hydrodynamical model, mentioned above,[7] which works reasonably well for medium and heavy nuclei. However the model developed by Steinwedel and Jensen has a serious limitation, which at that time was not considered to be important: though density fluctuations of separate proton and neutron fluids are allowed, the total density remains constant during the oscillation. Danos and Greiner admit themselves that "we neglect this nuclear property," i.e. nuclear compressibility, "and as a result lose the states corresponding to compression waves (ordinary waves)." From the modern perspective the assumption that the sound-wave state lies at a much higher energy than the dipole state on account of the incompressibility of nuclear matter is no longer acceptable. Presently (see the discussion and literature in Ref. 41) we are aware of the existence of an isoscalar dipole resonance where the nuclear compresibility plays a crucial role. Danos and Greiner considered in the DCM framework nuclei having large deformations and, assuming that rotational and vibrational motion is slow compared to giant resonance oscillations (adiabatic approximation), they showed that once quadrupole vibrations are included, the nuclear ellipsoid acquires a dynamic triaxiality and each of the two principal peaks is accompanied by a satellite carrying about 10% of its strength.

Once professor in Frankfurt, Greiner vastly extended the DCM framework. With Weber and Huber he published an exhaustive study in two parts dedicated to the application of the DCM to spherical nuclei and compared the result of calculations to experimental data showing thus that the structures in the gamma-absorbtion cross-section are surprisingly well described.[42,43] Together with Ligensa[44] he operated the dynamic coupling of the giant quadrupole oscillations with the RV modes and single-particle motions and obtained an excellent agreement to the experimental $E2$ absorbtion cross sections for ^{165}Ho and ^{159}Tb. They found that the giant electric quadrupole resonance has five components for these rare earth nuclei. The inclusion of both dipole and quadrupole giant resonances in DCM was published in the same year with another of his Frankfurt students (T. Urbas).[45] A formulation of the DCM in the particle–hole framework was operated in Refs. 46 and 47. While retaining from the original model the treatment of surface excitations, the dipole Hamiltonian and the dipole operator from the dipole–quadrupole Hamiltonians are replaced by corresponding operators in the one particle–one hole (1p–1h) subspace. The new interaction between the giant dipole resonance and the quadrupole oscillations leads to collective correlations.

Extensive tables of neutron and proton partial and total escape widths of various dipole states in several spherical nuclei were published in Ref. 48. They stressed the strong dependence of the widths on the angular momenta of the hole state of the nucleus and the emitted particle. A crucial assumption made in this paper was that the residual interaction determines the coupling of the single-particle continuum to the collective states of the DCM.

4.2. *Eigenchannel model of nuclear reactions*

In 1965 the main framework to calculate resonant nuclear states was the R-matrix method (for an early review see Ref. 49). As we have seen above, Greiner and Danos were foccusing on the giant resonances and at that particular time they were looking to study the excitation of these modes based on shell-model techniques. Therefore they proposed a new treatment[50,51] of nuclear reactions, which shared some common features with the R-matrix theory such as the separation from the total Hamiltonian H of a shell-model Hamiltonian $H_{\text{shell}} = \sum_i h_0(i)$, the division of the space into an internal (nuclear) region and an external (scattering) one, separated by the so-called surface channel of radius a_i in each channel i, as well as the diagonalization of H in a truncated set of eigenfunctions of H_{shell} with the exclusion of all configurations with more than one nucleon in continuum eigenstates (i.e. considering only 1p–1h excited states in the reacting nuclei). One should remind the reader that in the traditional R-matrix framework, the wave-function in the internal region is expanded into a set of solutions of $h_0(i)$ and subject to an arbitrary boundary condition (BC) at the matching radius (beyond which the nuclear potential dies out), which has the annoying consequence that the wave-function slope is discontinuous thus entails a slow convergence of the basis set. In its turn the eigenchannel method considers at a given excitation energy a complete set of scattering states (eigenchannels), which diagonalize simmultaneously H and the scattering matrix S, avoiding thus the wave-function discontinuity at the channel surface. This advantage over the R-matrix method was admitted even by some of the bitter criticizers of the new method.[52] In order to cure this unsatisfactory feature, the framework devised by Danos and Greiner introduces so-called "natural" BC, insuring thus a smooth transition between the inside and outside regions. Instead of choosing an arbitrary BC like in R-matrix theory, the eigenchannel procedure gives the BC parameter in each channel in terms of the known form of the channel radial wave-function. A trial value of real eigenphase δ is assumed at a given scattering energy E and therefore also the S-matrix is known in diagonal form ($S = e^{2i\delta}$). Next, the continuity of the logarithmic derivative at the matching radius is imposed in order to obtain the eigenvalues E_λ of the complete Hamiltonian. If none of E_λ is equal to E, the S-matrix is iterated until the equality $E_\lambda = E$ is reached. The supplementary computer-time required by the Danos and Greiner method compared to the standard R-matrix involves the computation and diagonalization of a $Nm \times Nm$ dimensional matrix where N is the number of basis states retained in each channel and m is the number of eigenphases (open channels); naturally the time required for this operation increases rapidly as the matrix dimension increases. After five decades such large-matrices computations are much easy to handle and the advantage over the standard R-matrix methods, obtained via the improved convergence of the scattering wave-function expansion, makes the eigenchannel method attractive again for resonant nuclear reaction calculations.

Applications of this method have been published by Greiner and collaborators in the next years. In Refs. 53 and 54 the photon absorbtion cross section of ^{16}O was calculated and concluded that the broad structures, e.g. the splitting of the main peak indicated by the experiment, are largely reproduced within the eigenchannel framework with the inclusion of 1p–1h channels. The fact that some of the secondary peaks and other fine structures in the experimental total cross section are not accounted by the theory was attributed to the absence of 2p–2h and higher configurations. An extension of the formalism to describe resonant nuclear reactions above the two-particle threshold was proposed with Danos in Ref. 55 for nucleon channels and with H. J. Weber[56] for cluster channels, but no numerical applications were given. A comprehensive review on the eigenchannel method was published in 1973.[57]

5. Ordentliche Professor in Frankfurt am Main and Further Developments in Nuclear Collective Dynamics

In the summer of 1964 Walter Greiner accepted the chair for theoretical physics at J. W. Goethe University in Frankfurt am Main as full professor. In the first years he pursued some of the directions described previously (new applications and further developments of the RVM, eigenchannel theory and the DCM) but he also initiated new ones. One of the most significative of these concerned the nuclear response to external electromagnetic fields. In Ref. 58 he investigated with Danos and Kohr the role played by the nucleus orientation with respect to the incident photon and electric field in the excitation of quadrupole modes.

He developed together with Hartmuth Arenhövel a general formalism of photon scattering by oriented nuclei and explicit formulae for the dipole radiation in terms of orientation were provided in reference to aligned ^{165}Ho nuclei.[59,60] It was concluded that the photon scattering by oriented nuclei could be a useful tool for direct observation of the tensor polarizability of nuclei.

With two of his students, Gotthard Gneuss and Ulrich Mosel, he proposed in 1968 a new treatment of the collective nuclear Hamiltonian that is able to describe simultaneously a pure spherical vibrator and a permanently deformed rotator in the basis of the five-dimensional oscillator.[61,62] While the Faessler–Greiner model assumes that the nucleus rotates or vibrates as a whole about an equilibrium deformation and therefore is not able to describe transitional nuclei, the model dubbed in the literature as Gneuss–Greiner (GGM) is capable of describing all types of nuclei, the quadrupole degree of freedom playing the prevailing role in the low-energy properties of even–even nuclei. The main purpose envisaged by Greiner in developing this model was to obtain a reliable tool that allows the calculation of spectra and transition probabilities of an even-even nucleus of arbitrary shape. In this way it was possible to deduce from experimentally known spectra the corresponding potential energy surface (with respect to the intrinsic nuclear coordinates β and γ), which can be compared with microscopic results that are obtained after more tedious calculations. They emphasized that a number of experimental level sequences, where the earlier collective models (Bohr–Mottelson, Jean–Wilets, Davydov–Filippov, RV) failed completely, could now be reproduced. A far-reaching goal of the "five-dimensional oscillator" approach was the calculation of potential energy surfaces of nuclei undergoing fission, especially in regard to the stability of superheavies. In the years to come Walter Greiner will dedicate his efforts to this problem.

Whereas the RV model was limited to small-amplitude oscillations around the ground state deformation, the model of Gneuss and Greiner explores wide regions of the quadrupole variable and consequently, anharmonicities in the potential energy, which are included up to the sixth order in the quadrupole coordinate $\alpha_{2\mu}$, plays an instrumental role (see the review paper of Gneuss and Greiner[63] and an up-to-date presentation in Ref. 25). A limitation in the GGM is represented by the large number of parameters and the slow convergence for large deformations due to the diagonalisation procedure in spherical basis. Despite of that, the model played a historical role in the development of nuclear collective models, the anharmonicities in the low-lying vibrational states receiving for the first time in the literature a first row seat.

The GGM was improved with the significative contribution of some of Greiner's students (especially P. O. Hess and J. Maruhn, M. Seiwert, L. V. Bernus) by extending its flexibility to high spins. This new version, called the General Collective Model (GCM) or Frankfurt Collective Model[64,65] contains as new features three terms in the kinetic energy instead of two as in the GGM, and the potential energy contains many additional terms of the type $\beta^n(\cos 3\gamma)^m$ with $2 \leq n \leq 6$ and $0 \leq m \leq 4$ such that GCM uses 20 parameters instead of 8 like in GGM. These parameters are fixed by fitting the data to low-lying energy levels, reduced $E2$ transition probabilities, and quadrupole moments. Applying the Frankfurt Model to isotopic chains of Pt, Os, W, two minima (one prolate and one oblate) in the potential energy surface were obtained in several instances instead of only one like in the GGM. The isotopes with two minima exhibit a backbending structure in the yrast band due to the crossing of the lowest bands in the two minima.

At the end of the period that concerns this biographical excourse, the first papers on the ion–ion potentials are published.[66,67] Major themes such as stability and formation of superheavy nuclei, nuclear molecules, elementary matter in strong fields that will attract the attention of Walter Greiner for the next decades are coming into play after this year.

6. Conclusions

This contribution was confined to a limited but very prolific period of Walter Greiner scientific career. The large majority of subjects evoked above were included in his magnum opus on Nuclear Models.[68] Even so, the best reference in gaining a profound understanding of Greiner's work remain the original articles.

Walter Greiner always tried to convey to his students and collaborators, that they should avoid to fall in the trap of "Trockene Physik" (dry physics) and rather venture in new and exciting problems whenever possible. This situation is particularly true in theoretical nuclear physics where always new subjects are prompted by new or foreseeable experiments, problems that require dedication, assiduity and intuition in order to be solved. Walter Greiner possessed these skills.

Acknowledgments

I am grateful to Prof. A. A. Raduta who read this manuscript, expressed critical remarks and drew my attention on relevant aspects related to Walter Greiner's work that should not fall into oblivion. I also owe to Prof. P. O. Hess gratitude for helping me to improve the fifth section of the present work. To Prof. A. Faessler I am indebted for some clarifications regarding the Freiburg period. I am also indebted to Mrs. Daniela Radulescu from ITP Frankfurt for providing me with an electronic copy of Walter Greiner Dissertation. She obtained this rare document due to the kindness of Prof. Carsten Greiner. This work has been financed by the national project PN III 5/5.1/ELI-RO, Project 04-ELI/2016 ("QLASNUC"), under the financial support of Institute for Atomic Physics - IFA.

References

1. https://genealogy.math.ndsu.nodak.edu
2. W. Schlögl and S. Flügge, *Physics Today* **51**(8), 77 (1998).
3. W. Greiner, Hans Marschall 70 Jahre alt, *Physikalische Blätter*, **39**, 410 (1983).
4. F. Beck and O. Scherzer: Wegbereiter der Elektronenoptik, *Phys. Blätter*, **39**, 50 (1983). M. Marko and H. Rose, The contributions of Otto Scherzer (1909–1982) to the development of the electron microscope, *Microsc. Microanal.* **16**, 366 (2010).
5. W. Greiner, Die Kernpolarisation in Mesonenatomen bei kugelsymmetrischen Mesonenzuständen, *Zeit. für Physik* **164**, 374 (1961).
6. L. N. Cooper and E. M. Henley, μ-mesonic atoms and the electromagnetic radius of the nucleus, *Phys. Rev.* **92**, 801 (1953).
7. H. Steinwedel and J. H. D. Jensen, Hydrodynamik von Kerndipolschwingungen, *Zeit. für Naturforschung* **5a**(8), 413 (1950).
8. E. Nuding, Kernpolarisation und Ebbe-Flut-Effekt beim μ-Meson-Atom, *Zeit. für Naturforschung* **12a**, 187 (1956).
9. W. Greiner and H. Marschall, Eine Abschätzung des von einem nichtkugelsymmetrischen Mesonenzustand hervorgerufenen Beitrags zur Kernpolarisation bei Mesonenatomen, *Zeit. für Physik* **165**, 171 (1961).
10. Y. N. Kim, *Mesic Atoms and Nuclear Structure* (North-Holland, Amsterdam, 1971).
11. S. Devons and I. Duerdoth, Muonic atoms, *Adv. Nucl. Phys.* **2**, 295 (1972).
12. W. Greiner and F. Scheck, Mesonenatomexperimente und Schallenstruktur, *Nucl. Phys. A* **41**, 424 (1968).
13. W. Pieper and W. Greiner, Nuclear dynamics and the X-ray spectrum of muonic atoms, *Nucl. Phys. A* **109**, 539 (1968).
14. F. W. Ruppert, Y. K. Chen and W. Greiner, The influence of nuclear excitations on the muonic X-ray spectrum of heavy spherical nuclei, *Nucl. Phys. A* **139**, 563 (1969).
15. A. Faessler and W. Greiner, Die Rotations-Vibrations-Wechselwirkung in deformierten Atomkernen, *Zeit. für Physik* **168**, 425 (1962).
16. A. S. Davydov, *Excited States of Atomic Nuclei* (Atomizdat, Moscow, 1967).
17. W. Greiner and A. Faessler, Vergleich der Davydov-Theorie mit dem Modell des achsial-symmetrischen Kreisels mit Rotations-Vibrations-Wechselwirkung für deformierte gg-Kerne, *Zeit. für Physik* **177**, 190 (1964).
18. A. Faessler, W. Greiner and R. K. Sheline, Rotation-vibration-interaction in deformed nuclei, *Nucl. Phys.* **70**, 33 (1965).
19. A. Faessler and W. Greiner, Zur potentiellen energie der γ-Schwingungen in deformierten gg-Kernen, *Zeit. für Physik* **170**, 105 (1963).
20. A. Faessler, W. Greiner and R. K. Sheline, Nuclear models and the Osmium isotopes, *Phys. Rev.* **135**, B591 (1964).
21. L. C. Biedenharn and P. J. Brussard, *Coulomb Excitation* (Clarendon Press, Oxford, 1965).
22. J. P. Davidson, *Collective Models of the Nucleus* (Academic Press, New York, 1968).
23. D. Rowe, *Nuclear Collective Motion* (Methuen, London, 1970).
24. W. F. Hornyak, *Nuclear Structure* (Academic Press, New York 1975).
25. A. A. Raduta, *Nuclear Structure with Coherent States* (Springer-Verlag, Heidelberg, 2015).
26. W. Greiner, Michael Danos — As I Remember Him, in *Proc. Symposium on Fundamental Issues in Elementary Matter*, 241 W. & E. Heraeus Seminar, ed. Walter Greiner (EP Systema, Debrecen, 2001), p. 415.

27. A. Faessler and W. Greiner, Zum kollektiven g_R-faktor in deformierten kernen, *Zeit. für Physik* **179**, 343 (1964).
28. W. Greiner, Magnetic dipole transitions and g_R-factors in deformed even-even nuclei, *Phys. Rev. Lett.* **14**, 599 (1965).
29. W. Greiner, Magnetic properties of even nuclei, *Nucl. Phys. A* **80**, 417 (1966).
30. A. A. Raduta and A. Faessler, A coherent state description of the shape phase transition in even-even Gd isotopes, *J. Phys. G: Nucl. and Part. Phys.* **31**(8), 873 (2005).
31. V. Maruhn-Rezwani, W. Greiner and J. Maruhn, Collective description of magnetic properties of even-even nuclei, *Phys. Lett. B* **57**, 109 (1975).
32. W. Greiner, Eine qualitative Diskussion der kollektiven potentiellen Energie in Quasispinmodell, *Zeit. für Physik* **172**, 386 (1966).
33. A. Faessler, W. Greiner and R. K. Sheline, The Coriolis anti-pairing and blocking effects in deformed nuclei, *Nucl. Phys.* **62**, 241 (1965).
34. W. Donner and W. Greiner, Octupole vibrations of deformed nuclei, *Zeit. für Physik* **197**, 440 (1966).
35. A. A. Raduta, Al. H. Raduta and A. Faessler, Positive and negative parity bands in pear-shaped nuclei, *J. Phys. G: Nucl. Part. Phys.* **23**, L49 (1997).
36. T. Guhr *et al.*, On the nature of low-lying electric dipole excitations in light and heavy deformed nuclei, *Nucl. Phys. A* **501**, 95 (1966).
37. W. Greiner and A. M. Green, Nuclear bremsstrahlung, *Nucl. Phys.* **49**, 481 (1963).
38. W. Greiner and M. Danos, The γ-absorption cross section of Holmium in the dynamic collective theory, *Phys. Lett.* **8**, 113 (1963).
39. W. Greiner and M. Danos, Dynamical theory of the nuclear collective model, *Phys. Rev.* **134**, B284 (1964).
40. M. Danos, W. Greiner and B. Kohr, The dynamic collective theory of odd-A-nuclei, *Phys. Rev.* **138**, B1055 (1965).
41. Ş. Mişicu, Interplay of compressional and vortical nuclear currents in overtones of the isoscalar giant dipole resonance, *Phys. Rev. C* **73**, 024301 (2005).
42. H. J. Weber, M. G. Huber and W. Greiner, Theorie der Riesenresonanzen in sphaerischen gg-kernen, *I. Zeit. für Physik* **192**, 182, II, 223 (1966).
43. M. G. Huber, M. Danos, H. J. Weber and W. Greiner, Collective treatment of the giant resonances in spherical nuclei, *Phys. Rev.* **155**, 1073 (1965).
44. R. Ligensa and W. Greiner, Dynamic collective theory of the quadrupole giant resonance, *Nucl. Phys. A* **92**, 673 (1967).
45. T. Urbas and W. Greiner, On the interaction of collective degrees of freedom in nuclei, *Zeit. für Physik* **196**, 44 (1966).
46. J. B. Seaborn, D. Drechsel, H. Arenhövel and W. Greiner, Collective correlations in the giant resonance of ^{60}Ni, *Phys. Lett.* **23**, 576 (1966).
47. D. Drechsel, J. B. Seaborn and W. Greiner, Collective correlations in spherical nuclei and the structure of giant resonances, *Phys. Rev.* **162**, 983 (1967).
48. R. Ligensa and W. Greiner, Nucleon escape widths of the giant dipole resonances in spherical nuclei, *Ann. Phys.* **51**, 28 (1969).
49. A. M. Lane and R. G. Thomas, R-matrix theory of nuclear reactions, *Rev. Mod. Phys.* **30**, 257 (1958).
50. M. Danos and W. Greiner, Shell-model treatment of nuclear reactions, *Phys. Rev.* **138**, B93 (1965).
51. M. Danos and W. Greiner, Eigenchannel theory for the nuclear reactions, *Phys. Rev.* **146**, 708 (1966).
52. C. Mahaux and H. A. Weidenmüller, Comparison between the R-matrix and eigenchannel methods, *Phys. Rev.* **170**, 847 (1966).

53. H. G. Wahsweiler, W. Greiner and M. Danos, Continuum nuclear structure of O^{16} in the eigenchannel reaction theory, *Phys. Rev.* **170**, 893 (1968).
54. H. G. Wahsweiler, W. Greiner and M. Danos, The 3^-continuum states of O^{16} in the eigenchannel reaction theory, *Phys. Rev. Lett.* **17**, 395 (1966).
55. M. Danos and W. Greiner, Treatment of nuclear reactions above the two-particle threshold, *Zeit. für Physik* **202**, 125 (1967).
56. H. J. Weber and W. Greiner, Treatment of cluster channels in nuclear reactions above the two-particle threshold, *Zeit. für Naturforschung* **24a**(5), 736 (1969).
57. R. F. Barrett, L. C. Biedenharn, M. Danos, P. P. Delsanto, W. Greiner and H. G. Wahsweiler, The eigenchannel method and related theories for nuclear reactions, *Rev. Mod. Phys.* **45**, 44 (1973).
58. M. Danos, W. Greiner and C. Byron Kohr, Static theory of the giant quadrupole resonance in deformed nuclei, *Phys. Rev.* **151**, 761 (1966).
59. H. Arenhövel and W. Greiner, The photon scattering by oriented nuclei, *Nucl. Phys.* **86**, 193 (1968).
60. H. Arenhövel, W. Greiner and M. Danos, The photonuclear effect in heavy deformed nuclei, *Phys. Rev.* **157**, 1109 (1967).
61. U. Mosel and W. Greiner, Investigation of the collective potential-energy surface, *Zeit. Für Physik* **217**, 256 (1968).
62. G. Gneuss, U. Mosel and W. Greiner, A new treatment of the collective nuclear Hamiltonian, *Phys. Lett.* **30**, 397 (1969).
63. G. Gneuss and W. Greiner, Collective potential energy surfaces and nuclear structure, *Nucl. Phys. A* **171**, 449 (1971).
64. P. O. Hess, M. Seiwert, J. A. Maruhn and W. Greiner, General collective model and its application to ^{238}U, *Zeit. für Physik* **296**, 147 (1968).
65. P. O. Hess, J. A. Maruhn and W. Greiner, The general collective model applied to the chains of Os and W-isotopes, *J. Phys. G: Nucl. and Part. Phys.* **7**, 737 (1981).
66. W. Scheid, R. Ligensa and W. Greiner, Ion-ion potentials and the compressibility of nuclear matter, *Phys. Rev. Lett.* **21**, 1479 (1968).
67. W. Scheid and W. Greiner, Quasimolecular optical nuclear potentials, *Zeit. für Physik* **226**, 364 (1969).
68. J. Eisenberg and W. Greiner, *Nuclear Theory I, Nuclear Models*, 3rd edn. (North-Holland, Amsterdam, 1988).

Photon Scattering off Nuclei

Hartmuth Arenhövel
Institut für Kernphysik, Johannes Gutenberg-Universität, D-55099 Mainz, Germany

Dedicated to the memory of Walter Greiner

The study of nuclear and subnuclear structure by means of photon scattering is outlined. Besides a brief exposition of the formalism a few illustrative examples are discussed.

1. Introduction

I would like to begin with a brief personal remark: It was during the fall of 1964 - I just had completed my diploma thesis in experimental physics at the University of Freiburg - , when Walter Greiner offered me to work on a PhD thesis in his new established theory group at the University of Frankfurt. As subject of the thesis he had proposed to investigate the structure of heavy deformed nuclei in the region of the giant resonances by means of photon scattering. By the end of 1965 Hans-Jürgen Weber and myself were Walter's first PhD-students to complete their PhD. Since then, my interest in the study of electromagnetic reactions on nuclei in general, e.g. photo absorption and scattering, electron scattering and meson production, has continued up to present times.

Electromagnetic reactions on nuclei provide an excellent tool to investigate nuclear structure. In addition, they also lead to valuable insights into the electromagnetic properties of the nuclear constituents, proton and neutron, for example, the electric and magnetic polarizabilities and electromagnetic form factors. In this context, photon scattering experiments are a particularly interesting source of information on off-shell properties of the nuclear constituents. On the other hand, genuine microscopic calculations of photon scattering cross sections are rather complicated since in principle the complete excitation spectrum has to be taken into account.

Here I will give a brief account of photon scattering reactions on nuclei with emphasis on my own work and that of my collaborators. It is not intended as a general review, rather a personal view on this interesting reaction which over many years has fascinated myself and which was and still is subject of my own research. Early reviews on nuclear photon scattering may be found in Refs. 1–3 and a more recent one in Ref. 4.

In the next section I will present a short summary of the basic scattering formalism, in particular the expansion of the scattering matrix in terms of generalized polarizabilities as basic quantities. It will be followed by a few illustrative examples, partly on earlier work on medium and heavy weight nuclei within the dynamic collective model of the giant dipole resonances and partly on the lightest nucleus, the deuteron, with emphasis on subnucleon degrees of freedom like meson exchange and isobar currents.

2. Formalism of Photon Scattering

I will briefly describe the formal features of the photon scattering process

$$\gamma_\lambda(k) + N_i(P_i) \longrightarrow \gamma'_{\lambda'}(k') + N_f(P_f)\,, \tag{1}$$

where an incoming photon with four momentum $k = (k_0, \vec{k})$ and circular polarization $\vec{e}_\lambda$ ($\lambda = \pm 1$) is scattered off a nucleus in the initial intrinsic state $|i\rangle$ with total four momentum $P_i = (E_i, \vec{P}_i)$ making a transition to a final intrinsic state $|f\rangle$ with total four momentum $P_f = (E_f, \vec{P}_f)$ while emitting a final photon with four momentum $k' = (k'_0, \vec{k}\,')$ and polarization $\vec{e}\,'_{\lambda'}$.

2.1. *The photon scattering amplitude*

Since the electromagnetic interaction is weak one makes a Taylor expansion with respect to the electromagnetic field A_μ up to second order, because at least two photons are involved in the scattering process. Thus in this lowest, i.e. second order in the e.m. coupling, the scattering amplitude is given by two terms, the contact or two photon amplitude (TPA or seagull) $B_{\lambda'\lambda}(\vec{k}\,', \vec{k})$, arising from the second order term of the Taylor expansion, and the resonance amplitude (RA) $R_{\lambda'\lambda}(\vec{k}\,', \vec{k})$ from the iterated linear interaction term. A diagrammatic illustration is shown in Fig. 1.

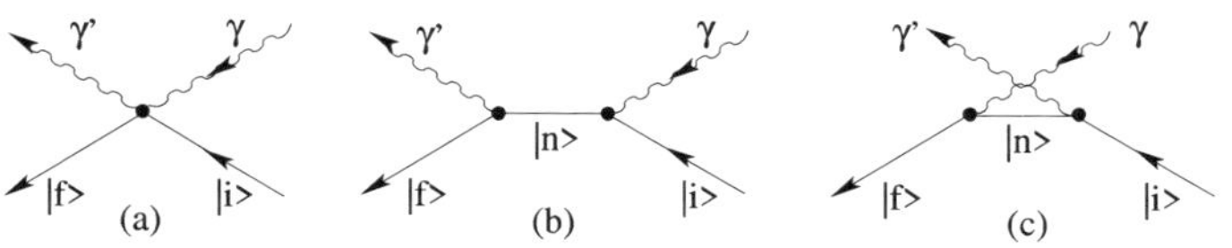

Fig. 1. Diagrammatic representation of photon scattering: The two-photon amplitude (a) and the resonance contribution (direct (b) and crossed (c)). To be read from right to left.

Accordingly, the total scattering amplitude is the sum of these two contributions

$$T^{fi}_{\lambda'\lambda}(\vec{k}\,', \vec{k}) = B^{fi}_{\lambda'\lambda}(\vec{k}\,', \vec{k}) + R^{fi}_{\lambda'\lambda}(\vec{k}\,', \vec{k})\,, \tag{2}$$

where the two-photon amplitude (diagram (a) of Fig. 1) has the form

$$B^{fi}_{\lambda'\lambda}(\vec{k}', \vec{k}) = \sum_{l,m=1}^{3} e'^{*}_{\lambda',l} \langle f|\widehat{B}_{lm}(\vec{k}\,', \vec{k})|i\rangle e_{\lambda,m}\,, \tag{3}$$

with $\widehat{B}_{lm}(\vec{k}\,', \vec{k}\,)$ as Fourier transform of the second order coefficient of the Taylor expansion $\widetilde{B}_{lm}(\vec{x}, \vec{y}\,)$, i.e.

$$\widehat{B}_{lm}(\vec{k}\,', \vec{k}\,) = -\int d^3x\, d^3y\, e^{i\vec{k}\,'\cdot\vec{x}}\widetilde{B}_{lm}(\vec{x}, \vec{y}\,)e^{-i\vec{k}\cdot\vec{y}}\,. \tag{4}$$

Assuming a nonrelativistic description, the overall center-of-mass (c.m.) motion can be separated and the Hamiltonian splits accordingly into an intrinsic part and the c.m. kinetic energy

$$H = H_{int} + \frac{\vec{P}^{\,2}}{2M_A}\,, \tag{5}$$

where $H_{int} = T_{int} + V$ denotes the Hamiltonian of the internal motion, $\vec{P}$ the total c.m. momentum and M_A the mass of the nucleus. An intrinsic state is denoted by $|n\rangle$ with intrinsic energy e_n. Then the resonance amplitude (RA) (diagrams (b) and (c) of Fig. 1) is given by the following matrix element between intrinsic states

$$\begin{aligned} R^{fi}_{\lambda'\lambda}(\vec{k}\,', \vec{k}) = \langle f|\, \Big[\, & \vec{e}^{\,\prime *}_{\lambda'}\cdot\vec{J}(-\vec{k}\,', 2\vec{P}_f + \vec{k}\,') \\ & \times G_{int}\Big(k_0 - \frac{\vec{k}\cdot(2\vec{P}_i + \vec{k})}{2M_A} + i\epsilon\Big)\,\vec{e}_\lambda\cdot\vec{J}(\vec{k}, 2\vec{P}_i + \vec{k}) \\ & + \Big(\vec{e}_\lambda \leftrightarrow \vec{e}^{\,\prime *}_{\lambda'}\,, k_\mu \leftrightarrow -\vec{k}^{\,\prime}_\mu\Big)\Big]|i\rangle\,, \end{aligned} \tag{6}$$

with the resolvent or propagator

$$G_{int}(z) = (H_{int} - e_i - z)^{-1}\,. \tag{7}$$

The current operator in Eq. (6)

$$\vec{J}(\vec{k}, \vec{P}) = \vec{\jmath}(\vec{k}) + \frac{1}{2M_A}\,\vec{P}\,\rho(\vec{k}) \tag{8}$$

consists of the intrinsic current density operator $\vec{\jmath}(\vec{k}\,)$ plus a term taking into account the convection current of the separated c.m. motion, where $\rho(\vec{k}\,)$ denotes the Fourier transform of the intrinsic charge density operator of the nucleus. The intrinsic charge and current density operators consist of a kinetic or one-body $(\rho_{[1]}, \vec{\jmath}_{[1]})$ and a two-body meson exchange part $(\rho_{[2]}, \vec{\jmath}_{[2]})$

$$\rho(\vec{k}\,) = \rho_{[1]}(\vec{k}\,) + \rho_{[2]}(\vec{k}\,)\,, \tag{9}$$

$$\vec{\jmath}(\vec{k}\,) = \vec{\jmath}_{[1]}(\vec{k}\,) + \vec{\jmath}_{[2]}(\vec{k}\,)\,, \tag{10}$$

with

$$\rho_{[1]}(\vec{k}) = \sum_l e_l\, e^{-i\vec{k}\cdot\vec{r}_l}\,, \tag{11}$$

$$\vec{\jmath}_{[1]}(\vec{k}\,) = \frac{1}{2M}\sum_l \Big(e_l\{\vec{p}_l, e^{-i\vec{k}\cdot\vec{r}_l}\} + \mu_l\vec{\sigma}_l\times\vec{k}\,e^{-i\vec{k}\cdot\vec{r}_l}\Big)\,. \tag{12}$$

Here, e_l and μ_l denote charge and magnetic moment of the l-th particle and $\vec{r}_l$, $\vec{p}_l$ and $\vec{\sigma}_l$ its internal coordinate, momentum and spin operators, respectively. The expressions for the corresponding exchange operators depend on the interaction model. At least in the nonrelativistic limit, the exchange contribution to the charge density vanishes (Siegert's hypothesis). Furthermore, also the TPA consists of a kinetic one-body contribution and a two-body exchange amplitude

$$\widehat{B}_{lm}(\vec{k}\,',\vec{k}\,) = \widehat{B}_{[1],lm}(\vec{k}\,',\vec{k}\,) + \widehat{B}_{[2],lm}(\vec{k}\,',\vec{k}\,)\,, \tag{13}$$

where the kinetic one-body operator is given by the sum of the individual proton Thomson scattering amplitudes

$$\widehat{B}_{[1],lm}(\vec{k}\,',\vec{k}\,) = -\frac{1}{M}\sum_j e_j^2\, e^{-i(\vec{k}-\vec{k}\,')\cdot\vec{r}_j}\,\delta_{lm}\,. \tag{14}$$

It is important to note that the splitting of the scattering amplitude into a resonance and a two-photon amplitude is gauge dependent. This gauge dependence is reflected in gauge conditions for the current and the two-photon amplitude which follow from the Gauge invariance of the electromagnetic interaction. In detail one finds the following gauge conditions for the e.m. operators

$$\vec{k}\cdot\vec{\jmath}(\vec{k}\,) = [H,\rho(\vec{k}\,)]\,, \tag{15}$$

$$\sum_l k_l' B_{lm}(\vec{k}\,',\vec{k}\,) = [\rho(-\vec{k}\,'),j_m(\vec{k}\,)]\,. \tag{16}$$

The first condition in (15), connecting the charge density with the current, describes current conservation, while the second relates the TPA to the commutator of charge and current densities. Separating the one-body and two-body (exchange) contributions, one finds the following conditions

$$\vec{k}\cdot\vec{\jmath}_{[1]}(\vec{k}\,) = [T,\rho_{[1]}(\vec{k}\,)]\,, \tag{17}$$

$$\vec{k}\cdot\vec{\jmath}_{[2]}(\vec{k}\,) = [V,\rho_{[1]}(\vec{k}\,)] + [T,\rho_{[2]}(\vec{k}\,)]\,, \tag{18}$$

$$\sum_l k_l' \widehat{B}_{[1],lm}(\vec{k}\,',\vec{k}\,) = [\rho_{[1]}(-\vec{k}\,'),j_{[1],m}(\vec{k}\,)]\,, \tag{19}$$

$$\sum_l k_l' \widehat{B}_{[2],lm}(\vec{k}\,',\vec{k}\,) = [\rho_{[1]}(-\vec{k}\,'),j_{[2],m}(\vec{k}\,)] + [\rho_{[2]}(-\vec{k}\,'),j_{[1],m}(\vec{k}\,)]\,. \tag{20}$$

Important consequences are the low energy limits[5,6]

$$\vec{\jmath}(0) = [H,\vec{D}]\,, \tag{21}$$

the Siegert theorem, where $\vec{D}$ denotes the unretarded dipole operator, and

$$\widehat{B}^{ii}_{[1],lm}(0,0) = -\frac{Ze^2}{M}\delta_{lm}\,, \tag{22}$$

$$\widehat{B}^{ii}_{[2],lm}(0,0) = -\langle i|[D_l,[V,D_m]]|i\rangle\,, \tag{23}$$

$$\widehat{R}^{ii}_{lm}(0,0) = \frac{NZe^2}{AM}\delta_{lm} - \widehat{B}^{ii}_{[2],lm}(0,0)\,, \tag{24}$$

resulting in the classical Thomson limit for the total nuclear scattering amplitude

$$T^{ii}_{\lambda'\lambda}(0,0) = -\vec{e}\,'^{*}_{\lambda'}\cdot\vec{e}_\lambda\frac{(Ze)^2}{AM}\,, \tag{25}$$

with the approximation $M_A = AM$.

To close this section, I would like to mention the optical theorem, which relates the forward elastic scattering amplitude to the total photo absorption cross section

$$\sigma_{tot}(k,\rho) = \frac{4\pi}{k}\,\mathrm{Im}\Big[\mathrm{Tr}\Big(\rho\, T^{ii}(\vec{k},\vec{k}\,)\Big)\Big]\,, \tag{26}$$

where ρ denotes the photon-nucleus polarization density matrix of the initial state.

2.2. *Generalized nuclear polarizabilities*

In the scattering process the incoming photon and the scattered one transfer various angular momenta via the e.m. multipole operators L and L' to the nucleus which can be coupled to a total angular momentum transfer J. It is very useful to expand the scattering amplitude with respect to this total angular momentum transfer. It leads to the concept of generalized polarizabilities $P^{(L'L)J}_{fi,\lambda'\lambda}(k',k)$, first introduced by Fano[7] for the case of pure $E1$ transitions and later generalized in Refs. 8–11. A review may be found in Ref. 1. A further generalization to the $(e,e'\gamma)$ reaction of electron scattering (virtual compton scattering) is given in Ref. 12.

The polarizabilities allow one to separate geometrical aspects related to the angular momentum properties from dynamical effects as contained in the strength of the various polarizabilities. The expansion of the total scattering amplitude in terms of these polarizabilities reads

$$\begin{aligned} T^{fi}_{M_f M_i,\lambda'\lambda}(\vec{k}\,',\vec{k}) = (-)^{I_f - M_i} \sum_{L',M',L,M,J,m} (-)^{L+L'}\widehat{J}^2 \begin{pmatrix} I_f & J & I_i \\ -M_f & m & M_i \end{pmatrix} \\ \times \begin{pmatrix} L & L' & J \\ M & M' & -m \end{pmatrix} P^{(L'L)J}_{fi,\lambda'\lambda}(k',k) D^{L}_{M,\lambda}(R) D^{L'}_{M',-\lambda'}(R')\,, \end{aligned} \tag{27}$$

where the abbreviation $\widehat{J} = \sqrt{2J+1}$ is used, and (I_i, M_i) and (I_f, M_f) refer to the angular momenta of the initial and final states and their projections on the quantization axis, respectively. Furthermore, R and R' describe the rotations which carry the quantization axis into the directions of the photon momenta $\vec{k}$ and $\vec{k}\,'$, respectively, and $D^{L}_{M,\lambda}(R)$ and $D^{L'}_{M',-\lambda'}(R')$ denote the corresponding rotation matrices in the convention of Rose.[13] The polarizabilities contain the dynamic properties of the system and depend on the absolute values of the photon momenta only, whereas the geometrical aspects, i.e. the angular dependencies, are contained in the rotation matrices.

The general definition of the polarizabilities is then obtained by the inversion of Eq. (27), i.e.

$$\begin{aligned} P^{(L'L)J}_{fi,\lambda'\lambda}(k',k) = \frac{(-)^{L'-L-I_f}}{8\pi^2}\widehat{L'}^2\widehat{L}^2 \sum_{M_f M_i M' M m} \begin{pmatrix} I_f & J & I_i \\ -M_f & m & M_i \end{pmatrix} \begin{pmatrix} L & L' & J \\ M & M' & -m \end{pmatrix} \\ \times \int dR' \int dR\, D^{L*}_{M,\lambda}(R) D^{L'*}_{M',-\lambda'}(R')\, T^{fi}_{M_f M_i,\lambda'\lambda}(\vec{k}\,',\vec{k})\,. \end{aligned} \tag{28}$$

One should note that J is bound by the multipole orders L and L' and the spins I_i and I_f of the initial and final nuclear states, respectively, i.e.

$$|L - L'| \leq J \leq L + L' \quad \text{and} \quad |I_i - I_f| \leq J \leq I_i + I_f \,. \tag{29}$$

For example, to elastic scattering off a spin-zero nucleus only the scalar polarizabilities ($J = 0$) contribute.

It is furthermore useful to classify the polarizabilities according to the total parity transfer in case that parity is conserved. This leads to the introduction of

$$P^J_{fi}(M^{\nu'}L', M^\nu L, k', k) = \frac{1}{4} \sum_{\lambda',\lambda=\pm1} \lambda'^{\nu'} \lambda^\nu \, P^{(L'L)J}_{fi,\lambda'\lambda}(k', k) \,. \tag{30}$$

In terms of these one has

$$P^{(L'L)J}_{fi,\lambda'\lambda}(k', k) = \sum_{\nu',\nu=0,1} \lambda'^{\nu'} \lambda^\nu P^J_{fi}(M^{\nu'}L', M^\nu L, k', k) \,, \tag{31}$$

where ν and ν' classify the type of multipole transition, i.e. $\nu = 0$ means electric ($M^0L = EL$) and $\nu = 1$ magnetic ($M^1L = ML$).

For parity conservation a simple selection rule follows

$$P^J_{fi}(M^{\nu'}L', M^\nu L, k', k) = 0 \,, \quad \text{if} \quad (-)^{L'+\nu'+L+\nu} \neq \pi_i \pi_f \,, \tag{32}$$

with π_i and π_f denoting the parities of intial and final states, respectively. A graphical visualization of the generalized polarizability is shown in Fig. 2 for the direct resonance term.

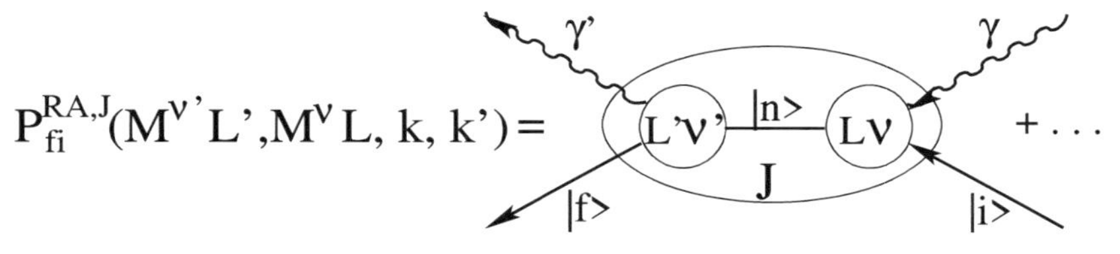

Fig. 2. Graphical representation of the contribution of the direct term of the resonance amplitude to the generalized polarizability $P^J_{fi}(M^{\nu'}L', M^\nu L, k', k)$, where the angular momentum transfer L of the incoming photon is coupled with the angular momentum transfer L' of the outgoing photon to a total angular momentum transfer J.

The polarizabilities $P^J_{fi}(M^{\nu'}L', M^\nu L, k', k)$ can be separated into a TPA and a resonance contribution

$$P^J_{fi}(M^{\nu'}L', M^\nu L, k', k) = P^{TPA,J}_{fi}(M^{\nu'}L', M^\nu L, k', k) + P^{RA,J}_{fi}(M^{\nu'}L', M^\nu L, k', k) \,, \tag{33}$$

where for the resonance amplitude one has

$$P^{RA,J}_{fi}(M^{\nu'}L', M^\nu L, k', k) = 2\pi(-)^{L+J} \frac{\hat{L}\hat{L}'}{\hat{J}} \times \langle I_f e_f || \Bigg(\Big[M^{\nu'[L']}(k') G_{int}(k_0 - \frac{\vec{k} \cdot (2\vec{P}_i + \vec{k})}{2M_A} + i\varepsilon) M^{\nu[L]}(k) \Big]^{[J]} + \Big\{ M^{\nu[L]}(k) \leftrightarrow M^{\nu'[L']}(k') \,, k_\mu \leftrightarrow -\vec{k}'_\mu \Big\} \Bigg) || I_i e_i \rangle \,. \tag{34}$$

The superscript $[J]$ indicates a spherical tensor of rank J, and "$[\dots]^{[J]}$" means that two spherical tensors are coupled to a spherical tensor of rank J. Furthermore, $M^{\nu'[L']}(k')$ denotes a standard electromagnetic current multipole operator[13]

$$M_m^{\nu'[L']}(k') = \begin{cases} \nu = 0 \quad \text{electric}: \int d^3x\, \vec{A}_m^{[L]}(E;k,\vec{x}\,)\cdot\vec{J}(\vec{x}\,)\,, \\ \nu = 1\, \text{magnetic}: \int d^3x\, \vec{A}_m^{[L]}(M;k,\vec{x}\,)\cdot\vec{J}(\vec{x}\,)\,, \end{cases} \tag{35}$$

with the multipole fields

$$\vec{A}_m^{[L]}(M;k,\vec{x}\,) = i^L j_L(kx)\vec{Y}_m^{(L1)[L]}(\hat{x})\,, \tag{36}$$

$$\vec{A}_m^{[L]}(E;k,\vec{x}\,) = \frac{1}{k}\vec{\nabla}\times\vec{A}_m^{[L]}(M;k,\vec{x}\,)\,. \tag{37}$$

The two-photon contribution to the polarizability is given by

$$P_{fi}^{TPA,J}(M^{\nu'}L', M^{\nu}L, k', k) = 2\pi(-)^{L+J+1}\frac{\hat{L}\hat{L}'}{\hat{J}}$$
$$\langle I_f e_f || \int d^3x\, d^3y \sum_{lm}\Big[A_l^{[L']}(M^{\nu'};k,\vec{x}\,)B_{lm}(\vec{x},\vec{y}\,)A_m^{[L]}(M^{\nu};k,\vec{y}\,)\Big]^{[J]}||I_i e_i\rangle\,. \tag{38}$$

The evaluation of the TPA contribution to the polarizabilities is straightforward once the TPA operator $B_{lm}(\vec{x},\vec{y}\,)$ is given.

For the resonance contribution, one finds by evaluating the reduced matrix element in standard fashion (see e.g. Ref. 14)

$$P_{fi}^{RA,J}(M^{\nu'}L', M^{\nu}L, k', k) = 2\pi(-)^{L+I_f+I_i}\hat{L}\hat{L}'$$
$$\times \sum\!\!\!\!\!\!\int_{e_n,I_n}\Bigg[\begin{Bmatrix} L & L' & J \\ I_f & I_i & I_n\end{Bmatrix}\frac{\langle I_f e_f||M^{\nu'[L']}(k')||I_n e_n\rangle\langle I_n e_n||M^{\nu[L]}(k)||I_i e_i\rangle}{e_n - e_i - k_0 + \frac{\vec{k}\cdot(2\vec{P}_i+\vec{k})}{2M_A} - i\varepsilon}$$
$$+(-)^{L+L'+J}\begin{Bmatrix} L' & L & J \\ I_f & I_i & I_n\end{Bmatrix}\frac{\langle I_f e_f||M^{\nu[L]}(k)||I_n e_n\rangle\langle I_n e_n||M^{\nu'[L']}(k')||I_i e_i\rangle}{e_n - e_i + k_0' - \frac{\vec{k}\,'\cdot(2\vec{P}_i-\vec{k}\,')}{2M_A} - i\varepsilon}\Bigg]\,. \tag{39}$$

Obviously, the calculation of the resonance part is more involved because of the summation over all possible intermediate states $|I_n e_n\rangle$ with angular momentum I_n and intrinsic energy e_n.

The optical theorem in (26) allows one to relate the imaginary part of the scalar polarizabilities of elastic scattering $P_{ii}^0(M^\nu L, M^\nu L, k, k)$ to the partial contribution $\sigma(M^\nu L)(k)$ of the multipole $M^\nu L$ to the total unpolarized photo absorption cross section $\bar{\sigma}_{tot}$

$$\begin{aligned}\bar{\sigma}_{tot}(k) &= \sum_L \Big(\sigma(EL)(k) + \sigma(ML)(k)\Big) \\ &= \frac{4\pi}{2k(2I_i+1)}\sum_{\lambda M}\operatorname{Im} T^{ii}_{\lambda\lambda MM}(\vec{k},\vec{k}) \\ &= \frac{4\pi}{k\hat{I}_i}\sum_L\frac{(-)^{L+1}}{\hat{L}}\operatorname{Im}\Big[P_{ii}^{RA,0}(EL,EL,k,k) + P_{ii}^{RA,0}(ML,ML,k,k)\Big]\,. \end{aligned} \tag{40}$$

From this relation follows

$$\mathrm{Im}[P^0_{ii}(M^\nu L, M^\nu L, k, k)] = \frac{k}{4\pi}(-)^{L+1}\widehat{I}_i\widehat{L}\sigma(M^\nu L)(k)\,. \tag{41}$$

An important property of the scattering amplitude is the low energy theorem[5] according to which up to terms linear in the photon momentum k the scattering amplitude is completely determined by global properties like charge, mass and magnetic moment. This means for the polarizabilities that in the limit $k = 0$ only the scalar E1-polarizability is nonvanishing, i.e.

$$P^J(E1, E1)|_{k=0} = -\delta_{J0}\,\widehat{I}_i\,\sqrt{3}\,\frac{e^2 Z^2}{M_A}\,, \tag{42}$$

with I_i as ground state spin. It corresponds to the Thomson scattering amplitude. Internal properties like static electric and magnetic polarizabilities contribute in the next nonvanishing order (k^2) only. An extension of this theorem by a more general low energy expansion of the polarizabilities has been discussed in Ref. 6 with the result that for electric multipole transitions EL with even L the contributions to $P^J(EL, EL)|_{k=0}$ up to order $2L-2$ vanish.

2.3. *The elastic scattering cross section*

The elastic scattering cross section for unpolarized photons and targets is given by

$$\frac{d\sigma_{elastic}}{d\Omega} = \frac{c(P_i, k, k')}{2(2I_i+1)} \sum_{\lambda,\lambda',M_i,M_f} |T^{ii}_{M_f M_i,\lambda'\lambda}(\vec{k}\,', \vec{k})|^2\,, \tag{43}$$

with a kinematic factor for collinear initial momenta

$$c(P_i, k, k') = \frac{E_i(k_0 + E_i - k'_0)}{(\widehat{k}' \cdot (k + P_i))^2}\,, \tag{44}$$

where $\widehat{k}' = k'/k'_0$. For elastic scattering in the c.m. frame ($\vec{k} + \vec{P}_i = 0$ and $k_0 = k'_0$) one has $c_{c.m.}(P_i, k, k') = (M_A^2 + k^2)/W^2$ with $W = k_0 + E_i$ as invariant mass.

In terms of the polarizabilities one finds for the unpolarized cross section[1]

$$\begin{aligned}\frac{d\sigma_{elastic}}{d\Omega} = \frac{c(P_i, k, k')}{2I_i+1} &\sum_{L',L,K',K,J}\ \sum_{\nu',\nu,\bar\nu',\bar\nu} P^J_{ii}(M^{\nu'}L', M^\nu L, k', k)\\ &\times P^J_{ii}(M^{\bar\nu'}K', M^{\bar\nu}K, k', k)^* g_J^{\nu' L'\nu L;\bar\nu' K'\bar\nu K}(\theta)\,,\end{aligned} \tag{45}$$

where the angular functions depend on the scattering angle θ only and are given by

$$\begin{aligned}g_J^{\nu' L'\nu L;\bar\nu' K'\bar\nu K}(\theta) = &\frac{(-)^J}{2}\widehat{J}^2(-)^{L+K+\nu'+\bar\nu'}\\ &\times \sum_j \widehat{j}^2(1+(-)^{L+K+j+\nu+\bar\nu})(1+(-)^{L'+K'+j+\nu'+\bar\nu'})\\ &\times \begin{pmatrix} L' & K' & j\\ 1 & -1 & 0\end{pmatrix}\begin{pmatrix} L & K & j\\ 1 & -1 & 0\end{pmatrix}\begin{Bmatrix} L & K & j\\ K' & L' & J\end{Bmatrix} P_j(\cos\theta)\,,\end{aligned} \tag{46}$$

with $P_j(\cos\theta)$ denoting a Legendre polynomial. For pure $E1$ transitions this expression simplifies considerably and one obtains with scalar, vector and tensor polarizabilities

$$\frac{d\sigma_{elastic}(E1)}{d\Omega} = \frac{c(P_i, k, k')}{(2I_i + 1)} \sum_{J=0}^{2} |P_{ii}^J(E1, E1)|^2 \, g_J^{E1}(\theta)\,, \tag{47}$$

where in an abbreviated notation the angular functions are

$$g_0^{E1}(\theta) = \frac{1}{6}\,(1 + \cos^2\theta)\,, \tag{48}$$

$$g_1^{E1}(\theta) = \frac{1}{4}\,(2 + \sin^2\theta)\,, \tag{49}$$

$$g_2^{E1}(\theta) = \frac{1}{12}\,(13 + \cos^2\theta)\,. \tag{50}$$

In this case the angular distribution is symmetrical around 90° in the c.m. system.

The generalization to polarized photons and oriented nuclei is presented in Refs. 1 and 15.

3. Applications

Now I will discuss several applications: (i) Heavy deformed nuclei within the dynamic collective model of Danos and Greiner, (ii) Photon scattering off ^{12}C for energies in the region of the first nucleon resonance, the $\Delta(1232)$, and finally (iii) photon scattering off the deuteron as a means to study subnuclear degrees of freedom, for example meson exchange currents or the static polarizabilities of the neutron.

3.1. *Photon scattering off complex nuclei in the giant dipole resonance region*

The giant dipole resonance (GDR) in medium and heavy weight nuclei can well be explained as a collective phenomenon in the framework of the hydrodynamical model of Steinwedel and Jensen as an oscillation of a proton fluid against a neutron fluid. According to this model, as pointed out independently by Danos and Okamoto,[16] the GDR will be split into two peaks for an axially symmetric deformed nucleus corresponding to different frequencies for oscillations along and perpendicular to the symmetry axis.

In view of the additional collective surface degrees of freedom, Danos and Greiner[17] proposed in 1964 a unified dynamic collective model of the giant resonances (DCM) which includes the coupling between the rotation-vibration surface degrees of freedom and the giant resonance d.o.f. leading to additional dynamic effects. A weak point of this approach is that it provides energies and strengths of the GDR states at discrete energies only, but not the shape nor the widths of the states. Usually a Lorentzian shape is assumed with some simple *ad hoc* model for the width, the parameters of which are used for a fit procedure to the absorption cross section.

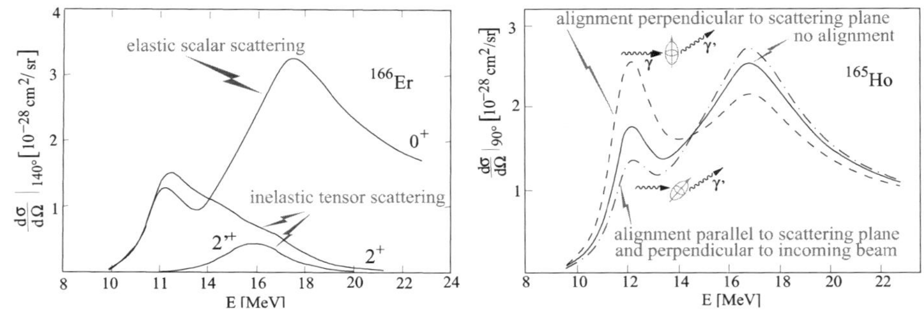

Fig. 3. Left panel: Calculated elastic and inelastic photon scattering cross sections at 140° for ^{166}Er (from Ref. 9). Right panel: Elastic photon scattering cross sections for ^{165}Ho (from Ref. 15): unoriented target: solid curve; aligned target: (a) perpendicular to scattering plane: dashed curve, (b) parallel to scattering plane and perpendicular to incoming photon beam: dash-dot curve.

As a result of this dynamic coupling, considerably strong dipole transition strengths from the GDR states to the low lying rotational and vibrational states appear leading to sizeable Raman scalar and tensor scattering into these low lying collective states (see Fig. 3, left panel). Indeed, such a Raman scattering has been measured for ^{238}U, ^{232}Th and ^{209}Bi by Jackson and Wetzel[18] although these authors found a significantly weaker Raman cross section (about 40%) than predicted by the DCM. A similar reduction of the inelastic DCM strenghts was found later for a series of vibrational medium weight nuclei by Bowles *et al.*[19] A possible explanation of this reduction could be a 10–15% nonresonant contribution of direct transitions into the continuum which would appear in the scalar but not in the tensor polarizability as has been discussed in Ref. 20.

Another interesting feature of the DCM is the fact that a deformed nucleus with a nonvanishing ground state spin becomes dynamically triaxial and thus optically anisotropic (nonvanishing elastic tensor polarization) and thus its absorption and scattering cross sections depend on the nuclear orientation. The reason for this feature is the fact that for a nonvanishing ground state spin $I \geq 1$ there is a tensor contribution to elastic scattering which, however, does not show up for an unoriented nucleus. This is shown on the right panel of Fig. 3 for ^{165}Ho having a ground state spin $I = 7/2$. Experimentally such a dependence of the photon absorption cross section of ^{165}Ho on the nuclear orientation was found by Ambler *et al.*[21]

Subsequently, the DCM was further developed in order to describe a much more general class of potential energy surfaces for the low energy collective d.o.f. allowing a unified description of the GDR for nuclei with quite different collective characteristics by Rezwani *et al.*[22,23] An example, taken from Ref. 22, is shown in Fig. 4 displaying the collective potential energy surface, the absorption strength and cross section and the elastic and inelastic scattering cross sections for three different types of nuclei, resembling a transition from a nucleus with an anharmonic vibrational character (A) to a strongly deformed rotational nucleus (C).

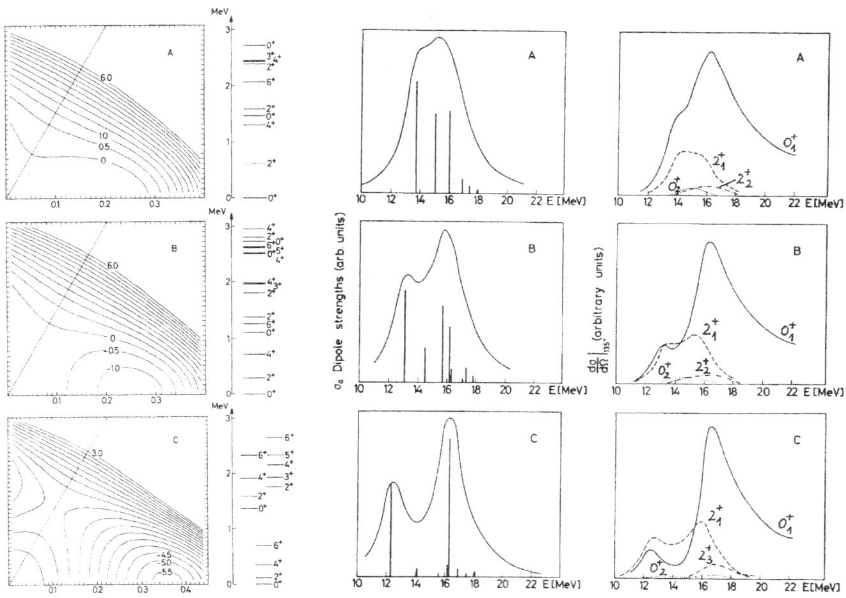

Fig. 4. Transition study from a vibrational (A) to a strongly deformed nucleus (C) (from Ref. 22): Left panels: Collective potential energy surfaces and low-energy spectra. Middle panels: Dipole strengths and γ-absorption cross sections. Right panels: Elastic and Raman γ-scattering cross sections to low lying collective states.

Case A represents a vibrational nucleus with strong anharmonicities, resulting in a shift and splitting of the surface two-phonon triplet. Furthermore, a slight axially symmetric deformation leads to a nonvanishing intrinsic quadrupole moment. Since the nucleus is easily deformable in β- and γ-directions, it appears dynamically triaxial resulting in a slight splitting of the dipole strength into three equally spaced states. Moreover, considerable inelastic tensor γ-scattering into the low-lying 2^+-states is observed.

The opposite situation of a strongly deformed nucleus is displayed by case C. It corresponds to a good axially symmetric rotator with clearly separated ground, β- and γ-rotational states. Accordingly the GDR is split into two distinct peaks, and strong Raman-scattering via the tensor polarizability into the 2^+-state of the ground state rotational band appears. The transition between these two extreme cases is represented by case B.

3.2. *Photon scattering off complex nuclei in the* $\Delta(1232)$ *region*

The influence of internal nucleon degrees of freedom constitutes an important field of research in nuclear and medium energy physics. In particular, the role of the lowest excited state of the nucleon, the $\Delta(1232)$ resonance, has been studied in the low energy domain[24] as well as in the energy region of real Δ excitation in pion production and photo absorption.

For photon scattering in the energy region of about 300 MeV the simplest model is a static approach assuming pure $M1$-scattering off the individual nucleons, whereas the nuclear structure is manifest only via the elastic form factor with respect to the momentum transfer. Then the scattering matrix has the simple form[25]

$$T^{fi}_{\lambda'\lambda} = \langle f| \sum_{l=1}^{A} t_{\lambda'\lambda}(l) e^{-i(\vec{k}-\vec{k}')\cdot\vec{r}_l} |i\rangle \,, \tag{51}$$

where the elementary scattering operator is given by

$$\begin{aligned} t_{\lambda'\lambda}(l) = c^2 \Big(& \frac{\vec{e}^{\,\prime *}_{\lambda'} \cdot [\vec{\sigma}^{\,l}_{N\Delta} \times \vec{k}\,'] \, \vec{e}_{\lambda} \cdot [\vec{\sigma}^{\,l}_{\Delta N} \times \vec{k}\,]}{M_\Delta - M_N - k_0 - i\Gamma_\Delta/2} \\ & + \frac{\vec{e}_{\lambda} \cdot [\vec{\sigma}^{\,l}_{N\Delta} \times \vec{k}\,] \, \vec{e}^{\,\prime *}_{\lambda'} \cdot [\vec{\sigma}^{\,l}_{\Delta N} \times \vec{k}\,']}{M_\Delta - M_N + k_0' - i\Gamma_\Delta/2} \Big) . \end{aligned} \tag{52}$$

The $N \to \Delta$ spin transition matrix $\vec{\sigma}^{\,l}_{N\Delta}$ is defined as in Ref. 24. Furthermore,

$$c = G^{\Delta N}_{M1} (M_\Delta + M_N)/(4 M_\Delta M_N) \,, \tag{53}$$

where M_N, M_Δ and Γ_Δ denote respectively nucleon and Δ mass and width. The latter and the magnetic transition strength $G^{\Delta N}_{M1}$ are fit to the experimental photo absorption cross section of the nucleon in the Δ resonance region.

In this simple approach nuclear structure enters only via the nuclear form factors of mass and spin (for more details see Ref. 25). A comparison of this approach with experiment for ^{12}C and ^{208}Pb is shown in Fig. 5.

It turns out, that the resulting calculated cross section largely underestimates the experiment by orders of magnitude for both nuclei. Collective effects in a more refined Δ-hole model for ^{12}C by Koch *et al.*[27] lead to a slight enhancement but the discrepancy remained essentially. However, Hayward and Ziegler[26] already had pointed out that the finite energy resolution with respect to the measured scattered photons of about 10% would lead to the inclusion of corresponding inelastic contributions in the experimental data.

Subsequently, such inelastic contributions have been studied in Ref. 25 within two simple models: (i) Inelastic scattering into 1-particle-1-hole excitations with excitation energies within an energy interval Δk, and (ii) a sum rule approach to the giant resonances as final states (see Ref. 25 for further details). The results of these two models are also shown in Fig. 5. Both approaches show an enhancement due to inelastic contributions, particularly strong for the 1p-1h approach. However, for ^{12}C still sizeable strength is missing and a more thorough theoretical treatment is needed.

3.3. *Photon scattering off the deuteron*

Now I would like to turn my attention to more fundamental studies with respect to the role of subnuclear degrees of freedom like meson exchange currents (MEC)

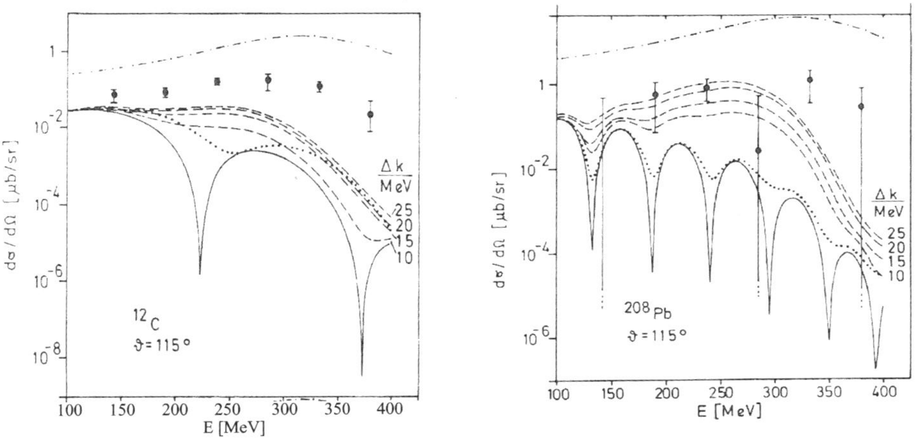

(a) ^{12}C (b) ^{208}Pb

Fig. 5. Elastic and quasielastic photon scattering cross sections at $\theta = 115°$ for ^{12}C and ^{208}Pb (from Ref. 25). Experimental data from Ref. 26. Solid curve: elastic scattering in static approach; dashed curves include inelastic contributions up to an excitation energy Δk; dotted curve: sum rule approach (see Ref. 25 for further details); dash-dotted curve: incoherent sum of elastic elementary $\gamma - N$ scattering cross section.

and internal nucleon degrees of freedom in terms of nucleon resonances (isobar currents IC) in electromagnetic reactions. Lightest nuclei present ideal laboratories for such studies and I will concentrate on the deuteron for which extensive studies on electromagnetic reactions exist.[2,3,28,29]

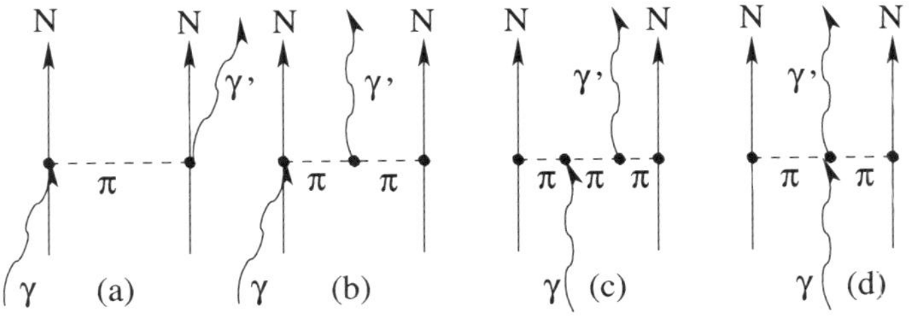

Fig. 6. Diagrammatic representation of pion exchange TPA contributions to the two-photon amplitude.

In photon scattering, meson exchange effects do not only appear via MEC in the resonance amplitude but also as additional contributions in the TPA as derived in Ref. 30 for a NN one-pion exchange potential (see Fig. 6 for the corresponding diagrams) as well as for isobar contributions.

It is intriguing, and indeed had been suggested, that diagram (d) of Fig. 6 allows the extraction of a density form factor of charged mesons in the nucleus (see references in Ref. 31). However, a careful analysis has shown that such an

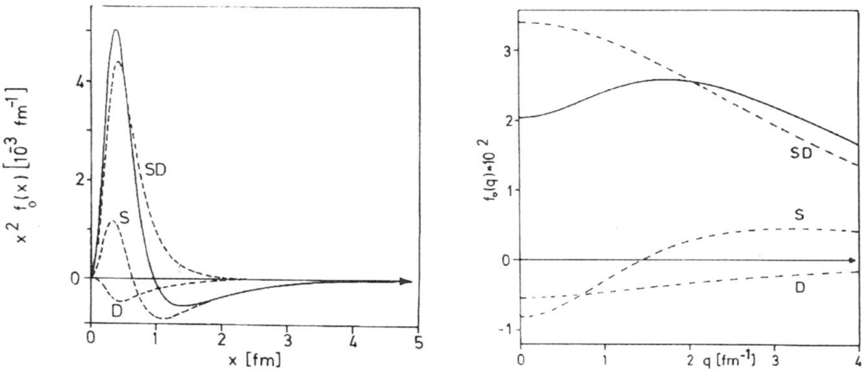

Fig. 7. Left panel: Monopole part of π-exchange transition density for the deuteron (solid curve) from the π-TPA diagram (d) of Fig. 6. Dashed curves show separate contributions from S- and D-states and $S-D$ interference. Right panel: Form factor of transition density displayed in the left panel, again with separate contributions (from Ref. 31).

interpretation is not possible.[31] It turns out that the TPA of an exchanged pion (diagram (d) of Fig. 6) reflects the form factor of a pion transition density between the two nucleons and not the virtual pion density inside the deuteron. This feature is illustrated in Fig. 7, where the monopole part of this transition density is displayed. Moreover, as already mentioned above, the diagrams in Fig. 6 are not by itself gauge invariant and thus not separately measurable.

A first realistic calculation of elastic deuteron photon scattering with inclusion of subnuclear effects has been carried out in Ref. 32 taking as reference frame the photon deuteron Breit frame, then $k' = k$. Furthermore, small contributions from the c.m. motion have been neglected. For unpolarized photons and deuterons only the scalar polarizability $P^0_{ii}(M^\nu L, M^\nu L, k, k)$ contributes to the elastic scattering cross section. The calculation of the resonance amplitude requires the summation over all excited energies and all possible intermediate states.

As detailed in Ref. 32, a subtracted dispersion relation for the individual polarizabilities of given multipolarity has been assumed using Eq. (41) for the imaginary part of the resonance amplitude (the TPA is real). Thus the real part of the scalar polarizability is determined from

$$Re(P^{RA,0}_{ii}(M^\nu L, M^\nu L, k, k)) - \delta_{\nu 0}\delta_{L1}\, P^{RA,0}_{ii}(E1, E1, 0, 0)$$
$$= (-)^{L+1}\widehat{L}\widehat{I}_i \frac{k^2}{2\pi^2} P \int_{k_{th}}^{\infty} dk' \frac{\sigma(M^\nu L)(k')}{k'^2 - k^2}\,. \quad (54)$$

Here, k_{th} denotes the threshold energy for photo absorption. Since the integral extends in principle up to infinity, one has to include also contributions from particle production above the corresponding threshold energy, e.g. from the total photo pion production cross section for energies above the pion production threshold. These were neglected limiting this approach to the low energy region. For the evaluation of the partial contributions $\sigma(M^\nu L)(k)$ to photo disintegration the Reid soft core potential had been used. Besides the one body current and TPA π-meson exchange currents also $N\Delta$ and $\Delta\Delta$ isobar configurations have been included.

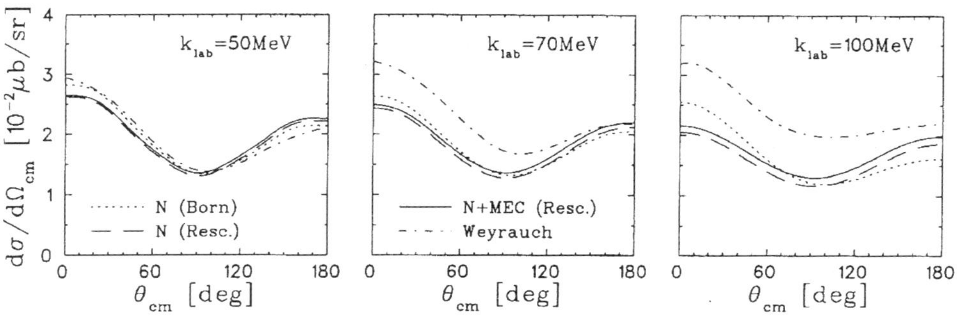

Fig. 8. Differential scattering cross sections for photon energies 50, 70 and 100 MeV from Ref. 33. Notation of curves: dotted (labeled "N (Born)") without FSI and MEC beyond Siegert, dashed ("N (Resc.)") including FSI but no MEC, and full ("N+MEC (Resc.)") with FSI and MEC. Dash-dot curves ("Weyrauch") represent a calculation with a separable interaction from Ref. 35.

A more recent realistic calculation from my group by Wilbois *et al.*[33] has been based on the solution of the off-shell NN-scattering matrix instead of using dispersion relations. Again inelasticities, which appear above pion production threshold, are excluded, limiting this approach also to low photon energies. For the deuteron bound state and the final state interaction (FSI) the realistic Bonn OBEPQ-B potential has been used. MEC contributions from pion- and rho-meson exchange to the current and the TPA amplitude have been included. For the electric transitions so-called Siegert operators have been used, incorporating thus implicitly the major part of MEC contributions.[34] Figure 8 displays the resulting differential scattering cross section at three photon energies, 50, 70, and 100 MeV. For comparison the results of another approach by Weyrauch[35] using a separable NN-interaction are also shown. Contributions from internal nucleon structure as manifest, e.g. in nucleon polarizabilities, were neglected in both calculations.

Comparing the dotted with the dashed curves one notes a sizeable increase from FSI with increasing energy, at forward angles a reduction and in the backward region an increase, thus reducing the strong asymmetry of the case without FSI considerably. The additional MEC effects beyond the Siegert operators are quite small for the lowest energy, but become more pronounced at 100 MeV, leading to an overall increase of the cross section. The increasing difference to Weyrauch's result[35] indicate that the separable interaction used in that work is not appropriate at higher energies.

A comparison with experimental data is shown in Fig. 9. One readily notes a systematic overestimation of the experimental results (dashed curves). An improved

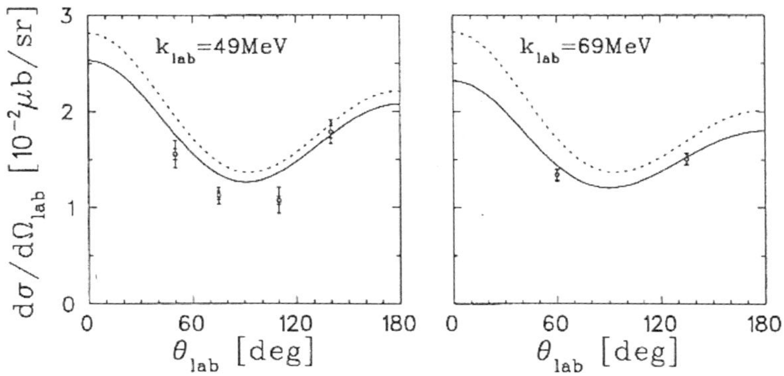

Fig. 9. Differential scattering cross sections for photon energies of 49 and 69 MeV (from Ref. 33). Experimental data from Ref 38. Solid curves: complete calculation with nucleon polarizabilities included. Dashed curves: without nucleon polarizabilities.

description is achieved if the influence of the internal nucleon structure in terms of nucleon polarizabilities is considered (full curves). The additional introduction of the free neutron and proton polarizabilities reduces the cross sections sizeably and gives a better description of the experiment. In view of the fact, that the neutron polarizabilities are not directly measurable, one turns the argument around in order to determine them from photon scattering off the deuteron. However, this procedure is not completely free from model dependencies.

Since then quite a few more theoretical investigations of this reaction using various approaches have been published,[36] where the major emphasis had been laid on the extraction of the neutron polarizabilities (for recent results see Ref. 37).

In fact, the deuteron is often used to determine internal neutron properties like polarizabilities and electromagnetic form factors, which otherwise are not available, since free neutron targets are absent. The question of off-shell modification of such internal nucleon properties is thereby left open with the hope that such effects are small since the deuteron is quite a loosely bound system.

Another interesting aspect which reactions on the deuteron offer is that the deuteron can also serve as a test ground for new theoretical methods. As an illustrative example, I will consider recent work based on the Lorentz integral transform (LIT) method (for a review of this method see Ref. 39). The LIT method is particularly suited for theoretical studies of reactions on complex nuclei. It allows the calculation of reactions without the need of determining the complex final scattering states, reducing this problem to the solution of a bound state equation. Therefore, as a test case for the application of the LIT method to photon scattering reactions, the LIT has been applied to the deuteron for a pure unretarded $E1$ radiation by Bampa *et al.*[40]

The resulting cross section at one energy is shown in Fig. 10 together with

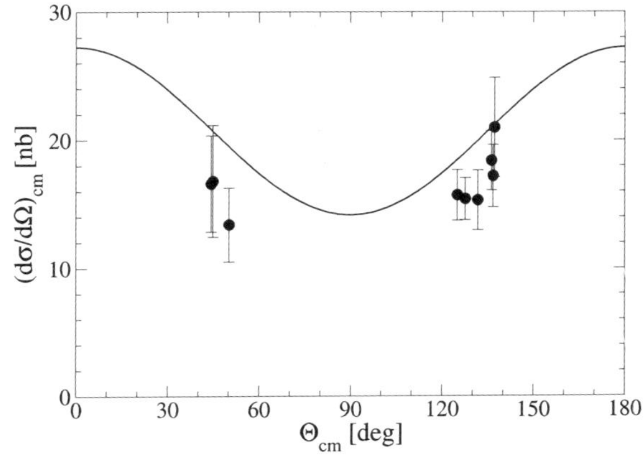

Fig. 10. Differential scattering cross section in the unretarded dipole approximation calculated with the Lorentz integral transform method for a photon energy of 55 MeV (from Ref. 40). Experimental data from Ref. 41.

recent experimental data. Nucleon polarizabilities were not included. The slight difference to the more complete calculation at 50 MeV in Fig. 8 is caused by the restriction to $E1$ radiation which is also the reason for the completely symmetric angular distribution around 90° (see Eq. (48)) in contrast to the slight asymmetry at 50 MeV in Fig. 8.

4. Conclusions

With these few examples, I hope to have convinced the reader, that nuclear photon scattering reactions has been and still is a wide and interesting field of research in nuclear structure studies from low energy collective properties to the influence of subnuclear degrees of freedom as manifest, e.g. in meson exchange and isobar currents. Hopefully, in the future more experimental results will be available, although the cross sections are quite small compared to hadronic reactions. Of particular interest are scattering experiments on complex nuclei at higher energies in the region of the first nucleon resonance, in order to study the behavior of a nucleon resonance in a nuclear environment.

References

1. H. Arenhövel and W. Greiner, *Prog. Nucl. Phys. Vol.* **10** (Pergamon Press, Oxford 1969), p. 167.
2. H. Arenhövel, *Workshop on Perspectives in Nuclear Physics at Intermediate Energies*, eds. S. Boffi, C. Ciofi degli Atti, and M. M. Giannini (World Scientific, Singapore, 1984), p. 97.
3. H. Arenhövel, *New Vistas in Electro-Nuclear Physics*, eds. E. L. Tomusiak, H. S. Caplan, E. T. Dressler (Plenum Press, New York 1986), p. 251.

4. M.-Th. Hütt, A. I. L'vov, A. I. Milstein, and M. Schumacher, *Phys. Rep.* **323**, 457 (2000).
5. J. L. Friar, *Ann. Phys.* (N.Y.) **95**, 1428 (1975), and references therein.
6. H. Arenhövel and M. Weyrauch, *Nucl. Phys. A* **457**, 573 (1986).
7. U. Fano, NBS Technical Note 83 (1960), reprinted in *Photonuclear Reactions*, eds. E. G. Fuller and E. Hayward (Dowden, Hutchinson & Ross, 1976), p. 338.
8. H. Arenhövel, PhD Thesis, University Frankfurt, 1965.
9. H. Arenhövel, M. Danos, and W. Greiner, *Phys. Rev.* **157**, 1109 (1967).
10. R. Silbar and H. Überall, *Nucl. Phys. A* **109**, 146 (1968).
11. R. Silbar, *Nucl. Phys. A* **118**, 389 (1968).
12. H. Arenhövel and D. Drechsel, *Nucl. Phys. A* **233**, 153 (1974).
13. M. E. Rose, *Elementary Theory of Angular Momentum*, (Wiley, New York, 1957).
14. A. R. Edmonds, *Angular Momentum in Quantum Mechanics*, (Princeton University Press, Princeton, 1957).
15. H. Arenhövel and W. Greiner, *Nucl. Phys.* **86**, 193 (1966).
16. M. Danos, *Nucl. Phys.* **5**, 23 (1958); K. Okamoto, *Phys. Rev.* **110**, 143 (1958).
17. M. Danos and W. Greiner, *Phys. Rev.* **134**, B284 (1964).
18. H. E. Jackson and K. J. Wetzel, *Phys. Rev. Lett.* **28**, 513 (1972).
19. T. J. Bowles *et al.*, *Phys. Rev. C* **24**, 1940 (1981).
20. H. Arenhövel, *Phys. Rev. C* **6**, 1449 (1972).
21. E. Ambler, E. G. Fuller, and H. Marshak, *Phys. Rev.* **138**, B117 (1965).
22. V. Rezwani, G. Gneuss, and H. Arenhövel, *Phys. Rev. Lett.* **25**, 1667 (1970).
23. V. Rezwani, G. Gneuss, and H. Arenhövel, *Nucl. Phys. A* **180**, 254 (1972).
24. H. J. Weber and H. Arenhövel, *Phys. Rep. C* **36**, 277 (1978).
25. H. Arenhövel, M. Weyrauch, and P.-G. Reinhard *Phys. Lett. B* **155**, 22 (1985).
26. E. Hayward and B. Ziegler, *Nucl. Phys. A* **414**, 333 (1984).
27. J. H. Koch, E. J. Moniz, and N. Ohtsuka, *Ann. Phys.* (N.Y.) **154**, 99 (1984).
28. H. Arenhövel and M. Sanzone, *Few-Body Syst. Suppl.* **3**, 1 (1991).
29. H. Arenhövel, W. Leidemann and E. L. Tomusiak, *Eur. Phys. J. A* **23**, 147 (2005).
30. H. Arenhövel, *Z. Phys. A* **297**, 129 (1980).
31. M. Weyrauch and H. Arenhövel, *Phys. Lett. B* **134**, 21 (1984).
32. M. Weyrauch and H. Arenhövel, *Nucl. Phys. A* **408**, 425 (1983).
33. T. Wilbois, P. Wilhelm, and H. Arenhövel, *Few-Body Syst. Suppl.* **9**, 263 (1995).
34. H. Arenhövel, *Z. Phys. A* **302**, 25 (1981).
35. M. Weyrauch, *Phys. Rev. C* **41**, 880 (1990).
36. M. I. Levchuck, *Few-Body Syst. Suppl.* **9**, 439 (1995); J. J. Karakowski and G. A. Miller, *Phys. Rev. C* **60**, 014001 (1999); M. I. Levchuck and A. I. L'vov, *Nucl. Phys. A* **674**, 449 (2000); *Nucl. Phys. A* **684**, 490 (2001); R. P. Hildebrandt, H. W. Grießhammer, T. R. Hemmert, and D. R. Phillips, *Nucl.Phys. A* **748**, 573 (2005); R. P. Hildebrandt, H. W. Grießhammer, and T. R. Hemmert, *Eur. Phys. J. A* **46**, 111 (2010).
37. L. S. Myers *et al.*, *Phys. Rev. C* **92**, 025203 (2015).
38. M. A. Lucas, Ph.D. thesis, University of Illinois at Urbana-Champaign, unpublished (1994).
39. V. D. Efros, W. Leidemann, G. Orlandini, and N. Barnea, *J. Phys. G: Nucl. Phys.* **34**, R459 (2007).
40. G. Bampa, W. Leidemann, and H. Arenhövel, *Phys. Rev. C* **84**, 034005 (2011).
41. M. Lundin *et al.*, *Phys. Rev. Lett.* **90**, 192501 (2003).

The QCD Phase Diagram from Statistical Model Analysis

Marcus Bleicher, Jan Steinheimer
Frankfurt Institute for Advanced Studies, Ruth-Moufang-Strasse 1, D-60438 Frankfurt, Germany

Reinhard Stock
Frankfurt Institute for Advanced Studies, Ruth-Moufang-Strasse 1, D-60438 Frankfurt, Germany
Institut fuer Kernphysik, Goethe Universitaet, D-60438 Frankfurt, Germany

Ideally, the Statistical Hadronization Model (SHM) freeze-out curve should reveal the QCD parton–hadron phase transformation line in the (T, μ_B) plane. We discuss the effects of various final state interaction phenomena, like baryon–antibaryon annihilation, core–corona effects or QCD critical point formation, which shift or deform the SHM freezeout curve. In particular, we present a method to remove the annihilation effects by quantifying them with the microscopic hadron transport model UrQMD. We further discuss the new aspects of hadronization that could be associated with the relatively broad cross-over phase transformation as predicted by lattice-QCD theory at low μ_B. That opens up the possibility that various observables of hadronization, e.g. hadron formation or susceptibilities of higher order (related to grand canonical fluctuations of conserved hadronic charges) may freeze out at different characteristic temperatures. This puts into question the concept of a universal *(pseudo-)critical* temperature, as does the very nature of a cross-over phase transformation.

1. Introduction

The phase diagram of strongly interacting matter, as addressed by the thermodynamics of Quantum Chromodynamics (QCD) theory, represents one of the open problems of the Standard Model of elementary interactions. Thermal QCD confronts matter at high energy density, as was prevailing during the early phases of the cosmological expansion but can also be investigated (albeit not at macroscopic scale) in collisions of heavy nuclei at relativistic energy. One thus creates a hot and compressed *fireball* in which, for an instant of time, the QCD confinement of partons into hadrons is overcome, creating a hot partonic QCD plasma state, the so-called Quark–Gluon Plasma (QGP). It is the goal, both of experiment and of

QCD thermodynamics, to elaborate the phase diagram of *fireball matter*. Its most prominent feature, the transition line between hadrons and partons, in the plane spanned by temperature T and baryochemical potential μ_B, is located in the non-perturbative sector of QCD. Here, the theory can (only) be solved on the lattice,[1] within the above variables T and μ_B, and has recently led to predictions of the parton–hadron boundary line.[2,3]

This line can also be addressed by experiment, in relativistic collisions of heavy nuclei where the primordial interaction volume is far from equilibrium at first; but after a small relaxation time (of order 1 fm/c, or even less) it settles onto the equilibrium (T, μ_B) plane at a temperature well above the *critical* (or pseudocritical) temperature T_c that corresponds to the phase transformation. Expansion and cooling then take the fireball down to the QCD phase boundary where hadronization occurs,[4] with the parameters (T_c, $\mu_{B,c}$) being preserved in the relative abundances (multiplicities) of the various created hadronic species.[4,5] Assuming, for the moment, that these multiplicities are conserved throughout the final hadron-resonance cascade expansion, their analysis in the framework of the Statistical Hadronization Model (SHM)[4–8] reveals the hadronization point, at which, in this ideal picture, the partons freeze-out into hadrons.[9–11] The hadron abundances stay essentially unobliterated throughout the ensuing hadron/resonance expansion evolution. The latter fixes flow observables, spectra and Bose correlations. So there are two separate, well spaced freeze-outs, the so-called hadro-chemical and the kinetic ones.

As the primordial temperature (baryochemical potential) shifts upward (downward) with increasing collision energy, an ascending sequence of experimental energies can, thus, map a sequence of hadronization points along the QCD parton–hadron boundary line. At the low AGS energies, in the domain of $\sqrt{s} = 5$ GeV, this investigation begins at μ_B of about 550 MeV, falling to about 250 MeV at top SPS energy ($\sqrt{s} = 17.3$ GeV), then on to the 20 to 100 MeV domain at RHIC energies ($\sqrt{s}$ from 60 to 200 GeV), and ending at baryochemical potential practically zero at the LHC energy of 2.76 TeV, in central collisions of Pb or Au projectiles.[6,7,12] It is here that we meet with the conditions prevailing in the big-bang expansion, thus recreating the cosmological phase transition to hadrons that occurred at a few microseconds time.

With these assumptions which of course need careful discussion and refinement, the subject of the present article, one of the main goals common to QCD theory and nuclear collision experiments — the parton–hadron boundary line — comes well within reach as we shall demonstrate in the following. We will describe the state of the art in applying the SHM to the hadron multiplicity results created at the AGS, SPS, RHIC and LHC (also discussing difficulties arising in the application of the statistical model), and compare the resulting hadronization curve with the corresponding lattice QCD calculations.

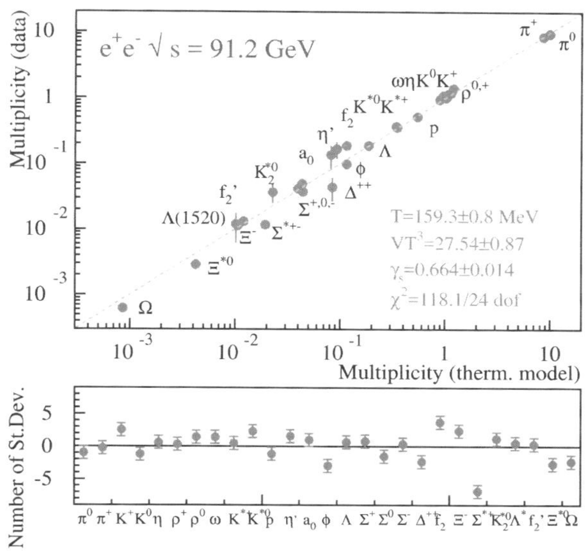

Fig. 1. Hadron production in electron–positron annihilation at LEP energy, confronted with canonical Statistical Hadronization Model calculations.[15]

Turning to a more detailed argumentation, we note that at present the identification of the QCD hadronization point with the emergence of a hadron/resonance population that is "born into equilibrium",[9] thus resembling an equilibrium canonical or grand-canonical Gibbs ensemble (the basis of the SHM[4,5]), remains a (plausible) conjecture. No analytic solution of the QCD Lagrangian exists in the non-perturbative domain. Two approaches nevertheless suggest the validity of this assumption. The first is a line of argument developed by Amati, Veneziano and Webber,[13,14] in an early attempt to understand hadron production in electron–positron annihilation. A QCD mechanism called *color preconfinement* occurring toward the end of the QCD shower evolution prompts colour neutralization into singlet *clusters* or resonances, that subsequently decay quantum mechanically onto the hadron/resonance mass spectrum. This decay is governed by phase space weights (Fermi's Golden Rule), thus the outcome appears to be born into equilibrium, a maximum entropy state preserving the initial quantum numbers and the size of available phase space, represented by temperature and baryochemical potential in the Gibbs ensemble. Indeed the hadron multiplicities from $e^+ - e^-$ annihilation can be well accounted for by the SHM. We show in Fig. 1 the LEP data at 91.2 GeV confronted with the canonical SHM,[15] revealing a hadronization temperature of 160 MeV. It agrees with Hagedorns limiting hadronic temperature as was predicted, already in 1975, right with the discovery of QCD, by Cabibbo and Parisi.[16] The second line of argument concerning hadronization equilibrium stems from the recent idea[17] to investigate the overlap between the Hadron Gas Model and lattice QCD with regard, e.g., of the energy and entropy densities as a function of temperature.

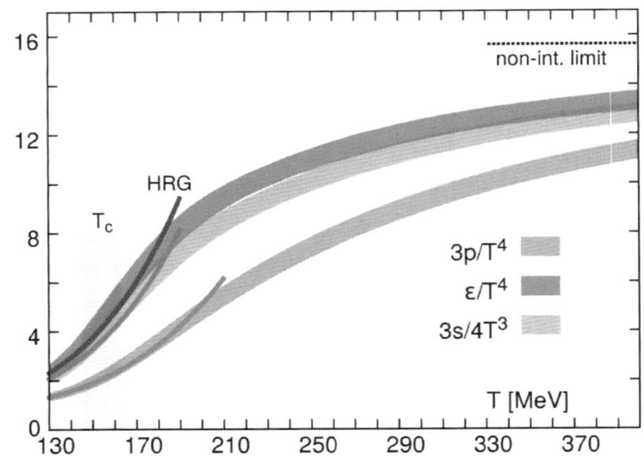

Fig. 2. Lattice QCD calculations for energy density and other quantities, confronted with Hadron-Resonance Gas model predictions for the same quantities.[18]

An example[18] is shown in Fig. 2 where we see the HRG curves intersect with the respective Lattice predictions in the vicinity of 165 MeV. That is below this temperature the degrees of freedom of an equilibrium HRG appear to take over from the lattice proper degrees of freedom. Note that the HRG and the SHM refer to the same hadron/resonance partition functions.

Next, we note that the results and predictions of lattice QCD are *exact* (in the sense of the model) only at zero baryochemical potential where the phase transition is a rapid cross-over. Toward finite μ_B various extrapolations are employed,[2,3] for example a Taylor expansion[19] the coefficients of which are, in fact, related to experimentally accessible higher order fluctuations of conserved quantities like baryon number and charge.[20,21] We shall return to this work later on but note, for now, that an interesting feature of the emerging phase boundary line at finite μ_B would be the intensely discussed critical point of QCD,[22,23] at which the cross-over nature of the parton–hadron transition would end giving way to a first order transition toward higher μ_B. This would mark, on the one hand, the end of the lattice Taylor expansion, by divergence, but on the other hand lead to measurable changes in the sequence of hadronization points[23] which are accessible via SHM analysis of the hadronic multiplicity distributions. This analysis could, therefore, not only ascertain the much debated existence of a critical point, but also serve to locate the phase transition line toward higher μ_B where lattice QCD is inapplicable, as of yet.

The arguments above tacitly assume that the nucleus–nucleus collision dynamics does indeed cross the phase transformation line, and settles above it before re-expansion of the hot and dense matter volume sets in, to approach the phase boundary. That is we are at collisional energies above the so-called "onset of deconfinement".[24] Estimates of the corresponding incident energy domain (in central, heavy nuclear collisions) remain uncertain at present, pointing to the $\sqrt{s}$ region from about 4 to 8 GeV. Below this (yet to be determined) energy the hadronic multiplicities would not stem from the QCD hadronization phase transition, which is responsible for the hadro-chemical equilibrium among the species' abundances.[4,5,9–11] Thus, unless another phase transition comes into play at high baryochemical potential,[25] we should expect observable changes in the hadronic freeze-out pattern, such as a sequential chemical freeze-out in inverse order of inelastic cross section (characteristic of a diluting hadron gas). We shall return to this question at the end of this article but state, for now, that we cannot yet arrive at clearcut conclusions concerning such features, due to a lack of data in the AGS energy domain, and below.

Turning to another important assumption of the above overall model considerations it was shown[26,27] that the simple picture of an instant, synchronous chemical freeze-out of all hadronic species, occurring directly at the hadronization phase transformation, requires some revision. Final state inelastic or annihilation processes need consideration owing to the high spatial particle density after hadronization. This effect is perfectly absent in elementary collisions such as $e^+ - e^-$ annihilation to hadrons where hadrons are born into the physical vacuum. The SHM was initially developed for such elementary collisions.[28] Turning to A+A collisions it was shown that a detailed investigation of final state modifications of the multiplicity distribution exhibits relatively weak effects on the bulk meson production, pions and kaons which embody more than 90% of the total hadronic output energy at the LHC energy, but is important via annihilation of baryons and antibaryons, notably p and pbar. These effects were quantified using the microscopic transport model UrQMD, in its so-called *hybrid* version[29] where a hydrodynamic expansion mode of the collisional volume is terminated by a simulation of the hadronization process via the Cooper–Frye mechanism. The emerging hadrons and resonances are then traced through the ensuing, final hadron-resonance cascade expansion phase of UrQMD, and its attenuating effects can thus be determined, species-wise. These final state modifications affect the outcome of the SHM analysis as we shall demonstrate in the next section; they need to be taken into account before concluding on the position of the QCD phase boundary line. Beyond studying the changes occurring in baryon and antibaryon multiplicity, and their effects on the SHM analysis[26,27,30] as further discussed below, other groups have also focused on the consequences of this final state annihilation for the hydrodynamic, notably the elliptic flow of various identified hadronic species.[31]

In the next section we shall discuss an approach to overcome the final state effects, by generating UrQMD-modification factors for the hadronic multiplicities that lead to a modified SHM analysis. Turning to SHM analysis of data at LHC, SPS and AGS energies in Sec. 3 we discuss the resulting new freeze-out curve[30] and its differences to the result of a *standard* SHM analysis, as well as to the often quoted smooth interpolation curve given by Cleymans *et al.*[32] In Sec. 4 we compare to the lattice QCD results for the parton-hadron phase boundary, confirming the lattice prediction[2,3] that the line is almost horizontal up to a rather high μ_B. We then turn to a discussion of the apparent tension between the SHM conclusion that $T_c = 164 \pm 5$ MeV and the recent conclusion from consideration of data, HRG and Lattice QCD results concerning higher order fluctuations of proton and charge multiplicity distributions, that appear to favor a lower transition temperature, of about 150–155 MeV.[33] Section 5 will give a short presentation of the issues concerning the SHM, i.e acceptance and stopping power effects, the core–corona model, canonical strangeness suppression, and new LHC $p+p$ collision results. A discussion section will end the paper in Sec. 6, where we will briefly return to the issues of critical point, and onset of deconfinement.

It remains to say that we will not re-iterate the theoretical formulation of the grand canonical SHM in this brief overview. Comprehensive presentations can be found in Refs. 4–6 and 8.

2. UrQMD Final State Effects and the Statistical Model Analysis

Combined microscopic–macroscopic models are among the most promising approaches to describe nucleus–nucleus collisions.[34] Initial state effects causing eventwise fluctuations of the dynamics can be installed and investigated separately, as is the case for the study of the ensuing hydrodynamic flow expansion, and for the matching to the final state hadron-resonance cascade that leads to decoupling. The latter aspect is our concern here.

In the hybrid UrQMD model employed[26,27] to assess the final state effects on hadronic multiplicities in A+A collisions, the hydrodynamic stage is a full (3+1)-dimensional ideal hydro model executed by the SHASTA algorithm,[35] employing the CH EOS from Ref. 36 which corresponds to a cross-over transition from a hadronic gas to the QGP at $\mu_B = 0$ that extends to the region of finite μ_B, relevant for all beam energies discussed, at present. The hydrodynamic evolution was stopped, generally speaking, once the energy density (temperature) in the system falls below a pre-set critical value. Then the Cooper–Fry equations were sampled on a defined hypersurface, in accordance with conservation of all charges as well as the total energy. Two different choices of the hypersurface have been employed in order to assess the sensitivity of the ensuing *afterburner* effects to the initialization procedure. The first choice was an effective iso-proper time hadronization,[37] implemented by freezing out in successive transverse slices, of thickness $dz = 0.2$ fm, whenever the last flow cell of the considered slice fulfills the freeze-out criterion: an energy density ϵ below five times the nuclear ground state energy density (i.e. about 750 MeV/fm^3). This procedure was carried out separately for each slice, and the particle vector information transferred to the cascade part of the UrQMD model.[29] The effect of final state interaction was then quantified by either stopping the calculation directly after hadronization, letting the produced hadronic and resonance species undergo all their strong decays as if in vacuum, thus establishing a fictitious

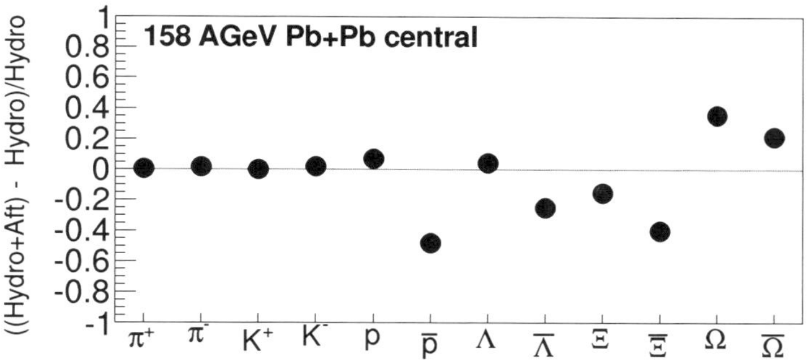

Fig. 3. Modification factors accounting for final state attenuation in the hadron/resonance expansion.[26,27]

multiplicity distribution ideally referring to the hadronization *point*. Alternatively, the final afterburner UrQMD stage was attached, and the multiplicity distribution at decoupling generated. For each hadronic species one could then extract a modification factor indicating the strength of the afterburner effects (a method introduced by Bass and Dumitru[34]). This factor got finally employed in the SHM analysis of the data (see Sec. 3).

In a further investigation[30] an alternative choice of the hadronization hypersurface was explored, corresponding to an isothermal termination of the hydro expansion stage of UrQMD. Here, one implements the switch to hadronic/resonance degrees of freedom locally, in each hydro-flow cell, once it falls below the pre-set energy density, or temperature, calculating the hypersurface element with a state of the art hypersurface finder.[38]

In the following we illustrate this overall procedure, and its main consequences as far as the UrQMD predictions for final state attenuation of hadronic abundances are concerned. The resulting modification factors are employed in the data analysis with the SHM (Sec. 3). For illustration of the phenomena we choose the case of central $Pb + Pb$ collisions at the top SPS energy, $\sqrt{s} = 17.3$ GeV (one of the data sets to be analyzed in Sec. 3). Figure 3 shows the modification factors[26]

$$M = \frac{yield\,(\text{with afterburner}) - yield\,(\text{at hadronization})}{yield\,(\text{at hadronization})} \tag{1}$$

for the hadronic species covered by the experiment. One sees the pion, kaon, proton, Lambda and Xi yields, i.e. the bulk output from hadronization, to be essentially unaffected at this incident energy, a remarkable result. Thus far the assumption of a synchronous freeze-out, directly at hadronization, as made in the standard version of the SHM,[4–7] is substantiated. On the other hand, antiprotons suffer a net loss of about 50%, and Anti-Lambda and Anti-Xi multiplicities are reduced by 25% to 40%. The Omega/Anti-Omega hyperons are less affected because they have no excited resonant states, and freeze out from the cascade evolution almost instantaneously, owing to their low total cross-section. There may also occur

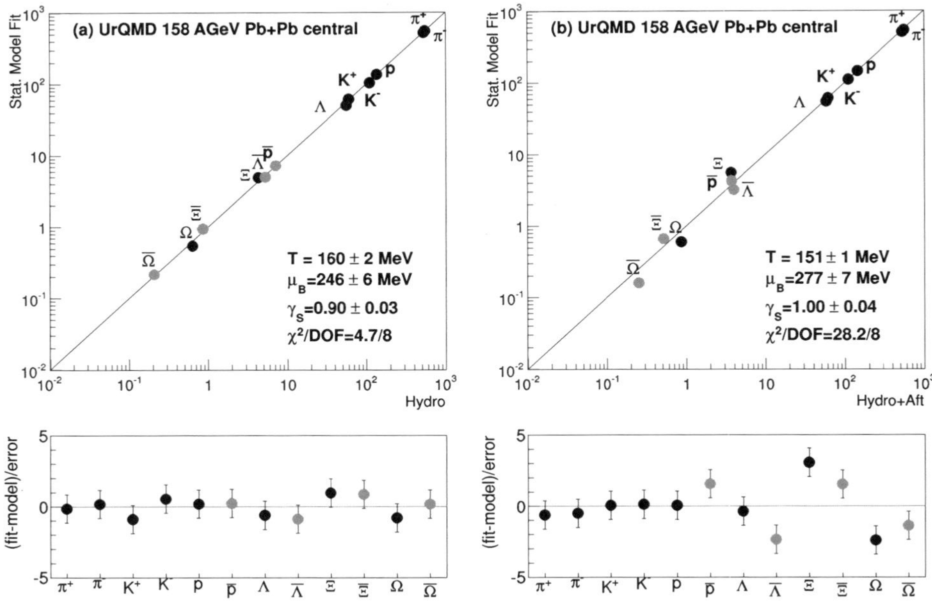

Fig. 4. UrQMD multiplicities fitted by the Grand Canonical model, fitted to the "data" right after hadronization, and to the results with attached UrQMD afterburner.[26]

contributions of baryon/antibaryon regeneration[39] which are only partially implemented in UrQMD. A new study of their importance is under way with the QHDS transport model.[40]

We see that in the UrQMD cascade evolution inelastic reaction channels are of no major importance whereas baryon–antibaryon annihilation channels do not freeze-out directly after hadronization. At the relatively low SPS energy, illustrated here, baryons are far more abundant than their anti-partners; thus their partial annihilation with antibaryons changes a minor fraction of their multiplicities, whereas the antibaryons are significantly diminished, fractionally. This pattern changes at LHC energies due to the near-perfect particle–antiparticle symmetry, prevailing there.[27] Under such conditions the antibaryons are equally affected.

These attenuations have a profound effect on the Statistical Model analysis. The loss of antibaryons is, clearly, a departure from the initial chemical equilibrium distribution which is reflected in a SHM analysis (described in Ref. 26) of the two sets of UrQMD multiplicities, obtained with, and without the afterburner stage.

This is shown in Fig. 4: the UrQMD multiplicities, taken as "data", have been fitted with the SHM. We see that, on the one hand, the SHM fit to the no-afterburner results is of excellent quality and reveals a hadronization temperature of 160 MeV. This is not surprising because the SHM fit merely responds to the chemical equilibrium multiplicity distribution imprinted by the Cooper–Frye hadronization, with

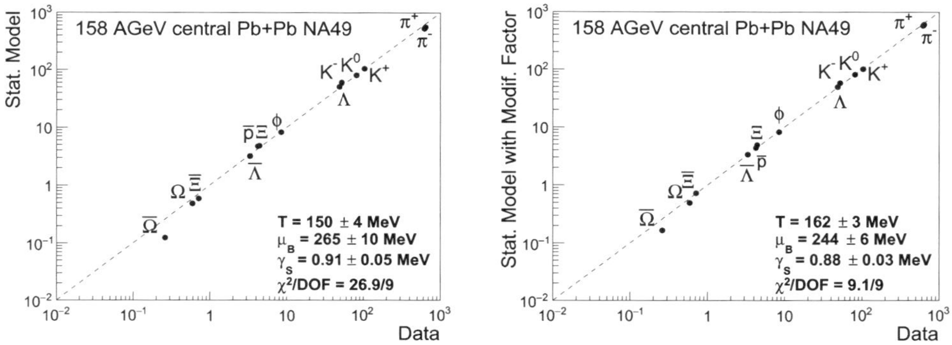

Fig. 5. Hadron multiplicities from central Pb+Pb collisions at top SPS energy, confronted with HRG fits employing the standard Grand Canonical approach, and the SHM corrected by UrQMD modification factors.[26]

its preset temperature value. On the other hand, the case with afterburner shows a very much deteriorated fit in the antibaryon sector, at a temperature reduced to 151 MeV. The apparent chemical freeze-out temperature thus drops down, significantly, due to the final state distortions of the multiplicity distribution. The traditional statistical model analysis thus needs revision. We will show this in the next section, turning to data analysis.

3. Data Analysis: The SHM Freeze-Out Curve

In this section we illustrate the UrQMD-modified Statistical Model analysis[30] that addresses five hadronic multiplicity data sets obtained, respectively, by ALICE[41] at the LHC energy of 2.76 TeV per nucleon pair, by NA49[42] at the SPS energies 17.3, 8.7 and 7.6 GeV, and by the AGS experiments E891, 802, 896, 877[43] at 4.85 GeV. All data refer to central collisions of Au+Au (at the AGS) and Pb+Pb (at SPS and LHC). Note that the new STAR data[44] from the RHIC BES program have not been included in this analysis because they are not yet corrected for antiproton feed-down from secondary weak antihyperon decays. In Fig. 5 we illustrate this analysis choosing the case of the top SPS energy data, central Pb+Pb collisions at 17.3 GeV. The outcome of the *traditional* SHM analysis[6] is compared with the version in which the SHM fit to the data is done with theoretical multiplicities modified by the final state attenuation factors from UrQMD (see Fig. 3). Two major effects are obvious: the deduced temperature increases significantly, from $T = 150$ MeV in the former, to $T = 162$ MeV in the latter case. Concurrently, the fit chi-square/dof drops dramatically, from about 2.7 to about 1.0. The UrQMD-reconstructed yield distribution, which aims to re-establish the distribution at hadronization, obviously meets with the grand canonical equilibrium hypothesis, implicit in the SHM approach. Recalling the analogous observations derived above, from the SHM study of UrQMD predictions for the multiplicity distribution (Fig. 4), we re-iterate the conclusion that the final state hadron/resonance cascade evolution (here represented by the UrQMD *afterburner*), generates a distortion of the hadro-chemical equilibrium yield distribution that is initially imprinted at hadronization. This is reflected in the unsatisfactory SHM fit quality, a feature also reported from the SHM analysis of Andronic *et al.*[12]

A short remark is in order here, to clarify the above use of the term equilibrium. First, it refers, not, to a global equilibrium of all aspects of the phase space distributions of the hadrons that emerge from the QCD hadronization phase transformation. It refers, only, to its first moment, the hadronic yield (multiplicity) distribution. Higher moments, such as collective, discrete hadron emission flow patterns (radial, directed, elliptic and higher flow orders) represent non-isotropic momentum space flow moments at any higher order, in the final state. Furthermore, if the hadronization process creates a hadronic species thermal equilibrium distribution (which is represented by the quasi-classical equilibrium grand canonical Gibbs ensemble underlying the Statistical Model) directly corresponding to the conditions prevailing at hadronization (chiefly the energy density/temperature), this equilibrium state, if it freezes-in directly at this point (as assumed in the traditional SHM analysis), is of course out of equilibrium immediately after the onset of expansive dilution and cooling. The yield distribution does, ideally, not adjust to the falling energy density but stays, frozen into the ongoing expansion, as we have seen in Fig. 3, to appear as the bulk hadronization output. The hadron/resonance expansion stage will adjust other, higher moments. We note that also the higher moments of conserved baryon number and charge multiplicity fluctuation may not become stationary (freeze out) at the very instant at which the number and entropy densities of the hadron/resonance degrees of freedom get fixed. We shall return to this consideration below, when discussing the recent work on fitting kurtosis and skewness data, and HRG predictions, to the corresponding results from lattice QCD. Thus there are two, or perhaps even three successive freeze-outs: the primordial hadronic species distribution reflects the energy/entropy densities prevailing at the onset of hadronization (with corrections that this study quantifies) whereas higher order multiplicity fluctuations may be imprinted throughout the duration of the hadronization period (which is finite in a cross-over transition). Finally, spectral and correlation features become stationary at the lower temperatures of decoupling from all strong interaction(kinetic freeze-out).

The results from Ref. 30 concerning the reconstruction of the hadronization line, in the plane of temperature T and baryochemical potential μ_B are summarized in Fig. 6. It shows the (T, μ_B) points at the five energies considered here, that result from SHM analysis of the corresponding hadron multiplicity data sets. The results of the standard SHM approach are compared to the UrQMD-modified SHM in both the hadronization modes described in Sec. 2, isothermal and slice-by-slice. The latter give almost indistinguishable results pointing to the conclusion that the detailed method employed for UrQMD hadronization has little influence on the effects of the ensuing cascade expansion stage, as far as multiplicities are concerned. Up to a baryochemical potential of about 370 MeV the points exhibit very little downward slope, unlike the results from standard SHM analysis which drop off more steeply. This invites comparison to the lattice QCD predictions for the curvature of the phase boundary line, a subject of the next section, as is the apparent abrupt downward turn toward higher μ_B (unfortunately documented by a solitary AGS point only). As far as the turn-off toward higher baryochemical potential is concerned, which may indicate the advent of new physics as discussed in the next section, we note here that this feature has been well recognized already in previous investigations of the SHM freeze-out, which did not employ any final state attenuation corrections. We show in Fig. 7 the state of the art as of 2006 by reproducing the results obtained by Andronic *et al.*[7] for the SHM freeze-out systematics as function of μ_B, gathering all results available at that time. Indeed, the authors note that the turn-off, corresponding to SHM investigations below a center-of-mass energy of about 8 GeV, is hardly compatible with any smooth interpolation.

Finally, we note here that the much discussed puzzle[12,41] of an apparent *non-thermal*, high pion to proton ratio observed in the ALICE data[41] at the LHC energy of 2.76 TeV (which resulted in an unsatisfactory SHM fit[12]) finds its resolution in the UrQMD modified approach illustrated above. Final state annihilation of nucleons and antinucleons reduces the p and $\bar{p}$ multiplicities while increasing the pion yield as each annihilation contributes 5 pions, on average. Both the p, $\bar{p}$ losses, and the additional pion gains are, of course, of non-thermal origin.

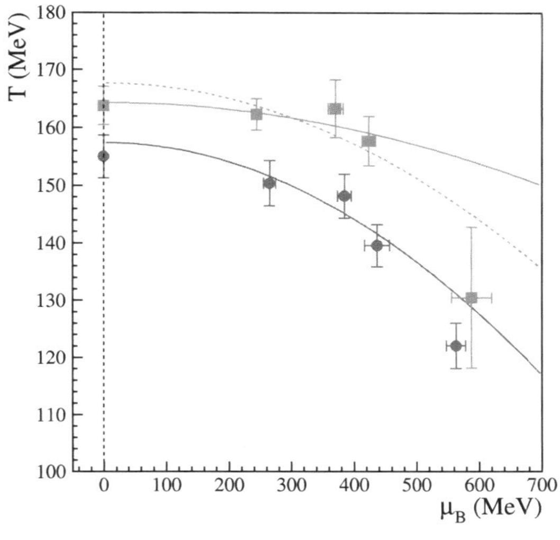

Fig. 6. Results of the UrQMD plus SHM approach, for hadronization points in the (T, μ_B) plane. The lower set of points represents a standard grand canonical SHM analysis.[30]

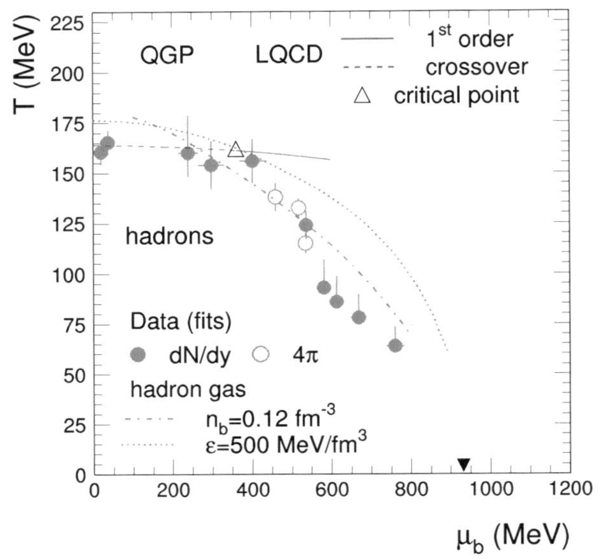

Fig. 7. Standard SHM analysis results for freeze-out points in the (T, μ_B) plane, covering energies from RHIC, via the SPS and AGS, down to SIS18.[7]

4. Comparison to Lattice QCD

We wish to discuss, in more detail, the observations described above.

(a) The small initial slope of the UrQMD modified SHM model points resembles the recent lattice predictions for the QCD transition line[2,3] which employ a quadratic ansatz,

$$T_{c,\mu_B} = T_{c,0}(1 - \kappa_{2,B})(\mu_B/T_{c,0})^2 \qquad (2)$$

reporting curvature values $\kappa_{2,B}$ between 0.007 in Ref. 2 and 0.015 in Ref. 3. The first four points are compatible with $T_{c,0} = 164$ MeV and $\kappa_{2,B} = 0.0048$.

(b) The steep drop-off in the μ_B domain above about 400 MeV may represent a manifestation of a fourth-order coefficient $-\kappa_{4,B}(\mu_B/T_{c,0})^4$ in the lattice Taylor expansion. This, then, would be of order 0.003 but we note that this estimate is based on a single data point (AGS) only. We cannot decide, with certainty, whether the QCD phase boundary moves downward, steeply, in this domain. Alternatively, the steep drop-off could signal a change in the collisional evolution, such as the advent, at high μ_B, of the hypothetical quarkyonic matter phase[25] interpolating between the domains of deconfined, and of hadron-resonance matter in the QCD phase diagram, thus shifting the final hadronization transition downward in temperature. Hadronization from a quarkyonic matter phase has not yet been considered theoretically so that we can not say with certainty that it would feature a similarly phase space dominated yield distribution over species. Finally, and more simply, we might witness the so-called "onset of deconfinement"[24] at the AGS energy, where the collisional volume does deconfine only partially, thus lacking a phase transition synchronized hadronic freeze-out.

(c) Obviously, the reconstructed hadronization points in Fig. 6 indicate a (pseudo-)critical temperature $T_{c,0}$ of 164 ± 5 MeV, at $\mu_B = 0$ and, likewise, a T_c above about 160 MeV in the entire μ_B interval up to about 370 MeV. This observation is, clearly, at odds with the recent consensus in lattice QCD studies about a lower $T_{c,0}$ value, in the vicinity of 150 to 155 MeV.[33] However, let us not rush for conclusions here. As we said above, higher order fluctuations (susceptibilities) might freeze out later than the zeroth-order multiplicity quantities (their overlap with lattice QCD thus occurring at a lower T) and, moreover, they might not be reliably accounted for by the noninteracting HRG model due to attractive short range interactions.[45] Thus the HRG-Lattice overlap temperature, as derived from the fluctuation/susceptibility observables, may not be a good approximation of $T_{c,0}$. Finally we recall that Fig. 2 did suggest a $T_{c,0}$ above 160 MeV from state of the art lattice calculations for the equation of state and the energy density — the quantities more closely related to hadron/resonance number density (that is of relevance in the SHM and also, of course, in the HRG) than, e.g., the Polyakov-loop slope turnover, or higher susceptibilities. Addressing the former observables the HRG-lattice overlap region rather suggested a $T_{c,0}$ above 160 MeV.

(d) The UrQMD-modified results represent a revision of the *hadronic freeze-out curve* that was obtained[32] by an interpolation of the (T, μ_B) values deduced from grand canonical SHM analysis of central A+A collisions, gathered from RHIC energies down to AGS and SIS18 energies.[4–7] This often shown curve also extrapolated to a $T_{c,0}$ of about 165 MeV. However, the various sets of hadron multiplicity data, available by 2006 for AGS, SPS and RHIC energies, had mostly not yet been corrected for experimentally unresolved feed-down contributions stemming from weak decays of the hyperons/antihyperons. This increased the apparent multiplicities for baryons/antibaryons, reflected, in the SHM analysis, by too high temperature assessments. The data sets used in Fig. 6 are properly corrected, thus the corresponding standard freeze-out curve moves below the 2006 result, by about 10 MeV as shown in Fig. 3. We note that, anyhow, this standard SHM result looses interest due to the annihilation processes in the final cascade evolution, which have, first of all, to be accounted for, by an analysis like the one described above. Whether or not the particular UrQMD approach represents the last word about this issue clearly remains to be clarified.

5. Issues of Concern in SHM Analysis

Let us begin this chapter with a general observation. The consistency of existing SHM analyses suffers from the plurality concerning details of the employed analysis formalism or, even, its basic formulation. We shall take up some such aspects in some more detail below but mention, for now, the options for or against, e.g., hadron eigenvolume corrections, strangeness correlation volumes or a strangeness fugacity, corona–core separation, light quark fugacity, to name just a few. Furthermore, the available data reflect different experimental acceptance conditions (both in rapidity space and in azimuthal coverage), and data sets comprise a varying sub-selection of hadronic species, the latter often a consequence of hyperon multiplicities falling strongly with incident energy. Thus, a set of multiplicities ranging from pions up to anti-Omega hyperons gets employed from LHC down to top SPS energies, but the rarer species fall away, gradually, toward lower energies. This is counterproductive if one tries to secure a systematic coverage of the overall T, μ_B dependence. Overall, we note that the ideal view represented by the standard grand canonical ensemble may not be universally congruent to the reality of A+A collisions,[6] and of their measurement within certain fixed rapidity windows (see below). Thus the data sets of multiplicities may reflect *tensions* that vary with energy; the acceptance problem, for example, disappears only with approaching boost invariance, i.e. at top LHC energy. And the annihilation corrections affect, significantly, only the antibaryons at low energy because of the large excess of B over $\bar{\text{B}}$. Whereas both B and $\bar{\text{B}}$ are attenuated at high energies.

Numerous further SHM analysis problems can be enumerated. An example is the question of how to treat the ϕ-meson: with strangeness suppression (creation from $K^+ - K^-$ pairs) or without(strangeness zero at hadronization). A further, new question: should the charmed hadrons be included in the general SHM fit(with a special fugacity factor as charm is not thermally produced), or should we expect a sequential hadronization of the heavier quarks[46]? We shall take up some of such questions below but end here showing a fairly complete selection of available standard SHM analyses,[47] from LHC down to SIS18 energies, in Fig. 8.[48] Quite expectedly the points exhibit considerable dispersion even though the results with UrQMD modification (Fig. 6), and the predictions of the Single Freezeout Model[48] have not been included here.

5.1. *Acceptance and stopping power effects in rapidity space*

The thermal rapidity spreading width of a single midrapidity fireball is proportional to $(T/m)^{1/2}$ for a hadron species of mass m.[49] For pions emitted at $T = 160$ MeV this amounts to the interval $-1.3 < y < 1.3$, whereas emitted Omega hyperons would occupy a much smaller interval, of about $-0.4 < y < 0.4$. Kaons fall in-between. Realistically, a fireball of 160 MeV temperature could get created at top SPS energy (recall Fig. 6), but here it features, in addition, collective radial

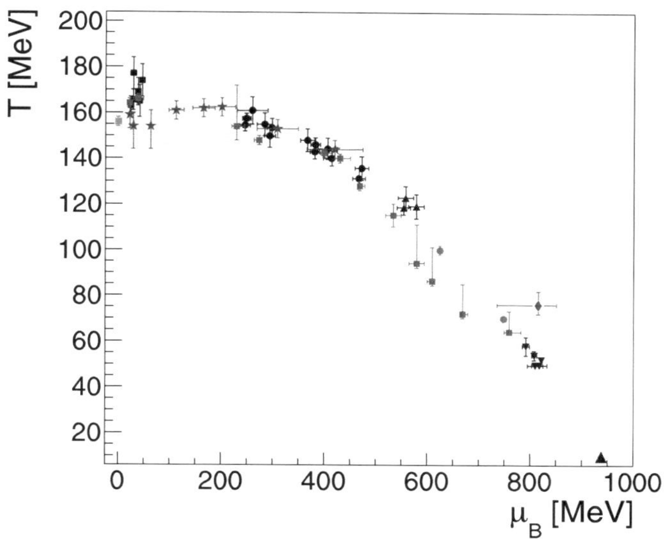

Fig. 8. Summary plot of present SHM analysis with various variants of the Statistical Hadronization Model, for the overall freeze-out curve in (T, μ_B) space.[46,47]

flow expansion which elongates the longitudinal rapidity distributions.[50] The pions thus acquire a width of about ± 1.5 units of rapidity whereas all heavier hadrons stay within an interval of unit rapidity. This is the STAR experiment acceptance, employed in the BES RHIC runs at low energy, down to the SPS regime where considerations such as above may be realistic. Such an acceptance thus creates an artificial strangeness enhancement, as reflected in the so-called Wroblewski ratio.[51] It counts the total number of s and sbar quarks, relative to the light quark number, recovered in the observed hadronic multiplicities. The pions that carry the major fraction of the light quarks partially leak out of the narrow acceptance. This creates *tensions* in the SHM analysis.

A remedy might be to widen the experimental acceptance. The resulting effects can be studied with data from fixed target experiments, such as NA49 at the SPS, and BRAHMS at RHIC. The former already reported a minimum at midrapidity in the net baryon rapidity distribution for central Pb+Pb collisions at 17.3 GeV center-of-mass energy.[52] This signals a minimum of the baryochemical potential which, in turn, is a consequence of the so-called stopping power of hadronic matter. The net baryon number density reflects the final distribution, in longitudinal phase space, of the valence quarks carried in by the colliding nuclei. In a Glauber-like picture the transport of valence quarks, from initial target/projectile rapidity positions inward toward midrapidity occurs via the successive binary collisions of nucleons (or their remnants) during the primordial interpenetration phase, which extract partons from the initial nucleon structure functions. This stopping mechanism spreads the bulk

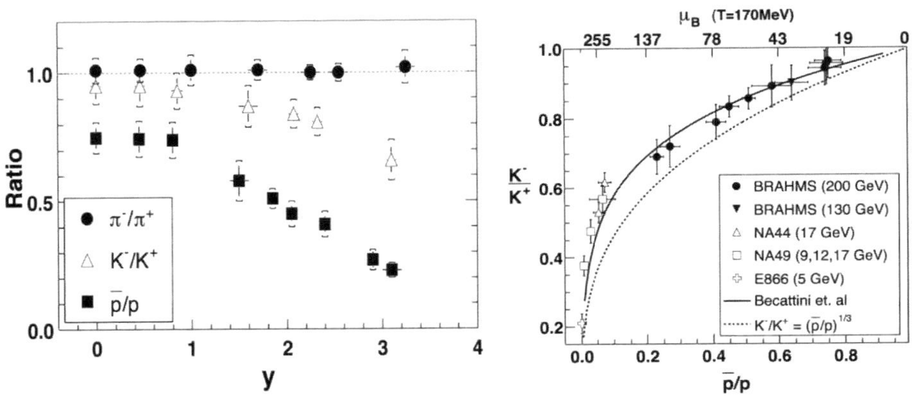

Fig. 9. Results of the RHIC BRAHMS experiment[53] for antihadron to hadron ratios at top RHIC energies. The left panel shows the rapidity dependence, the right confronts a correlation plot of K^-/K^+ vs. $\bar{p}/p$ with SHM calculations at constant temperature but with decreasing baryochemical potential.

of the valence quarks over about 2.5 units of rapidity in central mass 200 nuclear collisions. From top SPS energy onward, the rapidity gap between the initial target and projectile rapidities, increases beyond $dy = 6$, and the midrapidity region is increasingly devoid of valence quarks, thus μ_B falls toward zero here. This can be perfectly illustrated by the BRAHMS results[53] shown in Fig. 9. The left panel shows results for central Au+Au collisions at 200 GeV, exhibiting the rapidity dependence of the multiplicity ratios π^-/π^+, K^-/K^+ and $\bar{p}/p$. The latter two are proportional to $\exp(-2/3\ \mu_B/T)$ and $\exp(-2\ \mu_B/T)$, respectively, thus presenting a direct measure of μ_B. Obviously this stays near constant within $0 < y < 1$, the heavy hadron radiation width as argued above. However, a different physics seems to set in, in each successive further unit rapidity interval. No use in widening the acceptance in a SHM analysis that seeks to establish the prevailing $(T,\ \mu_B)$ point in the phase diagram, because μ_B increases, steeply, toward higher y where the valence quark density increases.

This is further demonstrate in the right panel of Fig. 9 where the correlation between K^-/K^+ and $\bar{p}/p$ is confronted with an interpolating SHM fit curve. The inferred μ_B shifts down from the 20 MeV domain, corresponding to a relatively valence quark free midrapidity domain, to about 140 MeV, smoothly interpolating, furtheron, to data points from the SPS domain.

To conclude about our above question of increasing the acceptance to secure the true Wroblewski ratio, we see that from top SPS to top RHIC energies this is not helpful because each successive unit rapidity window reflects a differently composed hadronization source. This is the consequence of the hitherto little explored stopping mechanism that governs interpenetrating hadronic matter. At top LHC energies, the valence quark domain recedes far from midrapidity, and the acceptance is of no concern in this domain where μ_B is practically zero, and T is constant (boost invariance). At low energies, on the other hand, the rapidity gap is so small that the energy and quark distributions fall mostly within a single fireball radiation width: conditions yet to be explored in detail.

5.2. *Core–corona effects and canonical strangeness suppression*

Glauber model calculations reveal that, even in central collisions, a small fraction of the target and projectile nucleons does not interact, at all, and a further about 10% fraction of the participant nucleons interacts only once. Toward peripheral collisions that latter fraction increases toward about 25%. These single, minimum bias nucleon–nucleon collisions exhibit canonical strangeness suppression.[6,8] The Wroblewski ratio of total strange to total nonstrange quarks in $p + p$ collisions is half that in central mass 200 A+A collisions.[54] In the SHM the transition from $p+p$ to A+A occurs as a change-over from a canonical to a grand canonical ensemble that occurs, gradually, with increasing N_{part}.[6,8] Singly strange/antistrange yields rise, by a factor of about two, relative to minimum bias $p+p$ at similar energy, and by factors of about 5 and 15, respectively, for multistrange hyperons.[54,55] The single collision fraction of A+A collisions stems from the diffuse surface regions of the nuclear Woods–Saxon nucleon density distributions. Thus one calls this the *corona effect*.[6,54,56] It reduces the strange to nonstrange hadron yield ratios in A+A collisions, as compared to a global grand canonical equilibrium population. One attempt to account for this strangeness undersaturation is to introduce a strangeness undersaturation parameter $\gamma(s)^S$ into the partition functions of the GC Gibbs ensemble;[6] $S = 1, 2, 3$ here. This is thus an additional fit parameter, in addition to the GC variables V, T and μ_B. It is found to amount to about 0.85 to 0.90 in central mass 200 collisions at SPS and RHIC energies,[6] increasing toward peripheral collisions.[57]

We can illustrate a typical interplay between several such second order corrections to the standard GC approach. The effect of a narrow midrapidity acceptance (see previous section) is expected to lead to an increase of the apparent Wroblewski ratio whereas the corona effect decreases it. The net effect might be a cancellation and, in fact, it was shown[6] that the NA49 data for central Pb+Pb collisions at top SPS energy (referring to a wider acceptance, of about $0 < y < 2$ required a $\gamma(s) = 0.85$ whereas $\gamma(s) = 1$ as the SHM fit was restricted to a $0 < y < 1$ acceptance interval.

5.3. *The origin of canonical strangeness suppression*

We address, next, a qualitatively different strangeness undersaturation effect, that governs elementary electron–positron annihilation to hadrons, and minimum bias $p+p$ collisions.[58,59] We have shown in Fig. 1 the canonical SHM fit to the LEP e^+–e^- annihilation data. It requires another, further strangeness suppression factor, $\gamma(s) = 0.66$. Unfortunately the same term is used here as in the GC SHM analysis of A+A collisions, confusing the issue as this suppression is of completely different origin. It can be understood from the Amati, Veneziano, Webber[13,14] model of color neutralization in elementary QCD hadronization that we referred to in Sec. 4. Here the primordial QCD partonic shower evolution is shown to end in a string-like configuration of successive complementary colour charges that locally neutralize into color singlet clusters, with varying invariant masses. They decay, independently, into hadrons and resonances, according to their respective phase space weights. The clusters are distributed over the entire longitudinal phase space, and thus they are, in general, not causally connected.[58,59] However, as shown by Becattini and Fries,[4] the Lorentz invariant single cluster multiplicities and net quantum numbers can be added into a single *equivalent global cluster*, whose volume is the sum of the proper individual cluster volumes, turning out to be large enough to make the canonical ensemble a good approximation.[60] However, the small volume , high microcanonical strangeness suppression (apparently due to the strange quark mass), prevailing in the individual cluster decays, gets transported into the equivalent global cluster strangeness content: a memory of the microscopic QCD dynamics leading to cluster formation. As the canonical SHM is ignorant of that microscopic pre-history it needs to be amended by an additional strangeness suppression parameter. It recalls that strangeness was produced in small, causally disconnected clusters, under high strangeness suppression.

From this, admittedly somewhat symbolic model, to hadron production from a QCD plasma, in A+A collisions: how does a QGP hadronize? Clearly, the first stage of the above schematic model, local color neutralization, has to be in common. We are not aware of any theoretical argument as to why this should not be a short-range, local process, imbedded into the overall collective expansion/cooling dynamics as the plasma approaches the critical or pseudocritical energy density domain. Furthermore it is plausible that the cluster density in phase space should now be about A times higher. So the phenomenon of isolated, single cluster hadronization that we addressed above under the symbolic term *causally not connected* should disappear, leading to a quantum number coherent large equivalent super-cluster decay, that is mirrored in the success of the Grand Canonical, large volume SHM description of the A+A hadron output. As we have seen from the BRAHMS results in Fig. 9, such fictitious superclusters can not extend over all longitudinal phase space, as successive intervals in rapidity space reveal a GC hadronization under different (T, μ_B) conditions. But, on the other hand, the genuine small cluster volume strangeness suppression, known from elementary (minimum bias) collisions, has disappeared here, hinting at a much larger, coherent decay system, in accord with the above schematic super-cluster picture.

In all, the reader will notice, however, that we employ here a highly symbolic terminology. Clusters, and super-clusters, are not yet known from genuine QCD calculations. They may be understood as an extrapolation from the concept of massive hadronic resonances. However, the formal three-step idea may reflect, in a minimal way, the three steps expected from fundamental QCD: color neutralization corresponding to the confinement transition, hadronic (cluster) mass generation symbolizing QCD chiral symmetry breaking, and the final quantum mechanical decay into on-shell hadrons/resonances representing the final tunneling effect, with its intrinsic phase space dominance, *à la* Fermi's Golden Rule, that makes the apparent hadro-chemical equilibrium plausible.

5.4. *Onset of Grand Canonical hadron production at the LHC*

It has been predicted, before the startup of the LHC, that at $\sqrt{(s)} = 7$ TeV the midrapidity production of hadrons in $p + p$ collisions might approach the grand canonical limit[60] of the SHM description. These expectations can now be checked with ALICE $p + p$ data[61] on strangeness production vs. multiplicity density $dN(ch)/d\eta$. In the cluster-terminology employed in the previous section the advent of Grand Canonical (GC) strangeness saturation would have to imply an increasing overlapp (or *causal connection*) among the microscopic singlet clusters, in longitudinal phase space. Looking at the midrapidity densities of charged hadrons we find $dN(ch)/d\eta = 3.3$ in LEP electron–positron annihilation (Fig. 1), and 6.0 in minimum bias $p + p$ at 7 TeV. The latter value[61] is quite in line with the $p + p$ systematics at lower energies.[54] As the former case did require the application of the strangeness suppression factor 0.66, characteristic of elementary collisions at such lower energies,[60] one would not expect a substantial, qualitative change in $p + p$ at the LHC as the cluster density increases by less than a factor of 2. However, the ALICE data shown in Fig. 10 imply that, on the contrary, significant strangeness enhancement (the reversal of strangeness suppression) occurs, with increasing midrapidity charged hadron density. Of course in the highest multiplicity bin, analyzed here, we have $dN_{ch}/d\eta = 21.3$, now in fact 6.5 times higher than in $e^+ - e^-$ annihilation at LEP. Also the enhancement effect increases strongly with ascending strangeness, quite resembling the strangeness hierarchical increase in the course of a transition from the canonical to the grand canonical ensemble.[8,54] However, we also see in Fig. 10 that the Omega to pi ratio, for example, grows by a further factor of about two, toward semi-central Pb+Pb collisions at 2.76 TeV. As the latter reaction should represent full GC strangeness saturation, we finally conclude that, even in the high multiplicity density windows of $p + p$ at LHC, we see no fully developed GC behavior but the onset of it. A GC SHM fit to these data would thus feature a non-unity, genuine $\gamma(s)$ factor.

A final remark concerning $p + p$ data as a reference for hadron production in A+A collisions, as expressed in the nuclear modification factor R_{AA}. Note that this is commonly defined as the ratio between A+A cross section and the corresponding minimum bias $p + p$ cross section scaled with the number of binary collisions, corresponding to the considered A+A centrality class. However, this ratio is only applicable at high transverse momentum, well above the domain of thermal bulk production. Here the hadron multiplicities scale with the volume of the emitting

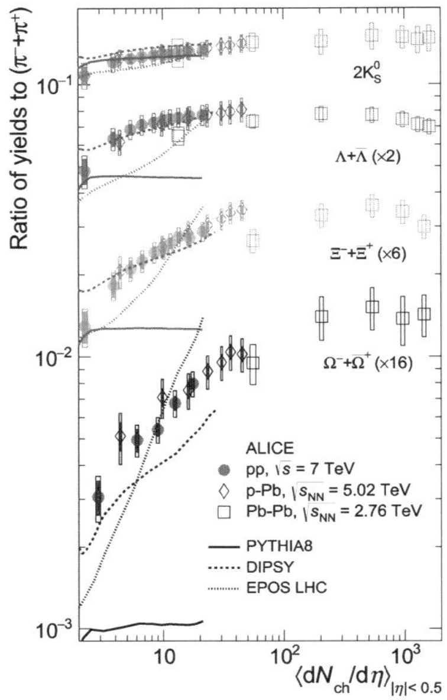

Fig. 10. ALICE results for $p+p$ collisions at LHC energy, $\sqrt{s} = 7$ TeV,[62] showing the multiplicity dependence of strange hadron to pion yields, matching to results from p+Pb and Pb+Pb collisions.

source, proportional to N_{part}, and not with the Glauber binary collision number that increases like $N_{\text{part}}^{4/3}$. At high p_t there should be no problem with employing R_{AA}.

6. Conclusions

The fundamental interest in Statistical Model analysis stems from the hypothesis that the derived freeze-out points in the (T,μ_B) phase diagram should, at modest μ_B, coincide with the QCD hadronization points, thus revealing the parton–hadron transformation line of QCD: one of the principal characteristics of thermal QCD, in the non-perturbative domain. It was the purpose of this article to demonstrate that SHM studies are converging toward this goal. One needs, however, to take care of numerous second-order concerns in the application of the grand canonical SHM. From among these, we have addressed final state annihilation corrections, experimental effects stemming from lack of feed-down corrections arising from unresolved, secondary vertex weak hyperon and antihyperon decays, and proper handling of acceptance and core–corona effects. Further problems require attention, such as finite hadron volume corrections[62] and the clarification of the relationship between the SHM and coalescence model approaches.[4,63] Finally the validity of the single freeze-out model[8,47] still awaits a clarification. Turning to the idea[17] to replace, or substitute the traditional SHM analysis by considering the overlap between Lattice

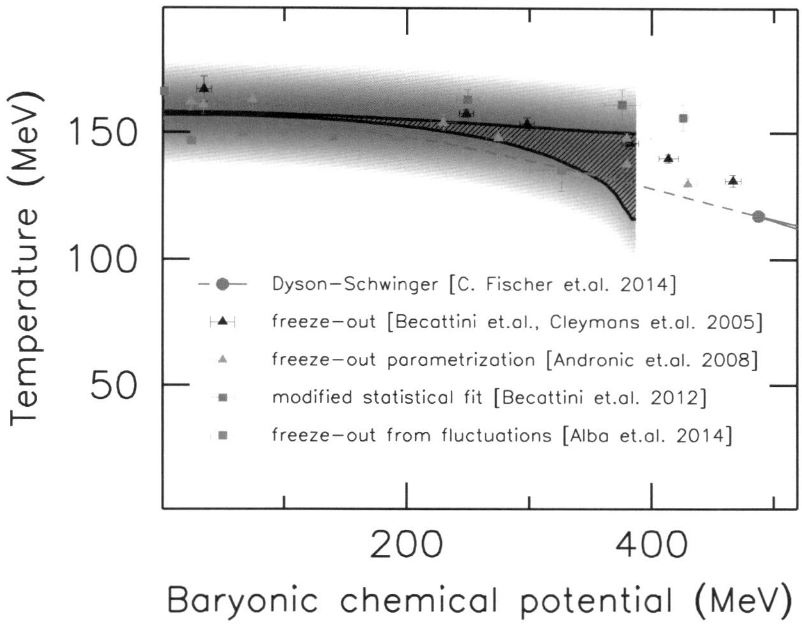

Fig. 11. The QCD phase diagram at finite baryochemical potential showing the domain of the crossover transition, confronted with various entries from SHM and Dyson–Schwinger analysis (see Ref. 66 for detail).

QCD and Hadron-Resonance Gas (HRG) model calculations: this should provide for a alternative method to locate the (pseudo-)critical temperature domain over which partonic degrees of freedom convert to the hadronic ones.[18] One can also analyze, directly, the first data concerning higher-order grand canonical fluctuations of conserved quantum numbers, such as skewness and kurtosis, in terms of Lattice predictions for higher-order susceptibilities.[19,20,33,64] We have argued that not all of the the various quantities, addressed in this framework (to which we could only provide a passing glance), should necessarily reveal a common freeze-out temperature,[65] given the broad band of temperatures encountered in a cross-over phase transformation as is encountered at modest μ_B. First-order quantities, like energy and entropy densities (closely related to the hadron number densities considered in the SHM), might freeze out sooner than fourth order susceptibilities. Finally the HRG approximation to the latter type of quantities may lack validity.[62] To wrap up on this new research field, of HRG-Lattice QCD-data analysis, and also the recent attempts in Lattice QCD to pin down the pseudo-critical line (or domain) in the QCD phase diagram, we show in Fig. 11 the result of a recent study of the phase diagram, by Bellwied *et al.*[66] The relatively broad domain of the cross-over transition in the (T, μ_B) plane, reaching up to limit of validity of the employed extrapolation method for finite μ_B domains, at about 400 MeV, is shown to envelope individual entries stemming from SHM and other approaches (see Ref. 66). We conclude that a detailed description of the QCD parton–hadron transition, far into the finite μ_B domain, appears within reach now, a formidable task commensurate to its fundamental importance. What remains, relatively much less explored, is the high μ_B domain. A rather abrupt turn-over occurs[7] at a μ_B of about 400 MeV, but the scarcity of data in this domain prevents us from answering several interesting questions concerning the high μ_B domain:

(1) If one considers the lattice extrapolations to apply, at all, to such high μ_B we might ascribe the the steepening slope to higher than quadratic terms in the lattice Taylor expansion.
(2) If the enigmatic critical point of QCD exists, at all, it should occur[16] at high μ_B, and influence the sequence, and position of the hadronization points (their dependence on the incident energy of the collision). This might occur, either, due to the onset, at the critical point, of a first order phase transition, or to the hypothetical focusing effect on the systems expansion trajectories.[17]
(3) A further untested suggestion would explain the rather steep fall-off toward AGS energy. If the so-called quarkyonic matter domain of QCD[19] exists, and sets in at the temperature of the critical point, one would expect a second QCD phase boundary line, to turn down, steeply, from the continuing deconfinement line, and become the site of hadronization.
(4) Finally, to the more trivial side, we might expect to turn away from the QCD transition line because the collisional volume does not (or not predominantly) enter the deconfined phase, to begin with. However, recent hadron transport model studies[67] consistently predict maximum energy densities well above one GeV/fm^3 to be reached at this energy, indicating an onset of deconfinement[18] to occur at lower energies.

We conclude that, neither the theoretical foundation, nor the approaches toward an experimental evidence, have been receiving even remotely such an attention recently as we witnessed in the attempts to work out the QCD phase diagram at modest μ_B, where Lattice QCD is applicable.

References

1. F. Karsch and E. Laermann, in *Quark-Gluon Plasma 3*, eds. H. C. Hwa and X. N. Wang (World Scientific, 2004), p. 1; Z. Fodor and S. D. Katz, *Landolt-Boernstein* **23** (2010) 227, arXiv:0908.3343.
2. G. Enrodi, Z. Fodor, S. D. Katz, and K. K. Szabo, *JHEP* **1104** (2011) 001; O. Kaczmarek *et al.*, *Phys. Rev. D* **83** (2011) 014504.
3. A. Bazavov *et al.*, arXiv:1509.05786.
4. F. Becattini and R. Fries, *Landolt-Boernstein* **23** (2010) 208, arXiv:0907.1031; R. Stock, arXiv nucl-th/0703050.
5. P. Braun-Munzinger, K. Redlich, and J. Stachel, in *Quark-Gluon Plasma 3*, eds. R. C. Hwa and X. N. Wang (World Scientific, 2004), p. 491.

6. F. Becattini, M. Gadzdicki, A. Keranen, J. Manninen and R. Stock, *Phys. Rev. C* **69** (2004) 024905.
7. A. Andronic *et al.*, *Nucl. Phys. A* **772** (2006) 167, and *Phys. Lett. B* **673** (2009) 142.
8. J. Rafelski and M. Danos, *Phys. Lett. B* **97** (1980) 167;
   A. Tounsi and K. Redlich, *J. Phys. G* **28** (2002) 2095;
   W. Florkowski, W. Broniowski and M. Michalec, *Acta Phys. Polon. B* **33** (2002) 761.
9. R. Stock, *Phys. Lett. B* **456** (1999) 761.
10. P. Braun-Munzinger, J. Stachel and C. Wetterich, *Phys. Lett. B* **598** (2004) 61.
11. U. Heinz and G. Kestin, arXiv nucl-th/0612105.
12. A. Andronic, P. Braun-Munzinger, K. Redlich and J. Stachel, *Nucl. Phys. A* **904-905** (2013) 535.
13. D. Amati and G. Veneziano, *Phys. Lett. B* **83** (1979) 87.
14. B. R. Webber, *Nucl. Phys. B* **238** (1984) 492.
15. F. Becattini, *Nucl. Phys. A* **702** (2002) 336.
16. N. Cabibbo and G. Parisi, *Phys. Lett. B* **59** (1975) 67.
17. F. Karsch, K. Redlich and A. Tawfik, *Phys. Lett. B* **571** (2003) 67.
18. A. Bozavov *et al.*, *Phys. Rev. D* **90** (2014) 094503, and *Phys. Rev. Lett.* **113** (2014) 072001.
19. F. Karsch, *J. Phys. G* **38** (2011) 124098;
    S. Mukherjee, *Nucl. Phys. A* **904-905** (2013) 873.
20. C. Ratti *et al.*, *Nucl. Phys.* **904-905** (2013) 802.
21. M. A. Stephanov, *J. Phys. G* **38** (2011) 124147.
22. C. Schmidt *et al.*, *Nucl. Phys. A* **904-905** (2013) 865;
    S. Datta, R. V. Gavai and S. Gupta, *ibid.* p. 883.
23. C. Nonaka and M. Asakawa, *Phys. Rev. C* **71** (2005) 044904;
    R. A. Lacey *et al.*, *Phys. Rev. Lett.* **98** (2007) 092301.
24. M. Gazdzicki, M. Gorenstein and P. Seyboth, *Acta Phys. Polon. B* **42** (2011) 307, arXiv:1006.1765 (hep-ph).
25. L. McLerran and M. Pisarski, *Nucl. Phys. A* **796** (2007) 83;
    A. Andronic *et al.*, *Nucl. Phys. A* **837** (2010) 65.
26. F. Becattini *et al.*, *Phys. Rev. C* **85** (2012) 044291, arXiv:1201.6349.
27. F. Becattini *et al.*, *Phys. Rev. Lett.* **111** (2013) 082302, arXiv:1212.2431.
28. R. Hagedorn, *Nuovo Cimento Suppl.* **3** (1965) 147.
29. H. Petersen *et al.*, *Phys. Rev. C* **78** (2008) 044901.
30. F. Becattini, J. Steinheimer, R. Stock and M. Bleicher, arXiv:1605.09694.
31. R. Snellings, arXiv:1411.7690;
    X. Zhu, F. Meng, H. Song and Y. X. Liu, arXiv:1501.0328.
32. J. Cleymans, H. Oeschler, K. Redlich and S. Wheaton, *Phys. Rev. C* **73** (2006) 034905.
33. H. T. Ding, *Nucl. Phys. A* **931** (2014) 52, and references therein.
34. S. Bass and A. Dumitru, *Phys. Rev. C* **61** (2000) 064909;
    D. Teaney and E. V. Shuryak, *Nucl. Phys. A* **698** (2002) 479.
35. D. H. Rischke, Y. Pursun and J. A. Maruhn, *Nucl. Phys.* **595** (1995) 383.
36. J. Steinheimer, S. Schramm and H. Stoecker, *Phys. Rev. C* **84** (2011) 045208.
37. J. D. Bjorken, *Phys. Rev. D* **27** (1983) 140.
38. P. Huovinen and H. Petersen, *Eur. Phys. J. A* **48** (2012) 171.
39. R. Rapp and E. V. Shuryak, *Phys. Rev. Lett.* **86** (2001) 2980.
40. A. Palmese, W. Cassing, E. Seifert, T. Steinert, P. Moreau and E. Bratkovskaya, *Phys. Rev. C* **94** (2016) 044912.
41. M. Floris *et al.*, *ALICE Coll.*, *Nucl. Phys. A* **931** (2014) 103.

42. T. Anticic *et al.*, *NA49Coll., Phys. Rev. C* **83** (2011) 014901, and references therein.
43. A. Ahmad *et al.*, *E891 Coll., Phys. Lett. B* **382** (1996) 35;
L. Ahle *et al.*, *E802 Coll., Phys. Rev. C* **60** (1999) 064901;
S. Albergo *et al.*, *E896 Coll., Phys. Rev. Lett.* **88** (2002) 062301;
J. Barette *et al.*, *E887 Coll., Phys. Rev. C* **62** (2000) 024901.
44. L. Kumar, *Nucl. Phys. A* **904-905** (2013) 256.
45. P. Alba *et al.*, *Phys. Rev. C* **92** (2015) 064910 and arXiv:1606.06542.
46. R. Bellwied *et al.*, arXiv:1305.6297.
47. A. Andronic, P. Braun-Munzinger and J. Stachel, *Phys. Lett. B* **673** (2009) 142;
J. Cleymans *et al.*, *Phys. Rev. C* **73** (2006) 034905;
J. Stachel *et al.*, *J. Phys. Conf. Ser.* **509** (2014) 012019;
G. Agakishiev *et al.*, *HADES Coll., Eur. Phys. J. A* **52** (2016) 178;
X. Lopez *et al.*, *FOPI Coll., Phys. Rev. C* **76** (2007) 052203.
48. T. Galatyuk, talk at *Erice Nucl. Coll. School* 2016.
49. U. Heinz, nucl-th/9810056.
50. H. Appelshaeuser *et al.*, *NA49Coll., Eur. Phys. J. C* **2** (1998) 611.
51. A. Wroblewski, *Acta Phys. Pol. B* **16** (1985) 379.
52. Ch. Blume *et al.*, *NA49Coll.*, nucl-ex/0701042.
53. I. G. Baerden *et al.*, *BRAHMS Coll., Phys. Rev. Lett.* **90** (2003) 102301.
54. R. Stock, *Landolt-Boernstein* **1-21A** (2008) 7.1; nucl-ex/0807.1610.
55. J. Adams *et al.*, *STAR Coll., Phys. Rev. Lett.* **92** (2004) 182301;
M. Lamont *et al.*, *STAR Coll., J. Phys. G* **32** (2006) 105.
56. K. Werner, *Phys. Rev. Lett.* **98** (2007) 152301.
57. F. Becattini *et al.*, *Phys. Rev.* **90** (2014) 054907.
58. P. Castorina and H. Satz, *J. Mod. Phys. E* **23** (2014) 1450019.
59. F. Becattini, P. Castorina, J. Manninen and H. Satz, *Eur. Phys. J. C* **56** (2008) 49.
60. F. Becattini and L. Ferroni, *Eur. Phys. J. C* **38** (2004) 225;
F. Becattini and G. Passaleva, *Eur. Phys. J. C* **23** (2002) 551.
61. F. Becattini *et al.*, *J. Phys. G* **38** (2011) 025002.
62. J. Adams *et al.*, *ALICE Coll.*, arXiv:1606.07424.
63. S. Mrowczynski, arXiv:1607.02267.
64. P. Braun-Munzinger, V. Koch, T. Schaefer and J. Stachel, *Phys. Repts.* **62** (2016) 76.
65. J. Steinheimer, V. Vovchenko, J. Aichelin, M. Bleicher and H. Stoecker, arXiv:1608.03737.
66. R. Bellwied *et al.*, arXiv:1507.07510.
67. S. Endres, H. van Hees and M. Bleicher, arXiv:1512.06549.

Why May Hydrodynamics Work for Classical Radiation Field?

Tamás S. Biró
H.A.S. Wigner Research Centre for Physics, Budapest, Hungary

I review what I learned from the late Professor Walter Greiner, not simply in physics. Albeit his effect on me and the evolution of my ideas was as a rule indirect, nonetheless relevant. As to the physics part, in this contribution I sketch some ideas about *why hydrodynamics works* in high energy collisions.

1. Introduction

Walter Greiner is no more among us. In his life I have heard different opinions on his role, some critical, a few supporting, but none of them was able to deny his importance and effective influence on nuclear physics of the late 20th century. As part of the generation coming after the "founding fathers" like Wigner, Neumann, Teller, nuclear physics became for the 1960s a mature field of research. Coming out from the secrecies under semi-militaristic influences the freedom of university research gave a fresh wind beneath the wings of nuclear physics. With the concurrence of the fast emerging field theory and accelerator particle physics in the background, in the 70s and 80s theoretical nuclear physics has developed some modern branches: in particular the seek for the Holy Grail of the quark–gluon plasma required a novel interdisciplinary approach between nuclear physics, many-body physics, quantum field theory, and yes, classical thermodynamics and theories of relativistic fluids. The 80s also have seen the start of lattice gauge theory calculations with ever since improving computer power in their background.

I have started my PhD studies in 1980 at the KFKI Research Institute in Budapest, with József Zimányi as my PhD advisor. He gave me the task to calculate the yield rate of strange quark–antiquark pairs in a thermal quark–gluon plasma. Since I was studying in the biophysics direction previously, I knew about phase transitions and thermal ensembles but I did not know a bit about Feynman graphs. So I had to calculate the $q\bar{q}$–$s\bar{s}$ process on the tree level, following formal rules I found in the Appendix of Itzyckson's and Zuber's textbook about field theory. The result was liked by my advisor who came to Frankfurt and lectured about it. This triggered Walter Greiner to squeeze his assistant, Johann Rafelski, somewhat: "See Johann, your idea about strangeness will not work." That initiated an avalanche of events, at those end a full calculation of strangeness production in the quark–gluon plasma, including the gluon fusion process, made the Rafelski–Muller–Koch author trio's calculation legendary. It took almost 40 years until Rafelski and myself could reconstruct all the events and could became friends without bad feelings. It is to be learned that concurrence should not make you blind for possible cooperation and learning from each other's work. It is the lighter way to slide into animosity, even in science.

A predecessor of these developments, which gave some realism to the very idea of looking for signs of a phase transition accompanied with bomb-like phenomena (deflagration, detonation, shock waves, after-burners), was emerging in heavy-ion physics already in the 70s, just before I have started my nuclear physics research as a student of József Zimányi, a Hungarian colleague and friend of Walter. This vocabulary was very much borrowed from hydrodynamical studies of bomb-explosions; a top secret issue even shortly before those dates. The very secret of "why does hydro works so well," in particular in systems which should have been described as quantum systems, remained as such until to date. We still do not posses a satisfactory answer to this question. The hydrodynamics approach is originally based on and often accompanied by a particle interpretation. Indeed its ingredients from general principles, derived from conservation of Noether charges, is incomplete: for a possible solution of the equations a relation between pressure and energy density, the $p(\epsilon)$ equation of state is needed. One may, of course, think that quarks and gluons and in the late stage of an ultrarelativistic collision the composite hadrons behave like particles. This would be a version of the impulse approximation for the quantum mechanical objects. But there is a talk even about the "flow of photons!" So how can the radiation field flow and satisfy hydrodynamics? In Refs. 1 and 2 we have circumlined some mathematical necessities which lead to the illusion of flow even in classical radiation patterns. I shall elaborate on these ideas in the next section.

There are some more memories about the influence of Walter Greiner on my life. Basically I worked and learned from several of his previous students: I have spent about a decade in Giessen, in Ulrich Mosel's Institute for Theoretical Physics. There I have met Professor Scheid, too. In those years happened to me that Berndt Müller invited me to work with him at Duke University, soon after he started there. Those were pioneering times which impregnated my later development: that I have to thank Berndt. I also met Walter's two sons in Giessen: Martin was a colleague at the parallel department of Professor Scheid, Carsten came to the Mosel group when I was there. With Carsten we co-authored quite a few papers, became friends and visited each other to collaborate on scientific projects in the sequel. Our generation keeps a part of the Frankfurt–Budapest connections until today. During these years the field theoretical studies became into the foreground of theoretical heavy-ion physics. Hydrodynamics, in the 90s with emphasis on the studies of the perfect fluid, eventually claimed to be found in RHIC experiments, then more recently strengthened by the inclusion of dissipative processes, kept its momentum and conquered the major part of theoretical efforts. It seems that the string theory motivated black hole estimate on the smallest possible shear viscosity, $\eta/s \geq 1/4\pi$ in Planck units, at a given entropy density, s, gave an extra momentum to the popularity of hydrodynamics. Although this estimate can be undergone, either by higher order corrections or by considering high density fermions.

Walter was always supportive towards international cooperation, not excluding any country or institution which behaved worthy for reaching scientific goals together. He promoted the connections of western scientists to all countries, including some which were considered on the "less free" part of the world. He had a special eye for Hungarians. He even donated a limerick about the three Hungarian physicist friends, István Lovas, József Zimányi and Judith Németh, to Acta Physica Hungarica. At an event before establishing the Journal of Heavy Ion Physics, as a new name for Acta Physica Hungarica, later called "New Series", by savoring at some Tokaj wine, Walter became part of the supporting International Editorial Committee. That was the event when — I suppose very rarely — he found some encouraging words for me in a private conversation among two of us, calling me (indirectly though) a "first class scientist". That gave me back a respectable amount of self-confidence and helped to work more enthusiastic after that.

Although I never was a student, an assistant or a co-author of Walter, I knew him well indirectly and was quite a few times exposed to his ideas about science, life and the world in general from a vicinity. I learned and agreed with his judgement that creativity and endeavor are much more important in science than popularity or position. Certainly, my own character and life-path might have also predestined me to like and create "heretic" theoretical models, but he gave me support on that. One of such heresies of mine is not to simply accept the success of hydrodynamics, but asking questions. Going beyond applications and testing the issue in extreme theoretical situations. In the rest of this contribution I would like to circumline one of these.

2. Illusory Temperature and Flow

In this section some simple mathematical formulas are presented demonstrating that temperature and flow, the two main ingredients in the vocabulary of the hydrodynamics picture of ultrarelativistic heavy-ion collisions, can be illusory. Illusory in the sense that although measurements, in particular observed spectra of extreme relativistic particles, kinematically behaving like photons, leave room for conjecturing a value of the temperature and a magnitude of a more or less scaling relativistic flow, their physical origin is not the presence of an energy reservoir and a flowing medium made of particles, like in the classical case. Such effects are also treated in Ref. 3.

We start with the Unruh temperature. This temperature is proportional to acceleration in Planck units $T = g/2\pi$. Although around its original publication a number of philosophical essays appeared about it as being a property of the quantum field theoretical vacuum, it can be viewed much more classically. One considers a point-like source of monochromatic radiation on a trajectory belonging to constant acceleration in the comoving proper frame. Such a trajectory is in general a Rindler trajectory:

$$x = \frac{1}{g}\Big(\cosh(g\tau) - 1\Big),$$
$$t = \frac{1}{g}\sinh(g\tau) \tag{1}$$

with τ proper time and using $c = 1$ units. The corresponding Minkowski velocity vector is given as

$$u^i = \frac{dx^i}{d\tau} = \big(\cosh(g\tau), \sinh(g\tau)\big) \tag{2}$$

and the acceleration vector as

$$a^i = \frac{du^i}{d\tau} = \big(g\sinh(g\tau),\, g\cosh(g\tau)\big). \tag{3}$$

The acceleration vector is spacelike with the invariant length $a_i a^i = -g^2$.

A monochromatic source emits a plane wave described by a complex amplitude as

$$\Phi(x,t) = \frac{1}{\sqrt{2\omega}}\,\mathrm{e}^{-i(\omega t - kx)}. \tag{4}$$

Photons exactly, extremely relativistic quanta of radiation approximately have the dispersion relation $\omega = k$. From it follows that the phase of the plane wave exhibits a light-cone behavior: $\varphi = \omega t - kx = \omega(t - x)$. With a fix ω frequency this is a monochromatic wave. An observer, moving with velocity v, encounters the coordinates $t = \gamma\tau$, $x = \gamma v\tau$, with the Lorentz factor $\gamma = 1/\sqrt{1-v^2}$. The phase looks like $\omega\gamma(1-v)\tau$ in this case, featuring a relativistic Doppler effect on the frequency of the emitter:

$$\omega(v) = \frac{d\varphi}{d\tau} = \omega\sqrt{\frac{1-v}{1+v}}. \tag{5}$$

For a constant proper acceleration path on the other hand the Doppler factor is continuously changing, and only an integrated effect occurs in the phase of the emitted and far away detected plane waves. With the Rindler trajectory settings in Eq. (1) $v = \tanh(g\tau)$ and the Doppler effected actual frequency is given by

$$\omega(\tau) = \omega\sqrt{\frac{1-v(\tau)}{1+v(\tau)}} = \omega\mathrm{e}^{-g\tau}. \tag{6}$$

The integrated phase up to the proper time τ becomes

$$\varphi(\tau) = \frac{\omega}{g}\left(1 - \mathrm{e}^{-g\tau}\right) + \varphi_0, \tag{7}$$

and the spectral density of the energy carried by the radiation with the sliding amplitude,

$$\mathcal{A}(f) = \frac{1}{\sqrt{2\omega}} \int_{-\infty}^{+\infty} e^{i(\varphi(\tau)+f\tau)}\, d\tau, \tag{8}$$

results as

$$f\, |\mathcal{A}(f)|^2 = \frac{1}{2\omega} \frac{\pi}{g} \frac{1}{e^{2\pi f/g} - 1}. \tag{9}$$

The last factor in the above formula reminds one to the Planck distribution and it contains the temperature-like factor $T = g/2\pi$. This is exactly the Unruh temperature, coinciding with the period of the Rindler trajectory analytically continued to imaginary proper time arguments, $\beta = 2\pi/g$, in $c = 1$, $\hbar = 1$, $k_B = 1$ units. This Unruh temperature can be used for deriving the Hawking–Bekenstein entropy, too, identifying the thermodynamical energy with the total mass inside an event horizon, and obtaining the proper acceleration at the coordinate singularity. In the Schwarzschild metric,

$$d\tau^2 = f(r)dt^2 - dr^2/f(r) - r^2 d\Omega^2, \tag{10}$$

with $f(r) = 1 - 2M/r$, the acceleration is given by $g = -f'(2M)/2$ leading to an Unruh temperature of $T = 1/8\pi M$. Clausius' formula for the entropy delivers

$$S = \int \frac{dM}{T} = 4\pi M^2 = \pi R^2, \tag{11}$$

a quarter of the horizon area.

Instead of getting involved with the question whether the Unruh temperature is purely an illusion, since there is no oven to heat any medium, no fluctuations of energy states, only a single frequency integrated over a history path which is particular, we demonstrate that much less exotic sources of radiation, like a classical accelerating point charge, are also able to generate a spectral illusion of thermicity and even a flow.

Considering the equivalent semiclassical photon distribution in a radiation stemming from a current represented by a charge q moving on the path $x(\tau)$, in the Fourier transformed picture,

$$J^i(k) = q \int e^{ik\cdot x(\tau)}\, u^i(\tau)\, d\tau, \tag{12}$$

one expects the following polarization summed spectrum

$$d^3N = \frac{1}{2k_0} \frac{d^3k}{(2\pi)^3} \sum_{a=1,2} \left| \epsilon^{(a)} \cdot J(k) \right|^2. \tag{13}$$

For leading on-shell contributions $k^2 = 0$. Although the formula (12) contains the velocity of the source, it can be re-arranged by partial integration in a form showing radiation only from an accelerating charge:

$$\epsilon \cdot J(k) = q \int_{\tau_1}^{\tau_2} e^{ik\cdot x(\tau)} \frac{d}{d\tau} \left(\frac{\epsilon \cdot u}{k \cdot u} \right) d\tau. \tag{14}$$

This is also the proper expression for a path fraction swept over in a finite proper time. Details of the calculation given in Refs. 1 and 2. For a constant co-moving acceleration path one obtains the amplitude of the radiation field as

$$\mathcal{A} = \frac{1}{k_\perp} \int_{\xi_1}^{\xi_2} e^{ik\cdot x(\tau)} \frac{d\xi}{\cosh^2(\xi-\eta)}, \tag{15}$$

where η is the rapidity of the emitted photon, while $\xi = g\tau$ is the rapidity of the moving source. Its integral for a Rindler trajectory, starting with the speed of light, following a straight path and ending with the lightspeed in the opposite direction, the intensity becomes

$$\mathcal{I} = \frac{4\alpha}{\pi} \ell^2 K_1^2(\ell k_\perp), \tag{16}$$

with $\ell = c^2/g$ characteristic length scale for the acceleration. The Bessel function, $K_1(x)$ has an exponential tail asymptotically for large transverse momenta of the radiation $k_\perp \gg \pi/\ell$. Interesting enough, such a result may be interpreted as being an integral over a rapidity like variable of K_2 Bessel functions, making the illusion of relativistic ideal gas particles sitting on a Bjorken scaling flow:

$$\frac{dN}{k_\perp dk_\perp d\eta d\varphi} = \frac{4\alpha\ell^2}{\pi} \int_{-\infty}^{+\infty} d\zeta\, K_2\big(2\ell k_\perp \cosh(\zeta-\eta)\big). \tag{17}$$

This result looks like a distribution of photons according to a Jüttner distribution, an ideal thermal ensemble sitting on a longitudinal flow. The effective temperature is $T_{\text{spectral}} = g/2 = \pi T_{\text{Unruh}}$.

Further numerical and analytical investigations in Ref. 2 have shown that for short sections of a constant acceleration path a Landau-like, while for long sections a Bjorken-like flow pattern is seen in the photon rapidity distribution, $dN/d\eta$. For light hadrons, like pions, the conclusion is similar; for them η is the pseudorapidity.

In summary, remembering Walter Greiner recalls conceptual and heretic ideas, however with the hope that such studies may help to reach a deeper understanding of more standard procedures, in particular to reveal why hydrodynamics, based on a flowing particle medium picture, also seems to work for phenomena and spectral features connected to extreme relativistic radiation. This work has been supported by the Hungarian National Bureau for Research Development and Innovation under the contract K123815.

References

1. T. S. Biró, M. Gyulassy and Z. Schram, Unruh gamma radiation at RHIC?, *Phys. Lett. B* **708**, 276 (2012).
2. T. S. Biró, Z. Szendi and Z. Schram, Illusory flow in radiation from accelerating charge, *Eur. Phys. J. A* **50**, 62 (2014).
3. T. S. Biró, *Is there a temperature? Conceptual Challenges at High Energy, Acceleration and Complexity*, Fundamental Theories of Physics 1014 (Springer, New York, 2011).

Chiral Symmetry Restoration and Deconfinement in Heavy-Ion Collisions

E. L. Bratkovskaya[1,2], W. Cassing[3], P. Moreau[2,4] and A. Palmese[3]
[1] *GSI Helmholtzzentrum für Schwerionenforschung GmbH, D-64291 Darmstadt, Germany*
[2] *Institute for Theoretical Physics, Johann Wolfgang Goethe Universität, D-60438 Frankfurt am Main, Germany*
[3] *Institut für Theoretische Physik, Universität Gießen, 35392 Gießen, Germany*
[4] *Frankfurt Institute for Advanced Studies, Johann Wolfgang Goethe Universität, D-60438 Frankfurt am Main, Germany*
[1] *E. Bratkovskaya@gsi.de*

We study the effect of the chiral symmetry restoration (CSR) on heavy-ion collisions observables in the energy range $\sqrt{s_{NN}}$ = 3–20 GeV within the Parton-Hadron-String Dynamics (PHSD) transport approach. The PHSD includes the deconfinement phase transition as well as essential aspects of the chiral symmetry restoration in the dense and hot hadronic medium, which are incorporated in the Schwinger mechanism for the hadronic particle production. The latter describes the formation of quark–antiquark pairs in the strong chromo-electric fields in analogy to e^+e^- production in supercritical electric fields. We adopt different parametrizations of the nuclear equation of state from the non-linear $\sigma - \omega$ model, which enter in the computation of the quark scalar density for the CSR mechanism, in order to estimate the uncertainty in our calculations. For the pion–nucleon Σ-term we adopt $\Sigma_\pi \approx 45$ MeV which corresponds to some 'world average'. Our systematic studies show that chiral symmetry restoration plays a crucial role in the description of heavy-ion collisions at $\sqrt{s_{NN}}$ = 3–20 GeV, realizing an increase of the hadronic particle production in the strangeness sector with respect to the non-strange one. Our results provide a microscopic explanation for the "horn" structure in the excitation function of the K^+/π^+ ratio: the CSR in the hadronic phase produces the steep increase of this particle ratio up to $\sqrt{s_{NN}} \approx 7$ GeV, while the drop at higher energies is associated to the appearance of a deconfined partonic medium.

1. Introduction

The formation of e^+e^- pairs in supercritical electric fields has been a major research topic of Walter Greiner and his collaborators[1,2] who pioneered the field of Quantum ElectroDynamics (QED) in strong fields. In analogy the formation of quark–antiquark pairs in strong chromo-electric fields is the major mechanism for the formation of matter from field energy as realized in the early stage of relativistic heavy-ion reactions. Contrary to QED the QCD vacuum is characterized by a large scalar quark condensate $\langle\bar{q}q\rangle \neq 0$ which is expected to melt at high baryon density and/or temperature and accordingly the QCD vacuum changes its structure to a perturbative vacuum at high energy density. Since the scalar quark condensate $\langle\bar{q}q\rangle \neq 0$ is an order parameter for chiral symmetry restoration (CSR), the quark–antiquark production mechanism should reflect the change of the vacuum in a hot or dense hadronic environment.

On the other hand, the main goal of heavy-ion collision (HIC) physics is the study of nuclear matter at high temperatures and/or high densities. In particular, the investigation of the nuclear matter phase diagram as a function of temperature and baryon chemical potential is of major interest. According to Quantum Chromo-Dynamics (QCD), at large temperatures and densities the hadrons cannot survive anymore as bound states and they dissolve forming the so-called Quark–Gluon Plasma (QGP). At Relativistic Heavy-Ion Collider (RHIC) energies for the first time the creation of a QGP, identified as an almost perfect fluid,[3] has been proven experimentally. Actually, the properties of this deconfined state of matter are still debated as well as the phase boundaries to the hadronic phase. In order to shed some light on this issue, many heavy-ion experiments are performed at the Super-Proton Synchrotron (SPS), RHIC, the Large Hadron Collider (LHC) and will be performed at the future Facility for Antiproton and Ion Research (FAIR) as well as the Nuclotron-based Ion Collider fAcility (NICA). The crucial challenge is to identify in the final particle distributions those signatures, which allow one to disentangle the QGP contribution that is impossible to observe directly or independently.

The strange particle production has always been suggested as one of the most sensitive observables that could spot out the creation of a QGP during the early stages of a HIC. The earliest suggested signature is the strangeness enhancement in A+A collisions with respect to elementary p+p collisions.[4,5] Later on, Gazdzicki and Gorenstein[6] proposed that a sharp rise and drop in the excitation function of the K^+/π^+ ratio (so-called "horn") should show up due to the appearance of a QGP phase at a center-of-mass energy $\sqrt{s_{NN}} \sim 7\,\mathrm{GeV}$.[a] Several statistical models[7–9] have succeeded in reproducing the trend of the experimental observation of the K^+/π^+ ratio and other strange to non-strange particle ratios, but they can provide only a statistical description of the HIC process and do not provide a dynamical explanation. On the other hand, there was no conclusive interpretation of the "horn" from dynamical approaches for HIC, like microscopic transport models.[10–12] Only recently, the Parton-Hadron-String Dynamics (PHSD), a transport approach describing HIC on the basis of partonic, hadronic and string degrees-of-freedom, obtained a striking improvement on this issue when including chiral symmetry restoration (CSR) in the string decay for hadronic particle production.[13–15]

[a]In this work we adopt natural units, hence $\hbar = c = 1$.

Apart from deconfinement the chiral symmetry restoration addresses another aspect of the QCD phase diagram in the (T, μ_B)-plane as an additional transition between a phase with broken and a phase with restored chiral symmetry. As in case of the QCD deconfinement phase transition, the boundaries of the CSR phase transition line are not well known. Lattice QCD (lQCD) calculations show that at vanishing baryon chemical potential $\mu_B = 0$ the CSR takes place at roughly the same critical temperature and energy density as the deconfinement phase transition which is a crossover. At finite baryon chemical potential lQCD calculations cannot be performed due to the fermion sign problem and one must rely on effective models (or extrapolations) in order to study the QCD phase transitions.[16–20] Different models support the idea that at finite chemical potential a partially restored phase is achieved before the deconfinement occurs.[21–23] In order to distinguish the two phases of such a transition, effective models use the scalar quark condensate $\langle \bar{q}q \rangle$ as an order parameter. As the baryon density and temperature increase, the scalar quark condensate $\langle \bar{q}q \rangle$ is expected to decrease from a non-vanishing value in the vacuum to $\langle \bar{q}q \rangle \approx 0$ which corresponds to CSR. Since $\langle \bar{q}q \rangle$ is not a measurable quantity, it is crucial to determine experimental observables which are sensitive to this condensate. Since long the dilepton spectroscopy has been in the focus in this respect since in a chirally restored phase the spectral functions of the the ρ- and the a_1-meson should become identical. However, no clear evidence has been achieved so far.[24] On the other hand, the strangeness production at Alternating-Gradient Synchrotron (AGS) and lower SPS energies was suggested to be a signature of CSR.[13]

In this contribution we report about a systematic analysis in Refs. 13–15 where a detailed study within the PHSD approach has been performed on effects of the CSR on final particle distributions in HICs in the energy range $\sqrt{s_{NN}} = 3$–$20\,\text{GeV}$. The PHSD is particularly suited to this aim since it includes both hadronic and partonic degrees-of-freedom[25] and has been extensively used to describe HIC observables from SPS to LHC energies.[26–28]

This work is organized as follows: In Sec. 2 we briefly recall the basic ingredients of the PHSD approach, while in Sec. 3 we illustrate the string fragmentation included in PHSD and the most recent extension of this particle production mechanism to incorporate essential aspects of chiral symmetry restoration. We also discuss the sensitivity of our results on the nuclear Equation of State (EoS) within the non-linear $\sigma - \omega$ model. In Sec. 4 we present the calculated results from PHSD — with and without the inclusion of CSR — for the particle rapidity spectra incorporating different nuclear equations of state in order to investigate the uncertainties of our approach. The excitation functions for the ratios K^+/π^+, K^-/π^- and $(\Lambda + \Sigma^0)/\pi$ — together with their uncertainties — complete Sec. 4 where a comparison with experimental data is performed wherever possible. We conclude this work with a summary in Sec. 5.

2. Reminder of the PHSD Transport Approach

The Parton-Hadron-String Dynamics (PHSD) is a microscopic covariant dynamical approach for strongly interacting systems in and out-of equilibrium.[25,26] It is a transport approach which goes beyond the quasi-particle approximation, since it is based on Kadanoff–Baym equations for the Green's functions in phase-space representation in first-order gradient expansion.[29,30] Including both a hadronic and a partonic phase as well as a transition between the effective degrees-of-freedom, PHSD is capable to describe the full time evolution of a relativistic heavy-ion collision. The theoretical description of the partonic degrees-of-freedom (quarks and gluons) is realized in line with the Dynamical-Quasi-Particle Model (DQPM)[30] which reproduces lQCD results in thermodynamical equilibrium and provides the properties of the partons, i.e. masses and widths in their spectral functions. In equilibrium the PHSD reproduces the partonic transport coefficients such as shear and bulk viscosities or the electric conductivity from lQCD calculations as well.[31]

An actual nucleus–nucleus collision in PHSD follows the following steps:

- Primary hard scatterings between nucleons take place and produce excited color-singlet states, denoted by "strings" (as described within the FRITIOF Lund model[32] and realized also in PYTHIA 6.4[33]). These strings decay into "prehadrons" with a formation time $\tau_f \sim 0.8$ fm/c and "leading hadrons", which originate from the string ends and can re-interact with hadrons almost instantly with reduced cross-sections (according to the constituent quark number).
- In case the local energy density is above the critical value of $\epsilon_c \sim 0.5$ GeV/fm^3, the deconfinement is implemented by dissolving the newly produced hadrons into the massive colored quarks/antiquarks and mean-field energy from the DQPM.
- Within the QGP phase, the partons (quarks, antiquarks and gluons) scatter and propagate in a self-generated mean-field potential. They are described as off-shell quasi-particles with temperature-dependent masses and widths, which are given by the DQPM.
- The expansion of the system is associated with a decrease of the local energy density and, once the local energy density becomes close to or lower than ϵ_c, the massive colored off-shell quarks and antiquarks hadronize to colorless off-shell mesons and baryons. The hadronization process is defined by covariant transition rates and fulfills the energy–momentum and quantum number conservation in each event.
- In the hadronic corona as well as in the late hadronic phase, the particles are still interacting and propagating. Elastic and inelastic collisions between baryons, mesons and resonances are implemented in PHSD and the corresponding backward reactions are included through detailed balance for all channels.

We further note that the pure hadronic phase in PHSD is equivalent to the Hadron-Strings Dynamics (HSD) model.[34] Accordingly, the comparison between PHSD and HSD calculations allows us to disentangle the role of the QGP phase in heavy-ion collisions. Even at low center-of-mass energies ($\sqrt{s_{NN}} \sim 3\,\text{GeV}$) the PHSD and the HSD results slightly differ due to the appearance of QGP 'droplets' in central cells of the system, which are characterized by high baryon and energy densities.

The PHSD approach has been tested for different colliding systems (p+p, p+A, A+A) in a wide range of bombarding energy, from AGS to LHC energies, and has been able to describe a large number of experimental observables, such as charged particle spectra, collective flow coefficients v_n as well as electromagnetic probes such as photons and dileptons.[31] More recently, it has been also shown to provide a microscopic description of the maximum in the K^+/π^+ ratio in central nucleus-nucleus collisions in Refs. 13 and 14.

3. String Fragmentation in PHSD

3.1. *Basic concepts*

The string formation and decay represents the dominant particle production mechanism in nucleus-nucleus collisions for bombarding energies from 2 *A*GeV to 160 *A*GeV. In PHSD, the primary hard scatterings between nucleons are described by string formation and decay in the FRITIOF Lund model.[32] A string is an excited color-singlet state, which is composed of two string ends corresponding to the leading constituent quarks of the colliding hadrons and a color flux tube (color-electric field) in between. As the string ends recede, virtual $q\bar{q}$ or $qq\bar{q}\bar{q}$ pairs are produced in the uniform color field by a tunneling process (described by the Schwinger formula[35]), causing the breaking of the string. This pair creation in the strong color field is in close analogy to the production of e^+e^- pairs in strong electric fields; a subject pioneered by Walter Greiner and his collaborators some decades ago.[1] Finally, the produced quarks and antiquarks recombine with neighboring partons to mesons or baryon–antibaryon pairs with formation time $\tau_f \sim 0.8$ fm/c (in the rest-frame of the string).

In the string decay, the flavor of the produced quarks is determined via the Schwinger formula,[33,35] which defines the production probability of massive $s\bar{s}$ pairs with respect to light flavor production ($u\bar{u}, d\bar{d}$) pairs:

$$\frac{P(s\bar{s})}{P(u\bar{u})} = \frac{P(s\bar{s})}{P(d\bar{d})} = \gamma_s = \exp\Big(-\pi\frac{m_s^2 - m_{u,d}^2}{2\kappa}\Big)\,, \tag{1}$$

with $\kappa \approx 0.176\,\text{GeV}^2$ representing the string tension and $m_{u,d,s}$ denoting the constituent quark masses for strange and light quarks. For the constituent quark masses $m_u \approx 0.35\,\text{GeV}$ and $m_s \approx 0.5\,\text{GeV}$ in the vacuum, the production of strange quarks is suppressed by a factor of $\gamma_s \approx 0.3$ with respect to the light quarks, which is the default setting in the FRITIOF routines. A further ingredient to fix the rapidity and transverse mass distributions for the newly produced hadrons (by recombination with neighboring partons) is the fraction of energy and momentum that they acquire from the decaying string. This is defined by the fragmentation function $f(x, m_T)$, which is the probability distribution for a hadron with transverse mass m_T to be produced with an energy-momentum fraction x from the fragmenting string:

$$f(x, m_T) \approx \frac{1}{x}(1 - x^a) \exp(-b\, m_T^2/x)\,, \tag{2}$$

where $a = 0.23$ and $b = 0.34\,\mathrm{GeV}^{-2}$ as reliable settings for p+p and p+A collisions. As becomes evident from Eq. (2) the meson m_T-scaling from string decay is included by default.

3.2. *Modeling of the chiral symmetry restoration*

In Refs. 13 and 14 the PHSD has been extended to include CSR in the string decay in a hadronic environment of finite baryon and meson density. Here we recall the main aspect of this extension which is based on the Hellman–Feynman theorem for the scalar quark condensate.[36] Accordingly, a linear decrease of the scalar quark condensate $\langle \bar{q}q \rangle$ — which is nonvanishing in the vacuum due to a spontaneous breaking of chiral symmetry — is expected with baryon density ρ_B towards a chiral symmetric phase characterized by $\langle \bar{q}q \rangle \approx 0$.[37,38] This decrease of the scalar quark condensate is expected also to lead to a change of the hadron properties with density and temperature, i.e. in a chirally restored phase the vector and axial vector currents should become equal;[39–43] the latter implies that, e.g. the ρ and a_1 spectral functions should become identical (as addressed above in the context of dilepton production). Since the scalar quark condensate $\langle \bar{q}q \rangle$ is not a direct observable, its manifestations should also be found indirectly in different hadronic abundances and spectra or particle ratios like K^+/π^+, $(\Lambda + \Sigma^0)/\pi^-$, etc. as advocated in Ref. 13.

In leading order the scalar quark condensate $\langle \bar{q}q \rangle$ can be evaluated in a dynamical calculation as follows:[44]

$$\frac{\langle \bar{q}q \rangle}{\langle \bar{q}q \rangle_V} = 1 - \frac{\Sigma_\pi}{f_\pi^2 m_\pi^2}\rho_S - \sum_h \frac{\sigma_h \rho_S^h}{f_\pi^2 m_\pi^2}\,, \tag{3}$$

where σ_h stands for the σ-commutator of the relevant mesons h, $\langle \bar{q}q \rangle_V$ represents the vacuum condensate, $\Sigma_\pi \approx 45\,\mathrm{MeV}$ is the pion–nucleon Σ-term and f_π and m_π are the pion decay constant and pion mass, respectively. Note, however, that the value of Σ_π is not so accurately known; a recent analysis points towards a larger value of $\Sigma_\pi \approx 59$ MeV[45,46] while actual lQCD results[47] suggest a substantially lower value. Accordingly, our following calculations — based on $\Sigma_\pi = 45$ MeV — have to be taken with some care although in some sense it represents a 'world average' (cf. Fig. 3 in Ref. 48). According to the light quark content, the Σ-term for hyperons is decreased by a factor of 2/3 for Λ and Σ hyperons and by a factor of 1/3 for

Ξ baryons. Furthermore, for mesons made out of light quarks and antiquarks, we use $\sigma_h = m_\pi/2$, whereas for mesons with a strange (antistrange) quark we consider $\sigma_h = m_\pi/4$. We mention here that improved results for the sigma-commutator for kaons can be obtained from chiral perturbation theory as in Ref. 49 and alternative assumptions for non-pseudoscalar mesons can be employed as, e.g. suggested by Cohen *et al.* in Ref. 36. In view of the subleading contributions of these mesons to the ratio in Eq. (3), we keep the simple estimates noted above for our present report and look forward to a clarification of the present tension between the results from lQCD and dispersive approaches.[50]

In Eq. (3), the quantities ρ_S and ρ_S^h denote the nucleon scalar density and the scalar density for a meson of type h, respectively. The scalar density of mesons h is evaluated in the independent-particle approximation as:

$$\rho_S^h(x) = \frac{(2s+1)(2\tau+1)}{(2\pi)^3} \int \mathrm{d}^3p \frac{m_h}{\sqrt{\mathbf{p}^2 + m_h^2}} f_h(x, \mathbf{p}) \,, \tag{4}$$

where $f_h(x, \mathbf{p})$ denotes the meson phase-space distribution ($x = (\mathbf{r}, t)$) and s, τ refer to the discrete spin and isospin quantum numbers, respectively. Moreover, the vacuum scalar condensate $\langle \bar{q}q \rangle_V = \langle \bar{u}u \rangle_V + \langle \bar{d}d \rangle_V \approx 2\langle \bar{u}u \rangle_V$ can be computed according to the Gell-Mann–Oakes–Renner (GOR) relation,[51,52]

$$f_\pi^2 m_\pi^2 = -\frac{1}{2}(m_u^0 + m_d^0)\langle \bar{q}q \rangle_V \,, \tag{5}$$

and gives $\langle \bar{q}q \rangle_V \approx -3.2\,\mathrm{fm}^{-3}$ for the bare quark masses $m_u^0 = m_d^0 \approx 7\,\mathrm{MeV}$. Finally, in Eq. (3) the nucleon scalar density ρ_S has to be determined in a suitable model with interacting degrees-of-freedom in order to match our knowledge on the nuclear EoS at low temperature and finite density. A proper (and widely used) approach is the non-linear $\sigma - \omega$ model for nuclear matter where ρ_S is defined as:

$$\rho_S(x) = \frac{d}{(2\pi)^3} \int \mathrm{d}^3p \frac{m_N^*}{\sqrt{\mathbf{p}_N^{*2} + m_N^{*2}}} f_N(x, \mathbf{p}) \,, \tag{6}$$

where m_N^* and p_N^* denote the effective mass and momentum, respectively, and $f_N(x, \mathbf{p})$ the phase-space occupation of a nucleon while the degeneracy factor is $d = 4$. In fact, in the non-linear $\sigma - \omega$ model the nucleon mass is modified due to the scalar interaction with the medium:

$$m_N^*(x) = m_N^v - g_s \sigma(x) \,, \tag{7}$$

where m_N^v denotes the nucleon mass in vacuum and $\sigma(x)$ is the scalar field which mediates the interaction between the nucleons and the medium with the coupling g_s. In order to calculate ρ_S, we need to determine the value of the scalar field $\sigma(x)$ at each space-time point x. This is done via the non-linear gap equation:[53,54]

$$\begin{aligned} m_\sigma^2 \sigma(x) + B\sigma^2(x) + C\sigma^3(x) &= g_s \rho_S(x) \\ &= g_s d \int \frac{\mathrm{d}^3p}{(2\pi)^3} \frac{m_N^*(x)}{\sqrt{\mathbf{p}^2 + m_N^{*2}}} f_N(x, \mathbf{p}) \,, \end{aligned} \tag{8}$$

since for matter at rest we have $\mathbf{p}^* = \mathbf{p}$. In Eq. (8) the self-interaction of the σ-field is included up to the forth order. The parameters g_s, m_σ, B, C are fixed in order to reproduce the values of the nuclear matter quantities at saturation, i.e. the saturation density, the binding energy per nucleon, the compression modulus, and the effective nucleon mass. Actually, there are different sets for these quantities that lead to slightly different saturation properties. We defer a discussion on the uncertainties of our results to the following Subsec. 3.3.

The main idea in Ref. 13 is to consider effective masses for the dressed quarks in the Schwinger formula (1) for the string decay in a hot and dense medium. The effective quark masses can be expressed in terms of a scalar coupling to the quark condensate $\langle \bar{q}q \rangle$ in first order as follows:

$$m_s^* = m_s^0 + (m_s^v - m_s^0)\frac{\langle \bar{q}q \rangle}{\langle \bar{q}q \rangle_V}, \tag{9}$$

$$m_q^* = m_q^0 + (m_q^v - m_q^0)\frac{\langle \bar{q}q \rangle}{\langle \bar{q}q \rangle_V}, \tag{10}$$

with $m_s^0 \approx 100\,\text{MeV}$ and $m_q^0 \approx 7\,\text{MeV}$ for the bare quark masses. In Eqs. (9) and (10) the effective masses decrease from the vacuum values with decreasing scalar condensate $\langle \bar{q}q \rangle$ to the constituent masses. This adaptation of the Schwinger formula in case of a hot and dense medium implies a modification of the flavor production factors in Eq. (1). In an actual nucleus-nucleus collision, PHSD incorporates a dynamical calculation of all these quantities for each cell in space-time:

- the scalar density ρ_S is determined by solving the gap equation (8) for the σ-field;
- the scalar condensate $\langle \bar{q}q \rangle$ is then computed via Eq. (3);
- the effective masses m_q^*, m_s^* are calculated according to Eqs. (9), (10) and plugged in the Schwinger formula (1) in order to compute the flavor production ratios for the string decay.

We stress that, once the nucleon scalar density ρ_S and Σ_π are fixed, there is no additional 'parameter' in the PHSD approach compared to the previous version that has been employed for a couple of years for the analysis of relativistic heavy-ion reactions.[31]

3.3. *Dependence on the nuclear equation of state*

In this subsection we analyze in more detail the flavor production ratios from the Schwinger formula in the presence of a hot and dense nuclear medium. As mentioned in the previous section there are different sets for the parameters g_s, m_σ, B, C in the gap equation (8). In fact, these parameters are fixed within the non-linear $\sigma - \omega$ model in order to reproduce empirical values of nuclear matter quantities at saturation (i.e. saturation density, binding energy per nucleon, compression modulus, effective nucleon mass etc.), but sizeable uncertainties remain with respect to the high density properties. In Table 1 we display the values of g_s, m_σ, B, C together with the vector coupling g_v, the vector meson mass m_v, the compression modulus K and the ratio between the effective and the bare nucleon mass m^*/m at saturation density for three sets commonly indicated as NL1, NL2 and NL3. The sets NL1 and NL3 have the same compression modulus K but differ in the effective mass m^*/m at saturation density whereas NL1 and NL2 have the same effective mass but differ in the compression modulus K. By comparing the results from NL1, NL2 and NL3 we will be able to explore separately the effects from the effective mass and compression modulus.

Table 1. Parameter sets NL1, NL2 and NL3 for the non-linear $\sigma - \omega$ model.

	NL1	NL2	NL3
g_s	6.91	8.50	9.50
g_v	7.54	7.54	10.95
B (1/fm)	−40.6	50.57	1.589
C	384.4	−6.26	34.23
m_σ (1/fm)	2.79	2.79	2.79
m_v (1/fm)	3.97	3.97	3.97
K (MeV)	380	210	380
m^*/m	0.83	0.83	0.70

In the context of the string decay, the most important parameter to focus on is the scalar coupling g_s, which is lower for the NL1 and NL2 set with respect to the corresponding values in the NL3 set. In Fig. 1 we show the dependence of the nucleon scalar density ρ_S on energy density ϵ in panel (a), the ratio between the scalar quark condensate and its value in the vacuum $\langle\bar{q}q\rangle/\langle\bar{q}q\rangle_V$ in panel (b), the light and strange quark effective masses m_q^*, m_s^* in panel (c), and the production probability of massive $s\bar{s}$ relative to light flavor production γ_s in panel (d). Note that an analogous dependence is observed with respect to the baryon density ρ_B since the energy density ϵ in leading order is just the nucleon mass times the baryon density. We find that all quantities plotted in Fig. 1 show practically identical results for NL1 (green dashed lines) and NL2 (thin orange lines) since the scalar density ρ_S essentially depends on the effective nucleon mass which is very similar for NL1 and NL2 when plotted as a function of the energy density ϵ. The results shown in Fig. 1 are obtained at vanishing temperature $T = 0$, but all the following considerations can be extended to a more realistic picture at finite temperature (meson density) and illustrate the consequences of CSR in the PHSD results for heavy-ion collisions. The energy density ϵ here is calculated within the non-linear $\sigma - \omega$ model by:

$$\epsilon = U(\sigma) + \frac{g_v^2}{2m_v^2}\rho_N^2 + d\int \frac{d^3p}{(2\pi)^3}\, E^*(\mathbf{p})\left(N_f(\mathbf{p}) + N_{\bar{f}}(\mathbf{p})\right), \tag{11}$$

with

$$E^*(\mathbf{p}) = \sqrt{\mathbf{p}^2 + m_N^{*2}}\,,$$

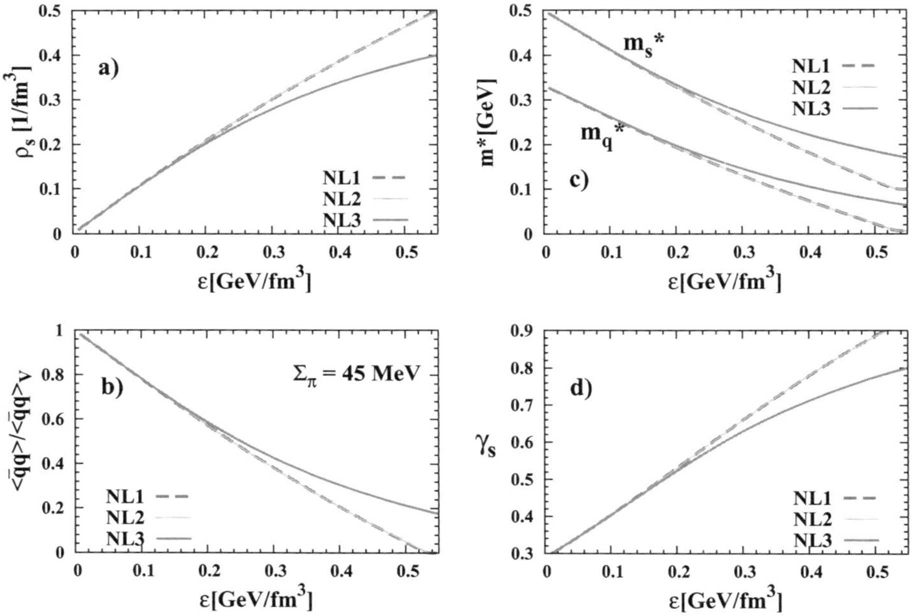

Fig. 1. The nucleon scalar density ρ_S (a), the ratio between the scalar quark condensate and its value in the vacuum $\langle \bar{q}q\rangle/\langle \bar{q}q\rangle_V$ (b), the light and strange quark effective masses m_q^*, m_s^* (c), and the production probability of massive $s\bar{s}$ relative to light flavor production γ_s (d) as a function of the energy density ϵ for the parameter sets NL3 (red solid lines), NL2 (thin orange lines) and NL1 (dashed green lines) at $T = 0$ and with $\Sigma_\pi = 45\,\text{MeV}$.

$$U(\sigma) = \frac{m_\sigma^2}{2}\sigma^2 + \frac{B}{3}\sigma^3 + \frac{C}{4}\sigma^4 ,$$

where ρ_N represents the nucleon density and $N_f(\mathbf{p})$ and $N_{\bar{f}}(\mathbf{p})$ are the particle/antiparticle occupation numbers at fixed momentum $\mathbf{p}$, respectively.

The scalar density ρ_S increases with increasing energy density ϵ as displayed in panel (a) of Fig. 1. We find a moderate sensitivity to the nuclear equation of state up to energy densities of ~ 0.5 GeV/fm^3 that are of relevance for the hadronic phase. In fact, the lines referring to the parameter sets NL1/NL2 and NL3 show a very similar behavior as a function of ϵ, but the NL3 (solid line) set is always characterized by lower values of the scalar density ρ_S relative to the NL1 or NL2 parametrization (dashed line). This is due to the larger value of the effective nucleon mass m_N^* in case of the NL1 and NL2 parameter sets. In panel (b) of Fig. 1 the ratio $\langle \bar{q}q\rangle/\langle \bar{q}q\rangle_V$ is presented as a function of ϵ. At $\epsilon = 0$ the scalar condensate corresponds to the vacuum value $\langle \bar{q}q\rangle_V$ and for fixed $\Sigma_N = 45\,\text{MeV}$ it decreases almost linearly with increasing energy density and almost vanishes for the critical energy density $\epsilon_c \approx 0.5$ GeV/fm^3. In this case, the order between NL1/NL2 and NL3 results is reversed: the NL3 parametrization for the nuclear EoS is associated to higher values of the scalar quark condensate with respect to the NL1 or NL2 sets. This feature can be easily explained looking at the definition of the ratio $\langle \bar{q}q \rangle / \langle \bar{q}q \rangle_V$ (3): at $T = 0$, there are no thermal mesons, thus the last term of the relation vanishes and the ratio is entirely fixed by the scalar density ρ_S; hence, higher values of ρ_S correspond to lower values of $\langle \bar{q}q \rangle / \langle \bar{q}q \rangle_V$. Therefore, the NL1 and NL2 parametrizations are characterized by lower values of the scalar quark condensate with respect to the NL3 parameter set. It follows that the light and strange quark effective masses m_q^*, m_s^* in panel (c) of Fig. 1 show a very similar dependence on the energy density. At vanishing energy density ϵ, the quark effective masses are equal to their vacuum values, $m_q \approx 0.33\,\text{GeV}$ and $m_s \approx 0.5\,\text{GeV}$; with increasing ϵ the quark masses decrease in line with the scalar quark condensate up to their bare values $m_s^0 \approx 100\,\text{MeV}$ and $m_q^0 \approx 7\,\text{MeV}$ for vanishing $\langle \bar{q}q \rangle / \langle \bar{q}q \rangle_V$. The decrease of both m_q and m_s is approximately linear in energy density where the slope associated to the light quark is flatter in comparison to the strange quark mass. Concerning the comparison between the different choices for the nuclear equation of state, we find also for these masses a non-negligible sensitivity and the same hierarchy as for the scalar quark condensate (the results associated to NL1/NL2 are always below the corresponding results for NL3).

We recall that the effective masses of the quarks enter the Schwinger formula (1) for the hadronic particle production via the string decays. In panel (d) of Fig. 1 the strangeness ratio γ_s is shown as a function of energy density for the two parameter sets. The factor γ_s increases from the vacuum case ($\gamma_s \approx 0.3$) with increasing energy density up to values of 0.8–0.9 for $\epsilon \approx \epsilon_c$. Thus the production of a $s\bar{s}$ pair relative to a light quark pair is no longer suppressed close to the phase boundary for CSR as it is in vacuum. The reason of this increase is the steeper decrease of the effective strange quark mass (with energy density) in comparison to the effective light quark mass as mentioned above. Furthermore, the NL1 and NL2 parametrizations give larger values of γ_s as the NL3 parametrization due to a faster change of the masses with ϵ (cf. panel (c)).

We stress that this scheme for CSR in the string decay mechanism can be applied not only to the light and strange quarks, but also to diquark combinations that are additionally produced in the fragmentation of the string and lead finally to baryon–antibaryon pairs. However, the modifications of the diquark ratios are found to be only very moderate (cf. Ref. 14). The variations of the ratio γ_s in Fig. 1(a), however, is limited to the hadronic phase with energy densities below ϵ_c. Above $\epsilon_c \approx 0.5\,\text{GeV/fm}^3$ strings cannot be formed anymore due to a vanishing string tension κ in the QGP and the hadrons dissolve to partonic degrees-of-freedom (and mean-field energy). Consequently, the s/u factor finally shows an increase for $\epsilon < \epsilon_c$ (associated with CSR) and in correspondence of $\epsilon \geq \epsilon_c$ it drops to the value $\sim 1/3$ (fixed by comparison with the strangeness production at RHIC and LHC energies observed experimentally). The energy dependence of the s/u ratio including the QGP phase is shown explicitely in Fig. 2 of Ref. 13. We recall that in the partonic phase the s/u ratio remains constant as a function of the energy density. As a result we can identify a "horn" structure in the s/u ratio as a function of ϵ, where the initial increase is related to chiral symmetry restoration in the hadronic phase and the subsequent sharp decrease is associated to the formation of the QGP. We thus expect that CSR modifies the particle abundances and spectra (especially in the strange particle sector) from heavy-ion collisions, where increasing energy densities ϵ in the overlap region can be achieved with increasing bombarding energy (at the same centrality of the collision).

4. Application to Nucleus–Nucleus Collisions

In this section we study observables from heavy-ion collisions in order to explore the traces of CSR in the dynamics. The following scenarios will be explored:

- default PHSD calculations without CSR;
- PHSD calculations including CSR with NL3 as parameter set for the nuclear EoS;
- PHSD calculations including CSR with NL1 as parameter set for the nuclear EoS;
- HSD calculations without CSR;
- HSD calculations including CSR with NL3 as parameter set for the nuclear EoS.

We recall that at AGS and SPS energies the strange particle production takes place at the early stages of the collision process as demonstrated in detail in Ref. 14 which is dominated by the string formation and decay in a nontrivial medium.

4.1. *Hadron rapidity distributions in central heavy-ion collisions*

In this subsection we show different hadron rapidity distributions from central heavy-ion reactions as well as the sensitivity of the results to the nuclear EoS. First, we compare the PHSD results with and without CSR at $E_{\rm Lab} = 10.7\,A$GeV in Fig. 2, to point out the general effect of this mechanism on the final particle rapidity distributions and in particular show the dependence on the parameter sets NL1 and NL3 for the non-linear $\sigma - \omega$ model for the nuclear EoS. The restoration of chiral symmetry gives an enhancement of the strange particle yields both for mesons and baryons. On the other hand, it produces a slight decrease in the number of pions at midrapidity due to the suppression of pions in the string decays in favor of strange hadrons. The proton rapidity spectra do not show any sensible variation, in fact the CSR (as implemented in PHSD) modifies essentially the chemistry of the newly produced particles in the string decay and has a minor impact on the dynamics of the nucleons, which in the string picture are associated to the string ends of the primary interactions in the system. The inclusion of the CSR is essential in order to correctly reproduce the strange particle rapidity spectra, as we can see

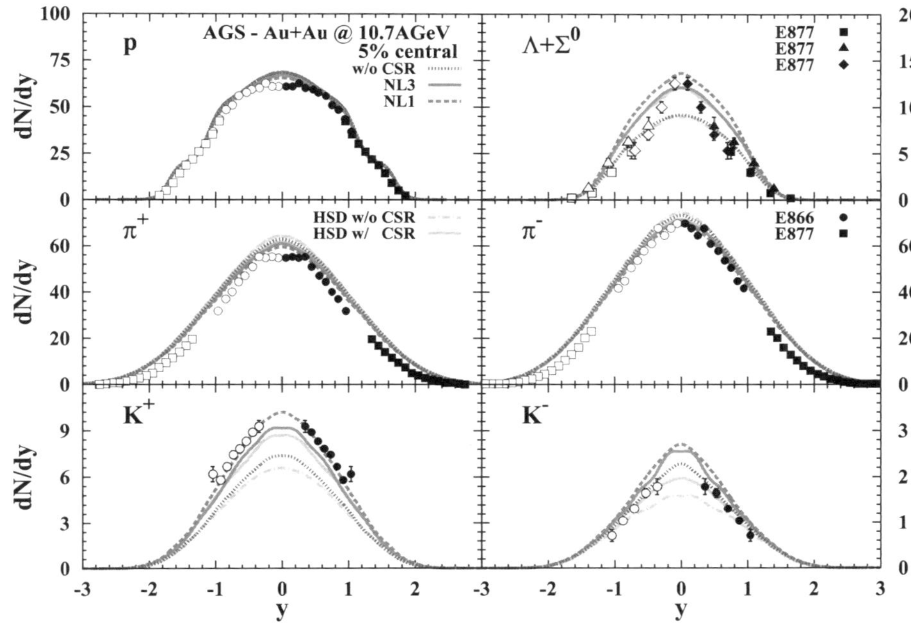

Fig. 2. The rapidity distribution of protons, $(\Lambda + \Sigma^0)$'s, pions and kaons for 5% central Au+Au collisions at 10.7 AGeV in comparison to the experimental data from Refs. 55 and 56. The solid (red) lines show the results from PHSD including CSR with NL3 parameters, the dashed green lines show the results from PHSD including CSR with NL1 parameters and the blue dotted lines show the result from PHSD without CSR. The solid and dashed orange lines present the results from HSD (without partonic phase) with and without CSR, respectively.

especially for $(\Lambda + \Sigma^0)$ hyperons and K^+ mesons. Furthermore, our calculations for the proton rapidity spectra are in good agreement with experimental observation.

Next, we discuss the results from PHSD with CSR using two different parametrization for the nuclear equation of state, i.e. NL3 and NL1. The general features of the strangeness enhancement hold for both parametrizations; in particular the NL1 set provides larger values for all strange particle rapidity spectra at midrapidity in line with the discussion of Fig. 1. The difference between the two parametrizations represents the uncertainty of our results related to CSR as implemented in PHSD. We stress that we do not tune the parameters of the equation of state to fit the data, but we employ different nuclear EoS to compute the scalar density (as explained in Subsec. 3.3) in order to explore the uncertainties of our approach.

In Fig. 3 the rapidity spectra of various hadrons at $E_{\text{Lab}} = 30\,A$GeV are shown. We find the same features as for $E_{\text{Lab}} = 10.7\,A$GeV concerning the strangeness enhancement and the comparison between the two parameter sets for the equation of state; the differences are slightly smaller at this energy. At midrapidity, both protons and pions are very slightly over-estimated in all explored scenarios which

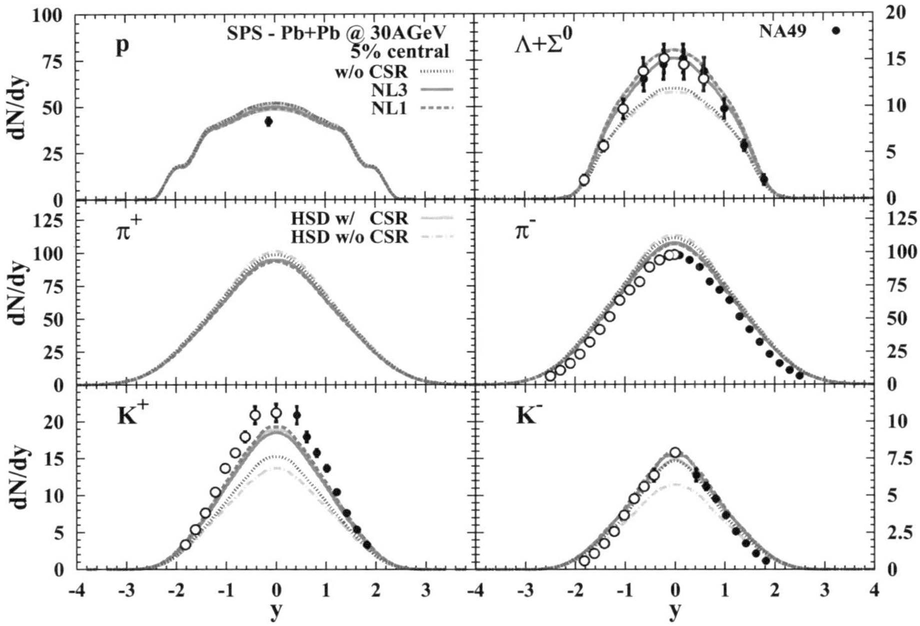

Fig. 3. The rapidity distribution of protons, $(\Lambda+\Sigma^0)$'s, pions and kaons for 5% central Au+Au collisions at 30 AGeV in comparison to the experimental data from Ref. 57. The coding of the lines is the same as in Fig. 2.

suggests that the nuclear stopping is still a bit overestimated. Finally, at larger collision energies (e.g. at $E_{\rm Lab} = 158\,A$GeV) the CSR does not play a significant role, since the dynamics is dominated by the QGP phase. Thus, there is no appreciable difference between the results with and without CSR for the two different EoS.[14]

4.2. *Strange particle abundances and ratios*

In this subsection we study the excitation function of the particle ratios K^+/π^+, K^-/π^- and $(\Lambda+\Sigma^0)/\pi$ at midrapidity from 5% central Au+Au collisions. In Fig. 4 we show the calculations for the following scenarios: the default PHSD without CSR (dotted blue line), PHSD including CSR with NL3 and NL1 as parameter sets for the nuclear EoS from the non-linear $\sigma-\omega$ model (solid red and dashed green lines, respectively), HSD (without partonic phase) with and without CSR (solid and dashed orange lines, respectively). The shaded area displays the uncertainties of our calculations from the two scenarios for the nuclear EoS since the results from the parameter set NL2 are always in between those from NL1 and NL3 (cf. Ref. 14). As described in Ref. 13, the inclusion of CSR in PHSD is responsible for the strong strangeness enhancement at AGS and low SPS energies. The experimental observations of the ratios K^+/π^+ and $(\Lambda+\Sigma^0)/\pi$ show the well-known "horn" structure, which is reproduced by the PHSD calculations with CSR. In fact, CSR

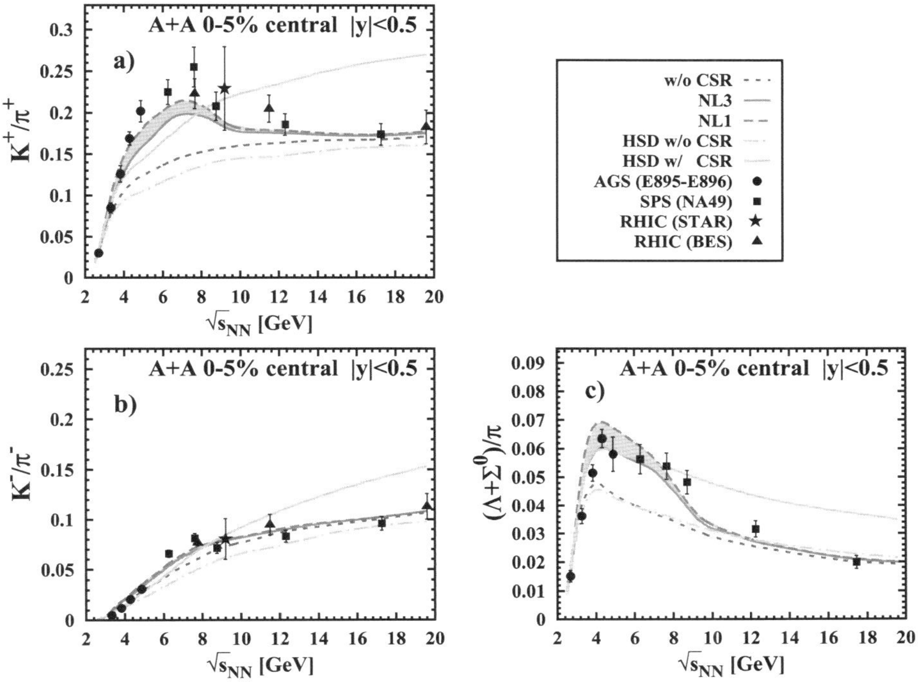

Fig. 4. The ratios K^+/π^+, K^-/π^- and $(\Lambda + \Sigma^0)/\pi$ at midrapidity from 5% central Au+Au collisions as a function of the invariant energy $\sqrt{s_{NN}}$ up to the top SPS energy in comparison to the experimental data from Refs. 58, 59 and 60. The grey shaded area represents the results from PHSD including CSR taking into account the uncertainty from the parameters of the $\sigma - \omega$ model for the EoS. The solid and dashed orange lines present the results from HSD (without partonic phase) with and without CSR, respectively.

gives rise to a steep increase of these ratios at energies lower than $\sqrt{s_{NN}} \approx 7\,\text{GeV}$, while the drop at larger energies is associated to the appearance of a deconfined partonic medium. As anticipated by the considerations in Subsec. 3.3, the NL1 parameter set produces a sharper peak both in the K^+/π^+ and in the $(\Lambda + \Sigma^0)/\pi$ excitation functions with a $\approx$ 10% maximum increase with respect to the NL3 result. We point out that even adopting different parametrizations for the $\sigma - \omega$ model, we recover the same "horn" feature. This supports the reliability of the CSR mechanism as implemented in the PHSD model.

At AGS energies, the energy dependencies of the ratios K^+/π^+ and $(\Lambda + \Sigma^0)/\pi$ are closely connected, since K^+ and Λ (or Σ^0) are mostly produced in pairs due to strangeness conservation. On the other hand, the excitation function of the K^-/π^- ratio does not show any peak, but it smoothly increases as a function of $\sqrt{s_{NN}}$. In fact, especially at AGS energies, the antikaon production differs substantially from the production of K^+ and Λ, which occurs dominantly via string formation. The antikaons are produced mainly via secondary meson-baryon interactions by

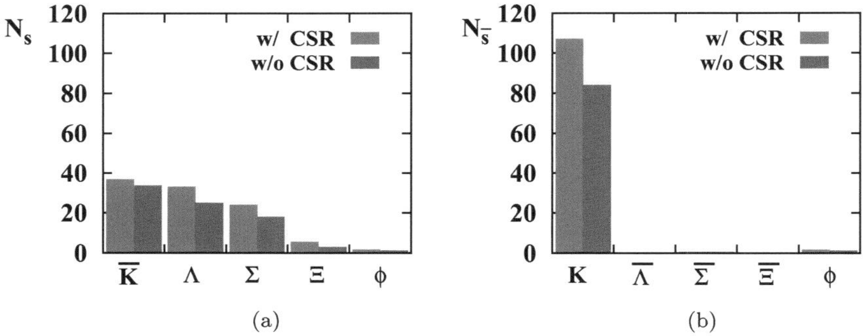

(a) (b)

Fig. 5. The final distribution of the strange (a) and antistrange quarks (b) from the Schwinger mechanism with respect to the hadrons in a central collision of Pb+Pb at 30 AGeV.

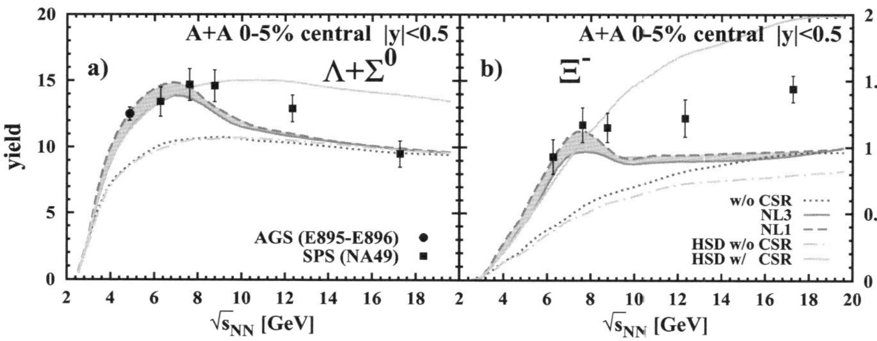

Fig. 6. The yields of $(\Lambda + \Sigma^0)$ and Ξ^- at midrapidity from 5% central Au+Au collisions as a function of the invariant energy $\sqrt{s_{NN}}$ up to the top SPS energy in comparison to the experimental data from Refs. 56 and 60. The coding of the lines is the same as in Fig. 4.

flavor exchange and their production is suppressed with respect to the Λ hyperons that carry most of the strange quarks. This is the reason why the inclusion of chiral symmetry restoration provides a substantial enhancement of the K^+/π^+ and $(\Lambda + \Sigma^0)/\pi$ excitation functions and a smaller change on the K^-/π^- ratio. This is illustrated in Fig. 5 which shows the final fate of the strange and antistrange quarks in a central collision of Pb+Pb at 30 A GeV. We also notice that there is no sizeable difference between the NL1 and NL3 results for the K^-/π^- ratio. At top SPS energies the strangeness is produced predominantly by the hadronization of partonic degrees-of-freedom, thus our results for all the ratios do not show an appreciable sensitivity to the nuclear EoS and the calculations with and without CSR tend to merge at $\sqrt{s_{NN}} \approx 20\,\text{GeV}$.

Finally, in Fig. 6 we present the yields of $(\Lambda+\Sigma^0)$ and Ξ^- at midrapidity from 5% central Au+Au collisions as a function of the invariant energy $\sqrt{s_{NN}}$ in comparison to the available data from Refs. 56 and 60. We recover a "horn" structure, similar to that shown in Fig. 4 for the energy dependence of the strange to non-strange particle ratios. A sensitivity on the nuclear model parametrizations persists at low energy, while in the top SPS energy regime the results corresponding to the different scenarios merge. The comparison with the available data at $\sqrt{s_{NN}} < 8\,\text{GeV}$ supports the validity of the CSR picture, while at larger energies we underestimate the experimental observations. We mention that this discrepancy is not due to the CSR mechanism, since it does not play an essential role in the high-energy regime as pointed out above.

5. Summary

In this contribution we have reported on the effects of chiral symmetry restoration (CSR) on observables from heavy-ion collisions in the energy range $\sqrt{s_{NN}} = 3$–$20\,\text{GeV}$ based on the studies in Refs. 13 and 14. Our results have been obtained within the Parton-Hadron-String Dynamics transport approach,[25] where essential aspects of CSR have been incorporated in the Schwinger mechanism for the string decay.[13] Since the PHSD approach includes both hadronic and partonic degrees-of-freedom and has been tested in a wide energy regime, it represents a powerful tool to study nucleus–nucleus collisions on a microscopic basis. The CSR, as implemented in PHSD, affects only the hadronic particle production and it does not imply modifications in the Quark–Gluon Plasma (QGP) phase. As already found in Ref. 13 the CSR induces an enhancement of the strange quark fraction γ_s produced via the string decay, while there are no sensible changes in the diquark production and accordingly in baryon–antibaryon production. The s/u ratio, as defined by the Schwinger formula (1), increases as a function of the energy density due to CSR and this is reflected in an enhancement of the strange particle abundances with respect to the non-strange ones. This has been observed explicitly in the particle spectra at AGS and lower SPS energies ($E_{\text{Lab}} = 10.7$ to $30\,A\text{GeV}$). On the other hand, at top SPS energies (e.g. $E_{\text{Lab}} = 158\,A\text{GeV}$) the results from PHSD with and without CSR merge, since the dynamics of the system is dominated by the QGP phase where CSR does not play a significant role.

We stress that our PHSD calculations provide a microscopic interpretation of the "horn" structure in the excitation function of the K^+/π^+ ratio in central Au+Au (or Pb+Pb) collisions. The steep rise of this ratio at AGS energies is associated to CSR, while the drop at higher SPS energies is due to the appearance of the QGP phase in an increasing volume of the interaction region. We have found an analogous energy dependence for the $(\Lambda+\Sigma^0)/\pi$ ratio, while the excitation function of the K^-/π^- ratio does not show any explicit peak. In general, the PHSD results obtained with the inclusion of CSR are in a good agreement with the available data for all observables analyzed, while calculations without CSR fail substantially.

In conclusion, our microscopic studies support the idea that CSR occurs in hadronic systems with high temperatures and densities before the deconfinement phase transition takes over. We suggest that the strange particle spectra and yields are suitable signatures to study the properties of CSR in HICs in future also as a function of system size and centrality (cf. Ref. 14).

Acknowledgments

The authors acknowledge inspiring discussions with J. Cleymans, M. Gazdzicki, M. Gorenstein, O. Linnyk, T. Song and thank the Helmholtz International Center for FAIR (HIC for FAIR), the Helmholtz Graduate School for Hadron and Ion Research (HGS-HIRe), the Helmholtz Research School for Quark Matter Studies in Heavy-Ion Collisions (H-QM), and the Bundesministerium für Bildung und Forschung (BMBF) for support. The computational resources have been provided by the LOEWE-CSC.

References

1. J. Reinhardt and W. Greiner, *Lect. Notes Phys.* **440**, 153 (1994).
2. J. Rafelski, J. Kirsch, B. Müller, J. Reinhardt, and W. Greiner, *FIAS Interdisc. Sci. Ser.* 211 (2017), ISBN 9783319441658.
3. P. Romatschke and U. Romatschke, *Phys. Rev. Lett.* **99**, 172301 (2007).
4. J. Rafelski and B. Müller, *Phys. Rev. Lett.* **48**, 106 (1982); *Phys. Rev. Lett.* **56**, 2334 (1986).
5. R. Stock, *J. Phys. G* **28**, 1517 (2002).
6. M. Gazdzicki and M. I. Gorenstein, *Acta Phys. Polon. B* **30**, 2705 (1999).
7. J. Cleymans *et al.* [NA49 Collaboration], *Phys. Lett. B* **615**, 50 (2005).
8. A. Andronic, P. Braun-Munzinger and J. Stachel, *Nucl. Phys. A* **772**, 167 (2006); *Phys. Lett. B* **673**, 142 (2009).
9. K. A. Bugaev *et al.*, *Europhys. Lett.* **104**, 22002 (2013).
10. J. Geiss, W. Cassing, and C. Greiner, *Nucl. Phys. A* **644**, 107 (1998).
11. E. L. Bratkovskaya *et al.*, *Phys. Rev. C* **69**, 054907 (2004).
12. E. L. Bratkovskaya, S. Soff, H. Stöcker, M. van Leeuwen, W. Cassing, *Phys. Rev. Lett.* **92**, 032302 (2004).
13. W. Cassing, A. Palmese, P. Moreau and E. L. Bratkovskaya, *Phys. Rev. C* **93**, 014902 (2016).
14. A. Palmese, W. Cassing, E. Seifert, T. Steinert, P. Moreau, and E. L. Bratkovskaya, *Phys. Rev. C* **94**, 044912 (2016).
15. E. L. Bratkovskaya, W. Cassing, P. Moreau, and A. Palmese, *Eur. Phys. J. A* **52**, 219 (2016).
16. P. Senger *et al.*, *Lect. Notes Phys.* **814**, 681 (2011).
17. C. S. Fischer, J. Luecker, and C. Welzbacher, *Phys. Rev. D* **90**, 034022 (2014).
18. C. S. Fischer, L. Fister, J. Luecker, and J. M. Pawlowski, *Phys. Lett. B* **732**, 273 (2014).
19. G. Eichmann, C. S. Fischer, and C. A. Welzbacher, *Phys. Rev. D* **93**, 034013 (2016).
20. T. K. Herbst, J. M. Pawlowski, and B.-J. Schaefer, *Phys. Rev. D* **88**, 014007 (2013).
21. L. McLerran and R. D. Pisarski, *Nucl. Phys. A* **796**, 83 (2007).
22. Y. Nambu and G. Jona-Lasinio, *Phys. Rev.* **122**, 345 (1961).
23. S. P. Klevansky, *Rev. Mod. Phys.* **64**, 649 (1992).
24. R. Rapp, *Adv. High Energy Phys.* **2013**, 148253 (2013).
25. W. Cassing and E. L. Bratkovskaya, *Nucl. Phys. A* **831**, 215 (2009).

26. E. L. Bratkovskaya, W. Cassing, V. P. Konchakovski and O. Linnyk, *Nucl. Phys. A* **856**, 162 (2011).
27. O. Linnyk *et al.*, *Phys. Rev. C* **92**, 054914 (2015); *Phys. Rev. C* **89**, 034908 (2014); *Phys. Rev. C* **88**, 034904 (2013); *Phys. Rev. C* **87**, 014905 (2013); *Phys. Rev. C* **85**, 024910 (2012); *Phys. Rev. C* **84**, 054917 (2011); *Nucl. Phys. A* **855**, 273 (2011).
28. V. P. Konchakovski *et al.*, *Phys. Rev. C* **85**, 044922 (2012); *Phys. Rev. C* **85**, 011902 (2012).
29. L. P. Kadanoff and G. Baym, *Quantum Statistical Mechanics*, Benjamin, New York, 1962.
30. W. Cassing, *Eur. Phys. J. ST* **168**, 3 (2009); *Nucl. Phys. A* **795**, 70 (2007).
31. O. Linnyk, E. Bratkovskaya, and W. Cassing, *Prog. Part. Nucl. Phys.* **87**, 50 (2016).
32. B. Nilsson-Almqvist and E. Stenlund, *Comp. Phys. Comm.* **43**, 387 (1987); B. Andersson, G. Gustafson, and H. Pi, *Z. Phys. C* **57**, 485 (1993).
33. T. Sjostrand, S. Mrenna and P. Z. Skands, *JHEP* **0605**, 026 (2006).
34. W. Cassing and E. L. Bratkovskaya, *Phys. Rep.* **308**, 65 (1999).
35. J. Schwinger, *Phys. Rev.* **83**, 664 (1951).
36. T. D. Cohen, R. J. Furnstahl, and D. K. Griegel, *Phys. Rev. C* **45**, 1881 (1992).
37. U. Vogl and W. Weise, *Prog. Part. Nucl. Phys.* **27**, 195 (1991).
38. M. C. Birse, *J. Phys. G* **20**, 1537 (1994).
39. V. Koch, *Int. J. Mod. Phys. E* **6**, 203 (1997).
40. J. V. Steele, H. Yamagishi, and I. Zahed, *Phys. Lett. B* **384**, 255 (1996); *Phys. Rev. D* **56**, 5605 (1997).
41. G. E. Brown, *Prog. Theor. Phys.* **91**, 85 (1987).
42. G. E. Brown, C. M. Ko, Z. G. Wu, and L. H. Xia, *Phys. Rev. C* **43**, 1881 (1991).
43. V. Koch and G. E. Brown, *Nucl. Phys. A* **560**, 345 (1993).
44. B. Friman, W. Nörenberg and V. D. Toneev, *Eur. Phys. J. A* **3**, 165 (1998).
45. J. M. Alarcon, J. Martin Camalich and J. A. Oller, *Phys. Rev. D* **85**, 051503 (2012).
46. M. Hoferichter, J. Ruiz de Elvira, B. Kubis, and U.-G. Meissner, *Phys. Rev. Lett.* **115**, 092301 (2015).
47. G. S. Bali *et al.* [RQCD Collaboration], *Phys. Rev. D* **93**, 094504 (2016).
48. Yi-Bo Yang *et al.*, *Phys. Rev. D* **94**, 054503 (2016).
49. J. Bordes, C. A. Dominguez, P. Moodley, J. Penarrocha, and K. Schlichter, *JHEP* **10**, 102 (2012).
50. M. Hoferichter, J. Ruiz de Elvira, B. Kubis, and U.-G. Meissner, *Phys. Lett. B* **760**, 74 (2016).
51. M. Bando, T. Kugo and K. Yamawaki, *Phys. Rep.* **164**, 217 (1988).
52. T. D. Cohen, R. J. Furnstahl, D. K. Griegel, and X. Jin, *Prog. Part. Nucl. Phys.* **35**, 221 (1995).
53. J. Boguta and A. R. Bodmer, *Nucl. Phys. A* **292**, 413 (1977).
54. A. Lang *et al.*, *Z. Phys. A* **340**, 287 (1991).
55. J. Stachel, *Nucl. Phys. A* **610**, 509C (1996).
56. S. Albergo *et al.*, *Phys. Rev. Lett.* **88**, 062301 (2002).
57. C. Alt *et al.* [NA49 Collaboration], *Phys. Rev. C* **77**, 024903 (2008); C. Alt *et al.* [NA49 Collaboration], *Phys. Rev. C* **73**, 044910 (2006); C. Alt *et al.* [NA49 Collaboration], *Phys. Rev. C* **78**, 034918 (2008).
58. J. L. Klay *et al.* [E895 Collaboration], *Phys. Rev. C* **68**, 054905 (2003).
59. B. I. Abelev *et al.* [STAR Collaboration], *Phys. Rev. C* **81**, 024911 (2010); M. M. Aggarwal *et al.* [STAR Collaboration], *Phys. Rev. C* **83**, 024901 (2011).
60. C. Alt *et al.* [NA49 Collaboration], *Phys. Rev. C* **78**, 034918 (2008).

The Physics Case for the $\sqrt{s_{NN}} \approx 10$ GeV Energy Region

J. Cleymans
UCT-CERN Research Centre and Physics Department,
University of Cape Town,
Rondebosch 7701, South Africa

To the memory of Professor Dr. Walter Greiner

There are indications that the beam energy region $\sqrt{s_{NN}} \approx 10$ GeV for heavy-ion collisions is an interesting one. The final state has the highest net baryon density at this beam energy. A transition from a baryon dominated to a meson dominated final state takes place around this beam energy. Ratios of strange particles to mesons show clear and pronounced maxima around this beam energy. The theoretical interpretation can be clarified by covering fully this energy region. In particular the strangeness content needs to be determined, data covering the full phase space (4π) would be helpful to establish the properties of this energy region.

1. Introduction

Heavy-ion collisions[1] at high energies produce a large numbers of secondaries. At the LHC the number of charged particles produced in Pb–Pb collisions at 5.02 TeV[2] is shown in Fig. 1, thus, including neutral particles, a total of approximately 30 000 particles is being produced on average in such a collision. It is natural to try a statistical–thermal model to analyze these. As it turns out such an analysis is useful for a very wide range of beam energies, stretching from 1 GeV all the way up to the highest energies available at the LHC. For such an analysis one has to keep in mind that a relativistic heavy-ion collision passes through several stages. At one of the later, hadronic, stages, the system is assumed to be dominated by hadronic resonances, on which the thermal model focuses. The identifying feature of the thermal model is that all the resonances as listed in Ref. 3 are assumed to be in thermal and chemical equilibrium. This assumption drastically reduces the number of free parameters and thus this stage is determined by just a few thermodynamic variables namely, the chemical freeze-out temperature T, the various chemical potentials μ determined by the conserved quantum numbers and by the volume V of

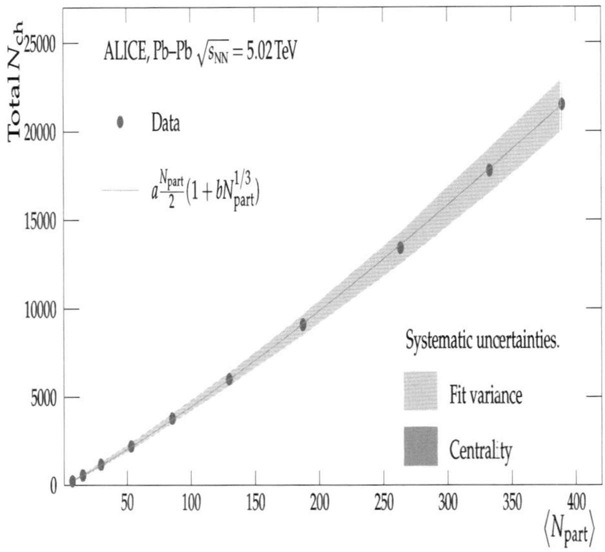

Fig. 1. Number of charged particles produced in a Pb–Pb collision as a function of beam energy as measured by the ALICE collaboration.[2]

the system. It has been shown that this description is also the correct one[4–6] for a scaling expansion as first discussed by Bjorken.[7]

In relativistic heavy ion collisions a new dimension was given to the model by the highly successful analysis of particle yields, leading to the notion of chemical equilibrium which is now a well-established one in the analysis of relativistic heavy ion collisions, see e.g. Refs. 8–10. In view of the success of chemical freeze-out in relativistic heavy ion collisions, much effort has gone into finding models that describe this chemical freeze-out, a comparison[11] of three parameterizations is shown in Fig. 2.

There are of course uncertainties in the thermal model, one of these is about the decays of resonances, another one is whether some resonances exist or not.[3] Particle yields are determined from:

$$N_i = \sum_j N_j \mathrm{Br}(j \to i).$$

Hence a lack of knowledge of branching ratios affects the quality of results obtained from the thermal model.

As an example, the final yield of π^+'s is given by

$$N_{\pi^+} = N_{\pi^+}(\text{thermal}) + N_{\pi^+}(\text{resonance decays})$$

and, depending on the temperature, over 80% of observed pions could be due to resonance decays. Hence the crucial importance of these decays. Various theoretical uncertainties have been recently discussed in Ref. 13.

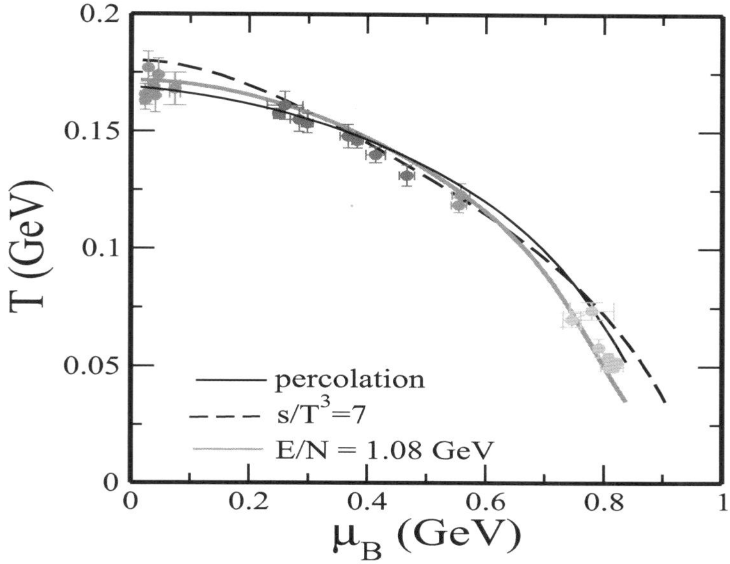

Fig. 2. Chemical freeze-out temperature T vs. the baryon chemical potential at different beam energies together with curves corresponding to a fixed ratio of energy per hadron divided by total number of hadrons in the resonance gas before decay of resonances.[11] Also shown are calculations based on the percolation model[12] and for a fixed value of the entropy density divided by T^3.

2. What Makes the Beam Energy $\sqrt{s_{NN}} \approx 10$ GeV Special?

2.1. *Maximum net baryon density*

The resulting freeze-out curve in the T–μ_B plane, shown in Fig. 2, can also be drawn in the temperature T vs net baryon density plane as was done in Ref. 14. The resulting curve is shown in Fig. 3. At very high beam energies the net baryon density is zero because equal numbers of particles and antiparticles are being produced while at low temperatures the net baryon density is very high. Figure 3 shows that the a clear maximum exists just below the $\sqrt{s_{NN}} = 10$ GeV beam energy region.

2.2. *Transition from a baryon dominated to a meson dominated final state*

A fairly good criterium for chemical freeze-outis the constant value of the entropy density divided $s/T^3 = 7$ ratio as can be seen from Fig. 2. The components that make up the entropy density are shown in Fig. 4, the change from a baryon-dominated to a meson-dominated final state also happens around a beam energy of $\sqrt{s_{NN}} \approx 10$ GeV.

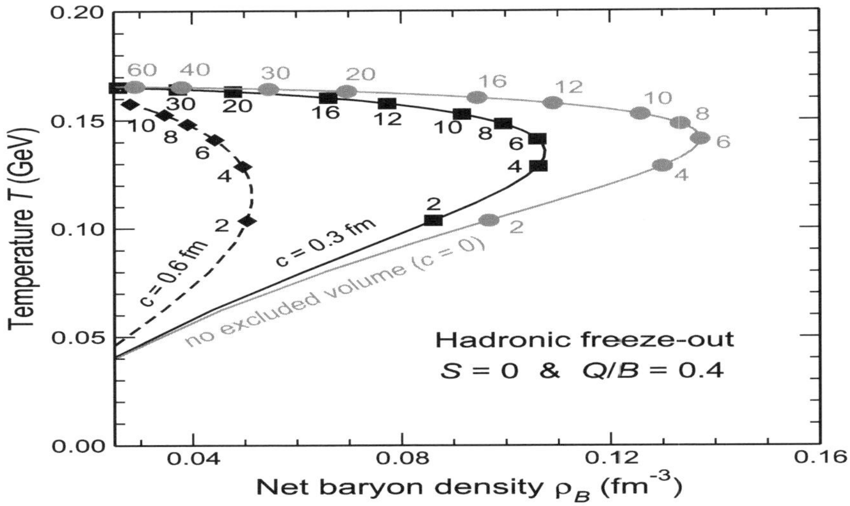

Fig. 3. The hadronic freeze-out line in the ρ_B–T phase plane as obtained from the values of μ_B and T that have been extracted from the experimental data in Ref. 11. The calculation employs values of μ_Q and μ_S that ensure $\langle S \rangle = 0$ and $\langle Q \rangle = 0.4\langle B \rangle$ for each value of μ_B.[14] Also indicated are the beam energies for which the particular freeze-out conditions are expected. The dependence on a hard-core radius is indicated.

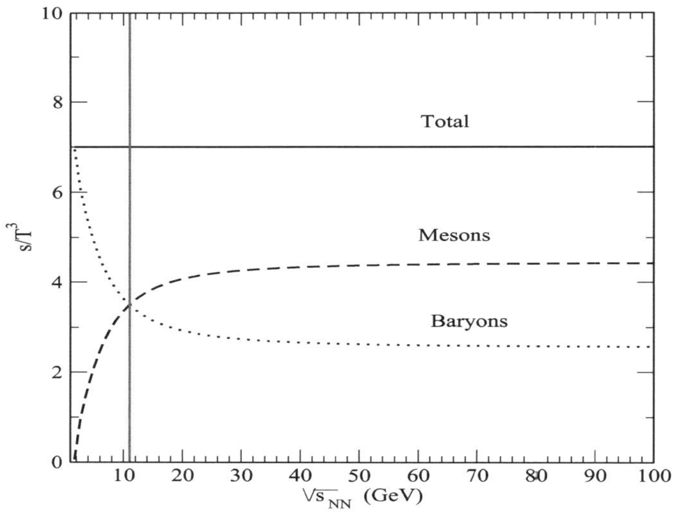

Fig. 4. The s/T^3 ratio calculated in the thermal model along the constant value consistent with chemical freeze-out. Also shown are the contributions from the mesons and the baryons.

2.3. *Ratios of strange hadrons to pions*

Despite the smoothness in the thermal freeze-out parameters as a function of beam energy, strong changes are observed in several particle ratios, e.g. the horn in the K^+/π^+ ratio and a similar strong variation in the Λ/π ratio.[15] These are not observed in p–p collisions, in Pb–Pb collisions they happen at a beam energy of around $\sqrt{s_{NN}} \approx 10$ GeV. Within the framework of thermal models this variation has been connected to a change from a baryon dominated to a meson dominated hadron gas.[16] The values of the K^+/π^+ and Λ/π^+ ratios[17] are shown in Fig. 5. From the lines of constant values for these ratios it can be seen that the maxima in the thermal model hug the chemical freeze-out line. It is also important to note that the maxima occur for different values of T and μ_B.

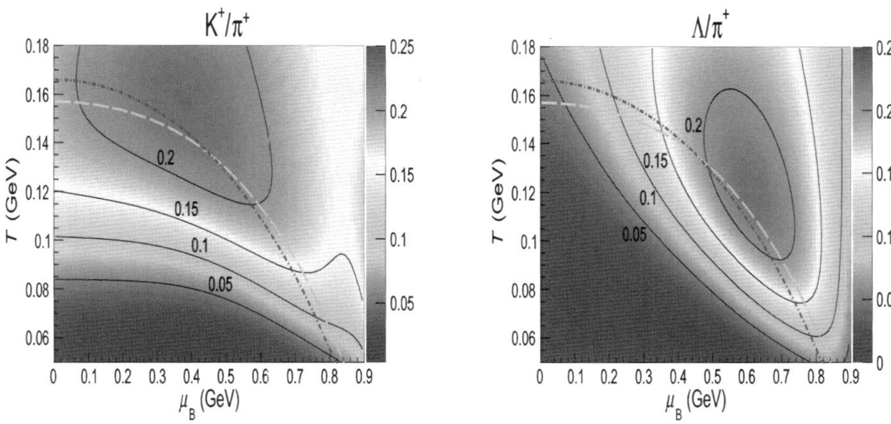

Fig. 5. Lines of constant values of the K^+/π^+ (left panel) and the Λ/π^+ (right panel) ratios in the T–μ_B plane showing a clear maximum in each ratio close to the boundary given by the chemical freeze-out line but in a different position.[17]

3. Conclusions

In the thermal model a change is expected as the hadronic gas undergoes a transition from a baryon-dominated to a meson-dominated gas. The strong variations seen in the particle ratios coincide with this transition. This transition occurs at a

- temperature $T = 151$ MeV,
- baryon chemical potential $\mu_B = 327$ MeV,
- energy $\sqrt{s_{NN}} = 11$ GeV.

There are thus several indications that the energy region around 10 GeV, covered by proposed new facilities, is an extremely interesting one. The theoretical interpretation can be clarified by covering this energy region. In particular the strangeness content needs to be determined, data covering the full phase space (4π) would be very helpful to determine the thermal parameters of a possible phase transition and the existence of a quarkyonic phase as has been discussed recently in Ref. 18.

References

1. H. Stoecker and W. Greiner, *Phys. Rept.* **137**, 277 (1986).
2. ALICE Collaboration, J. Adam *et al.*, *Phys. Lett. B* **772**, 567 (2017).
3. Particle Data Group Collaboration, C. Patrignani *et al.*, *Chin. Phys. C* **40**, 100001 (2016).
4. J. Cleymans and K. Redlich, *Phys. Rev. C* **60**, 054908 (1999).
5. W. Broniowski and W. Florkowski, *Phys. Rev. Lett.* **87**, 272302 (2001).
6. S. Akkelin, P. Braun-Munzinger and Y. M. Sinyukov, *Nucl. Phys. A* **710**, 439 (2002).
7. J. Bjorken, *Phys. Rev. D* **27**, 140 (1983).
8. STAR Collaboration, L. Adamczyk *et al.*, *Phys. Rev. C* (2017), `http://www.arxiv.org/abs/1701.07065`.
9. ALICE Collaboration, M. Floris *et al.*, *Quark Matter 2014*, in *The 24th International Conference on Ultrarelativistic Nucleus-Nucleus Collisions* (Darmstadt, Germany, 2014). pp. c103–c112.
10. HADES Collaboration, M. M. Lorenz *et al.*, *Sqm2015*, in *15th International Conference on Strangeness in Quark Matter* (Dubna, Russia, 2015). p. 012022.
11. J. Cleymans, H. Oeschler, K. Redlich and S. Wheaton, *Phys. Rev. C* **73**, 034905 (2006).
12. V. Magas and H. Satz, *Eur. Phys. J. C* **32**, 115 (2003).
13. L. M. Satarov, V. Vovchenko, P. Alba, M. I. Gorenstein and H. Stoecker, *Phys. Rev. C* **95**, 024902 (2017).
14. J. Randrup and J. Cleymans, *Eur. Phys. J. C* **52**, 218 (2016).
15. NA49 Collaboration, C. Alt *et al.*, *Phys. Rev. C* **77**, 024903 (2008).
16. J.Cleymans, H. Oeschler, K. Redlich and S. Wheaton, *Phys. Lett. B* **615**, 50 (2005).
17. H. Oeschler, J. Cleymans, B. Hippolyte, K. Redlich and N. Sharma, *Eur. Phys. J. C* (2017), `http://www.arxiv.org/abs/1603.09553`.
18. A. Andronic *et al.*, *Nucl. Phys. A* **837**, 65 (2010).

Untangling Simple Patterns in Intricate Atomic Nuclei

Jerry P. Draayer,[1] Kristina D. Launey,[1] Tomáš Dytrych,[1,2]
Alison C. Dreyfuss,[1] Grigor H. Sargsyan[1] and Robert B. Baker[1]

[1]*Department of Physics and Astronomy, Louisiana State University, Baton Rouge, LA 70803, USA*

[2]*Nuclear Physics Institute, Academy of Sciences of the Czech Republic, 250 68 Řež, Czech Republic*

In this review, we present a symmetry-guided strategy that utilizes exact as well as partial symmetries for enabling a deeper understanding of and advancing *ab initio* studies for determining the microscopic structure of atomic nuclei. These symmetries are critical toward untangling simple patterns within an overarching complexity that defines nuclei, and naturally provide a physically relevant basis. Such basis, for large-scale calculations, allows the model space size to be reduced through a very structured selection of the basis states, and guides extensions of the *ab initio* theory beyond current limitations. This is key to facilitating studies of the structure and reactions of isotopes across the nuclear chart, using QCD-inspired interactions and high performance computing (HPC) resources. This, in turn, can guide explorations of orderly patterns in nuclei and how they emerge from first principles.

1. Introduction

Pattern recognition and deviations thereof are key features in understanding complex structures and phenomena. For nuclear physics, this translates into the identification and exploitation of symmetries, exact and partial. The field of nuclear physics has advanced with successes of models proffered to describe regularly occurring patterns, such as the Liquid Drop Model for the description of fission phenomena, or the Geometric Collective Model of Greiner[a] for the depiction of collective modes, and early renditions of the Independent-particle Shell Model, all within the dawn of the field. Since then, various layers of complexity were added to these earliest models. However, the most promising of the more evolved theories could not be tested against a growing repository of high-quality experimental data until the advent of high-performance computing (HPC) resources in the late 90s that ushered in an era of computational nuclear physics.

[a]This review is dedicated to the memory of Professor Greiner in recognition of his important contributions to nuclear physics, especially for his seminal and sustained leadership in the development of collective models of nuclear structure and fission processes.

Today, aided by the availability of petascale computing resources and looking ahead to next-generation exascale facilities, a new dawn in our understanding of nuclear physics is unfolding before us. Theories advanced through the last decades of the 20th century can now be tested and refined against a vast body of data that probes more widely and deeply into the structure of atomic nuclei and the universe that resided within nucleons themselves. Again two primary branches of theoretical physics dominate the landscape: Quantum Chromodynamics (QCD) for gaining a much deeper understanding of the strong interaction, and *ab initio* nuclear theory, with the No-core Shell Model (NCSM)[1,2] being a successful example, that seeks to test and explore the effective interactions that bind nucleons into nuclei.

This review focuses on the *ab initio* symmetry-adapted no-core shell-model (SA-NCSM),[3] which utilizes the SU(3) symmetry and its embedding $Sp(3,\mathbb{R})$ group, with results that corroborate and are complementary to those enabled within the framework of the NCSM. These symmetries are used to facilitate *ab initio* applications to challenging lower *ds*-shell nuclei and reveal that bound states of light nuclei are dominated by high-deformation and low-spin configurations.[4] The applicable symmetries reveal the nature of collectivity in such nuclei and provide a description of bound states in terms of a relatively small fraction of the complete space when the latter is expressed in an $(LS)J$ coupling scheme with the spatial configurations further organized into irreducible representations of SU(3). That SU(3) plays a key role tracks with the seminal work of Elliott,[5] and is further reinforced by the fact that SU(3) also underpins the microscopic symplectic model,[6,7] which provides a microscopic framework for understanding collective phenomena in atomic nuclei.[8–10] Furthermore, we discuss the use of a schematic microscopic interaction in a no-core shell-model framework, which preserves the $Sp(3,\mathbb{R})$ symmetry and reduces to the Elliott SU(3) model[5] in the single-shell limit, augmented by the SU(3)-symmetry breaking spin-orbit interaction. We show that it reproduces characteristic features of the low-lying 0^+ states in ^{12}C, including the elusive Hoyle state,[11] as well as ground-state rotational bands in *p* and *sd*-shell nuclei (from Be to Si),[12–14] while further confirming the mechanism of shape coexistence.[15–18] An implication of the latter is that efforts to reproduce the structure of ^{12}C using a 'bottom up' *ab initio* effective interaction theory may benefit from 'top down' algebraic considerations that serve to expose emergent properties in terms of simple interaction forms that seem to dominate the structure of deformed nuclei.

Overview of approaches to nuclear structure — We start with a brief overview of the major theoretical efforts that underpin development of the SA-NCSM and/or have advanced our understanding of particle- and collectivity-driven phenomena (Table 1). For a complete list of approaches that have made substantial contributions, we refer the reader to the review articles Refs. 2, 7, 19, 20 and references therein.

In the 1950s, two simple models of nuclear structure were advanced that are complementary in nature, namely, the independent-particle model of Mayer and

Table 1. Major cornerstone theories in the development of two classes of nuclear structure models, starting with the Shell Model (SM) and the Collective Model (CM).

"Particle" Focus	"Collectivity" Focus
Shell Model, Goeppert-Mayer and Jensen (1950s)[21–23] — independent-particle model, spherical HO mean field $+l \cdot s + l^2$.	**Collective Model**, Bohr and Mottelson (1950s)[24] — descriptions in shape variables, β and γ (deformation, rotations, vibrations).
Nilsson Model (1955)[25] — deformed mean field (MF).	**Elliott** SU(3) **Model** (1958)[5] — valence shell model in SU(3)-adapted basis with SU(3)-breaking interactions including pairing.
Pairing Model ∘ Algebraic SU(2) and Sp(2) pairing, Racah (1940s), Flowers (1950s), Kerman (1960s).[26–28] ∘ Exact solutions to standard pairing in spherical/deformed MF.[29–31]	**Geometric Collective Model**, Greiner (1969) — with interactions in terms of β and $\cos 3\gamma$.[32,33]
	Interacting Boson Model, Iachello (1975) — algebraic, pairs approximated by bosons.[34,35]
***Ab Initio* No-core Shell Model**, Barrett, Vary, Navrátil and Maris (2000)[1,2] — uses realistic interactions (local/nonlocal; NN, $3N$, ...) and "horizontal" cutoff ($N_{\max}$). Complementary developments: GFMC,[36] CC method,[37] Lattice-EFT,[38] IM-SRG,[39] *Ab initio* DFT (UNEDF, 2005).[40]	**Symplectic** Sp(3, $\mathbb{R}$) **Model**, Rowe and Rosensteel (1980s)[6,7,10] — shell model in Sp(3, $\mathbb{R}$)-adapted basis and effective interactions, with "vertical" cutoff (by symplectic slices); successful reproduction of rotational bands and transition rates without effective charges.
***Ab Initio* Symmetry-adapted NCSM (SA-NCSM)**, Draayer, Dytrych and Launey (2007)[3,4] — *ab initio* NCSM with symmetry-adapted basis (any interaction) and no spurious center-of-mass motion; no effective charges.	

Jensen,[21–23] and the collective model of Bohr and Mottelson.[24] The first of these, which is microscopic in nature, recognizes that nuclei can be described by particles independently moving in a mean field, with the harmonic oscillator (HO) potential being a very good first approximation to the average potential experienced by each nucleon in a nucleus. The second of these, which is collective in nature, recognizes that deformed shapes dominate the dynamics. For example, deformed configurations are found to be important even in a nucleus such as ^{16}O, which is commonly treated as spherical in its ground state, but 20% of this state is governed by deformation; in addition, the lowest-lying excited 0^+ states in ^{16}O and their rotational bands are dominated by large deformation.[41] Bohr and Mottelson offered a simple but important description of nuclei in terms of the deformation of the nuclear surface and associated vibrations and rotations.

The seminal work of Elliott[5,42,43] focused on the key role of SU(3), the exact symmetry of the three-dimensional spherical HO, and is further discussed in Sec. 2.1. Within a shell-model framework, Elliott's model utilizes an SU(3)-coupled basis that is related via a unitary transformation to the basis used in the conventional shell model. The new feature here is that SU(3) divides the space into basis states of definite $(\lambda\,\mu)$ quantum numbers of SU(3) linked to the intrinsic quadrupole deformation,[44–46] together with orbital momentum L of associated rotations. For example, $(0\,0)$, $(\lambda\,0)$, and $(0\,\mu)$, describe spherical, prolate, and oblate deformation, respectively.

With an expanding body of experimental evidence that exposed prominent systematic features of nuclei, such as pairing gaps in energy spectra and enhanced electric quadrupole transitions within collective rotational bands, deformation modes were added to the independent-particle model to yield the Nilsson Model (deformed HO mean field);[25] pairing correlations were taken into account in various algebraic[26–28] and exact pairing models.[29–31] As noted in Table 1, a more complete Density Functional Theory (DFT) is a modern derivative theory of this general type, a self-consistent mean-field theory, that can incorporate correlation effects and can accommodate realistic interactions to achieve better predictive capabilities. For example, outcomes using this approach generally yield an excellent accounting of binding energies as well as near ground state phenomena across much of the nuclear landscape.[40]

Also noted in the Table 1 on the "Collectivity Focus" side, are two other complementary models that served to inform us of the importance of deformation and pairing; namely, the Geometric Collective Model (GCM)[32,33] advanced by Greiner and collaborators that used interactions in terms of β and $\cos 3\gamma$ with the goal to understand striking collective features of nuclei, and the intriguing Interacting Boson Model (IBM) of Iachello and associates.[34,35] The latter has offered an informative systematics based on a bosonic realization in terms of a common overarching U(6) algebraic structure and its physical subgroups, U(5) for pairing modes, SU(3)$\supset$SO(3) for rotations and O(6)$\supset$SO(3) for triaxial systems.

Another very significant advance is the microscopic symplectic model,[6,7,9] developed by Rowe and Rosensteel, which provides a microscopic formulation of the Bohr–Mottelson collective model and is a multiple-shell generalization of the Elliott SU(3) model (with a classical realization that underpins the dynamics of rotating bodies, e.g., the rotation of deformed stars and galaxies[47]). It has achieved a remarkable reproduction of rotational bands and transition rates without the need for introducing effective charges (for details, see Secs. 2.4 and 2.5). The significance of the symplectic Sp(3, $\mathbb{R}$) symmetry, with Sp(3, $\mathbb{R}$)$\supset$SU(3), for a microscopic description of a quantum many-body system of interacting particles naturally emerges from the physical relevance of its 21 generators, which are directly related to the particle momentum $(p_{s\alpha})$ and coordinate $(r_{s\alpha})$ operators, with $\alpha = x$, y, and z for the 3 spatial directions and s labeling an individual nucleon, and realize important observables. Namely, the many-particle kinetic energy $\sum_{s,\alpha} p_{s\alpha}^2/2m$, the HO potential, $\sum_{s,\alpha} m\Omega^2 r_{s\alpha}^2/2$, the mass quadrupole moment $Q_{(2\mathcal{M})} = \sum_s q_{(2\mathcal{M})s} = \sum_s \sqrt{16\pi/5} r_s^2 Y_{(2\mathcal{M})}(\hat{\mathbf{r}}_s)$ and angular momentum L operators, together with multi-shell collective vibrations and vorticity degrees of freedom for a description from irrotational to rigid rotor flows are all part of this symmetry. As an important consequence, the $Q \cdot Q$ interaction is $\mathrm{Sp}(3,\mathbb{R})$-preserving.[b]

Major progress in the development of realistic inter-nucleon interactions[48–51] along with the utilization of massively parallel computing resources (e.g., see Ref. 52) have placed *ab initio* large-scale simulations at the frontier of nuclear structure explorations. Several *ab initio* nuclear-theory approaches have been recently advanced, including Green's function Monte Carlo (GFMC),[36,53] no-core shell model (NCSM)[1,2,54–56] together with NCSM with a core[57] and importance truncation NCSM,[58] coupled-cluster method (CC),[37,59] lattice effective field theory,[60] in-medium SRG,[39,61] symmetry-adapted no-core shell model (SA-NCSM),[3] Monte Carlo NCSM,[62] and self-consistent Green's function.[63] *Ab initio* approaches build upon a 'first principles' foundation. They provide a long-missing link that bridges from the many-particle nucleus down to the fundamental blocks, namely, the properties of only two or three nucleons often tied to chiral symmetry-breaking patterns dictated by the underlying QCD. They also target predictions of experimentally inaccessible nuclear species far off the valley of stability. As such nuclei are often found key to understanding processes in extreme environments, from stellar explosions to the interior of nuclear reactors or fusion capsules, first-principle nuclear models have been and will be demonstrating a tremendous impact for advancing the frontiers in multiple branches of physics such as astrophysics, neutrino physics, and applied physics.[63–73]

The next-generation *ab initio* symmetry-adapted no-core shell model (SA-NCSM)[3] combines the first-principle concept of the NCSM with symmetry-guided considerations of the collectivity-driven models. The model and its recent findings are discussed in Secs. 2.3 and 3.

2. Shell-Model Theory — Symmetries of Strongly Interacting Particles and Organization of the Shell-Model Space

In its most general form, the nuclear shell model (SM),[2,74–76] a many-body "configuration interaction" (CI) method, solves the many-body Schrödinger equation for A particles,

$$H\Psi(\vec{r}_1, \vec{r}_2, \ldots, \vec{r}_A) = E\Psi(\vec{r}_1, \vec{r}_2, \ldots, \vec{r}_A), \tag{1}$$

for which the interaction and basis configurations are adopted as follows.

[b]An important $\mathrm{Sp}(3,\mathbb{R})$-preserving interaction is $Q \cdot Q = \sum_s q_s \cdot (\sum_t q_t)$, which realizes the physically relevant interaction of each particle with the total quadrupole moment of the nuclear system; furthermore, the expectation value of $Q \cdot Q$ in an SU(3) state is proportional to β^2, while for $Q \times Q \cdot Q$ it relates to $\beta^3 \cos 3\gamma$, providing a direct link to the interaction terms that enter in Greiner's Geometric Collective Model.

Interaction — The intrinsic non-relativistic nuclear plus Coulomb interaction Hamiltonian is defined as:

$$H = T_{\rm rel} + \sum_{i<j}^{A} (V_{NN})_{ij} + \sum_{i<j<k}^{A} (V_{3N})_{ijk} + \cdots + V_{\rm Coulomb}, \qquad (2)$$

where $T_{\rm rel}$ is the relative kinetic energy $T_{\rm rel} = \frac{1}{A}\sum_{i<j}\frac{(\vec{p}_i-\vec{p}_j)^2}{2m}$ (m is the nucleon mass), and the nucleon–nucleon NN (and possibly $3N$) interactions (see, e.g., Refs. 48–51, 77) are included along with the Coulomb interaction.

Basis configurations — A complete orthonormal basis ψ_i is adopted, such that the expansion $\Psi(\vec{r}_1, \vec{r}_2, \ldots, \vec{r}_A)$ in terms of unknown coefficients c_i, $\Psi(\vec{r}_1, \vec{r}_2, \ldots, \vec{r}_A) = \sum_k c_k \psi_k(\vec{r}_1, \vec{r}_2, \ldots, \vec{r}_A)$, renders Eq. (1) into a matrix eigenvalue equation, $\sum_{k'} H_{kk'} c_{k'} = E c_k$, where the many-particle Hamiltonian matrix elements $H_{kk'} = \langle \psi_k | H | \psi_{k'} \rangle$ are calculated for the given interaction (2). Typically, the basis is a finite set of antisymmetrized products of single-particle states (Slater determinants), referred to as a "model space", where the single-particle states of a three-dimensional spherical HO are used, $\phi_{\eta l j m t_z}(\vec{r}; b)$, where $\eta = 2n_r + l$, l is coupled to spin-$\frac{1}{2}$ to j, t_z distinguishes between protons and neutrons, and the oscillator length $b = \sqrt{\frac{\hbar}{m\Omega}}$ with oscillator frequency Ω. Such a basis allows for preservation of translational invariance of the nuclear self-bound system and provides solutions in terms of single-particle wave functions that are analytically known. With larger model spaces utilized in the shell-model theory, the eigensolutions converge to the exact values.

The NCSM treats all A particles as active, while the valence shell model[76,78] assumes a core of inactive particles and a subset of valence particles within the valence partially filled shell. In addition, if one can recognize approximate symmetries that survive within the nuclear dynamics, they can be used to help reduce the dimensionality of a model space to tractable sizes.[79] Such models use basis states that respect an approximate symmetry (symmetry-adapted basis), that is, possess good quantum numbers associated with the symmetry group (see Ref. 4 for various schemes for organization of the shell-model space).

2.1. *Elliott model*

The Elliott model[5,42,43] exploits SU(3), the exact symmetry of the three-dimensional spherical HO (see also Refs. 80–82). For SU(3)-symmetric interactions, the model can be solved analytically. But regardless whether a simple SU(3)-preserving interaction is used (see Figs. 5.1–5.6 of Ref. 83), or an SU(3)-symmetry breaking interaction (see Fig. 1 of Ref. 43), the results have revealed a striking feature, namely, the dominance of a few most deformed configurations. This has been shown for *sd*-shell nuclei, such as ^{18}Ne, ^{20}Ne, ^{22}Ne, ^{22}Mg, ^{24}Mg, and ^{28}Si, that have been known to possess a clear collective rotational structure in their low-lying states.[5,42,43] For heavier nuclei, pseudo-spin symmetry, discussed next, and an approximate "quasi-SU(3)" symmetry has been also suggested.[84] Furthermore, the pairing interaction

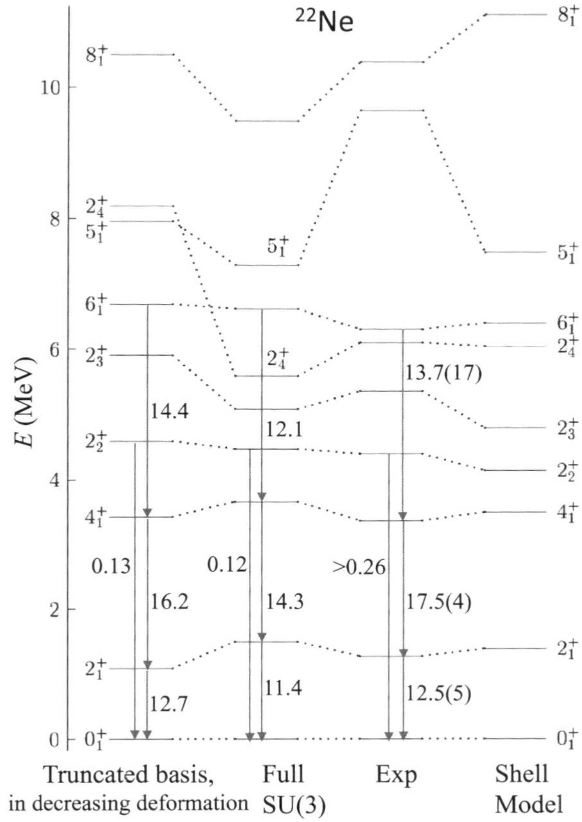

Fig. 1. Elliott's model with an SU(3)-preserving interaction + pairing in the *sd* valence shell for ^{22}Ne. $B(E2)$ transitions strengths (W.u.) are calculated for proton $e_{\text{eff}} = 1.3$ effective charge. Figure adapted from Ref. 85.

has been microscopically incorporated into the Elliott model where it breaks the SU(3) symmetry and mixes different $(\lambda\,\mu)$ configurations. It has been shown in Ref. 85 that using an SU(3)-symmetric interaction-plus-pairing yields results close to experiment and to the energies obtained using shell-model calculations in the full *sd* shell[86] (Fig. 1). It is remarkable that, even in the presence of pairing, comparable results have been obtained in a truncated model space that includes only about 10 most deformed configurations.

2.2. *Pseudo-spin symmetry for heavy nuclei*

For heavy nuclei ($A \gtrsim 100$), the discovery of the pseudo-spin symmetry[87,88] together with its fundamental nature[89–91] has established the pseudo-SU(3) model.[92] In particular, the microscopic origin of the pseudo-spin symmetry has been unveiled in Ref. 91, which has identified that the many-particle p-helicity operator generates a transformation to the pseudo-spin basis in heavy nuclei, while satisfying all other global symmetry requirements. Both mean-field and many-particle estimates demonstrate that in the helicity-transformed representation, the nucleons move in a finite-depth non-local potential with a reduced spin-orbit strength. Furthermore, the approximate independence of the single-nucleon spectrum in an infinite medium on the helicity transformation and the consistency of the microscopic estimates for the single-particle nuclear potentials with the Dirac–Brueckner calculations, allow one to connect the pseudo-spin symmetry to the one boson-exchange (OBEP) nature of the nucleon–nucleon NN interaction. Based on this link and because of the close relation (coincidence in the chiral symmetry limit) of the helicity and chirality operations, the goodness of pseudo-spin symmetry may be expected to increase with rising densities (or energy per particle) in hadronic systems, and actually yield to the chiral symmetry of massless hadrons in the high-energy region.[91]

The pseudo-spin scheme is an excellent starting point for a many-particle description of heavy nuclei, whether or not they are deformed. As for the SU(3) shell model, in many cases leading-irrep calculations (e.g., see Ref. 93) or mixed-irrep calculations (e.g., see Ref. 94) achieve good agreement with experimental data. The pseudo-SU(3) shell model provides a further understanding of the $M1$ transitions in nuclei such as the even–even $^{160-164}$Dy and $^{156-160}$Gd isotopes, specifically it reflects on the scissors and twist modes as well as the observed fragmentation (Fig. 2), that is, the break-up of the $M1$ strength among several levels closely clustered around a few strong transition peaks in the 2–4 MeV energy region[95] (for a detailed review on magnetic dipole excitations in nuclei, see Ref. 96). In medium-mass and heavy nuclei, where the pseudo-spin valence space is intruded by the highest-j orbit from the shell above, a major step towards understanding the significance of the intruder level is achieved by the pseudo-SU(3) plus intruder level shell model.[98]

Furthermore, the advantages of the symplectic Sp(3, $\mathbb{R}$) extension of the SU(3) model can be employed beyond the light nuclei domain towards a description of heavy nuclei in the framework of the pseudo-Sp(3, $\mathbb{R}$) shell model. For example, this model with a symmetry-preserving (in terms of the pseudo-Q and pseudo-L) two-, three-, and four-body interaction has been applied to ^{238}U with a focus on the energy spectrum and $B(E2)$ transition strengths from $J_i = 2, 4, 6, 8, 10$, and 12 to $J_i - 2$ within the ground-state rotational band.[99] In this study, the experimental $B(E2)$ values are remarkably well reproduced without the need of an effective charge. The results are comparable to a valence-shell pseudo-SU(3) model with effective charges, however, larger discrepancies have been observed for the higher-lying states.[99] While this pioneering work reveals the potential power of this model for heavy nuclear systems, it still remains not fully explored.

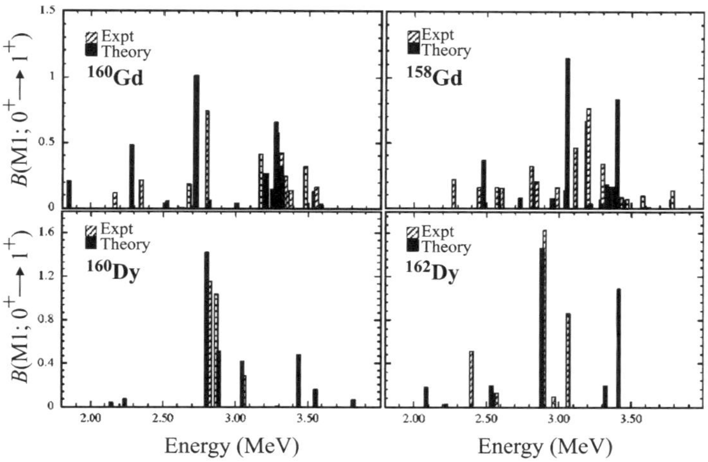

Fig. 2. Comparison of experimental (crosshatched bars)[97] and theoretical (solid bars) $M1$ strength distributions. In each case the eigenstates were determined by fitting parameters in the Hamiltonian to the experimental energy spectrum and associated $B(E2)$ transitions. Figure adapted from Ref. 95.

2.3. *Ab initio symmetry-adapted no-core shell model (SA-NCSM)*

The SA-NCSM[3] is a multi-shell generalization of the SU(3) scheme used in the Elliott model. It adopts the first-principle concept and is a no-core shell model carried forward in an SU(3) scheme. The many-particle basis states of the SA-NCSM are nuclear configurations of fixed parity, consistent with the Pauli principle, and truncated by a cutoff $N_{\max}$. The $N_{\max}$ cutoff is defined as the maximum number of HO quanta allowed in a many-particle state above the minimum for a given nucleus. For a given $N_{\max}$, the SA-NCSM many-particle basis states (Fig. 3) are constructed in the proton–neutron formalism, that is, we treat neutron and proton orbitals independently so total isospin is not conserved. For all possible distributions of protons $\{Z_0, Z_1, Z_2, \dots\}$ and neutrons $\{N_0, N_1, N_2, \dots\}$ over the major HO shells η ($\eta = 0, 1, 2, \dots$ for the s, p, ds, $\dots$ HO shell), limited by the number of HO quantum excitations up through $N_{\max}$, the $\mathrm{SU}(3)_\eta \times \mathrm{SU}(2)_{S_\eta}$ configurations are first enumerated for every major HO shell, following the $U(\Omega_\eta) \underset{\alpha_\eta}{\supset} \mathrm{SU}(3)_\eta$ reduction.[100] This is followed by inter-shell couplings, etc. Finally, the resulting proton and neutron configurations are coupled to good quantum numbers $(\lambda\,\mu)\kappa L$ of the $\mathrm{SU}(3)_{(\lambda\,\mu)} \underset{\kappa}{\supset} \mathrm{SO}(3)_L$ group chain, together with proton, neutron, and total intrinsic spins S_p, S_n, and S of the complementary SU(2) spin group. Each basis state in this scheme is thus labeled schematically as $|\vec{\gamma}\; [N(\lambda\,\mu)\kappa L; [S_p S_n]\, S]\, JM\rangle$, where N is the total number of HO excitation quanta and $\vec{\gamma}$ denotes additional quantum numbers needed to distinguish among configurations carrying the same $N(\lambda\,\mu)$ and $(S_p S_n)S$ labels. In this way, a complete shell-model basis is classified, with two

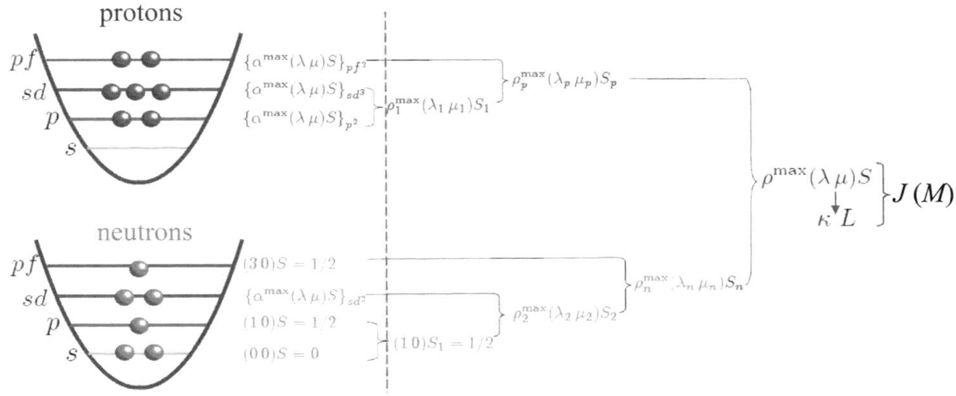

Fig. 3. (Color online) Example for an SU(3)-scheme basis state of good $(\lambda\,\mu)S$ and J. All additional quantum numbers needed to specify the basis state are shown in the illustration.

important features: (1) it can be down-selected to the physically relevant subspace, and (2) it allows for the spurious center-of-mass (CM) motion to be factored out exactly.[101–103]

The underlying principle behind the SA-NCSM kernel is an SU(3)-type Wigner–Eckhart theorem, which factorizes Hamiltonian matrix elements into the product of SU(3) reduced matrix elements and the associated SU(3) coupling coefficients. To compute the Hamiltonian matrix elements, the computational realization of the SA-NCSM, dubbed `LSU3shell`,[104] adopts state-of-the-art group-theoretical methods[100] and optimized numerical subroutines[105] for computing SU(3) coupling/recoupling coefficients. Recent developments and applications in the framework of the *ab initio* SA-NCSM are reviewed in Sec. 3.

2.4. *Symplectic model*

A significant breakthrough for the nuclear modeling is the microscopic symplectic model, developed by Rosensteel and Rowe.[6,7] In its simplest depiction,[16] the symplectic shell model is based on nucleons occupying HO shells with important correlations within each shell and between shells differing by $\pm 2\hbar\Omega$. The in-shell correlations are dominated by interactions of the quadrupole–quadrupole type, as first introduced by Elliott,[5,42] while the inter-shell correlations are of the giant monopole and giant quadrupole type. The inter-shell correlations enhance the electric quadrupole collectivity in such a way as to eliminate the need for effective charges. It is found that in many strongly deformed heavy nuclei the SU(3) quantum numbers, λ and μ, possess very large values, e.g., $\lambda \sim 100$ and $\mu \sim 10$.[106] This naturally leads to contraction of the SU(3) model to a rotor model.[8,107]

The symplectic model with $\mathrm{Sp}(3,\mathbb{R})$-preserving interactions have achieved a remarkable reproduction of rotational bands and transition rates without the need for introducing effective charges, while only a single $\mathrm{Sp}(3,\mathbb{R})$ irrep is used.[7,108] The symplectic model is used to construct rotational states for a rare-earth nucleus with microscopic wave functions. Analysis of the states in terms of their SU(3) content shows that SU(3) may be a relatively poor dynamical symmetry (mixing of many irreps) but an excellent *quasi-dynamical* symmetry for the model (the same combination of SU(3) irreps serves to uniquely characterize the intrinsic structure of a rotational band, independent of the L value).

Another successful extension to multiple shells has been achieved and applied to the ^{24}Mg ground-state rotational band,[109] where an interaction given as a polynomial in Q up through $(Q \cdot Q)^2$ was employed. Furthermore, a shell-model study in a symplectic basis that allows for mixing of Sp(3, $\mathbb{R}$) irreps due to pairing and non-degenerate single-particle energies above a ^{16}O core[10] has found that using only seven Sp(3, $\mathbb{R}$) irreps, which extend up through 15 HO shells, is sufficient to achieve a remarkable reproduction of the ^{20}Ne energy spectrum, as well as of $E2$ transition rates [Fig. 4(left)]. Recently, an Sp(3, $\mathbb{R}$)-based study using self-consistent arguments has been successful to give further insight into the cluster states of ^{16}O and shape-coexistence.[16]

Electron scattering form factors for transitions between low-lying states of ^{24}Mg have been calculated in the symplectic shell model using up to 4-body Q- and L-dependent interactions.[112] Parameters of the Hamiltonian were fit to reproduce measured energies and reduced transition probabilities in up to $N_{\max} = 20$ model spaces, with no adjustments made in obtaining the predicted form factors. The

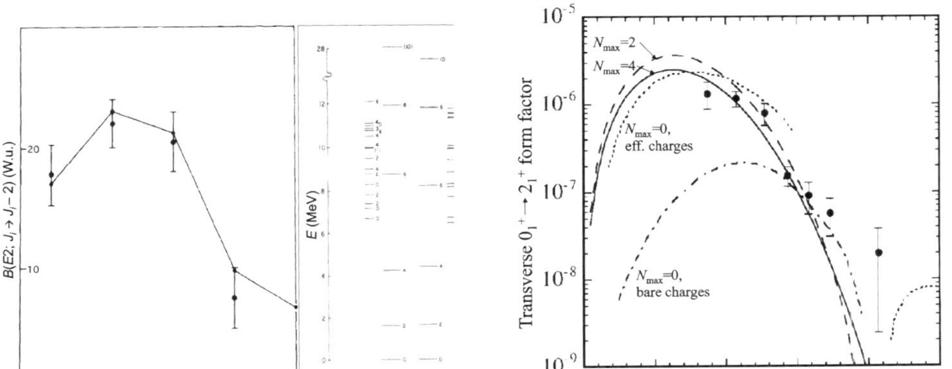

Fig. 4. (Left) Microscopic symplectic model with a set of effective single-particle energies, a $Q \cdot Q$-type interaction+pairing for ^{20}Ne. Left panel: Calculated $B(E2\downarrow)$ transition strengths *without effective charges* fall within the uncertainties of the corresponding experimental measurements. Right panel: Calculated energy spectrum for ^{20}Ne as compared to the experiment. Figure adapted from Ref. 10. (Right) Transverse $0_1^+ \rightarrow 2_1^+$ electron scattering form factors for ^{24}Mg in the symplectic model for $N_{\max} = 2$ and 4 *with no effective charges*, as compared to valence-shell calculations with bare and effective charges[110] and experiment.[110,111] Form factors are corrected for the center-of-mass motion and the finite-size effect of the nucleon. Figure adapted from Ref. 112.

symplectic form factors demonstrate excellent agreement with the available data [Fig. 4(right)], indicating that larger-$N_{\max}$ spaces and associated correlations play a substantial role in describing nuclear current and charge densities.

2.5. *No-core symplectic shell model (NCSpM) and the elusive Hoyle state*

Using the $Sp(3,\mathbb{R})$ scheme, the no-core symplectic shell model (NCSpM)[11] offers $N_{\max} = 12$–24 shell-model descriptions of low-lying states in deformed *sd*-shell nuclei (^{20}O, ^{20,22}Ne, and ^{20,22,24}Mg)[12] and of phenomena tied to giant monopole and quadrupole resonances,[13] as well as to collectivity and alpha-clustering in ^{12}Be (Ref. 14) and ^{12}C, and in particular, the challenging Hoyle state and its first 2^+ and 4^+ excitations.[11,113] While such ultra-large model spaces remain inaccessible by *ab initio* shell models, the NCSpM addresses a long-standing challenge,[114–116] namely, understanding highly-deformed spatial configurations from a shell-model perspective. Our present-day knowledge of various phenomena of astrophysical significance, such as nucleosynthesis, the evolution of primordial stars in the Universe, and X-ray bursts depends on reaction rates for the stellar triple-α process, which can considerably affect, e.g., results of core-collapse supernovae simulations and stellar evolution models, predictions regarding X-ray bursts, as well as estimates of carbon production in asymptotic giant branch (AGB) stars.[117] These rates, in turn, are greatly influenced by accurate measurements and theoretical predictions of several important low-lying states in ^{12}C, including the second 0_2^+ (Hoyle) state and its 2^+ excitation that has fostered long-lasting debate in experimental studies.[118–124] Further challenges relate to the α-cluster substructure of these states that has been explored within cluster-tailored[125–130] or self-consistent[131,132] framework, but has hitherto precluded a fully microscopic *ab initio* no-core shell-model description[133] (see, e.g., detailed reviews on the topic[134,135]). Only recently, first *ab initio* state-of-the-art calculations have been attempted using lattice effective field theory (EFT) with a coarse lattice grid.[60,136]

The physically relevant symplectic irreps are chosen among all possible symplectic $Sp(3,\mathbb{R})$ irreps within an $N_{\max}$ model space. For ^{12}C, these are 4 irreps extended up to $N_{\max} = 20$ of dimensionality of 6.6×10^3. For the interaction, the long-range part of the central NN force and a spin-orbit term are considered (better accounting of the symplectic symmetry mixing has been explored in Ref. 113 with preliminary results that use the bare JISP16 NN instead of the spin-orbit term). The interaction is augmented by $e^{-\gamma Q \cdot Q}$, with a single adjustable parameter γ that controls the contribution of the many-nucleon forces[148] (while short-range and tensor forces are considered important, they appear to be of secondary importance to the physics of the clusters, as suggested by the reasonably close agreement of the model outcome with experiment and *ab initio* results in smaller spaces[11,13]). Once the γ parameter is fixed by the order of the three low-lying 0^+ states in ^{12}C, excitation energies and other observables such as matter rms radii, electric quadrupole

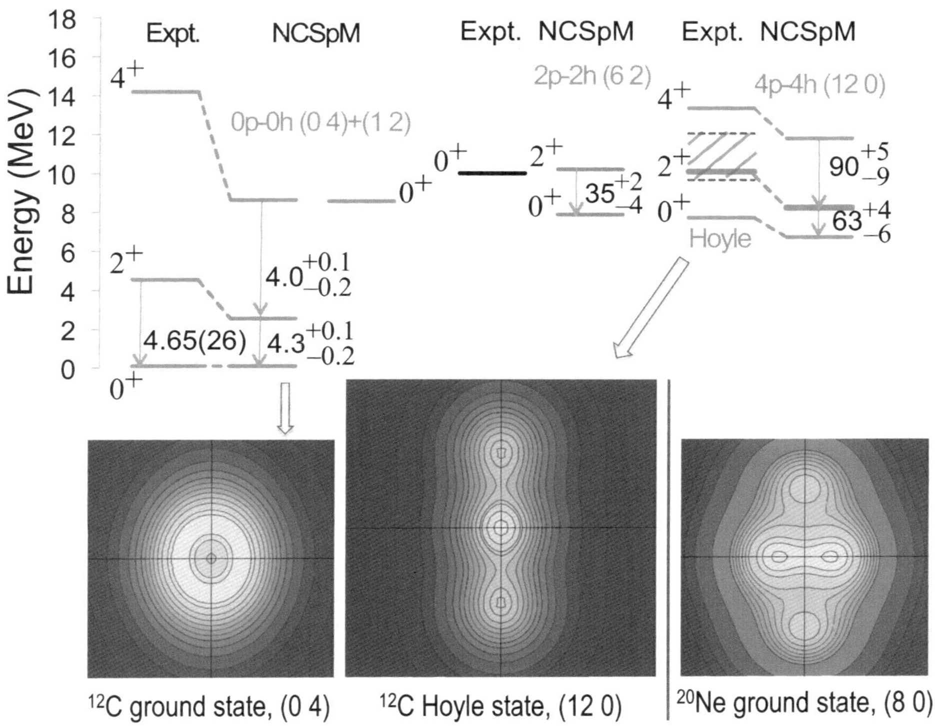

Fig. 5. (Color online) $N_{\max} = 20$ NCSpM energy spectrum of ^{12}C. Experimental data is from Refs. 120,123,137,138 (with a shaded area showing the energy range from Refs. 118–122). $B(E2)$ transition rates are in W.u.; theoretical uncertainties are estimated for a ±60% deviation of the Hoyle state energy. One-body density profiles *in the intrinsic frame* are shown for the 0^+_{gs} (for the most dominant symplectic irrep) and the Hoyle state in ^{12}C, and the ground state in ^{20}Ne. Figure adapted from Ref. 11.

moments and $E2$ transition rates (Fig. 5 and Table 2) for various p- and sd-shell nuclei are found, with no parameter adjustment, to be in a remarkable agreement with the experiment.[11,12,14] For ^{12}C, the NCSpM realizes a very reasonable $r_{\rm rms}$ for the ground state, and the Hoyle-state $r_{\rm rms}$ is found to lie close to a recent value deduced from experiment,[143] and as well tracks with the *ab initio* lattice EFT results at a leading order,[60] but differs considerably from predictions of cluster models, e.g., 3.4–4.3 fm.[126–128] Furthermore, the model yields a positive $Q_{2_1^+}$ very close to the experimental value, and a large negative one for the 2^+ and 4^+ states above the Hoyle state (Table 2), which indicates oblate (prolate) deformation for the ground-state (Hoyle-state) rotational band. This, together with the dominance of prolate $(\lambda\,\mu = 0)$ configurations in the Hoyle state (Fig. 6) and the 2^+ and 4^+ states above it, indicates a substantial prolate deformation for these states. Indeed, the one-body density profile for the Hoyle state very clearly supports an underlying α-particle,

Table 2. NCSpM point-particle rms matter radii and electric quadrupole moments for ^{12}C (for $N_{\max} = 20$) and *sd*-shell nuclei (for $N_{\max} = 12$)[11,12] compared to experimental (experimentally deduced) data. Experimentally deduced matter radii are summarized in Ref. 139; Q are taken from Refs. 140 and 141 for $A = 20$ and 22, respectively.

		matter $r_{\rm rms}$, fm		Q, $e\,{\rm fm}^2$	
		Expt.	NCSpM	Expt.	NCSpM
^{12}C	0^+_{gs}	$2.43(2)^a$	2.43(1)		
	2^+_1	$2.36(4)^{b*}$	2.42(1)	$+6(3)^d$	+5.9(1)
	4^+_1	–	2.41(1)	–	+8.0(3)
	0^+_2 (Hoyle)	$2.89(4)^{b*}$	2.93(5)		
	2^+ above 0^+_2	$3.07(13)^{c*}$	2.93(5)	–	−21(1)
	4^+ above 0^+_2	–	2.93(5)	–	−26(1)
	0^+_3	–	2.78(4)		
^{20}Ne	0^+_{gs}	$2.87(3)^e$	2.79		
	2^+_1 [4^+_1, 6^+_1]			−23(3)	−15.69 [−19.69, −21.05]
^{22}Ne	0^+_{gs}	–	2.82		
	2^+_1 [4^+_1, 6^+_1]			−17(3)	−14.90 [−19.22, −21.61]
^{20}Mg	0^+_{gs}	$2.88(4)^e$	2.73		
	2^+_1 [4^+_1]			–	−12.67 [−16.67]
^{22}Mg	0^+_{gs}	$2.89(6)^f$	2.82		
	2^+_1 [4^+_1, 6^+_1]			–	−17.88 [−23.07, −25.93]
^{24}Mg	0^+_{gs}	$2.97(12)^g$	3.03		
	2^+_1			$-16.6(6)^d$	−22.7

[a]Ref. 142; [b]Ref. 143; [c]Ref. 144; [d]Ref. 137, [e]Ref. 145; and [f]Ref. 146; and [g]Ref. 147.
*Experimentally deduced by model-dependent analyses of diffraction scattering.

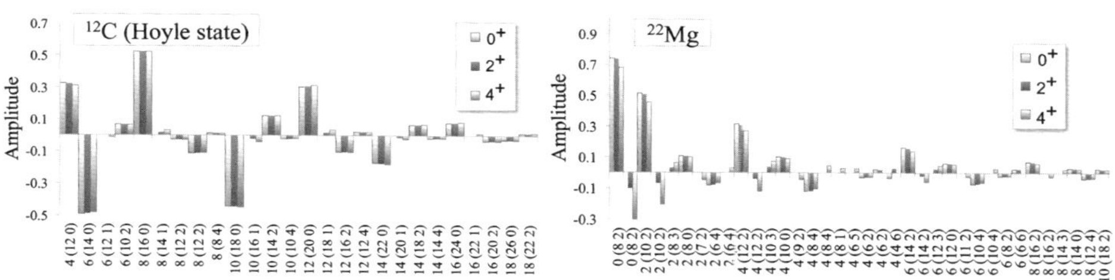

Fig. 6. (Color online) NCSpM amplitudes of relevant symplectic basis states (horizontal axis) labeled by SU(3)'s $N(\lambda_\omega\,\mu_\omega)$ for the 0^+ (blue) and the excited 2^+ (red) and 4^+ (green) states in the rotational band for (left) the Hoyle state in ^{12}C and (right) the ground state of ^{22}Mg.[12,14] Note that the Hoyle-state probability distribution peaks at $8\hbar\Omega$.

cluster-like picture (Fig. 5). We emphasize that this cluster structure is essentially very different from a simple α chain suggested by cluster models, as the clusters partially overlap (while the ground state exhibits a torus-like shape with a dip in the middle).

For intermediate-mass nuclei, we find that it is imperative that model spaces be expanded well beyond the current limits up through 15 major shells to accommodate particle excitations, which appear critical to highly deformed spatial structures (Figs. 5 and 6) and the convergence of associated observables, as detailed in Ref. 12.

3. Understanding Emergent Collectivity from First Principles with SA-NCSM

3.1. *Symmetry-guided concept*

The low-lying states in light nuclei exhibit orderly patterns, as illustrated in Fig. 7, that favor spatial configurations with strong quadrupole deformation and complementary low intrinsic spin values, a picture that is consistent with the nuclear symplectic model. This suggests a symmetry-guided basis selection philosophy that

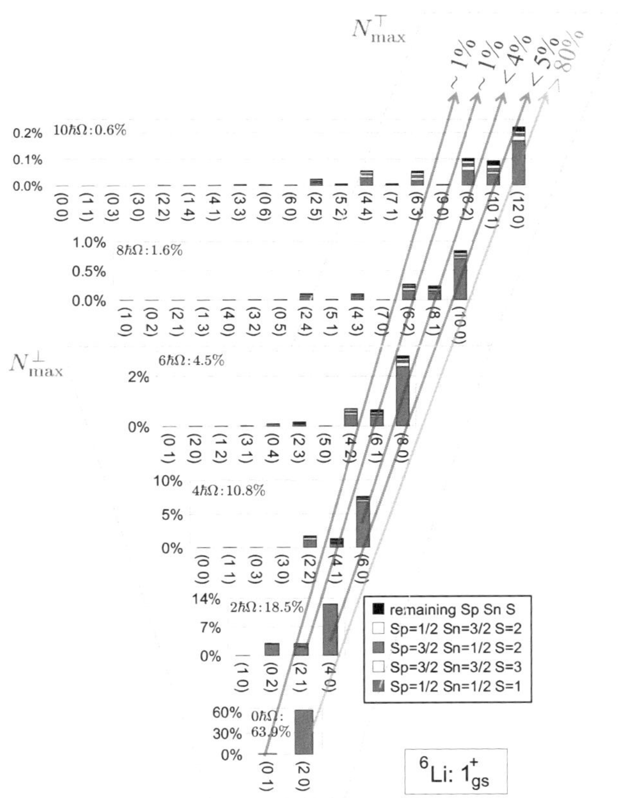

Fig. 7. (Color online) Probability distributions for proton, neutron, and total intrinsic spin components (S_pS_nS) across the Pauli-allowed ($\lambda\,\mu$) values (horizontal axis) for the calculated 1^+ ground state of ^{6}Li obtained for $N_{\max} = 10$ and $\hbar\Omega = 20$ MeV with the JISP16 bare interaction. The concentration of strengths to the far right within the histograms demonstrates the dominance of collectivity in the calculated eigenstates — this supports a *symmetry-guided* concept, which implies inclusion of all configurations up through $N^{\perp}_{\max}$, and a subset of deformation/spin configurations beyond this, up through $N^{\top}_{\max}$. The projection onto symplectic vertical slices (with probability $\geq 1\%$) is schematically illustrated for ^{6}Li by arrows and clearly reveals the preponderance of a single symplectic irrep (vertical cone). Figure adapted from Ref. 3.

allows the SA-NCSM to obtain results in much smaller spaces that are nearly indistinguishable from the complete basis counterparts. Specifically, the outcome supports a symmetry-guided concept,[4] a key feature of the SA-NCSM, namely, the relevant model space can be systematically selected, using a quantified cutoff, starting from the largest deformation/lowest spin within a low-$N\hbar\Omega$ subspace and associated symplectic excitations thereof (right sector of Fig. 7), and including ever smaller deformation until convergence of results is achieved. The SA-NCSM model spaces are thus characterized by a pair of numbers, $\langle N_{\max}^{\perp}\rangle N_{\max}^{\top}$ (schematically illustrated in Fig. 7), which implies inclusion of the full space up through $N_{\max}^{\perp}$, and a subset of deformation/spin configurations beyond this, up through $N_{\max}^{\top}$. The validity of the symmetry-guided concept can be illustrated with SA-NCSM results for p-shell nuclei obtained in model spaces, which are expanded beyond a complete $N_{\max}^{\perp}$ space ($N_{\max}^{\perp} = 2, 4, \ldots, 10$) by using a relatively few dominant intrinsic spin and deformation components up through $N_{\max}^{\top} = 12$ (Fig. 8).

Clearly, the results indicate that the observables obtained in the much smaller symmetry-guided selected spaces are excellent approximations to the corresponding $N_{\max} = 12$ complete-space counterparts. A crucial advantage of this symmetry-guided concept is that SA-NCSM can carry out investigations beyond the current $N_{\max}$ NCSM limits.

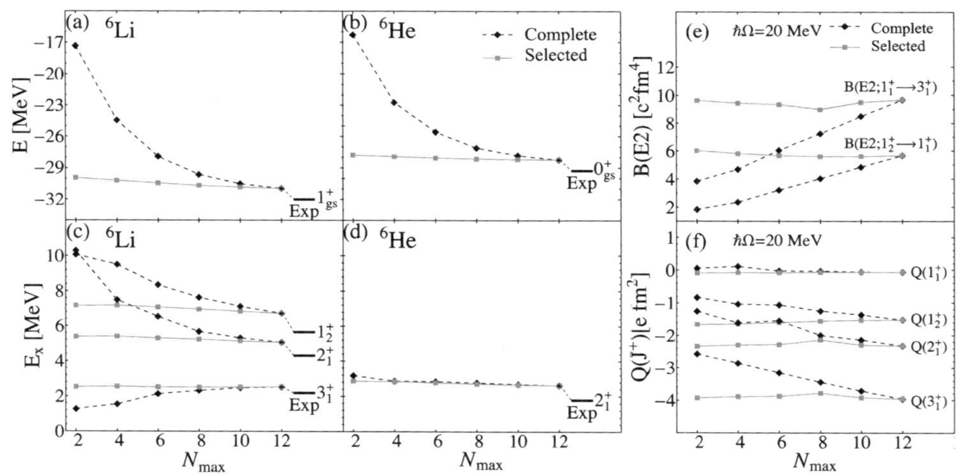

Fig. 8. (Color online) Energies of ^{6}Li and ^{6}He, (a)–(d), together with (e) $E2$ transition probabilities and (f) quadrupole moments for ^{6}Li, shown for the complete $N_{\max}$ model spaces (dashed black curves) and for the $\langle N_{\max}\rangle 12$ SA-NCSM selected model spaces (solid red lines). Results shown are for JISP16 and $\hbar\Omega = 20$ MeV. Note the relatively large changes in the results when the complete space is increased from $N_{\max} = 2$ to $N_{\max} = 12$ as compared to the nearly constant $\langle N_{\max}\rangle 12$ SA-NCSM outcomes. Figure adapted from Ref. 3.

The number of basis states used, e.g., for each ^{6}Li state, is only about 10–12% for $\langle 2\rangle 12$, $\langle 4\rangle 12$, $\langle 6\rangle 12$, 14% for $\langle 8\rangle 12$, and 30% for $\langle 10\rangle 12$ as compared to the number for the complete $N_{\max} = 12$ SA-NCSM model space. The net result is that calculations in the selected spaces require one to two orders of magnitude less computational time. Indeed, with the help of HPC resources, the use of such basis spaces in *ab initio* studies is manageable as well as expandable; that is, one expects to be able to extend the reach of our SA-NCSM scheme from applications that are doable today to the larger spaces and heavier nuclear systems of tomorrow, utilizing at each stage the accuracy and predictive power of the *ab initio* approach.

3.2. *Ab initio SA-NCSM — reaching sd-shell nuclei and beyond*

Ab initio SA-NCSM calculations for spectral and scattering properties of *p*-shell nuclei have been found to yield results that are in reasonably close agreement with known experimental data (Fig. 8 and Table 3).[3,149,150] Calculations are performed in the SU(3) scheme for selected model spaces that expand up to $N_{\max} = 12$, using the JISP16[151] or NNLO$_{\text{opt}}$[152] NN interactions, plus the Coulomb potential along with the Gloeckner–Lawson prescription[153] for eliminating spurious center-of-mass excitations. Indeed, the *ab initio* SA-NCSM results agree with those of other *ab initio* models, such as the NCSM (or complete-space SA-NCSM), the variational Monte Carlo (VMC)[53] and GFMC with AV18/UIX (Fig. 8 and Table 3). The results show good correlations among the different models with, perhaps, the exception of

Table 3. ^{6}Li binding energy (BE), excitation energies (E), electric quadrupole (Q) and magnetic dipole (μ) moments, point-nucleon proton (r_p) and matter (r_m) rms radii, as calculated in the $\langle 6\rangle 12$ SA-NCSM with the JISP16 NN interaction (for $\hbar\Omega = 20$ MeV) and compared to other *ab initio* approaches: the complete $N_{\max} = 12$ model space (or NCSM), as well as VMC and GFMC[53,154] using AV18/UIX interaction[48,155] (without contributions from two-body currents). Experimental results (Expt.) taken from Ref. 156 unless otherwise specified.

State		SA-NCSM	NCSM	VMC	GFMC	Expt.
1^+_{gs}	BE [MeV]	30.445	30.951	27.0(1)	31.2(1)	31.99
	rms r_p [fm]	2.112	2.125	2.46(2)		2.43^a
	rms r_m [fm]	2.106	2.119			2.35(3)^b
	Q [e fm^2]	−0.080	−0.064	−0.33(18)		−0.0818(17)
	μ [μ_N]	0.839	0.838	0.828(1)		0.822
3^+	E [MeV]	2.515	2.526	3.0(1)	2.7(3)	2.186
	rms r_m [fm]	2.044	2.063			
	Q [e fm^2]	−3.880	−3.965			
	μ [μ_N]	1.866	1.866			
2^+	E [MeV]	5.303	5.066	4.4(1)	4.4(4)	4.312
	rms r_m [fm]	2.180	2.204			
	Q [e fm^2]	−2.279	−2.318			
	μ [μ_N]	1.014	0.970			

^{a}Deduced from the ^{6}Li charge radius of 2.56(5) fm;[157] ^{b}From Ref. 158.

the smaller rms radii obtained with JISP16 and the larger magnitude of the $Q_{1^+_{\rm gs}}$ moment in ^{6}Li obtained with the VMC.

Of special note is the quadrupole moment of the ^{6}Li ground state. In addition to being in excellent agreement to experiment (Fig. 8 and Table 3), the SA-NCSM result for $Q_{1^+_{\rm gs}}$ reproduces its challenging sign and nearly vanishing magnitude. In fact, the *ab initio* SA-NCSM has unveiled that even the lightest of nuclei in their ground state exhibit considerable collectivity, as exemplified by the dominance of prolate deformation in the ^{6}Li wave function (Fig. 7) — nonetheless, the considerable contribution of $L = 0$ configurations ($\sim 87\%$) to $1^+_{\rm gs}$ is responsible for the small expectation value of the quadrupole moment, $L = 2$ operator.[3] Furthermore, as discussed in Ref. 149, we have demonstrated that the SA-NCSM space selection exhibits a relatively small effect on form factors and charge density.

Ab initio investigations of open-shell nuclei in the *sd*-shell region (Fig. 9) are now made possible by the SA-NCSM.[159] First SA-NCSM calculations for ^{24}Si, important for modeling and understanding x-ray burst nucleosynthesis, and for its mirror nucleus ^{24}Ne,[4,73,160] as well as for ^{18}Ne and odd-parity states of ^{19}Ne have been achieved starting with chiral interactions. Furthermore, the description of the ^{20}Ne ground-state rotational band is available now from first principles and reveals an interesting substructure formation. *Ab initio* SA-NCSM calculations are performed in the challenging $N_{\rm max} = \langle 2 \rangle 10$ model space. For example, for the 6^+ state there are 51×10^6 basis states in the SA-NCSM, which can be compared to the inaccessible $N_{\rm max} = 10$ NCSM model space of 4×10^{11} dimensionality required for the corresponding J-scheme calculations.

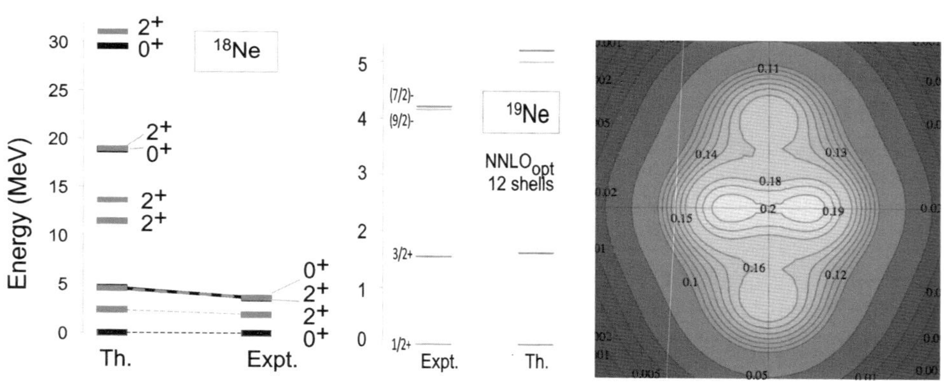

Fig. 9. (Color online) (Left) First SA-NCSM calculations with chiral potentials for ^{18,19}Ne in the ultra-large model space of 9–12 shells. (Right) *Ab initio* SA-NCSM density for the ^{20}Ne ground state in the challenging model space of 13 shells. Simulations use the NNLO$_{\rm opt}$ NN interaction[152] and $\hbar\Omega = 15$ MeV, and are performed on the Blue Waters system.

4. Summary and Outlook

We have reviewed exact and partial symmetries of atomic nuclei that have been long recognized, and only recently exposed from first principles and exploited in *ab initio* theories. Such partial symmetries include the deformation-related SU(3) and the symplectic $\mathrm{Sp}(3,\mathbb{R})$ for descriptions of shapes, rotations, and vibrations. We have focused on an open-core shell-model theory (starting with chiral potentials or an effective microscopic interaction) and the use of symmetries in large-scale nuclear simulations that permit symmetry mixing. In such calculations, symmetries are not used to restrict the nuclear dynamics, but rather to expose simple patterns and appropriate degrees of freedom to organize and down-select physically relevant spaces from what would otherwise be unwieldy at best or even prohibitively large model spaces. We have discussed an important new development that has demonstrated, for the first time, that observed collective phenomena in light and intermediate-mass nuclei emerge naturally from *ab initio* and microscopic considerations. The results, supporting experimental evidence, underscore the strong dominance of large quadrupole deformation and low intrinsic spin, and symplectic excitations thereof.

The findings reviewed here start to unveil new physics, namely, understanding the mechanism on how simple patterns in nuclei and a diversity of nuclear properties emerge from a fundamental level.

Acknowledgments

It is a pleasure to recall and acknowledge the important role Professor Greiner played in helping to advance the field of nuclear physics, more often than not challenging the status quo by proffering alternative views of reality. No theory escaped his critical scrutiny; he challenged young and old alike, stimulating students and colleagues to seek excellence in teaching and research. We also thank D. Rowe, G. Rosensteel, J. Wood, C. Bahri, J. P. Vary, and P. Maris for useful discussions. This work was supported by the U.S. NSF (ACI-1516338, ACI-1713690, OIA-1738287), Southeastern Universities Research Association (SURA), Louisiana State University (through its support of J.P.D.'s work with SURA), and the Czech Science Foundation under Grant No. 16-16772S. This work also benefitted from computing resources provided by Blue Waters, as well as the Louisiana Optical Network Initiative and HPC resources provided by LSU (`www.hpc.lsu.edu`). A portion of the computational resources were provided by the National Energy Research Scientific Computing Center and by an INCITE award from the DOE Office of Advanced Scientific Computing. T. D. acknowledges support from Michal Pajr and CQK Holding.

References

1. P. Navrátil, J. P. Vary and B. R. Barrett, Properties of ^{12}C in the *Ab Initio* Nuclear Shell Model, *Phys. Rev. Lett.* **84**, 5728 (2000).
2. B. Barrett, P. Navrátil and J. Vary, *Prog. Part. Nucl. Phys.* **69**, 131 (2013).
3. T. Dytrych, K. D. Launey, J. P. Draayer, P. Maris, J. P. Vary, E. Saule, U. Catalyurek, M. Sosonkina, D. Langr and M. A. Caprio, Collective Modes in Light Nuclei from First Principles, *Phys. Rev. Lett.* **111**, 252501 (2013).
4. K. D. Launey, T. Dytrych and J. P. Draayer, Symmetry-guided large-scale shell-model theory, *Prog. Part. Nucl. Phys.* **89**, 101 (2016), doi:10.1016/j.ppnp.2016.02.001.
5. J. P. Elliott, Collective Motion in the Nuclear Shell Model. I. Classification Schemes for States of Mixed Configurations, *Proc. Roy. Soc. A* **245**, 128 (1958).
6. G. Rosensteel and D. J. Rowe, Nuclear Sp(3, $\mathbb{R}$) Model, *Phys. Rev. Lett.* **38**, 10 (1977).
7. D. J. Rowe, Microscopic theory of the nuclear collective model, *Reports on Progr. in Phys.* **48**, 1419 (1985).
8. D. J. Rowe, Dynamical symmetries of nuclear collective models, *Prog. Part. Nucl. Phys.* **37**, 265 (1996).
9. G. Rosensteel and D. J. Rowe, *Ann. Phys. N.Y.* **126**, 343 (1980).
10. J. Draayer, K. Weeks and G. Rosensteel, Symplectic Shell-model Calculations for ^{20}Ne with Horizontal Configuration Mixing, *Nucl. Phys.* **A419**, 1 (1984).
11. A. C. Dreyfuss, K. D. Launey, T. Dytrych, J. P. Draayer and C. Bahri, Hoyle state and rotational features in Carbon-12 within a no-core shell model framework, *Phys. Lett. B* **727**, 511 (2013).
12. G. K. Tobin, M. C. Ferriss, K. D. Launey, T. Dytrych, J. P. Draayer and C. Bahri, Symplectic No-core Shell-model Approach to Intermediate-mass Nuclei, *Phys. Rev. C* **89**, 034312 (2014).
13. A. C. Dreyfuss, K. D. Launey, T. Dytrych, J. P. Draayer, R. B. Baker, C. M. Deibel and C. Bahri, Understanding emergent collectivity and clustering in nuclei from a symmetry-based no-core shell-model perspective, *Phys. Rev. C* **95** (2017).
14. K. D. Launey, T. Dytrych, J. P. Draayer, G. K. Tobin, M. C. Ferriss, D. Langr, A. C. Dreyfuss, P. Maris, J. P. Vary and C. Bahri, Symmetry-adapted No-core Shell Model for Light Nuclei, in *Proceedings of the 5th International Conference on Fission and properties of neutron-rich nuclei, ICFN5, November 4–10, 2012, Sanibel Island, Florida*, eds. J. H. Hamilton and A. V. Ramayya (World Scientific, Singapore, 2013), p. 29.
15. K. Heyde and J. Wood, *Rev. Mod. Phys.* **83**, 1467 (2011).
16. D. J. Rowe, G. Thiamova and J. L. Wood, Implications of Deformation and Shape Coexistence for the Nuclear Shell Model, *Phys. Rev. Lett.* **97**, 202501 (2006).
17. W. Kulp *et al.*, *Phys. Rev. C* **77**, 061301(R) (2008).
18. D. J. Rowe and J. L. Wood, *Fundamentals of Nuclear Models: Foundational Models* (World Scientific, Singapore, 2010).
19. J. Drut, R. Furnstahl and L. Platter, *Prog. Part. Nucl. Phys.* **64**, 120 (2010).
20. K. D. Launey (ed.) World Scientific Publishing Co. (2017). ISBN 978-981-3146-04-4. `http://www.worldscientific.com/worldscibooks/10.1142/10180`.
21. M. G. Mayer, *Phys. Rev.* **75**, 1969 (1949).
22. O. Haxel, J. H. D. Jensen and H. E. Suess, *Phys. Rev.* **75**, 1766 (1949).
23. M. G. Mayer, *Phys. Rev.* **78**, 16 (1950).
24. A. Bohr and B. R. Mottelson, *Nuclear Structure*. Vol. 1 (Benjamin, New York, 1969).

25. S. G. Nilsson, Binding states of individual nucleons in a strongly deformed field, *Mat. Fys. Medd. Dan. Vid. Selsk.* **29**(16), (1955).
26. G. Racah, *Phys. Rev.* **62**, 438 (1942).
27. B. H. Flowers, *Proc. Roy. Soc. (London)* **A212**, 248 (1952).
28. A. K. Kerman, *Ann. Phys.* **12**, 300 (1961).
29. R. W. Richardson, *Phys. Lett.* **3**, 277 (1963).
30. M. Gaudin, *J. Physique.* **37**, 1087 (1976).
31. X. Guan, K. D. Launey, M. Xie, L. Bao, F. Pan and J. P. Draayer, *Phys. Rev. C* **86**, 024313 (2012).
32. W. G. G. Gneuss, U. Mosel, *Phys. Lett. B* **30**, 397 (1969).
33. J. Eisenberg and W. Greiner, in *Nuclear Theory I: Nuclear collective Models* (North-Holland Publishing Company, Amsterdam, 1987).
34. A. Arima and F. Iachello, *Phys. Lett. B* **35**, 1069 (1975).
35. F. Iachello and A. Arima, *The Interacting Boson Model* (Cambridge University Press, 1987).
36. S. C. Pieper, K. Varga and R. B. Wiringa, Quantum Monte Carlo calculations of $A = 9, 10$ nuclei, *Phys. Rev. C* **66**, 044310 (2002).
37. M. Wloch, D. J. Dean, J. R. Gour, M. Hjorth-Jensen, K. Kowalski, T. Papenbrock and P. Piecuch, *Ab-Initio* Coupled-Cluster Study of ^{16}O, *Phys. Rev. Lett.* **94**, 212501 (2005).
38. D. Lee, *Ab initio* lattice effective field theory and Wigner's SU(4) symmetry, in *Emergent Phenomena in Atomic Nuclei from Large-Scale Modeling: A Symmetry-Guided Perspective*, ed. K. D. Launey (World Scientific, 2017).
39. K. Tsukiyama, S. K. Bogner and A. Schwenk, In-medium similarity renormalization group for nuclei, *Phys. Rev. Lett.* **106**, 222502 (2011).
40. S. Bogner, A. Bulgac, J. Carlson, J. Engel, G. Fann, R. Furnstahl, S. Gandolfi, G. Hagen, M. Horoi, C. Johnson, M. Kortelainen, E. Lusk, P. Maris, H. Nam, P. Navratil, W. Nazarewicz, E. Ng, G. Nobre, E. Ormand, T. Papenbrock, J. Pei, S. Pieper, S. Quaglioni, K. Roche, J. Sarich, N. Schunck, M. Sosonkina, J. Terasaki, I. Thompson, J. Vary and S. Wild, Computational nuclear quantum many-body problem: The UNEDF project, *Computer Physics Communications* **184**(10), 2235 (2013).
41. T. Dytrych, K. D. Sviratcheva, J. P. Draayer, C. Bahri and J. P. Vary, Ab initio symplectic no-core shell model, *J. Phys. G: Nucl. Part. Phys.* **35**, 123101 (2008).
42. J. P. Elliott, Collective Motion in the Nuclear Shell Model. II. The Introduction of Intrinsic Wave-Functions, *Proc. Roy. Soc. A* **245**, 562 (1958).
43. J. P. Elliott and M. Harvey, Collective Motion in the Nuclear Shell Model. III. The Calculation of Spectra, *Proc. Roy. Soc. A* **272**, 557 (1962).
44. G. Rosensteel and D. J. Rowe, *Ann. Phys. N.Y.* **104**, 134 (1977).
45. Y. Leschber and J. P. Draayer, *Phys. Letts. B* **190**, 1 (1987).
46. O. Castaños, J. P. Draayer and Y. Leschber, *Z. Phys. A* **329**, 33 (1988).
47. G. Rosensteel, Self-gravitating symplectic systems, *Astrophys. J.* **416**, 291 (1993).
48. R. B. Wiringa, V. G. J. Stoks and R. Schiavilla, *Phys. Rev. C* **51**, 38 (1995).
49. R. Machleidt, *Phys. Rev. C* **63**, 024001 (2001).
50. D. R. Entem and R. Machleidt, *Phys. Rev. C* **68**, 041001 (2003).
51. E. Epelbaum, *Prog. Part. Nucl. Phys.* **57**, 654 (2006).
52. D. Langr, I. Simecek, P. Tvrdik, T. Dytrych and J. P. Draayer, Adaptive-Blocking Hierarchical Storage Format for Sparse Matrices, in *Proceedings of the Federated Conference on Computer Science and Information Systems (FEDCSIS), 2012*, p. 545 (2012).
53. R. B. Wiringa and R. Schiavilla, *Phys. Rev. Lett.* **81**, 4317 (1998).

54. P. Maris, J. P. Vary and P. Navrátil, *Phys. Rev. C* **87**, 014327 (2013).
55. P. Navrátil, S. Quaglioni, I. Stetcu and B. R. Barrett, Recent developments in no-core shell-model calculations, *J. Phys. G: Nucl. Part.* **36**, 083101 (2009).
56. S. Baroni, P. Navrátil and S. Quaglioni, *Phys. Rev. Lett.* **110**, 022505 (2013).
57. A. F. Lisetskiy, B. R. Barrett, M. K. G. Kruse, P. Navratil, I. Stetcu and J. P. Vary, *Ab-initio* shell model with a core, *Phys. Rev. C* **78**, 044302 (2008).
58. R. Roth and P. Navrátil, *Phys. Rev. Lett.* **99**, 092501 (2007).
59. G. Hagen, T. Papenbrock, D. J. Dean and M. Hjorth-Jensen, *Phys. Rev. Lett.* **101**, 092502 (2008).
60. E. Epelbaum, H. Krebs, D. Lee and U.-G. Meissner, *Phys. Rev. Lett.* **106**, 192501 (2011).
61. S. K. Bogner, H. Hergert, J. D. Holt, A. Schwenk, S. Binder, A. Calci, J. Langhammer and R. Roth, Nonperturbative shell-model interactions from the in-medium similarity renormalization group, *Phys. Rev. Lett.* **113**, 142501 (2014).
62. T. Abe, P. Maris, T. Otsuka, N. Shimizu, Y. Utsuno and J. Vary, *Phys. Rev. C* **86**, 054301 (2012).
63. A. Cipollone, C. Barbieri and P. Navratil, Isotopic chains around oxygen from evolved chiral two- and three-nucleon interactions, *Phys. Rev. Lett.* **111**, 062501 (2013).
64. R. B. Wiringa and S. C. Pieper, Evolution of Nuclear Spectra with Nuclear Forces, *Phys. Rev. Lett.* **89**, 182501 (2002).
65. A. C. Hayes, P. Navratil and J. P. Vary, Neutrino-^{12}C scattering in the ab initio shell model with a realistic three-body interaction, *Phys. Rev. Lett.* **91**, 012502 (2003).
66. P. Maris, A. M. Shirokov and J. P. Vary, *Phys. Rev. C* **81**, 021301(R) (2010).
67. P. Navrátil and S. Quaglioni, Ab Initio Many-Body Calculations of the ^{3}H(d,n)^{4}He and ^{3}He(d,p)^{4}He Fusion Reactions, *Phys. Rev. Lett.* **108**, 042503 (2012).
68. G. Hagen, M. Hjorth-Jensen, G. R. Jansen, R. Machleidt and T. Papenbrock, Evolution of Shell Structure in Neutron-Rich Calcium Isotopes, *Phys. Rev. Lett.* **109**, 032502 (2012).
69. S. Bacca, N. Barnea, W. Leidemann and G. Orlandini, Isoscalar Monopole Resonance of the Alpha Particle: A Prism to Nuclear Hamiltonians, *Phys. Rev. Lett.* **110**, 042503 (2013).
70. G. Hupin, S. Quaglioni and P. Navrátil, A Predictive Theory for Elastic Scattering and Recoil of Protons from ^{4}He, *Phys. Rev. C* **90**, 061601 (2014).
71. E. Epelbaum, H. Krebs, T. A. Lähde, D. Lee, U.-G. Meißner and G. Rupak, Ab initio calculation of the spectrum and structure of ^{16}O, *Phys. Rev. Lett.* **112**(10), 102501 (2014), doi:10.1103/PhysRevLett.112.102501.
72. C. Romero-Redondo, S. Quaglioni, P. Navrátil and G. Hupin, ^{4}He+n+n continuum within an ab initio framework, *Phys. Rev. Lett.* **113**, 032503 (2014).
73. T. Dytrych, K. D. Launey and J. P. Draayer, Symmetry-adapted no-core shell model (YB140314, research review), in *McGraw-Hill Yearbook of Science and Technology* (McGraw-Hill, 2014).
74. P. Brussard and P. Glaudemans, *Shell-model Applications in Nuclear Spectroscopy* (North-Holland Publishing Company, Amsterdam, 1977).
75. I. Shavitt, The history and evolution of configuration interaction, *Molecular Physics* **94**, 3 (1998).
76. E. Caurier, G. Martinez-Pinedo, F. Nowacki, A. Poves and A. P. Zuker, The shell model as unified view of nuclear structure, *Rev. Mod. Phys.* **77**, 427 (2005).
77. D. Gazit, S. Quaglioni and P. Navratil, Three-Nucleon Low-Energy Constants from the Consistency of Interactions and Currents in Chiral Effective Field Theory, *Phys. Rev. Lett.* **103**, 102502 (2009). doi:10.1103/PhysRevLett.103.102502.

78. B. Brown, The nuclear shell model towards the drip lines, *Prog. Part. Nucl. Phys.* **47**, 517 (2001).
79. J. P. Draayer, Fermion Models, in *Algebraic Approaches to Nuclear Structure*, ed. R. Casten (Harwood Academic Publishers, 1992), chapter 7, p. 423.
80. M. Moshinsky, Wigner Coefficients for the SU_3 Group and Some Applications, *Rev. Mod. Phys.* **34**, 813 (1962).
81. M. Moshinsky, J. Patera, R. T. Sharp and P. Winternitz, Everything you always wanted to know about $SU(3) \supset O(3)$, *Ann. Phys. (N.Y.)* **95**, 139 (1975).
82. D. J. Millener, A note on recoupling coefficients for SU(3), *J. Math. Phys.* **19**, 1513 (1978).
83. M. Harvey, The Nuclear SU(3) Model, *Adv. Nucl. Phys.* **1**, 67 (1968).
84. A. Zuker, J. Retamosa, A. Poves and E. Caurier, *Phys. Rev. C* **52**, R1741 (1995).
85. C. Vargas, J. Hirsch and J. Draayer, *Nucl. Phys. A* **690**, 409 (2001).
86. B. Preedom and B. Wildenthal, *Phys. Rev. C* **6**, 1633 (1972).
87. K. T. Hecht and A. Adler, *Nucl. Phys. A* **A137**, 129 (1969).
88. A. Arima, M. Harvey and M. K. Shimizu, *Phys. Lett.* **30B**, 517 (1969).
89. J. Draayer, Dominance of Low Spin and High Deformation in *Ab Initio* Approaches to the Structure of Light Nuclei, in *Proceedings of Symmetries in Physics, Cocoyoc, Mexico, June 3–7, 1991*, eds. A. Frank and K. Wolf (Springer-Verlag, Berlin, 1991).
90. C. Bahri, J. Draayer and S. Moszkowski, Pseudospin symmetry in nuclear physics, *Phys. Rev. Lett.* **68**, 2133 (1992).
91. A. Blokhin, C. Bahri and J. Draayer, *Phys. Rev. Lett.* **74**, 4149 (1995).
92. R. R. Raju, J. Draayer and K. Hecht, *Nucl. Phys. A* **202**, 433 (1973).
93. J. Draayer and K. Weeks, *Phys. Rev. Lett.* **51**, 1422 (1983).
94. G. Popa, J. G. Hirsch and J. P. Draayer, *Phys. Rev. C* **62**, 064313 (2000).
95. T. Beuschel, J. P. Draayer, D. Rompf and J. G. Hirsch, *Phys. Rev. C* **57**, 1233 (1998).
96. K. Heyde, P. von Neumann-Cosel and A. Richter, Magnetic dipole excitations in nuclei: Elementary modes of nucleonic motion, *Rev. Mod. Phys.* **82**, 2365 (2010).
97. U. Kneissl, H. H. Pitz and A. Zilges, *Prog. Part. Nucl. Phys.* **37**, 349 (1996).
98. K. Weeks, C. Han and J. Draayer, *Nucl. Phys. A* **371**, 19 (1981).
99. O. Castaños, P. Hess, J. Draayer and P. Rochford, *Nucl. Phys. A* **524**, 469 (1991).
100. J. P. Draayer, Y. Leschber, S. C. Park and R. Lopez, Representations of U(3) in U(N), *Comput. Phys. Commun.* **56**, 279 (1989).
101. B. J. Verhaar, A Method for the Elimination of Spurious States in the Nuclear Harmonic Oscillator Shell Model, *Nucl. Phys.* **21**, 508 (1960).
102. K. T. Hecht, Use of SU(3) in Elimination of Spurious Center of Mass States, *Nucl. Phys. A* **170**, 34 (1971).
103. D. Millener, in *Group Theory and Special Symmetries in Nuclear Physics*, eds. J. Draayer and J. Janecke (World Scientific, Singapore, 1992), p. 276.
104. T. Dytrych, `LSU3shell` *code, Louisiana State University, available under the GNU General Public License at the git repository*, `http://sourceforge.net/projects/lsu3shell` (2013).
105. Y. Akiyama and J. P. Draayer, A User's Guide to FORTRAN Programs for Wigner and Racah Coefficients of SU(3), *Comput. Phys. Commun.* **5**, 405 (1973).
106. M. Jarrio, J. Wood and D. Rowe, *Nucl. Phys. A* **528**, 409 (1991).
107. D. Rowe, The fundamental role of symmetry in nuclear models, *AIP Conf. Proc.* **1541**, 104 (2013).
108. C. Bahri and D. J. Rowe, SU(3)quasi-dynamical symmetry as an organizational mechanism for generating nuclear rotational motions, *Nucl. Phys. A* **662**, 125 (2000).
109. D. Peterson and K. Hecht, *Nucl. Phys. A* **344**, 361 (1980).

110. A. Hotta, R. S. Hicks, R. L. Huffman, G. A. Peterson, R. J. Peterson and J. R. Shepard, Transverse isoscalar excitations in ^{24}Mg by 180° electron scattering, *Phys. Rev. C* **36**, 2212 (Dec 1987), doi:10.1103/PhysRevC.36.2212. URL `http://link.aps.org/doi/10.1103/PhysRevC.36.2212`.
111. G. C. Li, M. R. Yearian and I. Sick, High-momentum-transfer electron scattering from ^{24}Mg, ^{27}Al, ^{28}Si, and ^{32}S, *Phys. Rev. C* **9**, 1861 (May 1974), doi:10.1103/PhysRevC.9.1861. URL `http://link.aps.org/doi/10.1103/PhysRevC.9.1861`.
112. J. Escher and J. P. Draayer, Electron Scattering and Multi-$\hbar\omega$ Correlations, *Phys. Rev. Lett.* **82**, 5221 (1999).
113. K. D. Launey, A. C. Dreyfuss, J. P. Draayer, T. Dytrych and R. B. Baker, Emergence of cluster structures and collectivity within a no-core shell-model framework, *J. Phys.: Conf. Ser.* **569**, 012061 (2014).
114. P. J. Ellis and T. Engeland, *Nucl. Phys. A* **144**, 161 (1970).
115. T. Engeland and P. J. Ellis, *Nucl. Phys. A* **181**, 368 (1972).
116. Y. Suzuki and K. T. Hecht, Symplectic and Cluster Excitations in Nuclei: Evaluation of interaction matrix elements, *Nucl. Phys. A* **455**, 315 (1986).
117. H. O. U. Fynbo *et al.*, *Nature* **433**, 136 (2005).
118. M. Freer *et al.*, *Phys. Rev. C* **80**, 041303 (2009).
119. S. Hyldegaard *et al.*, *Phys. Rev. C* **81**, 024303 (2010).
120. M. Itoh *et al.*, *Phys. Rev. C* **84**, 054308 (2011).
121. W. R. Zimmerman, N. E. Destefano, M. Freer, M. Gai and F. D. Smit, *Phys. Rev. C* **84**, 027304 (2011).
122. A. R. Raduta *et al.*, *Phys. Letts. B* **705**, 65 (2011).
123. W. R. Zimmerman *et al.*, *Phys. Rev. Lett.* **110**, 152502 (2013).
124. D. J. Marin-Lambarri, R. Bijker, M. Freer, M. Gai, T. Kokalova, D. J. Parker and C. Wheldon, *Phys. Rev. Lett.* **113**, 012502 (2014).
125. Y. Kanada-En'yo, *Phys. Rev. Lett.* **81**, 5291 (1998).
126. Y. Funaki, A. Tohsaki, H. Horiuchi, P. Schuck and G. Ropke, *Phys. Rev. C* **67**, 051306 (2003).
127. T. Yamada and P. Schuck, *Eur. Phys. J. A* **26**, 185 (2005).
128. M. Chernykh, H. Feldmeier, T. Neff, P. von Neumann-Cosel and A. Richter, *Phys. Rev. Lett.* **98**, 032501 (2007).
129. D. T. Khoa, D. C. Cuonga and Y. Kanada-En'yo, *Phys. Letts. B* **695**, 469 (2011).
130. T. Neff and H. Feldmeier, *J. Phys.: Conf. Ser.* **569**, 012062 (2014).
131. L. Zamick, D. Zheng, S. J. Lee, J. A. Caballero and E. M. de Guerra, *Ann. Phys.* **212**, 402 (1991).
132. A. S. Umar, J. A. Maruhn, N. Itagaki and V. E. Oberacker, *Phys. Rev. Lett.* **104**, 212503 (2010).
133. R. Roth, J. Langhammer, A. Calci, S. Binder and P. Navrátil, *Phys. Rev. Lett.* **107**, 072501 (2011).
134. H. Horiuchi, K. Ikeda and K. Katō, Recent Developments in Nuclear Cluster Physics, *Prog. Theor. Phys. Suppl.* **192**, 1 (2012).
135. A. T. Y. Funaki, H. Horiuchi, *Prog. Part. Nucl. Phys.* **82**, 78 (2015).
136. E. Epelbaum, H. Krebs, T. Lähde, D. Lee and U.-G. Meißner, Structure and rotations of the Hoyle state, *Phys. Rev. Lett.* **109**, 252501 (2012). doi:10.1103/PhysRevLett.109.252501.
137. F. Ajzenberg-Selove and J. Kelley, *Nucl. Phys. A* **506**, 1 (1990).
138. M. Freer *et al.*, *Phys. Rev. C* **83**, 034314 (2011).
139. A. Ozawa, T. Suzuki and I. Tanihata, *Nucl. Phys. A* **693**, 32 (2001).

140. D. Tilley, C. Cheves, J. Kelley, S. Raman and H. Weller, *Nucl. Phys. A* **636**, 249 (1998).
141. R. Firestone, *Nuclear Data Sheets* **106**, 1 (2005).
142. I. Tanihata *et al.*, *Phys. Rev. Lett.* **55**, 2676 (1985).
143. A. N. Danilov, T. L. Belyaeva, A. S. Demyanova, S. A. Goncharov and A. A. Ogloblin, *Phys. Rev. C* **80**, 054603 (2009).
144. A. A. Ogloblin, T. L. Belyaeva, A. N. Danilov, A. S. Demyanova and S. A. Goncharov, *Eur. Phys. J. A* **49**, 46 (2013).
145. L. Chulkov *et al.*, *Nucl. Phys. A* **603**, 219 (1996).
146. T. Suzuki *et al.*, *Nucl. Phys. A* **630**, 661 (1998).
147. J. Vernotte, G. Berrier-Ronsin, J. Kalifa and R. Tamisier, *Nucl. Phys. A* **390**, 285 (1982).
148. R. LeBlanc, J. Carvalho, M. Vassanji and D. J. Rowe, *Nucl. Phys. A* **452**, 263 (1986).
149. T. Dytrych, A. C. Hayes, K. D. Launey, J. P. Draayer, P. Maris, J. P. Vary, D. Langr and T. Oberhuber, Electron-scattering form factors for ^{6}Li in the *ab initio* symmetry-guided framework, *Phys. Rev. C* **91**, 024326 (2015). doi:10.1103/PhysRevC.91.024326.
150. T. Dytrych, P. Maris, K. D. Launey, J. P. Draayer, J. P. Vary, M. Caprio, D. Langr, U. Catalyurek and M. Sosonkina, Efficacy of the SU(3) scheme for *ab initio* large-scale calculations beyond the lightest nuclei, *Comput. Phys. Commun.* **207** (2016). doi:10.1016/j.cpc.2016.06.006.
151. A. Shirokov, J. Vary, A. Mazur and T. Weber, Realistic Nuclear Hamiltonian: *Ab Exitu* Approach, *Phys. Lett. B* **644**, 33 (2007).
152. A. Ekström, G. Baardsen, C. Forssén, G. Hagen, M. Hjorth-Jensen, G. R. Jansen, R. Machleidt, W. Nazarewicz *et al.*, An optimized chiral nucleon-nucleon interaction at next-to-next-to-leading order, *Phys. Rev. Lett.* **110**, 192502 (2013).
153. D. H. Gloeckner and R. D. Lawson, *Phys. Lett. B* **53**, 313 (1974).
154. B. S. Pudliner, V. R. Pandharipande, J. Carlson, S. C. Pieper and R. B. Wiringa, *Phys. Rev. C* **56**, 1720 (1997).
155. B. S. Pudliner, V. R. Pandharipande, J. Carlson and R. B. Wiringa, *Phys. Rev. Lett.* **74**, 4396 (1995).
156. D. R. Tilley *et al.*, *Nucl. Phys. A* **708**, 3 (2002).
157. G. C. Li, I. Sick, R. R. Whitney and M. R. Yearian, *Nucl. Phys. A* **162**, 583 (1971).
158. I. Tanihata, *Phys. Lett. B* **206**, 592 (1988).
159. T. Dytrych, K. D. Launey, J. P. Draayer *et al.*, *to be submitted* (2015).
160. J. P. Draayer, T. Dytrych, K. D. Launey, A. C. Dreyfuss and D. Langr, Dominant Modes in Light Nuclei — Ab Initio View of Emergent Symmetries, in *Proceedings of the XXXVII Symposium on Nuclear Physics, January 6–9, 2014, Cocoyoc, Mexico*, J. of Phys.: Conf. Ser. Vol. 578, p. 012010 (2015).

Can One Determine the Neutrino Mass by Electron Capture?

Amand Faessler
Institute of Theoretical Physics, University of Tuebingen, 72076 Tuebingen, Germany
faessler@uni-tuebingen.de

Dedicated to the Memory of Walter Greiner (1935–2016)

There are three different methods used to search to determine the neutrino mass:

(1) The electron antineutrino mass can probably best be determined by the Triton decay.

(2) The neutrinoless Double Beta Decay yields information, if the neutrino is a Dirac or a Majorana particle. It can also determine the Majorana neutrino mass.

(3) Electron capture of an atomic bound electron by a proton in a nucleus $e+p \rightarrow n+\nu$ can give the mass of the electron neutrino.

This contribution discusses the possibility to determine the electron neutrino mass by electron capture. One expects the largest influence of the neutrino mass on this decay for a small $Q = 2.8$ keV ${}^{163}_{67}\text{Ho} + e \rightarrow {}^{163}_{66}\text{Dy} + \nu$. The energy of the Q value is distributed to the emitted neutrino and the excitation of the Dy atom. Thus the energy difference between the Q value and the upper end of the decay spectrum of Dy* is the electron neutrino mass. The excitation spectrum of Dy is calculated by one-, two- and three-electron hole excitations, and by the shake-off process. The electron wave functions are calculated self-consistently by the Dirac–Hartree–Fock approach for the bound and the continuum states. To extract the neutrino mass from the spectrum one must adjust simultaneously the neutrino mass, the Q value, the position,the relative strength and the width of the highest resonance. This fit is only possible, if the background is reduced relative to the present situation. The analysis presented here shows, that the determination of the electron neutrino mass by electron capture is difficult, but seems not to be impossible.

1. Introduction

Neutrinos are one of the least known particles, although they play a central role in weak interactions. We do not know the mass of the three neutrinos, electron, muon and tauon neutrinos (and antineutrinos). We do not know their hierarchy. Which one is the heaviest? The nature of the neutrinos, Dirac (neutrino different from the antineutrino) or Majorana (neutrino identical with the corresponding antineutrino) is still unknown. At the moment one searches with three different methods for the neutrino mass. The Tritium decay[1,3] can measure the antineutrino mass, while electron capture can determine the neutrino mass.[2,4–8] The neutrinoless Double Beta Decay can differentiate beween Dirac and Majorana neutrinos and can give the Majorana neutrino mass.[10–13]

Here I summarize the work on the determination of the electron neutrino mass by electron capture in ${}^{163}_{67}$Ho of the Tuebingen group and coworkers.[5,14–17] Electron capture is most sensitive for the neutrino mass for a small Q value: For the capture

$$^{163}\text{Ho} + e \to {}^{163}\text{Dy} + \bar{\nu} \tag{1}$$

the value obtained by a Penning trap measurement by Eliseev *et al.*[18] (see also Refs. 19–22) is:

$$Q = 2.833\,(30_{\text{stat}})\,(15_{\text{sys}})\,\text{keV}. \tag{2}$$

This means, that electron capture in ^{163}Ho is only possible from electron orbits bound less than $Q = 2.8$ keV. So capture is only possible from $3s_{1/2}$, $M1$ with a binding energy 2.128 keV and less bound electron states. In addition the electron orbits in Ho, from which an electron can be captured, must have an overlap with the nucleus to react with a proton. This is only the case for $(n > 2)s_{1/2}$ and the lower relativistic amplitude of the $(n > 2)p_{1/2}$ states. Thus a relativistic treatment of the electron wave functions is essential. The energy of the Q value after capture is distributed over the excitation energy of the Dy atom (one-, two-, three-hole, ... excitations), the emitted neutrino and the recoil of the Ho nucleus. The recoil of the Ho nucleus can be neglected and we have to consider the distribution of Q to the excitation of Dy and to the neutrino only. The excited Dy decays by X-rays and Auger electron emissions. The sum of these energies can be measured by a bolometer. The maximum bolometer energy yields then the minimum neutrino energy, which is the electron neutrino rest mass. Thus the neutrino mass can be determined by the difference of the Q value and the upper end of the bolometer spectrum. The excitation energy of the Dy* is calculated in the sudden approximation. The excitation of the Dy* is then given by the overlap with Ho* with an electron hole in the orbit of the capture with the different excited configurations $|D\rangle$ in the daughter Dy*. Lets estimate before we start the detailed calculations roughly an upper limit for the probability of exciting an electron in Dy* into the continuum (shake-off). We start with the normalization of the many electron one-hole state in Ho, from which the electron is captured and use the completeness relation:

$$\begin{aligned} 1 &= \langle P, (n, \ell_{1/2})^{-1}, \text{Ho}\,|P, (n, \ell_{1/2})^{-1}, \text{Ho}\rangle \\ &= \sum_{D,\text{Dy}} \langle P, (n, \ell_{1/2})^{-1}, \text{Ho}|D,\ \text{Dy}\rangle\langle L, \text{Dy}|P, (n, \ell_{1/2})^{-1}, \text{Ho}\rangle \end{aligned}$$

$$= |\langle P, (n, \ell_{1/2})^{-1}, \text{Ho}|P', \text{Dy}\rangle|^2 + \sum_{D \neq P} |\langle P, (n, \ell_{1/2})^{-1}, \text{Ho}|D, \text{Dy}\rangle|^2. \tag{3}$$

"P" stands here for configurations in the Ho parent (not to be mixed up with the large relativistic electron continuum $P_E(r)$ or bound state $P_b(r)$ amplitudes, which normally have a l.c. index) and "D" for a configuration in the Dy* daughter. The star indicates excited states. The first term in the last line of Eq. (3) is a diagonal overlap squared and thus the probability, that the electron Dy configuration $|P'\rangle$ with 66 electrons has a hole for the same quantum numbers from which the electron is captured in Ho (contributions to one-hole excitations). The prime reminds, that DHF electron orbits of Dy must be used. To get rid of the many electrons, which are practically not involved in the capture process and for which the wave functions have an overlap $\langle nlj, \text{Ho}|nlj, \text{Dy}\rangle \approx 0.999$ between Ho and Dy, we use the Vatai approximation[23,24] and put these overlaps for this first estimate to unity. For the final results in Secs. 2–5 we calculate these overlaps correctly and do not use the Vatai approximation.[23,24] For the first term in the last line of Eq. (3), which is the probability, that the configuration $|P, (n, \ell_{1/2})^{-1}, \text{Ho}\rangle$ is "not changing" in Dy, one has to consider the $(2j+1) = 2$ magnetic substates for the $ns_{1/2}$ and $np_{1/2}$ states, from which electrons can be captured. This yields for the first term in the last line of Eq. (3) an exponent $2(2j+1) = 4$. The last term of Eq. (3) contains the non-diagonal overlaps between Ho and Dy configurations and is the probability for shake-up and for the shake-off excitations in Dy* configurations $|D\rangle$ (different from the one-hole configurations $|P'\rangle$, contained in the first term of the last line of Eq. (3)) by moving electrons up into free bound states (shake-up) or into the continuum (shake-off).

The diagonal overlaps (first term of the last line of Eq. (3)) are very close to unity and are typically $\langle nlj, \text{Ho}|nlj, \text{Dy}\rangle = 0.999$, since the electron wave functions do only change minimal from $_{67}$Ho to $_{66}$Dy. To get a feeling how large the shake-off probability can roughly be, we use for a first estimate for the diagonal overlap as a measure for the deviation from the Vatai approximation $\langle P, (nlj)^{-1}, \text{Ho}|P', (nlj)^{-1}, \text{Dy}\rangle = 0.999$ (see Table 1). The probability to excite Dy* into a shake-off state is then less than one minus the probability, that in Dy the configuration $|P'\rangle$ is not changed. The probability to excite two-, three- and many-hole states (shake-up) and to excite continuum states (shake-off) is one minus the probability, that the configuration $|P\rangle$ of the excited Ho is not changed in Dy*:

$$\begin{aligned} \text{Prob}_{\text{shake-off}} &\leq [1.0 - \langle (n, \ell, j)^{-1},\ \text{Ho}|(n, \ell, j)^{-1},\ \text{Dy}\rangle^{2\cdot(2j+1)}] \\ &\approx [1.0 - 0.999^4] = 0.004 \equiv 0.4\%. \end{aligned} \tag{4}$$

Thus the shake-off process should be less than 0.4% relative to the one-hole probability. If perturbaton theory is used as in Ref. 8 to calculate this overlap of Ho and Dy states, one easily has a 10% error or more: $\langle (n, \ell, j)^{-1},\ \text{Ho}|(n, \ell, j)^{-1}, \text{Dy}\rangle \approx 0.9$:

$$\begin{aligned} \text{Prob}_{\text{shake-off}} &\leq [1.0 - 0.90^4] \approx 0.34 \equiv 34\%; \\ &34\% \text{ of the 1-hole states.} \end{aligned} \tag{5}$$

Table 1. Overlaps of the 3s and 4s wave functions of Ho with Dy. DHF (Dirac–Hartree–Fock) is the result of the self-consistent, relativistic calculation of this work. $1 - \langle \text{DHF} \rangle^4$ gives the probability, that Dy* is excited beyond the one-hole states including shake-up and shake-off. Thus the the shake-off must be less than 0.4% of the one-hole excitation probabilities in disagreement with Ref. 8. "Coulomb" shows the overlaps with non-relativistic Coulomb wave functions and effective charges $Z(3s)_{\text{eff}} = 54.9$ and $Z(4s)_{\text{eff}} = 43.2$ used by DeRujula and Lusignoli.[8] Here the Coulomb wave functions are calculated in Ho and Dy separately. In Refs. 8 and 29 the Coulomb wave functions in Dy are calculated with perturbation theory from Ho using expression (6) as perturbation. The diagonal overlaps show the right order of magnitude, but the non-diagonal elements of the overlap are completely wrong. Even the orthogonality of 3s and 4s in the same atom is not fulfilled.

−−−−−−−−	DHF	$1 - \langle \text{DHF} \rangle^4$	Coulomb	$1 - \langle \text{Coul} \rangle^4$
$\langle 3s, \text{Ho} \| 3s, \text{Dy} \rangle$	0.99940	0.00239	0.99932	0.00271
$\langle 4s, \text{Ho} \| 4s, \text{Dy} \rangle$	0.99909	0.00363	0.99848	0.00607
$\langle 3s, \text{Ho} \| 4s, \text{Dy} \rangle$	−0.01982	—	0.56828	—
$\langle 4s, \text{Ho} \| 3s, \text{Dy} \rangle$	0.02067	—	0.56817	—
$\langle 3s, \text{Ho} \| 4s, \text{Ho} \rangle$	0.0	—	0.56857	—
$\langle 3s, \text{Dy} \| 4s, \text{Dy} \rangle$	0.0	—	0.56952	—

Thus with such an error in the overlap the shake of probability can be about 34% of the one hole states and by about a factor 85 larger than a more correct estimate. This result requires very good electron wave functions for a reliable description of the capture process. DeRujula and Lusignoli[8] use non-relativistic pure Coulomb wave functions for the electrons in Ho, which do not allow capture from $np_{1/2}$. The effect of the partial screening of the nuclear Coulomb potential by the electron cloude is considered by effective charges. Since in Ref. 8 the perturbation approach is used to calculate the Dy* wave functions, this requires orthonormal basis functions. Thus Ref. 8 uses for each capture process, e.g. from $3s_{1/2}$, the same screening factor for all single electron orbits needed in the perturbation approach for this special capture process. This means the screening is paradoxically the same for inner and for outer electron shells for a specific capture process. Starting from these "bad" wave functions for Ho Ref. 8 calculates the bound and continuum wave functions for Dy* with the perturbation:

$$H'(r) = +1/r - \int d^3r_1 \, |\varphi_{3s/4s}(\vec{r}_1)|^2 / |\vec{r} - \vec{r}_1|. \tag{6}$$

The first term takes into account, that in Dy one has one proton less compared to Ho. The minus sign of the p–e interaction yields the plus sign in Eq. (6). The second term originates from the electron hole in the $3s_{1/2}$ or the $4s_{1/2}$ state, which are the only hole states considered in Ref. 8. Equation (3) serves as a lever to enlarge a small change in the overlap into a large change for the shake-up and shake-off probabilities. A 10% change in the overlap can produce an enlargement of almost a factor 100 for the shake-up and shake-off probabilities. Thus one needs very good electron wave functions, else the capture probabilities are not reliable.

We use here self-consistent Dirac–Hartree–Fock (DHF) wave functions calculated specifically for Ho* and Dy*.[25–27] The continuum electron wave functions in Yb* are determined[28] in the self-consistent $Z - 1 = 65$ electron potential orthogonal to all 65 bound electrons in Yb. The improvements relative to Inteman and Polok[29] and to DeRujula and Lusignoli[8] are:

- The sudden approximation[5,7,14–17] with self-consistent DHF wave functions for the Dy atom is used to determine the electron capture probability and not the less reliable first order perturbation theory.[8,29]
- We do not use for $^{163}_{67}$Ho non-relativistic screened Coulomb functions,[8,29] but calculate all electron wave functions by the relativistic, self-consistent DHF approach[25–27] with full antisymmetrization. Among many other advantages the electron orbitals are in this way all orthogonal.
- The wave function of the bound electrons in Dysprosium are again determined self-consistent and relativistic by DHF[25–27] even allowing for $3s$ and $4s$ hole states. In Refs. 29 and 8 the electron orbitals for the daughter Dy are calculated in first order perturbation theory (6) from the Coulomb functions of Ho.
- The continuum s-wave electron for shake-off in Dy is calculated relativistically in the self-consistent DHF potential of the self-consistent 65 electron core, under the condition, that the continuum s-orbitals are orthogonal the the bound s-orbits in Dy.
- Numerical stability is tested carefully. For the continuum electron wave functions in Dy for the radial coordinate 250 up to more than 700 mesh points were used depending on the energy. The integration over the continuum electron energies for the shake-off electron are performed from 0 to $Q = 2.8$ [keV] with 417 mesh points. Integrations for the norms, the overlaps and the integration over the shake-off in the continuum were done in parallel with the Trapez rule (error $\propto$ second derivative), the Kepler–Simpson rule (error $\propto$ fourth derivative), the Bode–Boole rule (error $\propto$ sixth derivative) and the Weddle rule. From the points of stability and accuracy the Bode–Boole's rule turned out to be the most reliable. All the calculations were done in double precision.
- The DHF overlaps of $\langle 3s, \mathrm{Ho}|3s, \mathrm{Dy}\rangle = 0.99940$ and $\langle 4s, \mathrm{Ho}|4s, \mathrm{Dy}\rangle = 0.99909$ (see Table 1) limit in the Vatai approximation[23,24] the 2-hole probability including the shake-off process, which requires a second hole, to 0.24% and 0.36% of the 1-hole excitations. An error of 10% in calculating the single orbital overlaps between Ho and Dy due to first order perturbation theory[8,29] estimated again with the Vatai approximation can increase the shake-off probability by two orders of magnitude. Equation (3) serves as lever to produce from a small uncertainty of the single electron overlaps a large increase of the shake-off probability. If one does not use the Vatai approximation and puts all electron orbital overlaps of Ho with Dy

to 0.999, the definite upper limit (including 1- and 2-hole and shake-off excitations) for shake-off is 12% relative to the 1-hole states. The norm gives without Vatai no restriction for the shake-off with an error of 10% for the $\langle n, \ell, j, \mathrm{Ho}|n, \ell, j, \mathrm{Dy}\rangle$ single electron overlaps.

- In this work the different 1-hole, 2-hole and shake-off contributions are taken from the theory without adjusting them in different ways to fit the experiment. In Ref. 8 the authors write:
  "Our estimate of the height of the N1(4s)O1(5s) shakeup peak is a factor ≈ 2.5 too low. It is possible to correct in similarly moderate ways the other contributions such as to agree with the data."

2. Description of Electron Capture

The spectrum of the the decay of the excited $^{163}\mathrm{Dy}^*$ after electron capture in $^{163}\mathrm{Ho}$ (Refs. 4, 8 and 5) assuming Lorentzian shapes for the decay resonances are:

$$\frac{d\Gamma}{dE_c} \propto \sum_{i=1,\ldots,N_\nu} (Q-E_c)\cdot U_{e,i}^2 \cdot \sqrt{(Q-E_c)^2 - m_{\nu,i}^2}$$
$$\times \left(\sum_{f=f'} \lambda_0 B_f \frac{\Gamma_{f'}}{2\pi} \frac{1}{(E_c - E_{f'})^2 + \Gamma_{f'}^2/4} \right.$$
$$+ \sum_{f=f';p'<F;q_b'>F} \lambda_0 B_{f,p'<F;q_b'>F} \frac{\Gamma_{f',p'}}{2\pi} \frac{1}{(E_c - E_{f',p'})^2 + \Gamma_{f',p'}^2/4}$$
$$\left. + \int dk_{q'} \lambda_0 B_{f,p'<F;q_c'>0} \frac{\Gamma_{f',p',q'}}{2\pi} \frac{1}{(E_c - E_{f',p',q'})^2 + \Gamma_{f',p',q'}^2/4} \right). \quad (7)$$

The factor in front of the bracket is the same as for the single beta decay. It reflects the phase space of the emitted neutrino (1). The three expressions in the bracket of Eq. (7) describe the decay of the 1-hole f' excitations in Dy*, of the 2-hole excitations f', p' with shake-up of p' to q' into a bound orbit and of 2-holes f', p' states with one electron p' moved to q' into the continuum (shake-off). The last term has as the others a dimensionless strength $dk_{q'} \cdot B_{f',p'<F,q'>o}$. The transformation to an integral over the energy yields the factor (21). Different neutrino mass eigenstates $|i\rangle$ are mixed into the electron neutrino flavor states by $U_{e,i}^2$. The Q value is given by Eq. (2) from the ECHo collaboration.[19,21,30–32] The excitation energy of the final Dy* is E_c. The energy $Q - E_c$ is carried away by the neutrino. B_f, $B_{f,p'<F;q_b'>F}$ and $B_{f,p'<F;q_c'>0}$ are the overlap and exchange corrections for the 1-hole, the bound 2-hole and the shake-off 2-hole states. λ_0 contains the nuclear matrix element squared.[33] Equation (7) yields the spectrum in arbitrary units and must be normalized at one energy to the data. We use the N1, $4s_{1/2}$ peak for the normalization. $E_{f'}$, $E_{f',p'}$ and $E_{f',p';q'>0}$ are the 1-hole, the 2-hole shake-up and the 2-hole shake-off excitation energies in Dysprosium (see

Tables 2 and 3). $\Gamma_{f'}$, $\Gamma_{f',p'}$ and $\Gamma_{f',p';q'>0}$ are the widths of the one- and two-hole states and the two-hole states with shake-off in Dysprosium.[14–17]

Table 2. Important electron binding energies in $^{163}_{67}$Ho.[34] Electrons can only be captured from orbitals overlapping with the nucleus. This restricts capture to $ns_{1/2}$ and in a relativistic treatment to the lower amplitude of $np_{1/2}$. Energy conservation requires, that the Q-value $Q = 2.8$ keV must be larger than the binding energy of the electron captured in Ho. This is the case for $3s_{1/2}$ and higher levels.

$n\ell_j$	Notation for electron orbit	Binding energy E_b [keV]
$1s_{1/2}$	$K1$	55.618
$2s_{1/2}$	$L1$	9.394
$2p_{1/2}$	$L2$	8.918
$3s_{1/2}$	$M1$	2.128
$3p_{1/2}$	$M2$	1.923
$4s_{1/2}$	$N1$	0.4324
$4p_{1/2}$	$N2$	0.3435

Table 3. Electron binding energies and width of two-hole states in $^{163}_{66}$Dy, which contribute to s-wave shake-off. Energy conservation requires, that the Q-value $Q = 2.8$ keV must be larger than the two–hole binding energy plus the energy of the electron in the continuum. The shake-off contributions for the 2-hole states start at the 2-hole binding energy in the bolometer spectrum as a function of E_c. The width includes only the contribution from the decay of the 2-hole states, but not the escape width of the continuum electron.

$n_1\ell_{j,1}, n_2\ell_{j,2}$	2-hole Binding Energy E_b [keV]	Width [keV]
$3s_{1/2}, 4s_{1/2}$	2.4742	0.0264
$4s_{1/2}, 4s_{1/2}$	0.8414	0.0108
$4s_{1/2}, 5s_{1/2}$	0.4583	0.0107
$4s_{1/2}, 3p_{1/2}$	2.2692	0.0114
$5s_{1/2}, 3p_{1/2}$	1.8861	0.0114
$3s_{1/2}, 4p_{1/2}$	2.3853	0.0186
$4s_{1/2}, 4p_{1/2}$	0.7525	0.0107
$5s_{1/2}, 4p_{1/2}$	0.3776	0.0106

The overlap and exchange corrections B_f, $B_{f,p'<F;q'_b>F}$ and $B_{f,p'<F;q'_c>0}$ are calculated in second quantisation. This yields automatically the antisymmetrisation and the exchange terms. To obtain these quantities we need:

$$|G\rangle = a_1^\dagger a_2^\dagger a_3^\dagger ... a_Z^\dagger |0\rangle. \tag{8}$$

The creation and anihilation operators of self-consistent DHF electron orbits in Dy $a_i'^\dagger$ and a_i' are characterized by a prime. The corresponding operators in Ho are unprimed

$$|A'_{f'}\rangle = a_1'^\dagger a_2'^\dagger ... a_{f'-1}'^\dagger a_{f'+1}'^\dagger ... a_Z'^\dagger |0\rangle. \tag{9}$$

The antisymmetrized two-hole state in Dy with shake-off is:

$$|A'_{p',f':q'>0}\rangle = a_1'^{\dagger} a_2'^{\dagger} ... a_{f'-1}'^{\dagger} a_{f'+1}'^{\dagger} ... a_{p'-1}'^{\dagger} a_{p'+1}'^{\dagger} ... a_Z'^{\dagger} a_{q'>0}'^{\dagger} |0\rangle. \tag{10}$$

The probability to form a two-hole shake-off state is proportional to:

$$P_{f',p';q'>0} = |\langle A'_{f',p';q'}|a_i|G\rangle|^2 \tag{11}$$

The relative shake-off probability normalized to the $3s$ one hole excitation is:

$$B_{f',p';q'>0} = \frac{|\psi_f(R)\langle A'_{f',p':q'<0}|a_f|G\rangle|^2}{|\psi_{3s1/2}(R)|^2} = P_{f',p';q'>0} \cdot \frac{|\psi_f(R)|^2}{|\psi_{3s1/2}(R)|^2}. \tag{12}$$

Normally the wave function of the captured Ho electron is taken for the nuclear matrix element at the origin. Here we take this electron wave function at the nuclear radius. Due to the weight r^2 of the integration this is a better choice

$$\begin{aligned} P_{f',p';q'>0} &= |\langle 0|a'_{q'} a'_Z ... a'_{p'+1} a'_{p'-1} ... a'_{f'+1} a'_{f'-1} ... a'_{1'} \cdot a_f \cdot a_1^{+} ... a_Z^{+}|0\rangle|^2 \\ &= |\langle A'_{p',f'<F;q'>F}(2 \text{ holes})|a_f|G\rangle|^2 \approx |\langle q'_{>0}|p_{<F}\rangle \cdot \prod_{k=1..Z\neq f,p} \langle k'|k\rangle|^2. \end{aligned} \tag{13}$$

To evaluate the probability for the shake-off process the excitations are restricted to s-waves

$$P_{f',p'} = \sum_{q'>F} |\langle p_{<F,\mathrm{Ho}}|q'_{>F,\mathrm{Dy}}\rangle\langle q'_{>F,\mathrm{Dy}}|p_{<F,\mathrm{Ho}}\rangle| \cdot \prod_{k=k'<F_{Dy}\neq f,p} |\langle k'_{\mathrm{Dy}}|k_{\mathrm{Ho}}\rangle|^2. \tag{14}$$

3. The Dy Continuum Wave Functions

The Dy continuum wave functions for the shake-off require the following:

- The continuum electrons move in the Coulomb potential of the Dy nucleus $V_{\text{nucleus}} = -66/r$ and of the self-consistent cloude of 65 core electrons allowing for the different empty states (Fig. 1).
- The relativistic continuum electron wave functions have to be orthogonal to all bound orbits in Dy. We determine them with the code of Salvat *et al.*[28] and Schmidt orthogonalization (Fig. 2).
- These continuum functions[28] are normalized to delta functions in wave numbers $2\pi \cdot \delta(k - k')$. To integrate over the excitation energy in the continuum (7), the continuum wave functions have to be normalized to energy delta functions $\delta(E - E')$.

The self-consistent Coulomb field for the shake-off electron is in atomic units:

$$V_{\text{shake-off}}(r) = \frac{-66}{r} + \sum_{k \text{ occupied } e} g'_k \int d^3 r \cdot \frac{|\varphi_k(\vec{r}')|^2}{|\vec{r} - \vec{r}'|}. \tag{15}$$

g'_ks are the number of bound electrons in the self-consistent occupied orbits $|k\rangle = |n, \ell, j\rangle$ of Dy. To determine the potential for the shake-off electrons in Dy, one

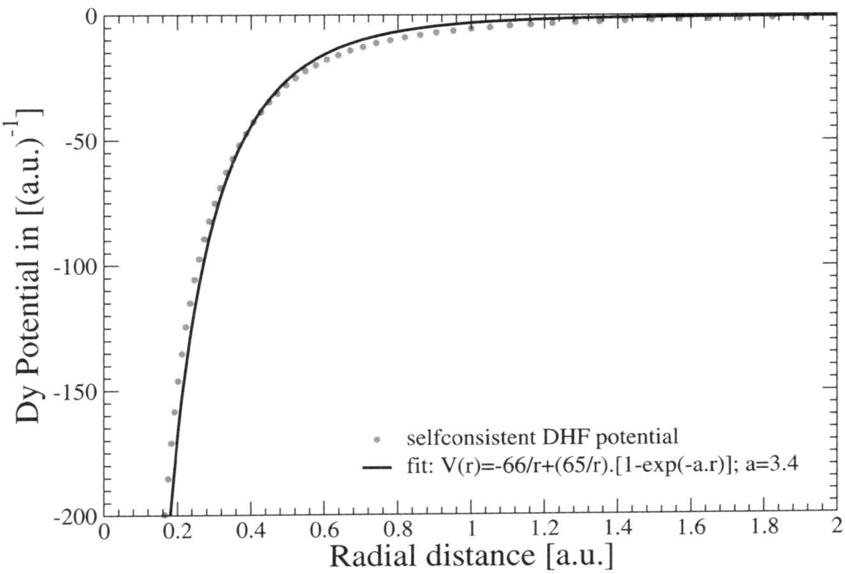

Fig. 1. Self-consistent DHF potential in Dy* with Z = 66 for a continuum electron (No 66) of 65 electrons (dimension: 1/length = atomic units) in the core and analytic approximation with $a = 3.4$ (see Eq. (17)).

needs the occupied self-consistent DHF orbitals $P_k(r)$ and $Q_k(r)$ (for examples of $P(r)$ see Fig. 2):

$$\varphi_k(\vec{r}) * r = (P_k(r); Q_k(r)), \tag{16}$$

$$V_{\text{shake-off}} \approx -\frac{66}{r} + \frac{65}{r} \cdot [1 - \exp(-a \cdot r)]\ [1/(\text{length} = \text{a.u.})]; \quad a = 3.4. \tag{17}$$

We adjust an analytic expression (17) to the DHF potential (see Fig. 1). The self-consistent Dy potential for the 66th electron in the continuum is shown in Fig. 1 for Dy.

The wave functions are normalized with the WKB approximation to:

$$\lim_{r \to \infty} P_k(r) \ = \ 2 \cdot \sin\left(kr - \ell \cdot \frac{\pi}{2} - \eta \ln 2kr + \Delta + \delta\right). \tag{18}$$

The normalization of the continuum Dirac wave function is treated in "Relativistic Electron Theory" (M. E. Rose[35]) or "Relativistic Quantum Mechanics" (Walter Greiner[36]). We follow here this recommendation of Perger and Karrighattam on their page 394.[37] In the asymptotic expression (18) η is the Sommerfeld parameter, Δ the Coulomb phase shift and δ takes into account deviations from a pure Coulomb potential. The electron energy in the continuum can be due to energy conservation not larger than the Q value of 2.8 keV minus the excitation energy of the two hole state. So for shake-off with capture from $3s$ the energy of the electron in the continuum must be less than 0.758 keV (extreme non-relativistic). The most

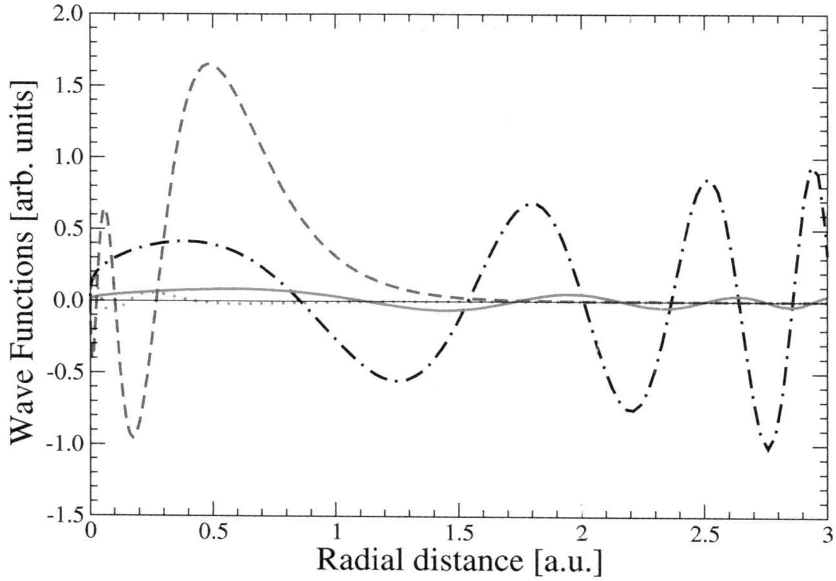

Fig. 2. The electron wave function $P_E(r)$ (dashed-dot) and $Q_E(r)$ (solid) (16) in the Dy continuum at 50 [Hartree] = 1.36 keV. The continuum wave functions P_E and Q_E are normalized to $\int dr[P_E(r)^2+Q_E(r)^2] = 2\pi\cdot\delta(k-k')$. The 4s bound state in Ho is normalized to unity (dashed for $P_{4s}(r)$ and dotted for $Q_{4s}(r)$): $\int dr[P_{4s}(r)^2 + Q_{4s}(r)^2] = 1$. The overlap $\langle(ns)_{-1},\ \text{Ho}|E,\ s,\ \text{Dy}\rangle$ squared is proportional to the shake-off process as a function of energy. The continuum wave functions P_E and Q_E are dimensionless, while the bound states $P_{4s}(r)$ and $Q_{4s}(r)$ are in atomic units $[(\text{a.u.})^{-1/2}]$.

important second hole is $4s1/2$ and thus for the two holes in 3s and 4s the integration over the shake-off electrons is restricted from zero to (2.800 − 2.474) keV = 0.236 keV. For capture from the two $4s1/2$ states the binding energy limits the integration in Eq. (7) to an upperlimit of 2.800 − 0.841 = 1.959 keV.

$$\int_{r=0\ to\ \infty} dr\cdot 2\cdot\sin\left(kr-\ell\cdot\frac{\pi}{2}-\eta\ln 2kr+\Delta+\delta\right) \times 2\cdot\sin\left(k'r-\ell\cdot\frac{\pi}{2}-\eta\ln 2kr+\Delta+\delta\right)\approx 2\pi\cdot\delta(k-k'). \tag{19}$$

The wave number is connected with the relativistic and non-relativistic energies by the equations:

$$E_{\text{rel}}^2 = c^2\hbar^2k^2+m^2c^4 \rightarrow\ c^2k^2+c^4\ \text{(in atomic units)};\quad E_{n-\text{rel}}=\frac{1}{2}k^2;$$
$$k=\alpha\sqrt{E_{n-rel}(E_{n-\text{rel}}+2\cdot c^2)};\text{with } c=1/\alpha=137.035999 \text{ in [atomic units]}. \tag{20}$$

The transformation from the asymptotic wave number normalization $P_k(r)$ to the asymptotic energy delta function normalization is:

$$P_E(r)=P_k(r)\cdot\sqrt{\frac{1}{2\pi}\cdot\frac{1}{c}\cdot\frac{\sqrt{k^2+c^2}}{k}}\approx P_k(r)\cdot\frac{1}{\sqrt{2\cdot\pi\cdot k}}, \tag{21}$$
$$\delta(E[\text{Hartree}]) = \delta(36.74932386\cdot E[\text{keV}]) = 0.027121138506\cdot\delta(E[\text{keV}]).$$

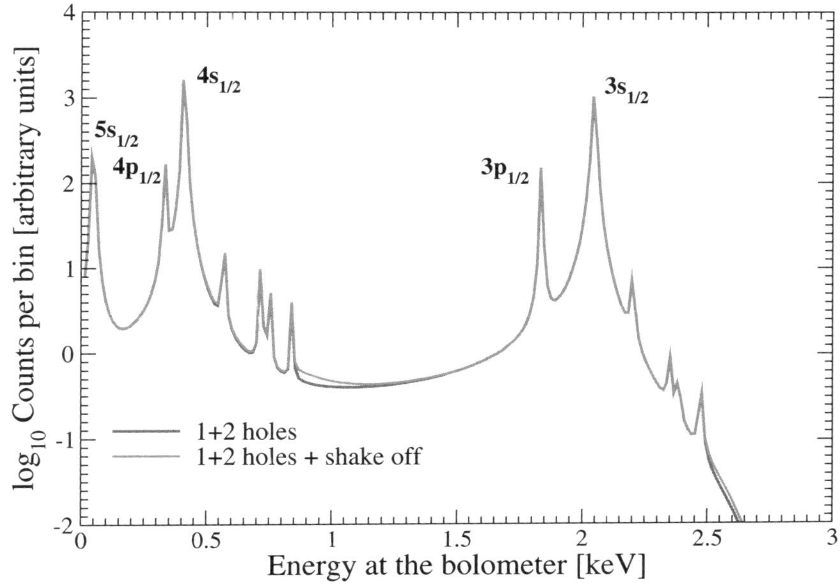

Fig. 3. Theoretical results in arbitrary units of the sum of the one- and two-hole deexcitations compared to the sum of the one-, two-hole and the shake-off deexcitation as measured by the bolometer spectrum (7). The arbitrary units are adjusted to the experimental N1, $4s_{1/2}$ 1-hole peak (see Fig. 10). The nature of the one-hole states are indicated. The two-hole peaks are by about two orders of magnitudes smaller than the one hole peaks. Shake-off can almost not been seen.

For shake-off one has to calculate the overlap of the bound Ho electron orbitals $|p\rangle$ with the continuum wave functions in Dy $|q'\rangle$, i.e. $\langle q'_{>0}|p_{<F}\rangle$ (13) (see Fig. 2). For this one expands the configuration of Ho after capture of the bound electron $|b, \text{Ho}\rangle$ with now the same number of protons as Dy (but not a Dy eigenstate) into the complete set of configurations $|D, \text{Dy}\rangle$ in Dy including the continuum:

$$
\begin{aligned}
|P,\ (b)^{-1}, \text{Ho}\rangle &= \sum_{D \neq P, \text{bound}} a_D \cdot |D,\ \text{Dy}\rangle + \int_{0 \text{ to } \infty} dE' \cdot a(D, E')|D,\ E', \text{Dy}\rangle, \\
a_D &= \langle D, \text{Dy}|(b)^{-1}|P,\ (b)^{-1}, \text{Ho}\rangle, \\
a(D, E'') &= \langle D, E'', \text{Dy}|P, (b)^{-1}, \text{Ho}\rangle \\
&= \int_{0 \text{ to } \infty} dE' \cdot a(D, E') \cdot \langle D, E'', \text{Dy}|D, E', \text{Dy}\rangle. \qquad (22)
\end{aligned}
$$

The probability forming a specific hole state $|k', \text{Dy}\rangle$ in Dy in a bound orbit after capture of the electron $|b, \text{Ho}\rangle$ is proportional to $|\langle k', \text{Dy}|P,\ (b)^{-1}, \text{Ho}\rangle|^2$ and and for the continuum $|E'', \text{Dy}\rangle$ to $|\langle E'', \text{Dy}|P,\ (b)^{-1}, \text{Ho}\rangle|^2$.

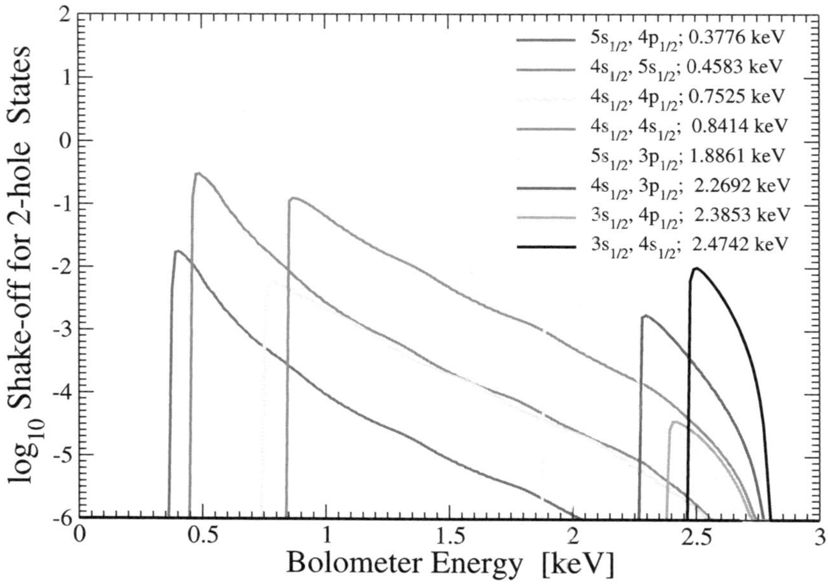

Fig. 4. Shake-off contributions for different 2-hole excitations in Dy normalized for the experimental bolometer spectrum (see Fig. 10) to the N1, $4s_{1/2}$ peak. Increasing the energy E_c of the bolometer spectrum the Q value $= 2.8$ keV is first used to excite the two hole state. So the shake-off contribution for the bolometer spectrum starts as function of E_c with the 2-hole binding energy. Energy conservation yields an upper limit of $Q = 2.8$ keV for the bolometer spectrum. To integrate over the continuum energy of the shake-off electron (7) we divided the interval $\langle 0.0\ ;\ 2.8\ keV \rangle$ into 417 mesh points. From the left to the right with increasing bolometer energy E_c the start of the different shake-off contributions are indicated in the figure. The energy difference between Q and E_c is carried away by the neutrino, which can not contribute to the bolometer spectrum.

4. Shake-off Probabilities

The overlap between the bound Ho states $ns_{1/2}$ and $np_{1/2}$ with $n \geq 3$ and the continuum wave functions in Dy squared gives the probability for shake-off. We restrict this work to s-wave shake-off, which is expected to be the largest contribution. The s-wave shake-off is given by the overlap $\langle n \geq 3, s_{1/2}, \mathrm{Ho}|E,\ s, \mathrm{Dy}\rangle$ squared. In the spectrum with 1-hole, 2-hole and shake-off in Figs. 3 and 5 the shake-off contribution is hardly visible. Here and also in Ref. 8 only the decay width of the 2-hole states are included. Three hole states can be neglected.[16] The electron in the continuum has an escape width, which is not included.

Figure 3 shows the logarithmic spectrum of the 1-hole, the 2-hole and the s-wave shake-off contributions. The shake-off contributions are calculated for the different 2-hole states listed in Table 3. The two-hole spectrum is about two orders of magnitude smaller than the one-hole states. The shake-off spectrum can hardly be seen on this scale in the total spectrum. Compared to the one-hole peaks it is at least two orders smaller. The integration over the continuum electron energy (7) is done by the Bode method using 417 mesh points. Shake-off is proportional to the

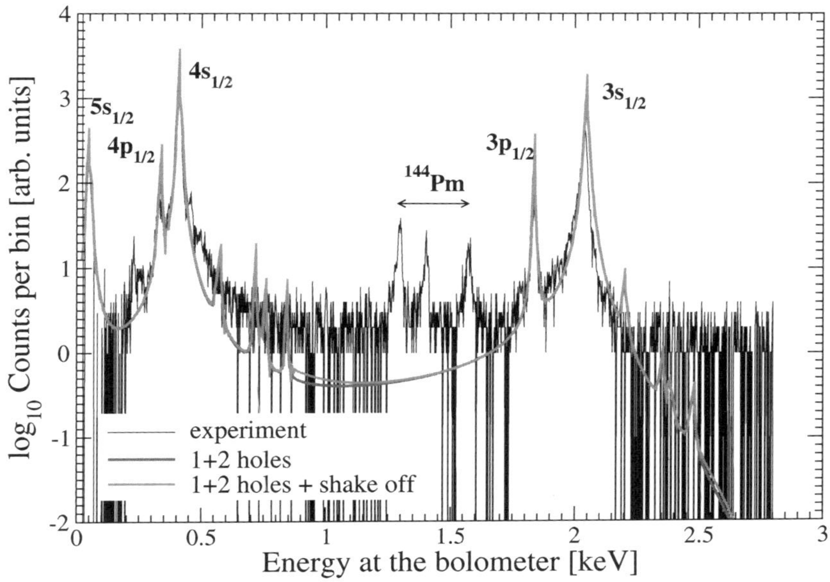

Fig. 5. Experimental and theoretical results of the sum of the one- and two-hole deexcitations and the sum of the one-, two-hole and the shake-off deexcitation for the bolometer spectrum (7). The experimental data are from the ECHo collaboration[21] and Ref. 19. The two theoretical spectra are adjusted to experiment at the $N1$, $4s_{1/2}$ peak. The nature of the one hole states are indicated. The two-hole peaks are by about two orders of magnitudes smaller than the one hole peaks. The shake off contributions can hardly been seen in this scale. Some bins contain no experimental counts, thus the $\log_{10}$ for these experimental values are minus infinity. To fit the 1931 experimental points for the bolometer energy of 0.0 to 2.8 keV, the theoretical spectrum of 200 mesh points had to be interpolated to the data points for this figure. Figure 3 contains the 200 original theoretical results without interpolation for the bolometer spectrum over E_c between 0.0 and 2.8 keV. The interpolation is normally very good (compare Figs. 3 and 5) but can be difficult at some sharp minima and maxima.

square of the overlap $\langle$Ho-bound$|$Dy-continuum$\rangle$. The 2-hole states contributing to s-wave shake-off are listed in Table 3. The shake-off contributions of the 2-hole states as function of the bolometer energy E_c is starting from the 2-hole binding energy up (see Table 3). The two main contributions originate from $4s_{1/2}$, $5s_{1/2}$ starting at 0.4583 keV and from $4s_{1/2}$, $4s_{1/2}$ starting at 0.8414 keV. The $\log_{10}$ contributions from $3s_{1/2}$, $4p_{1/2}$ starting at 2.33853 keV (see Table 3) are extremely small. The energy difference $Q - E_c$ is carried away by the neutrino and and does not show in the bolometer.

5. Summary

In this contribution to the memorial volume for Walter Greiner I summarized our work on the determination of the electron neutrino mass by electron capture[5,7,14–17] in $^{163}_{67}$Ho to $^{163}_{66}$Dy. The energy of the Q value $= 2.80 \pm 0.08$ keV is distributed to the excitation of Dy and the emission of a neutrino. The recoil of ^{163}Ho can be neglected. The deexcitation as X-rays and Auger electrons of Dy can be measured in a bolometer. The upper end of the spectrum requires the minimum energy of the neutrino. Thus the difference of Q minus this upper limit is the rest mass of the electron neutrino. For the excitation of Dy* we alowed one-hole excitations,[5] one- and two-hole excitations[17] (shake-up), three-hole excitations[16] (shake-up) and the excitation into the electron continuum[7] in Dy* (Shake-off). The excitation of Dy* is calculated by the overlap squared of Ho* with a hole in the orbit, from which the electron is captured, e.g. $3s_{1/2}, 3p_{1/2}, 4s_{1/2}, ...$, and the excited configurations in Dy*. Equations (3) and (4) show, that a small change of this overlap of about 10% can produce a change of the shake-up or shake-off by about a factor 100. Equations (3) and (4) serve as lever for large shake-up and shake-off effects. Inspite of this lever the estimate of (4) shows, that shake-up must be less the 0.4% of the probability for the one-hole excitations. The 1-hole and 2-hole electron configurations play the important role for the Dy* excitation. The three hole excitations and the excitation of a larger number of holes can be neglected.[16] The total 2-hole (shake-up) and continuum (shake-off) excitation probability is given in the sudden plus the Vatai[23,24] approximation by unity minus the overlap squared between Ho* minus the captured electron and Dy* configurations with the number of electrons in the exponent with the same caputure orbit quantum numbers in Dy as in Ho (3) and (4)

$$(1.0 - \langle \text{Ho}^*, n, \ell, j | \text{Dy}, n, \ell, j \rangle^{2(2j+1)}). \tag{23}$$

These Ho-Dy overlaps have in self-consistent relativistic Dirac–Hartree–Fock values of about 0.999 and even closer to unity (see Table 1). This yields the above already mentioned total 2-hole plus shake-off probability of about 0.4%. This estimate is only very approximate. (The final results in Secs. 2–5 are all calculated without the Vatai approximation.) Important is the result, that a small uncertainty for these electron overlaps between Ho and Dy produces a large increase for the 2-hole and the shake-off probability, which might be about two orders of magnitude for shake-off. Thus it is not permitted to use very approximate wave functions[8] for Ho* and Dy*. We describe the bound states in Ho and Dy by the Dirac–Hartree–Fock approach[25–27] even including occupations in Dy due to different hole states. The s-wave continuum wave functions in Dy are determined with the Dirac equation in the self-consistent potential[28] (see Figs. 1 and 2). The energy of the capture orbit in Ho and the excitations of Dy* including also the continuum states are limited by energy conservation to the Q value minus the excited one-hole states in Ho* and Q minus the excitation energies in Dy*. For example, the $3s_{1/2}$, $4s_{1/2}$ 2-hole state limits the upper bound of the continuum energy contributions to $2.8 - 2.4742 = 0.3258$ keV. So this contribution is very small. One of the two main contributions comes from the 2-hole state $4s_{1/2}$, $4s_{1/2}$ with a binding energy of 0.841 keV and thus an upper limit of the continuum energy of $2.8 - 0.841 = 1.959$ keV. A second large contribution is built on the 2-hole state $4s_{1/2}$, $5s_{1/2}$ with the binding energy 0.4398 keV. Thus the upper limit of the shake-off contributions in the continuum integration (7) is the

$2.8 - 0.4398 = 2.3602$ keV for the integral over the shake-off continuum electron. We prepared two different computer programs both calculating the s-wave shake-off to test the two codes against each other. All the calculations are done in double precision and for the critical integrations we use parallel the Trapez, the Kepler–Simpson, the Bode and the Weddle rules to test the accuracy. The numbers given are the ones from the Bode rule. (Trapez is not reliable enough.) The contributions from the shake-off process are small (see Figs. 3–5). The widths for the shake-off states include only the values from the 2-hole excitations as in Ref. 8. In reality one has to include the escape width of the electron in the continuum, which could perhaps even be larger than the contribution of the 2-hole states.
In summary this work shows, that one has not to worry about the three hole excitations and the shake-off process in $^{163}Dy*$ in the determination of the neutrino mass from electron capture in ^{163}Ho.
The remaining discrepancies between theory and experiment, e.g. the slope above the 1-hole state $4s_{1/2}$ $(N1)$, are probably due to configuration mixing not included here. Finally we want to stress, that the accuracy needed to extract the neutrino mass can not be obtained by theoretical calculation alone. One must fit simultaneously the neutrino mass, the Q value, the excitation energy of the highest resonance, their width and their strength at the upper end of the spectrum to extremely accurate data. For this the present background of the experimental spectrum has to be drastically reduced. The measurement must probably be done in an underground lab to protect against cosmic rays. This reduces the background as a first measurement in the Modane underground lab in the Fréjus tunnel has shown.[38] If the background can be reduced by one to two orders, it should be enough to adjust only the two parameters, the Q value and the neutrinos mass, to the upper end of the spectrum for the determination of the electron neutrino mass.

Acknowlegments

I want to thank Christian Enss, Loredana Gastaldo and Fedor Šimkovic for their collaboration in earlier works[5,14–17] on this topic.

References

1. G. Drexlin, V. Hannen, S. Mertens, C. Weinheimer, *Adv. High Energy Phys.* **2013**, 293986 (2013), arXiv 1307.0101v1 (2013).
2. A. Faessler, E. Huster, O. Kraft, F. Krahn, *Z. Phys.* **238**, 352 (1970).
3. E. W. Otten, C. Weinheimer, *Rep. Prog. Phys.* **71**, 086201 (2008).
4. A. De Rujula, arXiv 1305.4857v1 [hep-ph] 21 May 2013, unpublished. A. De Rujula, M. Lusignoli, *Phys. Lett. B* **118**, 429 (1982).
5. A. Faessler, L. Gastaldo, F. Šimkovic, *J. Phys. G* **42**, 015108 (2015) and arXiv: 1407.6504 [nucl-th] (2014).
6. R. G. H. Robertson, arXiv: 1411.2906v1 [nucl-th] (11 Nov 2014).
7. A. Faessler, arXiv1501.04338 and to be published.

8. A. De Rujula, M. Lusignoli, *JHEP* **1605** (2016) 015; arXiv: 1601.04990 [hep-ph].
9. A. Faessler, R. Hodak, S. Kovalenko, F. Šimkovic, arXiv 1304.5032v3 (2013) and *European Physical Journal Web of Conferences*.
10. A. Faessler, *Nucl. Phys. Proc. Suppl.* **188**, 20-24 (2009).
11. F. Šimkovic, A. Faessler, V. Rodin, P. Vogel, J. Engel, *Phys. Rev. C* **77**, 045503 (2008).
12. P. Vogel, J. Engel, *Phys. Rev. C* **77**, 045503 (2008).
13. F. Avignone, S. R. Elliott, J. Engel, *Rev. Mod. Phys.* **80**, 481-516 (2008).
14. A. Faessler, L. Gastaldo, F. Šimkovic, arXiv: 1611.00325 [nucl-th].
15. A. Faessler, F. Simkovic, *Phys. Scripta* **91**, 043007 (2016).
16. A. Faessler, Ch. Enns, L. Gastaldo, F. Šimkovic, *Phys. Rev. C* **91**, 064302 (2015).
17. A. Faessler, F. Šimkovic, *Phys. Rev. C* **91**, 045505 (2015).
18. S. Eliseev, K. Blaum, M. Block, S. Chenmarek, H. Dorrer, Ch. E. Duellmann, C. Enss, P. E. Filiamin, L. Gastaldo, M. Goncharov, U. Koester, F. Lautenschlaeger, Yu. N. Novikov, A. Rischka, R. X. Schuessler, L. Schweighard, A. Tuerler, *Phys. Rev. Lett.* **115**, 062501 (2015).
19. P. C.-O. Ranitzsch *et al.*, *Jour. of Low Temp. Phys.* **167**, 1004 (2012) and private communication.
20. L. Gastaldo, P. Ranitzsch, F. von Seggern, J. P. Porst, S. Schafer, C. Pies, S. Kempf, T. Wolf, A. Fleischmann, C. Enss, A. Herlert, K. Johnston, *Nucl. Inst. Methods A* **711**, 150-159 (2013) and arXiv: 1206.5647[physics.ins-det] (2012).
21. K. Blaum, A. Doerr, C. E. Duellmann, K. Eberhardt, S. Eliseev, C. Enss, A. Faessler, A. Fleischmann, L. Gastaldo, S. Kempf, M. Krivoruchenko, S. Lahiri, M. Matai, Yu. N. Novikov, P. C. Ranitzsch, F. Šimkovic, Z. Scusc, M. Wegner, arXiv 1306.2655v1 [physics.ins-det] (2013), unpublished.
22. P. C. Ranitzsch, C. Hassel, M. Wegner, S. Kempf, A. Fleischmann, C. Enss, L. Gastaldo, A. Herlert, K. Johnston, arXiv: 1409.007v1 (2014).
23. E. Vatai, *Nucl. Phys. A* **156**, 541 (1970).
24. E. Vatai, *Nucl. Phys. A* **402**, 1 (1983).
25. I. P. Grant, *Adv. Phys.* **19**, 747 (1970).
26. J. P. Desclaux, *Comp. Phys. Com.* **9**, 31-45 (1975).
27. A. L. Ankudinov, S. I. Zabinsky, J. J. Rehr, *Comp. Phys. Com.* **98**, 359-364 (1996).
28. F. Salvat, J. M. Fernandez-Varea, W. Williamson, *Comp. Phys. Com.* **90**, 151-168 (1995).
29. R. L. Intemann, F. Polock, *Phys. Rev.* 157, 41 (1967).
30. J. U. Anderson *et al.*, *Phys. Lett. B* **398**, 72 (1982).
31. F. Gatti *et al.*, *Phys. Lett. B* **398**, 415 (1997).
32. G. Audi, A. H. Wapstra, C. Thibault, *Nucl. Phys. A* **279**, 337 (2003).
33. W. Bambynek, H. Behrens, M. H. Chen, B. Crasemann, M. L. Fitzpatrick, K. W. D. Ledingham, H. Genz, M. Mutterer, R. L. Intemann, *Rev. Mod. Phys.* **49**, 109 (1977).
34. G. P. Williams, *X-Ray Data Booklet*, Lawrence Berkeley National Laboratory, Table 1-1.
35. M. E. Rose, *Relativistic Electron Theory* (John Wiley and Sons, New york, 1961).
36. W. Greiner, *Relativistische Quantenmechanik* (Band 6; Harri Deusch, Frankfurt, 1987).
37. W. F. Perger, V. Karighattam, *Comp. Physics Com.* **66**, 392 (1991).
38. L. Gastaldo, private information.

Open and Hidden Charm in My Collaboration with Walter Greiner

Mark I. Gorenstein

Bogolyubov Institute for Theoretical Physics, 03680 Kiev, Ukraine
Frankfurt Institute for Advanced Studies, D-60438 Frankfurt am Main, Germany

During 2000–2003 we published several papers devoted to charm and charmonium production in collaboration with Walter Greiner. In this presentation I am trying to recall the main steps of these studies.

PACS numbers: 25.75.Gz, 25.75.Ag, 21.65.Mn

1. Introduction

Walter Greiner played a great role in the development of nuclear physics, education of students and young scientists, and international scientific cooperation. His support was extremely important for me and changed crucially my life. I met Walter Greiner for the first time in 1987. This happened at the international conference in the USA. I was about 40 years old but it was only my second trip abroad which become possible due to the Gorbachjev's "Perestrojka". Walter Greiner came to me after my talk and invited me to visit his Institute of Theoretical Physics at Frankfurt University. During the next few months I managed to obtain a permission for this visit and came to Frankfurt for three months. For the next 30 years Frankfurt became my second native city after Kiev. Due to Greiner's support I spent in Frankfurt on the whole about 10 years.

My friendship with Walter was very important for me. I felt myself much stronger because this great man was my friend. Walter manifested his friendship in many different ways. Let me give one example. In 2007 our Kiev Institute organized an international conference in Crimea, during which one of the events was devoted to celebrating my birthday. When Walter knew about this event, he decided to come to the conference. It was a difficult decision. The conference was not so much important for Walter, the trip from Frankfurt to Crimea was not an easy one, and Walter was not more a young man. Nevertheless, Walter came and stayed for two days. He gave a talk and congratulated me on my birthday. Not only me but all my Ukrainian colleagues were very much impressed by that generous move of Walter Greiner.

2. J/ψ Suppression in Nucleus–Nucleus Collisions

Charmonium production in hadronic and nuclear collisions is usually considered to be composed of three stages: (1) creation of a $c\bar{c}$ quark pair, (2) formation of a bound $c\bar{c}$ state, and (3) subsequent interaction of this bound state with the surrounding matter. The first process is usually calculated within perturbative QCD, whereas modeling of non-perturbative dynamics is needed to describe the last two stages. The interaction of the bound $c\bar{c}$ state with matter causes suppression of the finally observed charmonium yield relative to the initially created number of bound $c\bar{c}$ states. This initial number is assumed to be proportional to the number of Drell–Yan lepton pairs, which then allows for the experimental study of the charmonium suppression pattern. It was proposed that the magnitude of the measured suppression in nuclear collisions can be used as a probe of the state of high density matter created at the early stage of the collision.[1] The suppression of the J/ψ yield observed in proton–nucleus (p+A) and light nuclei collisions at the CERN SPS was considered to be caused by interactions with nuclear nucleons.

The charmonium states J/ψ and ψ' were measured in nucleus–nucleus (A+A) collisions at CERN SPS by the NA38 and NA50 Collaborations. This experimental program was motivated by a suggestion to use the J/ψ as a probe of the state of matter created in the early stage of the collision. In this approach a significant suppression of J/ψ production relative to Drell–Yan lepton pairs is predicted when going from peripheral to central Pb+Pb interactions at 158 A GeV. The rapid increase of the suppression (anomalous suppression) observed when going from peripheral to central Pb+Pb collisions was attributed to the formation of a quark–gluon plasma (QGP).[1] This theoretical suggestion was confirmed by the data and it was declared as the most important signature of the QGP when the discovery of the new state of matter was announced in 2000 at CERN.

Alternative interpretations however were still under discussion. That time was rather hot in the experimental and theoretical studies of the possible QGP signals in central Pb+Pb collisions. I worked in Frankfurt and had the advantage of discussion with Walter Greiner new suggestions concerning the charm and charmonium production. Around 2000, almost each week the controversial ideas appeared. Walter did not like to elaborate suggestions of other people. Instead, he generated new ideas and encouraged all his colleagues and collaborators to follow their own ways — the more unexpected the better. Walter had a gift of prescience, he felt of importance in many out-of-the-way ideas much earlier than the surrounding scientific community.

3. Statistical Production of J/ψ

It was found that the mean multiplicity of J/ψ mesons increases proportionally to the mean multiplicity of pions when proton–proton (p+p), p+A, and A+A collisions at CERN SPS energies are considered. Note that the scaling of the ratio,

$$\frac{\langle J/\psi \rangle}{\langle h^- \rangle} = \text{const}(A) \ , \tag{1}$$

is valid also for central Pb+Pb collisions at the CERN SPS. In the standard picture of the J/ψ production based on the hard creation of $c\bar{c}$ pairs and the subsequent suppression of the bound $c\bar{c}$ states, the observed scaling behavior of the J/ψ multiplicity appears to be due to an 'accidental' cancelation of several large effects. This motivated our effort to find an alternative production mechanism of J/ψ mesons that would explain the experimental data in a natural way. It was argued that the scaling property (1) of the J/ψ multiplicity could be understood assuming that the dominant fraction of J/ψ mesons was produced directly at hadronization according to the available hadronic phase space.[2] For a long time statistical models are used to describe hadron multiplicities in high-energy collisions. Thermal hadron production models have been successfully used to fit the data on particle multiplicities in A+A collisions at the CERN SPS energies. Due to the large number of particles, the grand canonical formulation is used for modeling high-energy heavy-ion collisions. An impressive success of the statistical model applied to hadron multiplicities in elementary interactions at high energy was also reported. However, in the latter case the use of a canonical formulation of the model, which assures exact conservation of the conserved charges, is necessary. The temperature parameter which characterizes the available phase space for the hadron production is found in these interactions to be 160–180 MeV. It does not show any significant dependence on the type of reaction and on the collision energy. Based on the above facts, we formulated a hypothesis that a dominant fraction of the J/ψ mesons produced in hadronic and nuclear collisions at the CERN SPS energies was created at hadronization according to the available hadronic phase space.[2] J/ψ mesons are neutral and unflavored, i.e., all charges conserved in the strong interaction (electric charge, baryon number, strangeness and charm) are equal to zero for this particle. Therefore, its production is not affected by the conservation laws of quantum numbers. For sufficiently high collision energies, the effect of the strict energy–momentum conservation in the statistical model formulation can be neglected. Consequently, the J/ψ production can be calculated in the grand canonical ensemble (GCE) approximation and, therefore, its multiplicity is proportional to the volume, V, of the matter at hadronization. Thus, the statistical yield of J/ψ mesons at hadronization is given by

$$\langle J/\psi \rangle = (2j+1) \ V \ \left(\frac{T m_{J/\psi}}{2\pi}\right)^{3/2} \exp\left(-\ \frac{m_{J/\psi}}{T}\right) \ , \tag{2}$$

where $j = 1$ and $m_{J/\psi} = 3097$ MeV are the spin and the mass of the J/ψ meson, and T is the hadronization temperature. The previously mentioned results of the analysis of hadron yield systematics in elementary and nuclear collisions within the statistical approach indicate that the hadronization temperature is the same for different colliding systems and collision energies. This reflects the universal feature of the hadronization process. The total entropy of the produced matter is proportional to its volume. As most of the entropy in the final state is carried by pions, the pion multiplicity is also expected to be proportional to the volume of the hadronizing matter. Thus, the scaling property (1) follows directly from the hypothesis of statistical production of J/ψ mesons at hadronization and the universality of the parameter T.

The statistical J/ψ multiplicity (2) depends on two parameters, T and V. In general the calculation of the hadron yields in the statistical model should take into account the conservation of charges and resonance feeddown contributions. However, a simple way to estimate the temperature parameter in Eq. (2) from the experimental data is possible, provided that we find a second hadron which has the properties of the J/ψ meson, i.e., it is neutral, unflavored and stable with respect to strong decays. The best candidate is the η meson. The multiplicity of η mesons seems to obey also the scaling properties (1). Thus, the measured J/ψ to η ratio can be compared to the ratio calculated using Eq. (1):

$$\frac{\langle J/\psi\rangle}{\langle\eta\rangle} = 3\left(\frac{m_{J/\psi}}{m_\eta}\right)^{3/2} \exp\left(-\frac{m_{J/\psi} - m_\eta}{T}\right), \tag{3}$$

where $m_\eta = 547$ MeV is the mass of the η meson. This leads to an estimate of the hadronization temperature, $T = 170$ MeV. The maximal systematic error on T was estimated of about 10 MeV. The high sensitivity of the estimate of the T parameter by using the ratio is due to the large difference between mass of the J/ψ and the η mesons. The J/ψ production in hadronic and nuclear collisions at the SPS energy could be therefore understood assuming that a dominant fraction of J/ψ mesons was produced at hadronization according to the available hadronic phase-space. The estimate of the hadronization temperature based on J/ψ multiplicity, $T = 170$ MeV, agreed well with the values of the temperature parameter obtained from the analysis of the hadron yield systematics in A+A collisions. The consequences of the statistical model[2] are: (1) J/ψ yields are not sensitive to the state of high density matter created at the early stage of A+A collisions because their production takes place at hadronization. (2) The yield of J/ψ mesons is independent of the production of open charm, which is carried mainly by the D mesons in the final state.

4. Statistical Coalescence Model

The statistical approach, formulated in Ref. 2, assumes that J/ψ mesons are created at hadronization according to the available hadronic phase-space. In this model the J/ψ yield is independent of the open charm yield. The model offers a natural explanation of the proportionality of the J/ψ and pion yields and the magnitude of the J/ψ multiplicity in hadronic and nuclear collisions. The statistical coalescence model[3] was introduced for the charmonium production. Similar to the statistical model,[2] the charmonium states are assumed to be formed at the hadronization stage. However, they are produced as a coalescence of created earlier $c\bar{c}$ quarks and therefore the multiplicities of open and hidden charm hadrons are connected in the coalescence model. The charm quark–antiquark pairs are assumed to be created at the early stage of A+A collision and the average number of $c\bar{c}$ pairs, $Nc\bar{c}$, is fixed by the model consideration based on the hard scattering approach. The estimated number $Nc\bar{c}$ seems to be larger than the equilibrium hadron gas (HG) result. This requires the introduction of a new parameter in the HG approach — the charm enhancement factor γ_c (it was denoted as g_c in Ref. 3). This is analogous to the introduction of strangeness suppression factor γ_s in the HG model, where the total strangeness observed is smaller than its thermal equilibrium value. Within this approach the open charm hadron yield is enhanced by a factor γ_c and charmonium yield by a factor γ_c^2 in comparison with the equilibrium HG predictions. The thermal HG calculations in Ref. 3 were done within the GCE formulation. However, the validity of the GCE results are not valid for the charm hadron yield as the total number of charm hadrons is expected to be smaller than unity even for the most central Pb+Pb collisions at SPS energies. Therefore, exact charm conservation within the canonical ensemble (CE) should be imposed. Assuming the presence of the charm enhancement factor γ_c, the statistical coalescence model within the CE was formulated as[4]

$$N_{c\bar{c}} \;=\; \frac{1}{2}\gamma_c\, N_O\, \frac{I_1(\gamma_c N_O)}{I_0(\gamma_c N_O)} \;+\; \gamma_c^2\, N_H \;, \tag{4}$$

where I_0 and I_1 are the modified Bessel functions. Just these Bessel functions are consequences of the CE charm conservation. The GCE hadron multiplicities N_j are calculated as

$$N_j \;=\; d_j\, V \exp\left(\frac{\mu_j}{T}\right) \left(\frac{m_j T}{2\pi}\right)^{3/2} \exp\left(-\,\frac{m_j}{T}\right) \;, \tag{5}$$

where d_j and μ_j are the degeneracy factor and chemical potential of the jth hadron species, respectively. The quantities N_O and N_H in Eq. (4) are calculated as sums of N_j (5) over all hadronic states with open and hidden charm, respectively. The Boltzmann approximation and condition $m_j \gg T$ valid for all charm and charmonium states are used in Eq. (5). Therefore, the baryonic number, strangeness, and electric charge of the HG system are treated according to the GCE, but charm is considered in the CE formulation.

The HG model fit to the hadron yields data in Pb+Pb collisions at 158 A GeV leads to $T \cong 170$ MeV and $\mu_B \cong 250$ MeV. For a fixed number of participants N_p the volume V is defined from $N_p = V n_B(T, \mu_B)$, where n_B is the baryonic density calculated in the GCE HG model.

Such a consideration differs from that of Ref. 3 at three points. First, we use the exact charm conservation within the CE. Second, in our calculations we take into account all known particles and resonances with open and hidden charm. Third, we proceed with Eq. (4) in the reverse way. As the $\langle J/\psi\rangle$ multiplicities can be extracted from the NA50 data on Pb+Pb collisions at 158 A GeV for different values of N_p, we start from the requirement:

$$\langle J/\psi \rangle \;=\; \gamma_c^2 \left[N_{J/\psi} \;+\; \mathrm{Br}(\psi')N_{\psi'} \;+\; \mathrm{Br}(\chi_1)N_{\chi_1} \;+\; \mathrm{Br}(\chi_2)N_{\chi_2} \right] \;, \tag{6}$$

where $N_{J/\psi}$, $N_{\psi'}$, N_{χ_1}, and N_{χ_2} are given by Eq. (5) and $\mathrm{Br}(\psi') = 0.54$, $\mathrm{Br}(\chi_1) = 0.27$, and $\mathrm{Br}(\chi_2) = 0.14$ are the branching ratios of the excited charmonium states into J/ψ. Equation (4) is then used to calculate the value of $N_{c\bar{c}}$. This value is considered as a prediction of the statistical coalescence model. From the J/ψ multiplicity data in Pb+Pb collisions at 158 A GeV, the open charm yield was predicted: $N_{c\bar{c}} = 0.5$–0.6 in central ($N_p = 360$) collisions. An uncertainty of this prediction is mainly because of uncertainties in different compilations of the $\langle J/\psi \rangle$ data. Note that the perturbative QCD estimates suggested values of $N_{c\bar{c}}^{\mathrm{dir}} = 0.15$–$0.3$ in central Pb+Pb collisions at 158 A GeV (a value of $N_{c\bar{c}} = 0.17$ was used as input in Ref. 3). Assuming that $N_{c\bar{c}}^{\mathrm{dir}}$ scales as N_p^{α}, a value of $\alpha = 1.6$–1.7 was found.[4] This value is again larger than $\alpha = 4/3$ expected in the hard-collision model. Although the values of $\langle J/\psi \rangle$, N_O, and γ_c are rather sensitive to the temperature parameter, the model predictions for $N_{c\bar{c}}$ were found to be essentially unchanged.

Different aspects of the statistical coalescence model were discussed in Refs. 5–9. As an example let us consider the ratio

$$R \;=\; \frac{\langle J/\psi \rangle}{N_{c\bar{c}}} \;. \tag{7}$$

From Eq. (4) one finds at $N_{c\bar{c}} \ll 1$ (small collision energies and/or peripheral A+A collisions)

$$R \sim \frac{1}{V} \sim \frac{1}{\langle N_\pi \rangle} \sim \left(\sqrt{s_{NN}}\right)^{-1/2} N_p^{-1} \;. \tag{8}$$

On the other hand, at $N_{c\bar{c}} \gg 1$ (large collision energies and central A+A collisions) the ratio R increases with centrality,

$$R \sim \frac{N_{c\bar{c}}}{V} \sim \frac{N_{c\bar{c}}}{\langle N_\pi \rangle} N_p^{1/3} \;. \tag{9}$$

It also increases with collision energy, as an increase of $N_{c\bar{c}}$ with $\sqrt{s_{NN}}$ within the perturbative QCD is much stronger than the increase of the pion multiplicity.

5. Transverse Momentum Spectra of Charmed Hadrons

It was suggested[10] within statistical model of J/ψ production that the kinetic freeze-out of J/ψ and ψ' mesons takes place directly at hadronization. Therefore, these charmonium states carry information on the flow velocity of strongly interacting matter just after the transition from the QGP to the HG. Based on the measured J/ψ and ψ' spectra in Pb+Pb collisions at 158 AGeV the mean transverse collective flow velocity of hadronizing matter was estimated as $\bar{v}_T = 0.2$. On the other hand, the results of transport model calculations suggested that the transverse momentum spectrum of Ω was only weakly affected at the late stage of expansion even for central Pb+Pb collisions at the top SPS energy. This is because of the small hadronic cross section and large mass of the Ω hyperon.

If the above are correct for both charmonia and Ω, their p_T spectra should be simultaneously reproduced using the same hydrodynamic parameters, T and $\bar{v}_T$. It was demonstrated in Ref. 11 that such a description is indeed possible. Assuming kinetic freeze-out of matter at constant temperature T, the transverse mass m_T spectrum ($m_T = \sqrt{m_i^2 + p_T^2}$) of i-th hadron species (with mass m_i) in cylindrically symmetric and longitudinally boost invariant fluid expansion can be approximated as

$$\frac{dN_i}{m_T\, dm_T} \sim m_T \int_0^R r dr K_1\left(\frac{m_T \cosh y_T}{T}\right) I_0\left(\frac{p_T \sinh y_T}{T}\right) \tag{10}$$

where $y_T = \tanh^{-1} v_T$ is the transverse fluid rapidity, R is the transverse system size, K_1 and I_0 are the modified Bessel functions. The spectrum (10) is obtained under assumption that the kinetic freeze-out occurs at constant longitudinal proper time $\tau = \sqrt{t^2 - z^2}$, where t is the time and z is the longitudinal coordinate. Thus, the freeze-out time t is independent of the transverse coordinate r.

In order to calculate (10), the function $y_T(r)$ has to be given. A linear flow profile, $y_T(r) = y_T^{\max}\, r/R$, is often assumed in phenomenological fits. The numerical calculations justify this assumption. For heavy hadrons analyzed here the condition $m_i \gg T$ is always satisfied and, therefore, the asymptotic form for large arguments of $K_1(x) = x^{-1/2}\exp(-x)$ can be used in Eq. (10). At SPS energies typical values of v_T are small ($v_T^2 \ll 1$) and consequently $\cosh y_T \cong 1 + v_T^2/2$ and $\sinh y_T \cong vT$. The experimental m_T-spectra are usually parametrized with a function:

$$\frac{dN_i}{m_T\, dm_T} \sim \sqrt{m_T}\, \exp\left(-\frac{m_T}{T^*}\right) , \tag{11}$$

by introducing an effective temperature T^*. When Eq. (11) is considered as an approximation of Eq. (10), the inverse slope T^* in the limiting case of low p_T reads:

$$T^* \cong T + \frac{1}{2} m_i \bar{v}_T^2 , \tag{12}$$

where $\bar{v}_T$ is the average transverse collective velocity. The condition of the validity of Eq. (12), $p_T \bar{v}_T \ll T$, is too restrictive for heavy hadrons, e.g., for $T = 170$ MeV and $\bar{v}_T = 0.2$ it requires of $m_T - m_i$ to be much smaller than 0.2 GeV for Ω and than 0.1 GeV for J/ψ. A good approximation of Eq. (10) at $m_T - m_i < m_i$ can be obtained by replacing the v_T distribution in Eq. (10) by its average value $\bar{v}_T$ and by using large argument K_1 asymptotic:

$$\frac{dN_i}{m_T\, dm_T} \sim \sqrt{m_T}\, \exp\left[-\frac{m_T(1 + \bar{v}_T^2/2)}{T}\right] I_0\left(\frac{p_T\, \bar{v}_T}{T}\right) . \tag{13}$$

A simultaneous fit with Eq. (13) of the transverse momentum spectra of Ω, J/ψ, and ψ' in central Pb+Pb collisions at collision energy 158 A GeV leads to $T = 170$ MeV and $\bar{v}_T$. This gives a strong support of the hypothesis of statistical nature of J/ψ and ψ' production and their kinetic freeze-out occurring directly at hadronization.

The kinetic freeze-out parameters of pions were extracted from the analysis of single- and two-pion spectra measured for central Pb+Pb collisions at 158 A GeV by the NA49 Collaboration. The results are: T = 120 MeV and $\bar{v}_T = 0.55$, i.e., they are very different than those obtained here from the analysis of heavy hadron spectra. A decrease of temperature and an increase of transverse flow velocity with time is a general property of expanding systems. Note also that the increase of $\langle p_T \rangle$ with the centrality of the collision as well as with particle mass was observed for J/ψ and ψ'. These are characteristic features of hydrodynamics. In contrast to J/ψ and ψ' mesons, the m_T-spectra of the Drell–Yan pairs (dileptons with invariant mass $M > 4.2$ GeV) do not show this type of hydrodynamical behavior.

6. Open Charm Enhancement as a Signal of Deconfinement

The standard theoretical picture assumes that the average number of $c\bar{c}$ pairs created in A+A collision is connected with the probability to create a $c\bar{c}$ pair in a nucleon–nucleon (N+N) collision:

$$N_{c\bar{c}}^{\mathrm{AA}} = N_{\mathrm{coll}} \, N_{c\bar{c}}^{\mathrm{NN}} \, . \tag{14}$$

Note that in high-energy collisions, the charm production is dominated by the creation of hadrons with open charm. The charmonium states correspond to a tiny fraction of the total $N_{c\bar{c}}$ yield and can be safely neglected in our consideration. In the framework of the statistical coalescence model an enhancement of the open charm by the factor of about 2–4 over the direct extrapolation of Eq. (14) is needed to explain the data on the J/ψ multiplicity. It was suggested in Refs. 12 and 13 that this enhancement may appear due to the broadening of the phase space available for the open charm because of the presence of strongly interacting medium. There is an essential difference between charmed hadron production and hard dilepton production (the Drell–Yan process): created c and $\bar{c}$ can and even have to interact with the surrounding quarks and gluons to be transformed into observed charmed hadrons.

To get an intuitive picture of possible medium effects, let us start from the open charm production in e^+e^- annihilation. The $c\bar{c}$ pair created at the first stage, hadronizes into observed particles. The hadronization has non-perturbative nature. Its dynamics can be qualitatively understood in the framework of the string picture. When the distance between c and $\bar{c}$ reaches the range of the confinement forces, a string connecting these colored objects is formed. If the e^+e^- center-of-mass (c.m.) energy $\sqrt{s}$ (equal to the invariant mass of $c\bar{c}$ pair $M_{c\bar{c}}$) lies above the corresponding charmed meson threshold (equal to $2m_D$) c and $\bar{c}$ break the string into two (or more) pieces, so that the final state contains a charmed hadron pair (and possibly a number of light hadrons). However, when the e^+e^- c.m. energy exceeds the heavy quark threshold ($\sqrt{s} > 2m_c$) but lies below the corresponding charmed meson threshold

($\sqrt{s} < 2m_D$), the string can not be broken and no open charm hadron pair can be formed.

A possible source of the open charm enhancement in A+A collisions with respect to the direct extrapolation from p+p data may be the broadening of the phase space available for the open charm due to the presence of the hadron and/or the quark-gluon medium. Let a $c\bar{c}$ pair with an invariant mass below the open charm threshold, $2M_D = 3.7$ GeV, be created in a p+p collision. In the vacuum it cannot hadronize to an open charm hadron pair and thus it has to be transformed into non-charmed states to respect energy-momentum conservation. In A+A collisions, however, such a pair has a chance to shift its invariant mass above the open charm threshold due to secondary interactions with hadron and/or quark–gluon medium, thus resulting in additional open charm production. The strongest enhancement of the open charm production should be expected in the quark–gluon plasma. Due to Debye screening, c and $\bar{c}$ behave like free particles. Therefore, almost all subthreshold $c\bar{c}$-pairs created in the hard collisions hadronize into open charm mesons and baryons at the later stage of the A+A reaction. This effect can be substantial at moderate c.m. energies per nucleon pair, while at higher energies the threshold effects becomes less important. The above mechanism does not alter the production of direct Drell–Yan pairs: both the threshold effects and the interaction with the nuclear medium are negligible in that case.

There are of course essential differences between the open charm hadron production in e^+e^- annihilation and in N+N collisions. In N+N collisions, when no deconfined medium is expected, the created $c\bar{c}$ pair can interact with the spectator partons and has therefore a chance to form open charm hadron pair even if its primary invariant mass was insufficient for this process. Moreover, contrary to e^+e^- annihilation, most of $c\bar{c}$ pairs are created in the color octet state and therefore they have to interact with the spectators to form a color neutral final state. Instead of breaking the string, c and $\bar{c}$ can form hadron states by means of coalescence with light spectator (anti-)quark.

In Ref. 13 a rough estimation was made for the upper bound of the possible enhancement for charmed hadrons due to the color deconfined medium. An upper bound factor of about 5–6 was found for open charm enhancement at SPS energy. This means that the enhanced production of open charm hadrons by the factor 2–4 can be explained by the influence of the deconfined medium.

Acknowledgments

I am very thankful to my great friend Walter Greiner. This paper is a small part of my enduring memories of him.

References

1. T. Matsui and H. Satz, *Phys. Lett. B* **178**, 416 (1986).
2. M. Gazdzicki and M. I. Gorenstein, *Phys. Rev. Lett.* **83**, 4009 (1999).
3. P. Braun-Munzinger and J. Stachel, *Phys. Lett. B* **490**, 196 (2000).
4. M. I. Gorenstein, A. P. Kostyuk, H. Stoecker, and W. Greiner, *Phys. Lett. B* **509**, 277 (2001).
5. M. I. Gorenstein, A. P. Kostyuk, H. Stoecker, and W. Greiner, *Phys. Lett. B* **524**, 265 (2002).
6. M. I. Gorenstein, A. P. Kostyuk, L. McLerran, H. Stoecker, and W. Greiner, *J. Phys. G* **28**, 2151 (2002).
7. A. P. Kostyuk, M. I. Gorenstein, H. Stoecker, and W. Greiner, *Phys. Lett. B* **531**, 195 (2002).
8. A. P. Kostyuk, M. I. Gorenstein, H. Stoecker, and W. Greiner, *J. Phys. G* **28**, 2297 (2002).
9. A. P. Kostyuk, M. I. Gorenstein, H. Stoecker, and W. Greiner, *Phys. Rev. C* **68**, 041902 (2003).
10. K. A. Bugaev, M. Gazdzicki, and M. I. Gorenstein, *Phys. Lett. B* **523**, 255 (2001).
11. M. I. Gorenstein, K. A. Bugaev and M. Gazdzicki, *Phys. Rev. Lett.* **88**, 132301 (2002).
12. M. I. Gorenstein, A. P. Kostyuk, H. Stoecker, and W. Greiner, *J. Phys. G* **27**, L47 (2001).
13. A. P. Kostyuk, M. I. Gorenstein, and W. Greiner, *Phys. Lett. B* **519**, 207 (2001).

Dark Matter Compact Stars in Pseudo-Complex General Relativity

D. Hadjimichef, G. L. Volkmer, R. O. Gomes, and C. A. Zen Vasconcellos
Instituto de Física
Universidade Federal do Rio Grande do Sul (UFRGS)
Av. Bento Gonçalves 9500, Porto Alegre, RS, Brazil
CEP 91501-970, Caixa Postal 15051
dimiter@if.ufrgs.br, volkmer_@hotmail.com, rosana.gomes@ufrgs.br,
cesarzen@cesarzen.com

In the theory of pseudo-complex General Relativity (pc-GR), the field equations have an extra term, associated with the nature of spacetime, of repulsive character, which is believed to halt the gravitational attractive collapse of matter distributions in the evolution process of compact stars. According to Walter Greiner and Peter Hess, this additional term arises from micro-scale phenomena due to vacuum fluctuations which simulate the presence of dark energy in the Universe. In this contribution we explore the presence of this additional term and propose a toy model consisting of dark matter, represented by Standard Model Gauge Singlet WIMPs (Weakly-Interacting Massive Particles) having their origin in a portal model, held together by the presence of the gravitational interaction and superimposed to the (repulsive) background of dark energy, as predicted by pc-GR. The combination of these two ingredients, gravitational attraction and dark energy repulsion, allows in thesis the hydrostatic equilibrium condition of this non-luminous exotic star to hold. Solving the corresponding field equations and the TOV equations, and assuming that the fluid components interact only gravitationally, we determine the hydrostatic equilibrium equations of the star composed only by dark matter and dark energy, a rare compact object formed solely by exotic content. We then analyze the corresponding results obtained for the equation of state and for the mass–radius relations and we then determine the maximum mass of the exotic star for different parameter configurations.

1. Introduction

In 1933 Fritz Zwicky in his studies of the Coma cluster of galaxies[1] used the Virial Theorem, which relates the average kinetic energy of a physical system with its total potential energy, to infer the gravitational mass of the Coma cluster. Assuming only gravitational interactions and Newtonian gravity ($F \sim 1/r^2$), from the Virial Theorem, the following relation between the average kinetic energy ($\langle T_C \rangle$) and the average potential energy ($\langle U_C \rangle$) of the Coma cluster then holds:

$$\langle T_C \rangle = -\frac{1}{2}\langle U_C \rangle .$$

From the (at that time) known results of velocity dispersions of the individual galaxies of the Coma cluster, which allow the determination of the average kinetic energy of the system, Zwicky has found the average potential energy of the system and the total mass of the cluster as $4.5 \times 10^{13} M_{\odot}$. Zwicky also observed roughly 1000 nebulae in the cluster, so he calculated the average mass of the cluster as $4.5 \times 10^{10} M_{\odot}$. From the comparison of the predicted result for the gravitational mass of the system with the total sum of the bright and luminous masses of the stars and the gas contained in the cluster, he found an unexpected result: the values of the gravitational mass and the luminous mass of the Coma cluster were very different. And even more unexpectedly, he concluded that the mass of the luminous matter was not sufficient to keep the cohesion of the cluster since this difference corresponded to several times less than the inferred gravitational mass. He then postulated the presence in the galaxy of a mysterious kind of non-luminous matter, dark matter (DM), responsible for the integrity of the system. To add yet another nebulous ingredient to this concept, the presence of dark energy (DE) represents the most accepted hypothesis to explain observations since the 1990s indicating that the Universe is expanding at an accelerated rate.[2]

The presence of DM in the Universe was long considered a controversial idea, lacking phenomenology and observations that would enable its wide acceptance by the world scientific community. Presently, the assumption that the composition of the Universe is dominated by this form of non-luminous matter that affects the motion and overall properties of galaxies is widely accepted in modern cosmology.

The paradigmatic change for the acceptance by the scientific community of this remaining unsolved problem is fundamentally due to indirect evidences and observations that include, among others, the study of galaxy clusters, in special of galactic rotation curves, of the cosmic microwave background, of the large-scale structure of the Universe, of gravitational lens and so on. Despite this, scientists are still pushing forward to measure direct evidences of dark matter (see for instance Ref. 3).

However, the enigma of the nature of dark matter and dark energy remains unsolved. Still one of the most evasive and fascinating mysteries in physics, the problem of the dark matter/dark energy in the Universe stirs the imagination of most astronomers, cosmologists and particle physicists, that are convinced that the total mass-energy content of the Universe is composed by 4.9% of ordinary matter, at least 26.8% is due to some non-luminous form of matter (dark matter) and 68.3% is some unknown form of energy (dark energy).[4]

Over the years, dark matter studies have split into two major branches: one as an astrophysical phenomenon and the other as a particle physics problem. In the particle physics branch, the dark matter problem appears as an extension of the Standard Model (SM) of particles. Its search has introduced many candidates that converge to a variety and plausible *zoo* of exotic particles, namely the so-called

Weakly-Interacting Massive Particles (WIMPs).[5] In the *dark matter zoo* many different types of particles have been proposed and their properties theoretically studied, but experimentally there has been no conclusive evidence of particle dark matter to the present days. If particle searches fail to reveal a new physics, then one can ask: could dark matter be some form of exotic matter only coupled to gravity?

In this context, the astrophysical realm stands out as an unique testing ground for dark matter and dark energy models, where they could be both present in stellar structures formed by ordinary matter. And moreover, in this contextualization, the laws of General Theory of Relativity (GR) represent crucial elements for a proper analysis of the impact of these degrees of freedom on astrophysical objects.

General Theory of Relativity has deeply modified our understanding of the structure of spacetime and represents a beautiful achievement of applied mathematics. In Albert Einstein's General Theory of Relativity, gravity is a manifestation of spacetime being warped, bent or stretched due to the presence of matter and energy.[6] The theory has verified predictions like the perihelion shift of Mercury, frame dragging effects and more recently gravitational waves.[7]

Despite the success of General Theory of Relativity, modifications and extensions of this theory, with different motivations, have been proposed since its original formulation.[8,9] Einstein himself, aiming a unification between gravity and electromagnetism, has proposed an extension of the GR metric tensor to a complex field representation.[10,11]

The association of complex numbers with Group Theory allowed the mathematician Sophus Lie to generalize complex numbers to hypercomplex numbers.[12] The theory of hypercomplex numbers has found several applications in physics. For example, L. Sobrero applied the functions of a particular four-dimensional hypercomplex number to the theory of elasticity.[13] Before the work of Lie in Group Theory, Carl Friedrich Gauss observed that the generalization of real and complex numbers cannot be performed without losing some peculiarities of the algebraic operations involving these numbers: for instance, one must either renounce to the commutative property of the product of these numbers (quaternions lies in this case) or encounter the peculiarity that the product between some non-zero numbers may be equal to zero.[12] These products are called *zero divisors* or *generalized zeros*. The *pseudo-complex* numbers (sometimes also called hyperbolic numbers) fit the second case.

Developments in pseudo-complex numbers were stimulated due to its association with the Lorentz's group of Special Relativity.[12] In Ref. 14 a detailed consistent extension of General Relativity is made using pseudo-complex (pc) variables. The use of pc variables has a critical advantage that they allow the only algebraic extension without the presence of ghosts and tachyons solutions. An analysis of neutron stars within a pc-formalism is given in Ref. 14.

It was shown in Ref. 14, using pc-GR theory, that the resulting GR field equations include an extra term associated to the nature of spacetime itself and which is believed to halt the collapse of matter distributions avoiding for instance the standard GR predictions of black hole creation as the final stage of the evolution of compact stars. According to the authors, the physical origin of this additional term may arise from micro-scale phenomena where vacuum fluctuations could become considerable under certain conditions.[14]

In this contribution we propose a toy model consisting of a kind of *non-conventional* dark matter, represented by SM fermion gauge singlets,[15] having their origin in a *portal model*, held together only by the presence of the gravitational force, superimposed to the background of the pc-GR that contains this additional term, of repulsive nature, that mimic the presence of dark energy in the Universe.[a] The combination of these two ingredients in our approach would allow in thesis the hydrostatic equilibrium of the SM matter to hold forming an exotic composition: a type of unconventional and non-luminous star composed only by dark matter and dark energy, a rare compact object formed solely by exotic content.

2. The Pseudo-Complex General Relativity

In this section we briefly review the Pseudo-Complex General Relativity. Most of the material presented in this section is based on the publication "Pseudo-Complex General Relativity and Neutron Stars",[16] by Walter Greiner and Peter Hess.

In pseudo-complex General Relativity, real coordinates are algebraically extended to pseudo-complex ones in the form

$$x^\mu \to X^\mu \;=\; x^\mu + I y^\mu \quad \{x, y \in \mathbb{R},\; I \notin \mathbb{R};\; I^2 = 1\}\,; \tag{1}$$

in this expression, unlike the usual definition of a complex number, which can be written as a real number multiplied by the imaginary unit i, which obeys the property $i^2 = -1$, the corresponding pseudo-imaginary unity obeys $I^2 = 1$. In this expression, x^μ represent the usual spacetime coordinates in General Relativity and y^μ represent in turn the corresponding pseudo-imaginary components. Thus, the set (x^μ, y^μ) defines the coordinates of a pseudo-complex manifold.

An alternative way to express these coordinates is by introducing a new mathematical representation basis

$$\sigma_\pm = \frac{1}{2}\,(1 \pm I)\;, \tag{2}$$

which leads to an equivalent representation of the pseudo-complex coordinates, namely

$$\begin{aligned} X^\mu &= X^\mu_+ \sigma_+ + X^\mu_- \sigma_-\;; \\ X^\mu_\pm &= x^\mu \pm y^\mu\,. \end{aligned} \tag{3}$$

[a]Our objective in this work is to study the repulsive effects in the system due to the presence of the additional term of pc-GR described in Section 2. For this reason, in order to concentrate our analysis on the effects on the stellar stability due to the presence of this term, we will assume in the calculations that the SM-fermions are non-interacting.

One can verify that the quantities $\sigma_\pm$ are idempotent,[b] i.e. they obey the condition $\sigma_\pm^2 = \sigma_\pm$ and more importantly they present the following property $\sigma_+\sigma_- = 0$. This property indicates that, when an element proportional to σ_+ is multiplied by another one proportional to σ_-, their product may be equal to zero, even when they are individually different from zero, which implies the existence of a *zero-divisor*, i.e. the composition of two non-zero elements such that the product of a non-zero element by some other non-zero element results in the zero cardinal, following in mathematics the definition of a ring or a semi-group with the zero element. In this context, the authors call our attention to the important fact that, in pc-GR, the pseudo-complex variables do not form a field but rather a ring. The advantage of these new definitions is that mathematical manipulations can be performed in each zero divisor component, independently to the other one, thus broadening the possibilities of formal treatment of General Relativity equations under extreme physical conditions that generate spacetime singularities.

These features lead to the proposition to define a new GR formulation, in the contextualization of the pseudo-complex variables (pc-GR), considering two new branches, one for each zero-divisor component,[17–19] with the pseudo-complex metric tensor (assumed to be symmetric) defined as:

$$g_{\mu\nu}(X) = g^+_{\mu\nu}(X_+)\sigma_+ + g^-(X_-)\sigma_- \, . \tag{4}$$

Similarly, a pseudo-holomorphic function (the analogous of holomorphic functions in the complex case) may be represented in the pseudo-complex basis, as

$$f(X) = f(X_+)\sigma_+ + f(X_-)\sigma_- .$$

The metric functions $g^\pm_{\mu\nu}(X_\pm)$ are obtained by varying the standard action of GR, namely $S_\pm = \sqrt{-g_\pm}\mathcal{R}_\pm$, where $g_\pm$ is the determinant of the metric $g^\pm_{\mu\mu}$ and $\mathcal{R}_\pm$ represents the Riemann scalar in each component. The pseudo-complex relations for other physical quantities can be found in a similar fashion.

An important aspect to be considered in this stage of formulating a new theory for pc-GR is that, by using the standard condition for the variational principle of the least action to this formulation,

$$\delta S = \delta S_+\sigma_+ + \delta S_-\sigma_- \, ,$$

and making the two terms above, $\delta S_+ = 0$ and $\delta S_- = 0$, i.e. making the two terms independently equal to the zero cardinal, $\delta S_\pm = 0$, leads to two independent theories of GR.

In order to cure this *pathology* and obtain a new theory with two complementary non-independent branches, we should consider instead the definition of "zero divisors" of pseudo-complex numbers which have only one component in the σ_+ or σ_-

[b]Idempotence, as is well known, is the mathematical algebraic property of certain operators that can be applied indeterminate amounts of times in a functional form without changing the result of the initial application. From the algebraic point of view, idempotent algebraic elements remain invariant when raised to a positive integer power. In the description of properties of physical systems, the use of idempotents operators allows to describe in a consistent way conservation laws and symmetries associated to these systems.

sectors and allow this way non-trivial solutions to the equation $X \cdot Y = 0$. For this reason, this generalized notion of zero was used to formulate a modified variational principle[20] (in the context of field theories), namely

$$\delta S = \delta S_+ \sigma_+ + \delta S_- \sigma_- \in \text{zero divisor} = \lambda \sigma_\pm \,, \tag{5}$$

i.e., a modified variational least action principle which is proportional to the idempotent basis elements, more precisely to either σ_+ *or to* σ_-. Such elements are defined as we have seen in the context of zero-divisors and have zero norm, which justifies the denomination of *generalized zeros*. This modification is equivalent to add in S_- an additional term, whose variation just gives $-\lambda\sigma_-$.

It was shown in Ref. 14, within the pc-GR theory, that the resulting GR field equations include an extra term associated to the nature of spacetime itself. This term is believed to halt the collapse of matter distributions avoiding for instance the standard GR predictions of black hole creation as the final stage of the evolution of compact stars. In order to proceed within a familiar GR framework, similarly to the original formulation, we represent this term by an energy-momentum tensor $(T^\Lambda)_{\mu\nu}$:

$$G_{\mu\nu} = 8\pi (T^\Lambda)_{\mu\nu} + 8\pi (T^m)_{\mu\nu} \,; \tag{6}$$

thus, in this expression $(T^\Lambda)_{\mu\nu}$ represents the dark energy contribution and $(T^m)_{\mu\nu}$ is the energy-momentum tensor for matter. As previously mentioned, the physical origin of this additional term may arise from micro-scale phenomena where vacuum fluctuations could become considerable under certain conditions.[14]

In the following, we consider that the term $(T^m)_{\mu\nu}$ above describes the presence of *non-conventional* dark matter, represented by SM fermion gauge singlets.

3. Dark Matter: Higgs Portal Model

In the present, we consider that the dark matter sector is a renormalizable extension of the SM: a hidden sector consisting of the Standard Model gauge singlets (a singlet Dirac fermion χ and a singlet scalar S).[15] A vector boson V_μ, was added to this model (*dark massive photon*) and studied, revealing extremely small pair production cross sections in pair production reactions $e^- e^+ \to VV$.[21]

The singlet scalar interacts with the SM sector through the triple and quartic scalar interactions. The singlet fermion interacts with the SM only via the singlet scalar. The model is defined by the following Lagrangian density

$$\mathcal{L} = \mathcal{L}_{SM} + \mathcal{L}_\chi + \mathcal{L}_S - g_\varphi \, S \, \bar{\chi} \, \chi + \mathcal{L}_I \,; \tag{7}$$

$$\mathcal{L}_\chi = \bar{\chi} \, (i \not{\partial} - m_{\chi_0}) \chi \,; \tag{8}$$

$$\mathcal{L}_S = \frac{1}{2} \partial_\mu S \partial^\mu S - V_S \,; \tag{9}$$

$$\mathcal{L}_I = -\lambda_1 \Phi^\dagger \Phi S - \lambda_2 \Phi^\dagger \Phi S^2 \,, \tag{10}$$

where $\mathcal{L}_{SM}$ is the Lagrangian density for the SM; Φ the SM Higgs and S a scalar field connecting with the dark sector; χ is the dark fermion singlet; the potential in $\mathcal{L}_S$ is defined as

$$V_S = \frac{m_0^2}{2} S^2 + \frac{\lambda_3}{3!} S^3 + \frac{\lambda_4}{4!} S^4 \,. \tag{11}$$

The form of V_S clearly indicates the possibility of spontaneous broken symmetry in the model, which is parameterized by introducing $\langle S \rangle = x_0$. Considering spontaneous symmetry breaking in both sectors, we write the SM Higgs field as

$$\Phi = \frac{1}{\sqrt{2}} \begin{pmatrix} 0 \\ v_0 + h(x) \end{pmatrix} , \tag{12}$$

with

$$S = x_0 + \varphi \,,$$

and where $\mathcal{L}$ contains a quadratic form in h and φ, indicating that this is not a mass-eigenstates theory.

There remains the possibility that the interaction between the SM and the dark sectors occurs only through the Higgs scalar boson field; thus,the present model corresponds to a class of theories known as *Higgs portal dark matter*.[22]

Writing

$$\mathcal{V} = V_S + V(\Phi^\dagger \Phi) \,, \tag{13}$$

where the SM Higgs potential is

$$V(\Phi^\dagger \Phi) = -\mu^2 \Phi^\dagger \Phi + \lambda \, (\Phi^\dagger \Phi)^2 \,, \tag{14}$$

and demanding that

$$\left. \frac{\partial \mathcal{V}}{\partial h} \right|_{v_0, x_0} = \left. \frac{\partial \mathcal{V}}{\partial S} \right|_{v_0, x_0} = 0 \,, \tag{15}$$

one obtains

$$m_0^2 = -\frac{\lambda_3}{2} x_0 - \frac{\lambda_4}{6} x_0^2 - \lambda_2 v_0^2 - \frac{\lambda_1 v_0^2}{2 x_0} \,, \tag{16}$$

and

$$\mu^2 = \lambda v_0^2 + x_0 (\lambda_1 + \lambda_2 x_0) \,. \tag{17}$$

The neutral scalar states h and φ are defined through

$$H_0 = \frac{1}{2} \Big(v_0 + h \Big) \,, \tag{18}$$

and

$$S = x_0 + \varphi \,, \tag{19}$$

and are mixed to yield the mass matrix given by[15]

$$M_{hh}^2 = \left. \frac{\partial^2 \mathcal{V}}{\partial h^2} \right|_{h=\varphi=0} = 2 \lambda v_0^2 \,; \tag{20}$$

$$M_{ss}^2 = \frac{\partial^2 \mathcal{V}}{\partial \varphi^2}\bigg|_{h=\varphi=0} = \frac{\lambda_3 x_0}{2} + \frac{\lambda_4}{3} x_0^2 - \frac{\lambda_1 v_0^2}{2x_0}\,; \tag{21}$$

$$M_{hs}^2 = M_{sh}^2 = \frac{\partial^2 \mathcal{V}}{\partial \varphi \partial h}\bigg|_{h=\varphi=0} = \Big(\lambda_1 + 2\lambda_2 x_0\Big) v_0\,. \tag{22}$$

The two physical Higgs fields h_1 and h_2 corresponds to mixtures of the φ and h fields and the mass eigenstates are obtained from

$$h_1 = \varphi\,\sin\theta + h\,\cos\theta\,; \tag{23}$$

$$h_2 = \varphi\,\cos\theta - h\,\sin\theta\,; \tag{24}$$

the mixing angle θ on the other hand is defined as

$$\tan\theta = \frac{y}{1+\sqrt{1+y^2}}\,, \tag{25}$$

with

$$y = \frac{2M_{hs}^2}{M_{hh}^2 - M_{ss}^2}\,. \tag{26}$$

The masses of the fields h_1 and h_2 are

$$m_{1,2}^2 = \frac{M_{hh}^2 + M_{ss}^2}{2} \pm \frac{M_{hh}^2 - M_{ss}^2}{2}\sqrt{1+y^2}\,. \tag{27}$$

In this formulation, the SM Higgs is represented by the h_1 field with mass $m_1 = 125$ GeV.

4. Modified TOV Equations

As pointed out in Ref. 14 we have no argument to exclude the accumulation of dark energy around mass concentrations. The work of M. Visser[23] has shown that vacuum fluctuations are created by the presence of mass and that the greater the value of the mass, the higher are the intensities of vacuum fluctuations. The accumulated dark energy resulting from these vacuum fluctuations might have a role in phenomena like the gravitational collapse of bound objects and will be included in our toy model.

For simplicity we consider a spherically symmetric and static metric.

$$ds^2 = -e^{\nu(r)}dt^2 + e^{\mu(r)}dr^2 + r^2\Big(\sin^2(\theta)d\phi^2 + d\theta^2\Big)\,. \tag{28}$$

Moreover, in the following, both $(T^\Lambda)_{\mu\nu}$ and $(T^m)_{\mu\nu}$ will be treated as isotropic perfect fluids,

$$T_{\mu\nu} = \Big(\epsilon + p\Big)u_\mu u_\nu + p g_{\mu\nu}\,,$$

where p is the pressure and ϵ is the energy density of the fluid, and u_μ are the components of the four velocity of the fluid. Solving the modified field equations (6) and assuming that the fluid components interact only gravitationally, the hydrostatic TOV equilibrium equations can be written as:[14]

$$\frac{dp_m}{dr} = -\left(\frac{\epsilon_m(r) + p_m(r)}{r\Big(r - 2m_m(r) + 2m_\Lambda(r)\Big)}\right)\Big(m_m(r) - m_\Lambda(r) + 4\pi r^3\Big[p_\Lambda(r) + p_m(r)\Big]\Big)\,; \tag{29}$$

$$\frac{dp_\Lambda}{dr} = -\left(\frac{\epsilon_\Lambda(r) + p_\Lambda(r)}{r\Big(r - 2m_m(r) + 2m_{\Lambda i}(r)\Big)}\right)\Big(m_m(r) - m_\Lambda(r) + 4\pi r^3\Big[p_\Lambda(r) + p_m(r)\Big]\Big)\,, \tag{30}$$

where:

$$m_m(r) = 4\pi \int_0^r r'^2 \epsilon_m(r')dr'\,, \tag{31}$$

and

$$m_\Lambda(r) = -4\pi \int_0^r r'^2 \epsilon_\Lambda(r')dr'\,; \tag{32}$$

in these expressions, $m_m(r)$ and $m_\Lambda(r)$ represent the mass functions. In the following, we replace the standard matter contribution by the SM Fermi dark matter and analyze the subsequent characteristics of the exotic compact object composed of dark matter and dark energy.

5. Equations of State

In principle, outside the star one shall impose a condition to the character of the dark energy fluid: we may require for instance that the *Null Dominant Energy Condition* (NDEC)[24] is satisfied. In many cases some of the energy conditions may be not satisfied.[23] For example, if negative energy densities of the system are possible, this result would violate the *Weak Energy Condition* (WEC). The NDEC is less stringent and reads

$$T_{\mu\nu}l^\mu l^\nu \geq 0\,, \tag{33}$$

where l^μ are the components of a null-vector, and

$$W^\mu = T^\mu_\nu l^\nu \tag{34}$$

is a non-spacelike vector. This leads to the conditions

$$\begin{aligned} \epsilon_\Lambda + p_\Lambda &\geq 0 \\ \mid \epsilon_\Lambda \mid &\geq \mid p_\Lambda \mid\,, \end{aligned} \tag{35}$$

which are satisfied also for a negative dark energy density.

Inside the star, the numerical implementation of the system of differential equations demands the specification of the dark matter equation of state and its relation to the Λ-sector. Following Ref. 14 we will consider an isotropic fluid model for both the dark matter and the dark energy densities and we assume a linear coupling between ϵ_m and ϵ_Λ:

$$\epsilon_\Lambda = \alpha \epsilon_m . \tag{36}$$

To simulate our simple toy model, we must first fix $\theta = 0$ in equations (23) and (24), decoupling the physical Higgs from the dark Higgs, closing this way the Higgs portal. This choice also decouples dark matter from ordinary matter and transforms the χ fermionic-singlet system to a non-interacting Fermi gas of WIMPs. There is a residual interaction of the χ singlet with the dark Higgs. However, after symmetry breaking in the dark sector, the χ singlet acquires a mass $m_\chi = m_{\chi 0} + g_\varphi x_0$. We shall treat m_χ as a free parameter, studying the impact of different χ-mass values on the exotic stellar quantities.

The equation of state for a free gas of dark fermions at zero temperature $p_m(\varepsilon_m)$ can be calculated via explicit expressions for the energy density

$$\begin{aligned} \varepsilon_m &= \frac{1}{\pi^2}\int_0^{k_F} dk\, k^2 \sqrt{m_\chi^2 + k^2} \\ &= \frac{m_\chi^4}{8\pi^2}\left[(2z^3 + z)(1+z^2)^{\frac{1}{2}} - \sinh^{-1}(z)\right] \equiv m_\chi^4\, \bar{\varepsilon}_m \,; \end{aligned} \tag{37}$$

and pressure

$$\begin{aligned} p_m &= \frac{1}{3\pi^2}\int_0^{k_F} dk\, \frac{k^4}{\sqrt{m_\chi^2 + k^2}} \\ &= \frac{m_\chi^4}{24\pi^2}\left[(2z^3 - 3z)(1+z^2)^{\frac{1}{2}} + 3\sinh^{-1}(z)\right] \equiv m_\chi^4\, \bar{p}_m \,. \end{aligned} \tag{38}$$

In terms of the natural units,[c] the quantities $\bar{\epsilon}_m$ and $\bar{p}_m$ are dimensionless; above we have defined the relativity parameter $z = k_F/m_\chi$. The central pressure of the Λ-term is assumed to be related to the central energy density of the WIMPs, ϵ_0, through $p_{\Lambda_C} = |\alpha|\, \epsilon_0$, with a standard value $\epsilon_0 = 141$ MeV/fm^3.

6. Numerical Implementation

The system of equations (29)–(32) is integrated using a standard procedure, a fourth-order Runge-Kutta algorithm and as usual an "integration stop" condition, i.e., the integration of the equations ceases when the dark matter pressure vanishes. Although, in principle, the Λ-component does not vanish at the surface of the compact stellar object, for simplicity in this work we are not concerned with the exterior region of the exotic star.

[c] We assume the system of natural unites based on the condition $G = \hbar = c = 1$ applied to the universal physical constants G, $\hbar$ and c.

The range of m_χ is limited to a few values considered consistent by detection experiments that perform the direct search of dark matter. For instance, the CoGeNT Experiment[25] uses a single, 440-gram, high-purity germanium crystal cooled to liquid nitrogen temperatures in its measurements. The CoGeNT detector has an advantage of a very low energy threshold (0.5 keV) which allows to search for nuclear recoil events due to dark matter particles of relatively low mass, $m_\chi = 7$ to 10 GeV. Another important collaboration is the XENON100 Dark Matter Project.[26] The XENON100 detector has conducted several data runs but no dark matter signal was ever observed above the expected background, leading to the most stringent limit on the spin independent WIMP-nucleon cross section, with a minimum at 2.0×10^{-45} cm^2 for a 55 GeV WIMP mass.[26] In our calculation we shall restrict the values of m_χ to 1, 50 and 100 GeV.

7. Results

The results of our calculation are shown in the tables and figures below. In Table 1 we show the parameterization in our approach: central energy density (ε_{mC}) and central pressure (p_{mC}) of dark matter and central energy density ($\varepsilon_{\Lambda C}$) and central pressure ($p_{\Lambda C}$) associated to dark energy for $m_\chi = 1, 50, 100$ GeV. In Table 2 we show the adopted values for the minimum ($\varepsilon_{mC}^{[i]}$) and maximum ($\varepsilon_{mC}^{[f]}$) values of the central energy density ε_{mC} of dark matter for the central pressure of dark energy given by $p_{\Lambda c} = |\alpha| \cdot \varepsilon_0$, for $m_\chi = 1, 50, 100$ GeV. Tables 3–5 show the results of our calculations for the total mass (M_m) and radius (R) of the exotic star for, respectively, $m_\chi = 1$ GeV, 50 GeV, and 100 GeV, for different values of the α parameter that measures the role of pc-GR in the conformation of the exotic stellar mass. Figure 1 shows the equation of state of dark matter. The figure shows the dimensionless pressure $\bar{p}$ versus the dimensionless energy density $\bar{\varepsilon}$ for a free Fermi gas at zero temperature. Figures 2–4 show dark matter profiles for energy density (ε_m), pressure (p_m), total dark matter mass inside the star (m_m) and maximum star mass (M_m) for different values of the coefficient α with the dark fermion mass equal respectively to $m_\chi = 1, 50, 100$ GeV. The energy density variation shown in Table 2 was used to generate the graphs of $M_m \times R(km)$, which are the graphs located at the bottom right position of Figs. 2–4.

Table 1. Parameterization in our approach: central energy density (ε_{mC}) and central pressure (p_{mC}) of dark matter and central energy density ($\varepsilon_{\Lambda C}$) and central pressure ($p_{\Lambda C}$) associated with dark energy for $m_\chi = 1, 50, 100$ GeV.

m_χ (GeV)	ε_{mC} (MeV/fm^3)	p_{mC} (MeV/fm^3)	$\varepsilon_{\Lambda C}$ (MeV/fm^3)	$p_{\Lambda C}$ (MeV/fm^3)
1	2.82×10^3	2.62×10^2	$\alpha \cdot \varepsilon_{mC}$	$\lvert\alpha\rvert \cdot \varepsilon_0$
50	2.00×10^{10}	1.96×10^9	$\alpha \cdot \varepsilon_{mC}$	$\lvert\alpha\rvert \cdot \varepsilon_0$
100	3.00×10^{11}	2.87×10^{10}	$\alpha \cdot \varepsilon_{mC}$	$\lvert\alpha\rvert \cdot \varepsilon_0$

Table 2. Adopted values for the minimum ($\varepsilon^{[i]}_{mC}$) and maximum ($\varepsilon^{[f]}_{mC}$) central energy density ε_{mC} of dark matter for the central pressure of dark energy given by $p_{\Lambda c} = |\alpha| \cdot \varepsilon_0$.

m_χ (GeV)	$\varepsilon^{[i]}_{mC}$ (MeV/fm^3)	$\varepsilon^{[f]}_{mC}$ (MeV/fm^3)
1	1200	3200
50	7×10^9	2.5×10^{10}
100	3×10^{10}	4.5×10^{11}

Table 3. Total mass (M_m) and radius (R) of the exotic star for $m_\chi = 1$ GeV for different values of the α parameter that measures the role of pc-GR in the conformation of the exotic stellar mass. In the calculations we have used de values for ε_{mC} and $p_{\Lambda C}$ as given in Table 1.

Solution	M_m ($M_\odot$)	R (km)
GR	0.627	8.11
$\alpha = -0.1$	0.66	7.92
$\alpha = -0.3$	0.75	7.83
$\alpha = -0.5$	0.90	7.96
$\alpha = -0.7$	1.14	8.28
$\alpha = -0.9$	1.54	8.84

Table 4. Total mass (M_m) and radius (R) of the exotic star for $m_\chi = 50$ GeV for different values of the α parameter that measures the role of pc-GR in the conformation of the exotic stellar mass. In the calculations we have used de values for ε_{mC} and $p_{\Lambda C}$ as given in Table 1.

Solution	M_m ($\times 10^{-4} M_\odot$)	R (10^{-3} km)
GR	2.51	3.19
$\alpha = -0.1$	2.73	3.17
$\alpha = -0.3$	3.33	3.22
$\alpha = -0.5$	4.30	3.35
$\alpha = -0.7$	6.00	3.58
$\alpha = -0.9$	9.63	4.00

8. Discussion of the Results and Conclusions

As the equation of state is a function of the relativity parameter z, we can compute the EoS parametrically in a tabular form for a defined interval of z. Figure 1 shows the behavior of the dimensionless pressure of dark matter as a function of its dimensionless energy density; the figure corresponds to a double logarithmic plot.

Table 5. Total mass (M_m) and radius (R) of the exotic star for m_χ = 100 GeV for different values of the α parameter that measures the role of pc-GR in the conformation of the exotic stellar mass. In the calculations we have used de values for ε_{mC} and $p_{\Lambda C}$ as given in Table 1.

Solution	M_m ($\times 10^{-5} M_\odot$)	R (10^{-3} km)
GR	6.27	0.81
$\alpha = -0.1$	6.84	0.80
$\alpha = -0.3$	8.38	0.82
$\alpha = -0.5$	10.8	0.85
$\alpha = -0.7$	15.2	0.91
$\alpha = -0.9$	24.6	1.02

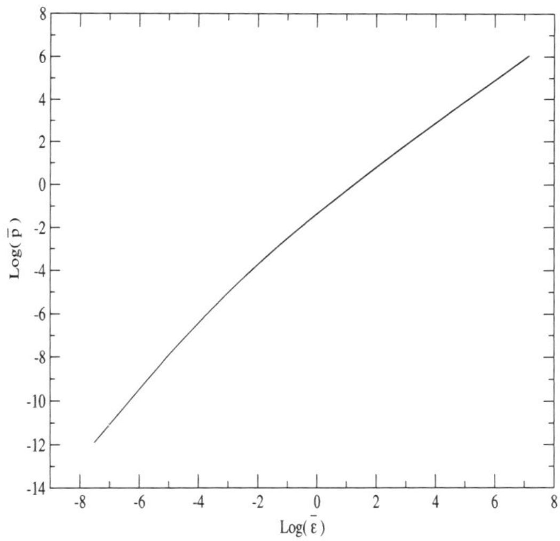

Fig. 1. The equation of state of dark matter. The figure shows the dimensionless pressure $\bar{p}$ versus the dimensionless energy density $\bar{\varepsilon}$ for a free Fermi gas at zero temperature. The two limits, the non-relativistic one for $z \ll 1$ and the ultra-relativistic one, for $z \gg 1$ corresponds respectively to the lower part of the figure (the larger slope at small values) and to the upper part of the figure (smaller slope at larger values).

In general, the equation of state of a compact star can not be expressed in a simple polytropic formulation of the type $p = \varepsilon^\gamma$ with a constant γ. However, a polytropic formulation may be assumed for two limits of the relativity parameter z (see Ref. 27 for an interesting discussion of these points):

(1) in the non-relativistic limit, $z \ll 1$, we get a polytropic law with $\gamma = 5/3$ (see Ref. 27), i.e.

$$p \sim \varepsilon^{5/3};$$

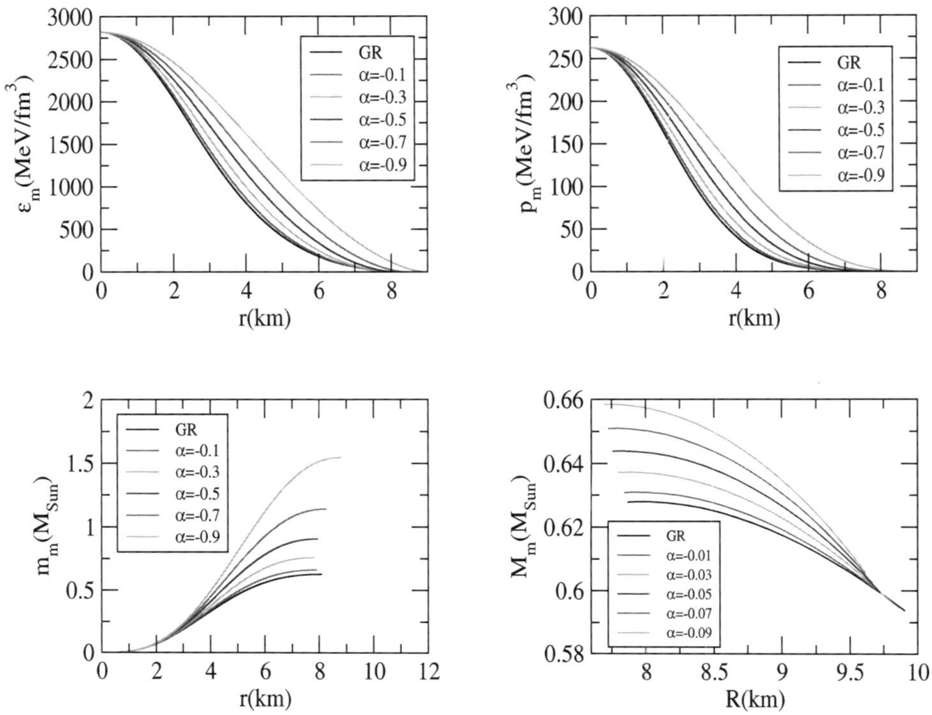

Fig. 2. Dark matter profiles for energy density (ε_m), pressure (p_m), total dark matter mass inside the star (m_m) and maximum star mass (M_m) for different values of the coefficient α. For the parameterization of the central energy density of the dark matter ε_{mC} and the central pressure $p_{\Lambda C}$ of the Λ-term see Table 1; the dark fermion mass is equal to $m_\chi = 1$ GeV.

(2) in the ultra-relativistic limit, $z \gg 1$, and the dark matter equation of state approaches a polytrope with $\gamma = 1$, i.e.

$$p = \frac{1}{3}\varepsilon$$

(see Ref. 27).

Our result in Fig. 1 exhibit the two different slopes: for small (non-relativistic) and large (ultra-relativistic) values of the dark matter energy density.

We compare our results of exotic compact stars composed of dark matter (SM fermions) and dark energy with the results of compact stars consisting of a free Fermi gas composed of fermions with arbitrary masses from Refs. 27 and 28. Scaling solutions are derived in Refs. 27 and 28 for a free and for an interacting Fermi gas and tested by numerical calculations. Although the authors of these references do not consider the effects of the presence of dark energy due to vacuum fluctuations, associated with the additional term of pc-GR, the comparison with their results only

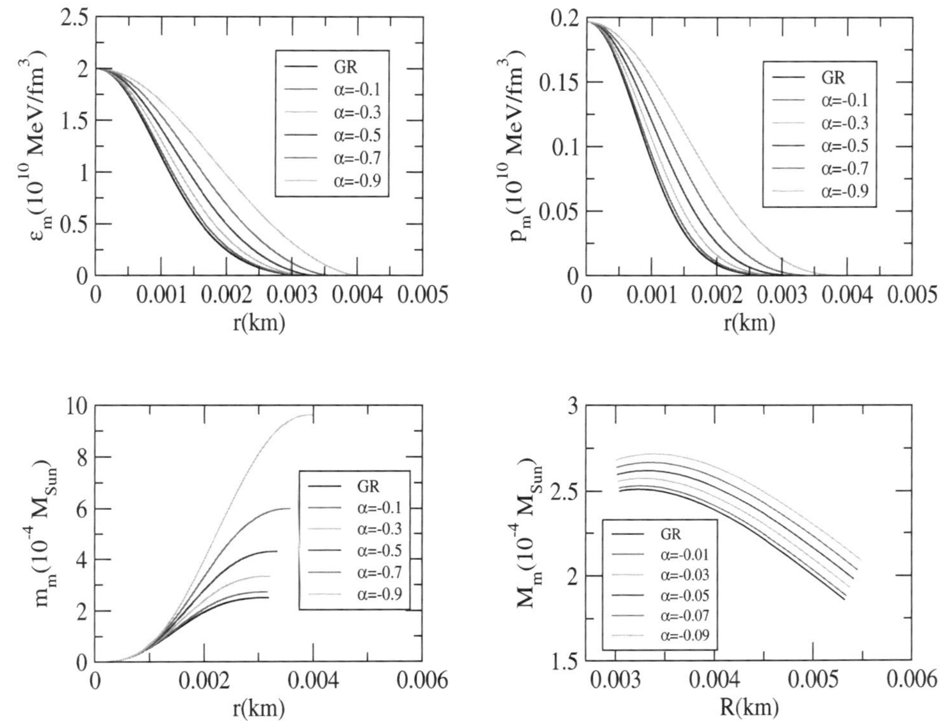

Fig. 3. Dark matter profiles for energy density (ε_m), pressure (p_m), total dark matter mass inside the star (m_m) and maximum star mass (M_m) for different values of the coefficient α. For the parameterization of the central energy density of dark matter ε_{mC} and the central pressure $p_{\Lambda C}$ of the Λ-term see Table 1; the dark fermion mass is equal to $m_\chi = 50$ GeV.

with respect to the Einstein Theory of General Relativity shows that our results are in tune. Moreover, the concordance of our results with the corresponding results of the aforementioned authors demonstrates the formal consistency of our approach.

Concerning the maximum mass of the exotic star, it is important to mention that gravitational compression produces a kind of packaging of the SM fermions so that only the repulsive components of the dark energy are relevant in our formulation to generate the stellar stability. In an ideal Fermi gas model applied to the description of global properties of an exotic star, that is, in a model in which the SM fermions are free and therefore non-interacting, the degeneracy pressure would represent, in Einstein's Theory of General Relativity, the only counterpoint to the gravitational force.

However, in this work we seek to analyze only the role in the stellar stability of the additional term of the pseudo-complex General Relativity that arises from the micro-scale phenomena due to vacuum fluctuations which simulate the presence of dark energy in the Universe. For this reason, in our preliminary calculations, we

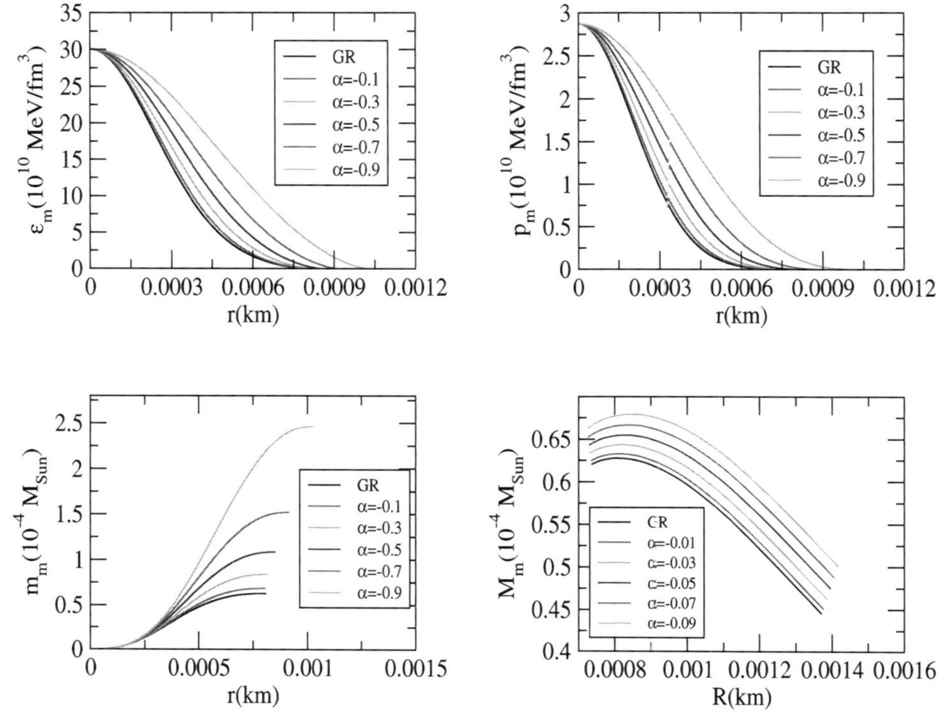

Fig. 4. Dark matter profiles for energy density (ε_m), pressure (p_m), total dark matter mass inside the star (m_m) and maximum star mass (M_m) for different values of the coefficient α. For the parameterization of the central energy density of dark matter ε_{mC} and the central pressure $p_{\Lambda C}$ of the Λ-term see Table 1; the dark fermion mass is equal to $m_\chi = 100$ GeV.

consider that the Standard Model fermion gauge singlets are held together only by the presence of gravitational interaction and at the same time are superimposed to the repulsive background of dark energy so that the combination of these two ingredients allow the hydrostatic equilibrium condition of this exotic star to hold.

On the other hand, even when we consider the idealized limiting case of a star composed of incompressible matter, therefore, with uniform density, the structure of the Tolman–Oppenheimer–Volkoff equations imposes an upper limit for the stellar mass above which gravitational collapse occurs. Using the typical jargon of this area, a state equation is then said to be more rigid or less rigid (less smooth or softer). For example, a state equation, A, is more rigid (less smooth) than another, B, less rigid (softer) if the internal pressure associated with A at any density value is greater than the internal pressure corresponding to B. Obviously, when we consider two stars, A and B, whose internal structures are described respectively by the state equations A and B, the state equation o of star A corresponds to a higher internal pressure compared to the internal pressure of the star B, for any density value. Given a greater internal pressure, the star A can withstand a greater gravitational compression than the star B and therefore the first one must have a mass greater than the second one, since the intensity of the gravitational force is directly associated to the stellar mass. Using again the typical jargon of the area, we say that the star A has a bigger mass than the star B.

In the search of an understanding of the maximum mass values of a compact star it becomes essential to understand its mass-radius relationship. In this sense, it is well known from Ref. 29, that a crude estimation of the mass and radius of a star can be made near the limit of gravitational collapse having $R = 2M_\star$. This will be an estimate of the maximum mass possible for the exotic compact star. In Ref. 29 it is considered that gravity packs up the fermions, which in this case are nucleons, up to the repulsive nuclear core, of $r_0 \approx 0.5 \times 10^{-13}$ cm. In our study the χ fermions are not nucleons, which, in principle, should allow a greater packing to smaller volumes as compared to nucleons. From Ref. 29,

$$M_\star \approx A\, m_\chi \quad ; \quad R \approx r_0\, A^{1/3}\,, \tag{39}$$

where A is the number of fermions. Combining Eqs. (39) with $R = 2M_\star$, one obtains the basic equations for the estimation of the number of SM fermions, mass and radius of the exotic star

$$A = \left(\frac{r_0}{2\, m_\chi}\right)^{3/2} \quad ; \quad R \approx r_0\, A^{1/3} \quad ; \quad M_\star = R/2\,, \tag{40}$$

written as a function of r_0. Table 6 shows a constant r_0, while in Table 7, r_0 decreases as m_χ increase. From a comparison of Tables 6 and 7 with the TOV tables, one can conclude that the TOV simulation has revealed that dark matter fermions χ can be packed to smaller distances than ordinary nuclear matter in usual neutron stars.

Table 6. Constant r_0.

m_χ (GeV)	r_0 (cm)	A	R (km)	$M_\star$ ($M_\odot$)
1	0.5×10^{-13}	3.00×10^{57}	7.2	2.4
50	0.5×10^{-13}	8.50×10^{54}	1.02	0.34
100	0.5×10^{-13}	3.00×10^{54}	0.72	0.24

Table 7. Variable r_0.

m_χ (GeV)	r_0 (cm)	A	R (km)	$M_\star$ ($M_\odot$)
1	0.5×10^{-13}	3.00×10^{57}	7.2	2.4
50	0.5×10^{-15}	8.50×10^{51}	1.02×10^{-3}	3.4×10^{-4}
100	0.25×10^{-15}	1.06×10^{51}	0.25×10^{-3}	8.59×10^{-5}

Using the typical jargon of this area, the exotic star with $m_\chi = 1$ GeV is more rigid (less smooth) than the exotic star with $m_\chi = 50$, or 100 GeV and may be as rigid as conventional neutron stars.

These results open the doors to future studies that will include the effects of the presence of dark matter (and dark energy) in describing the physical properties of compact stars using models that consider the fundamental baryonic octect interacting through light and heavy meson fields in order to describe the strong interaction in the short and long range regimes as well as the phase transition regime of compact stars that consider the presence of quarks and gluons. Research work along those lines are presently in progress in our research group.

References

1. F. Zwicky, *Helvetica Phys. Acta* **6**, 110 (1933).
2. J. Frieman, M. Turner, and D. Huterer, *Annual Review of Astronomy and Astrophysics*, **46**, 385 (2008).
3. As announced, in 2014, by the U.S. Department of Energy and National Science Foundation of upcoming dark matter experiments: LZ, SuperCDMS-SNOLAB, and ADMX-Gen2 (www.symmetrymagazine.org/article/july-2014/us-reveals-its-next-generation-of-dark-matter-experiments).
4. P. A. R. Ade *et al.*, *Astro. & Astroph.* **571**, A1 (2014).
5. G. Bertone, D. Hooper, and J. Silk, *Phys. Rep.* **405**, 279 (2005).
6. A. Das and A. DeBenedictis, *The General Theory of Relativity: A Mathematical Exposition* (Springer, N.Y., 2012).
7. C. A. Z. Vasconcellos (ed.), *Centennial of General Relativity: A Celebration* (World Scientific Publishing, Singapore, 2016).
8. Y. Fujii and K. ichi Maeda, *The Scalar-Tensor Theory of Gravitation* (Cambridge University Press, Cambridge, 2003).
9. E. Papantonopoulos (ed.), *Modifications of Einstein's Theory of Gravity at Large Distances* (Springer, New York, 2015).
10. A. Einstein, *Ann. Math. Second Ser.* **46**, 578 (1945).
11. A. Einstein, *Rev. Mod. Phys.* **20**, 35 (1948).
12. F. Catoni, D. Boccaletti, R. Cannata, V. Catoni, E. Nichelatti, and P. Zampetti, *The Mathematics of Minkowski spacetime: With an Introduction to Commutative Hypercomplex Numbers* (Frontiers in Mathematics, Birkhäuser Basel, 1 ed., 2008).
13. L. Sobrero, *Theorie der Ebenen Elastizität unter Benutzung eines Systems Hyperkomplexer Zahlen. Hamburger Mathematische Einzelschriften*, B. G. Teubner (1934).
14. P. O. Hess, M. Schäfer, and W. Greiner, *Pseudo-Complex General Relativity* (Springer, Switzerland, 2016).
15. K. Y. Lee, Y. G. Kim, and S. Shin, *Singlet Fermionic Dark Matter*, *JHEP05*, 100 (2008).
16. P. Hess, I. Rodríguez and W. Greiner, in *Nuclear Physics: Present and Future, Part of the series FIAS Interdisciplinary Science Series*, ed. Walter Greiner (Springer, Germany, 2015), p. 255.
17. P. O. Hess and W. Greiner, *Int. J. Mod. Phys. E* **18**, 51 (2009).
18. G. Caspar, T. Schönenbach, P. O. Hess, M. Schäfer, and W. Greiner, *Int. J. Mod. Phys. E*, **21**, 1250015 (2012).

19. P. O. Hess and W. Greiner, in *Centennial of General Relativity: A Celebration*, ed. C. A. Zen Vasconcellos (World Scientific Publishing, Singapore, 2016), p. 97.
20. F. P. Schuller, M. N. R. Wohlfarth, and T. W. Grimm, *Class. Quant. Grav.* **20**(1), 22 (2003).
21. D. Hadjimichef, *Massive Dark Photons in a Higgs Portal Model, AIP Conf. Proc.* **1693**, 050012 (2015); [arXiv:1601.00293].
22. L. Lopez-Honorez, T. Schwetz, and J. Zupan, arXiv:1203.2064 [hep-ph] (2012).
23. M. Visser, *Phys. Rev. D* **54**, 5116 (1996).
24. S. Carroll, *Spacetime and Geomety: An Introduction to General Relativity* (Addison-Wesley, Boston, 2004).
25. For the CoGeNT Experiment see the official website: http://cogent.pnnl.gov.
26. For the XENON Experiment see the official website: http://xenon.astro.columbia.edu/XENON100_Experiment/.
27. G. Narain, J. Schaffner-Bielich, and I. N. Mishustin, *Phys. Rev. D* **74**, 063003 (2006).
28. P. Mukhopadhyay and J. Schaffner-Bielich, *Phys. Rev. D* **93**, 083522 (2016).
29. N. K. Glendenning, in *Compact Stars, Nuclear Physics, Particle Physics, and General Relativity* (Springer, New York, 2000).

Some Aspects of Nuclear Structure

J. H. Hamilton, A. V. Ramayya and E. H. Wang
Physics Department, Vanderbilt University, Nashville, TN 37235, USA

The one phonon γ-vibrational bands in ^{154}Gd are compared with the theories existing at that time (1960) showing the best agreement with the rotation–vibration model of Faessler and Greiner. Nuclear shape coexistence was first discovered in both ^{72}Se and $^{184-188}$Hg with the Frankfurt Generalized Collective Model reproducing the two different shapes. In recent years experimental evidences have been found for both one, two, and possibly three phonon γ bands in the A = 103–110 region including the first such bands in odd N-even Z and even N-odd Z nuclei. Cluster radioactivity predicted by Sandulescu, Poenaru and Greiner has been observed. Maximum cluster radioactivity is seen in the zero neutron emission in cold binary and ternary fission of ^{252}Cf. Finally, recent super heavy element discoveries of the new elements 115, 117 and 118 are described. The half-lives in 117 which is closest in N to N = 184 on the Island of Stability support the theoretical predictions of the long-lived island.

1. Introduction

Gamma ray spectroscopy with heavy ions has developed into a vast field of research with many exciting possibilities. The collision between heavy nuclei can be roughly classified into four different categories determined by the impact parameter b, or the corresponding relative angular momentum. This paper will describe the intertwining of Professor Greiner's theoretical work and our experimental studies from the past through the present in several different areas of research from γ-vibrational bands in deformed nuclei, through nuclear shape coexistence and the generalized collective model on to super heavy nuclei.

2. Gamma-Vibrational Bands

Professor Faessler and Greiner[1] first introduced the rotation vibrational model to explain the deviations of the β and γ band properties from those of the Bohr–Mottelson model of deformed nuclei.[2] These deviations are explained in terms of band mixing calculations as shown in Table 1.[3] Better agreement with the Faessler–Greiner Model was found in our early work on these B(E2) branching ratios from these bands in ^{154}Gd,[4] as seen in Table 2.

Table 1. Experimental and theoretical ratios of reduced E2 transition probabilities from members of the β bands in ^{152}Sm and ^{154}Gd. Data are taken from Ref. 3. Here Z_0 denotes the mixing parameter[5] between the ground and the β band extracted from experimental B(E2) ratios.

Nucleus	(I'→I1)/(I'→I2)	B(E2;I1→I2)/B(E2;I3→I4)			
		Experiment	$-Z_0 \times 10^2$	Theory	
				B-M with mixing[a]	Davidson[b]
^{152}Sm	(2+'→4+)/(2+'→2+)	2.47±0.88	1.2±1.5	5.36±0.61	4.28
	(2+'→0+)/(2+'→2+)	0.19±0.05	8.0±1.1	0.33±0.04	0.38
	(2+'→4+)/(2+'→0+)	13.0±5.2	4.5±1.2	16.2±3.8	11.3
^{154}Gd	(2+'→4+)/(2+'→2+)	2.97±0.75	2.0±1.2	6.75±0.49	4.34
	(2+'→0+)/(2+'→2+)	0.12±0.03	9.8±0.8	0.25±0.03	0.37
	(2+'→4+)/(2+'→0+)	25.4±8.0	6.5±0.49	27.0±4.7	11.7
	(4+'→2+)/(4+'→4+)	0.11±0.11	4.9±1.1	0.004±0.01	0.24

a: The predictions of the Bohr–Mottelson model are corrected for band mixing through the use of weighted averages of the Z_0 values in column 4. For ^{154}Sm this average is $-(5.2 \pm 0.7) \times 10^{-2}$; for ^{154}Gd, $-(6.7 \pm 0.5)10^{-2}$.
b: Predictions result from asymmetric rotor model with $\gamma = 10.73°$, $\mu = 0.40$ for ^{152}Sm and $\gamma = 11.62°$, $\mu = 0.40$ for ^{154}Gd.

Table 2. Comparison of experimental ratios of reduced E2 transition probabilities from β and γ bands of ^{154}Gd to various model predictions. Data are taken from Ref. 4.

(I1→I2)/(I3→I4)[a]	B(E2;I1→I2)/B(E2;I3→I4)				
	Experiment	Theory			
		Adiabatic B-M[b]	R-V[c]	D-F[d]	D-C[e]
(4'→4)/(4'→2)	6.41±0.97	2.94	2.63		9.20
(3'→4)/(3'→2)	1.05±0.06	0.40	1.00		1.01
(2'→2)/(2'→4)	7.34±0.95	20.0	5.89		8.94
(2'→2)/(2'→0)	2.30±0.12	1.43	2.04	2.78	2.28
(2'→0)/(2→0)	0.035±0.014				0.035
(2''→4)/(2''→2)	3.15±0.30	1.8	3.0		4.34
(2''→0)/(2''→2)	0.12±0.02	0.7	0.33		0.37
(2''→4)/(2''→0)	25.2±3.7	2.6	9.1		11.6
(4''→2)/(4''→4)	0.09±0.06	1.1	0.17		0.24
(2''→0)/(2→0)	0.013		0.042	0.044	0.036

a: Primes denote a member of γ band, double primes a member of the β band and no primes of the ground state band.
b: Adiabatic symmetric-rotor model of Bohr and Mottelson.
c: Rotation–vibration model of Faessler, Greiner and Sheline.
d: Adiabatic asymmetric-rotor model of Davydov and Filippov.
e: Davydov–Chaban nonadiabatic asymmetric-rotor model.

3. Shape Coexistence in Complex Nuclei

Nuclear shape coexistence, the coexistence of energy levels built on two different nuclear shapes in the same nucleus, was first predicted by Hill and Wheeler.[6] Then Morinaga[7] observed high energy excited states in ^{16}O built on two quite different shapes, Greiner[8] introduced nuclear shape coexistence to explain the excited bands in ^{110}Cd. Then Soloviev[9] and Kumar[10] called for searches to be made for shape coexisting energy levels.

In 1974, while we were studying ^{16}O on ^{58}Ni, we discovered overlapping of energy levels of different shapes in ^{72}Se.[11] Later when we were studying decay of Tl to Hg isotopes at UNISOR, we discovered also shape coexistence in ^{184}Hg[12] built on near spherical and well deformed shapes. While some theoretically opposed the suggestion of shape coexistence in such complex nuclei, these nuclei are now textbook examples of shape coexistence as seen in Eisenberg and Greiner,[13] Figs. 1 and 2. The theoretical potential energy calculation in Fig. 1 for ^{72}Se[11] were carried out by two of Greiner's former students in the Generalized Collective Model introduced by Gneuss, Mosel and Greiner.[14] The overlapping near spherical and well deformed band seen in A = 182 to 188 for mercury nuclei far off stability are shown in Fig. 2. Nuclear shape coexistence is now seen throughout the Periodic Table.

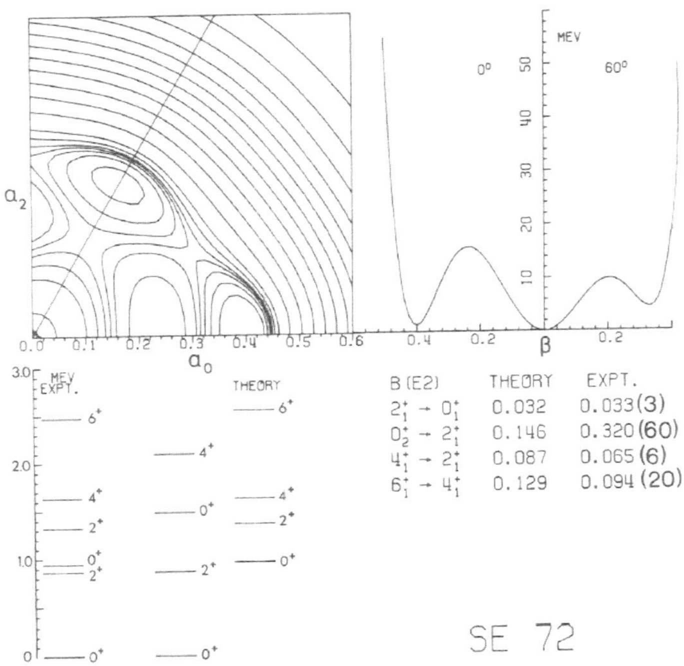

Fig. 1. First example of shape coexistence in a complex nucleus with calculations by the Frankfurt group.[11,13]

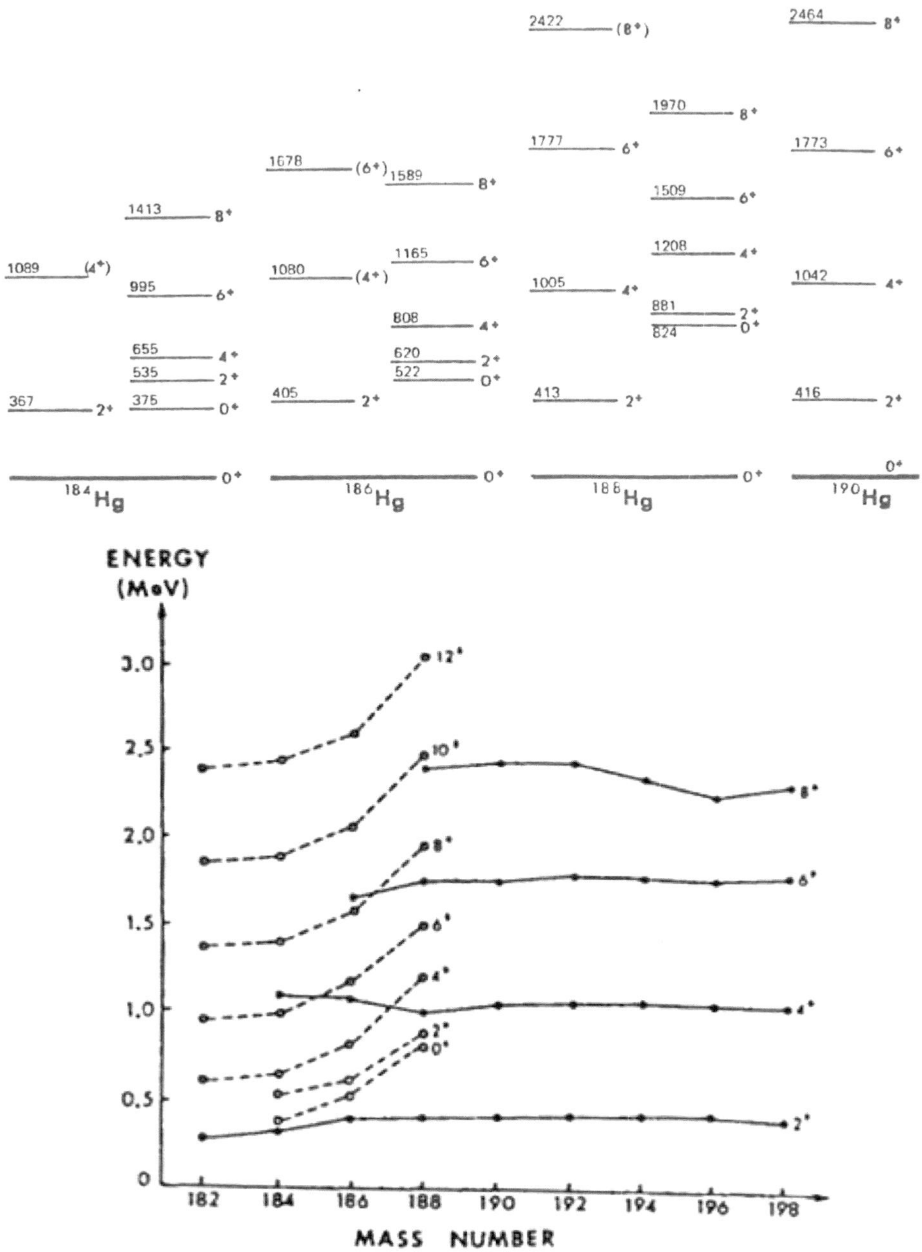

Fig. 2. First evidence for shape coexistence in heavy elements, Vanderbilt-UNISOR.[12]

4. Multi-phonon Gamma-Vibrational Bands

Gamma-vibrational bands have continued through the present to be important both experimentally and theoretically. We found the first evidence for a one and two phonon γ-vibrational band in an odd-A nucleus in ^{105}Mo [15] and in odd-Z nucleus in ^{103,105}Nd.[16] We have experimentally identified for the first time the one and two phonon γ-vibrational bands in odd-N-even-Z and even-N, odd-Z nuclei ^{103,105,107}Mo, ^{103,105}Nb and ^{107,109}Tc. An example of these results for ^{105}Nb is shown in Fig. 3. Also in ^{103}Nb one and two phonon γ bands and a third band candidate for a three phonon γ band are found which are essentially identical with their band heads within 50–85 keV of each other in ^{103,105}Nb. Note a band with all the properties of a three phonon band also are seen in this nucleus for the first time. Figure 4 shows a comparison of the odd-A and even-A Mo nuclei where γ-vibrational bands are seen. The energies of the one and two phonon γ bands in the odd-A and even-A are remarkably consistent in $^{103-108}$Mo to provide significant tests of microscopic theories.

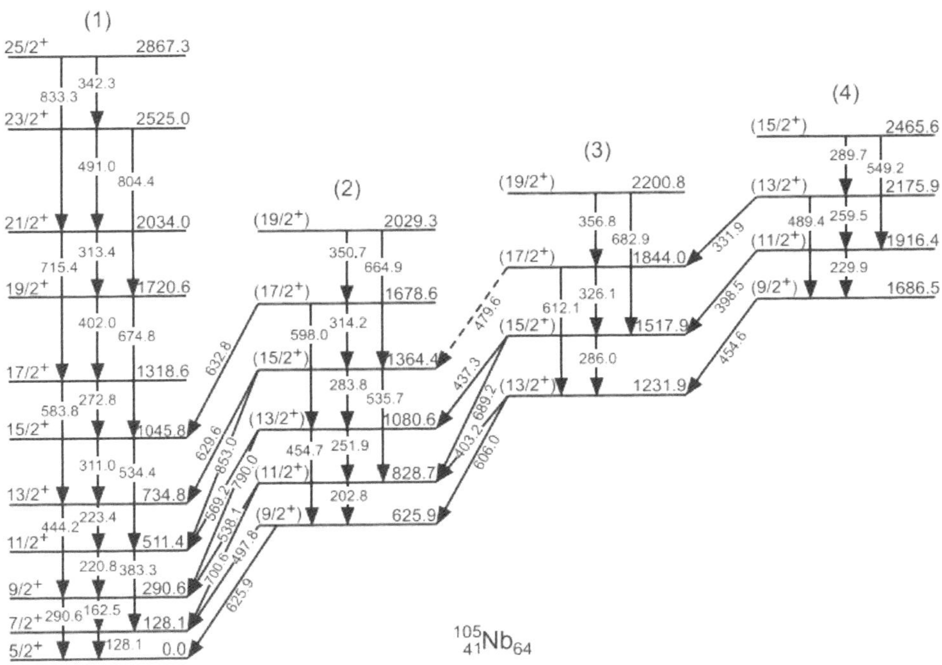

Fig. 3. Example of one and two and possibly a third γ-vibrational band.[16]

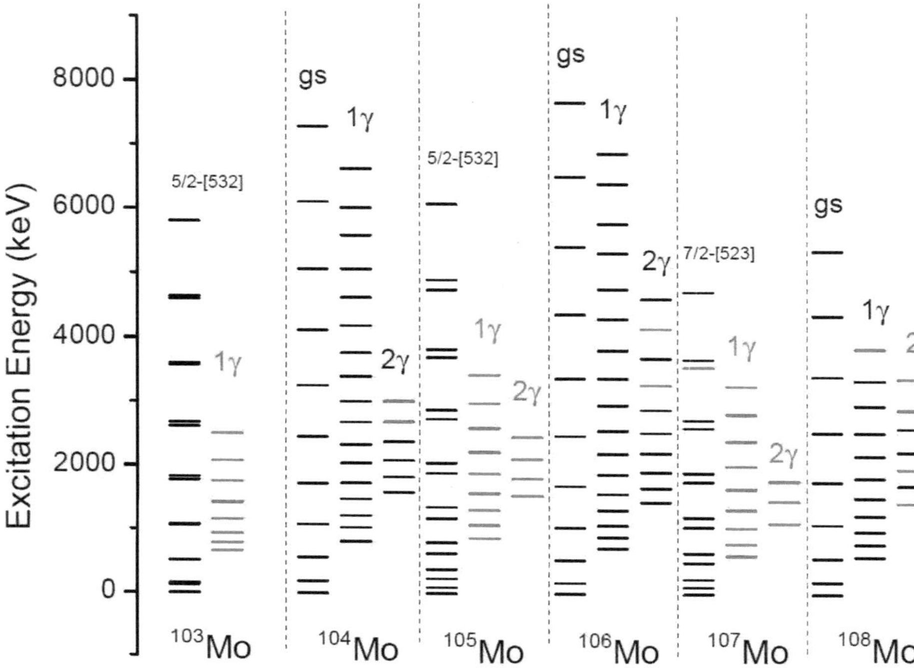

Fig. 4. Systematic comparison of the one and two phonon γ-vibrational bands in $^{103-108}$Mo. Such bands in ^{103,107}Mo were newly identified. The red levels are our new data.

5. Cluster Radioactivity and Cold Binary and Ternary Fission

We carried out an experiment with ^{252}Cf at the Lawrence Berkeley National Laboratory (LBNL). A 62-μCi ^{252}Cf source was sandwiched between two Fe foils of thickness 10 mg/cm^2. A 7.62 cm diameter plastic (CH) ball was used to absorb the beta rays and conversion elections, and partially moderate neutrons. A total of 101 Ge detectors of Gammasphere were used. The data were sorted into 5.7×10^{11} γ-γ-γ and higher fold γ events (3D cube) and 1.9×10^{11} γ-γ-γ-γ and higher fold γ coincident events (4D cube). These γ coincident data were analyzed by the RADWARE software package.

Sandulescu and Greiner[17] predicted in 1977 the new phenomena of cluster radioactivity where clusters like ^{14}C are emitted in decays to lead nuclei. The wide variety of such clusters are described in their review of the experimental and theoretical work on cluster radioactivity.[18] Together they published a paper entitled "Exotic Nuclear Decay of ^{223}Ra by Emission of ^{14}C Nuclei"[19] to point out the theoretical prediction before the experimental observing.

Then we discovered zero neutron emissions of binary spontaneous fission into ^{148}Ba + ^{104}Mo, ^{146}Ba–^{106}Mo, ^{144}Ba–^{108}Mo, ^{148}Ca–^{104}Zr, (Fig. 5) the ultimate (largest) cluster radioactivity co-authored with Prof. Greiner.[20] We followed this

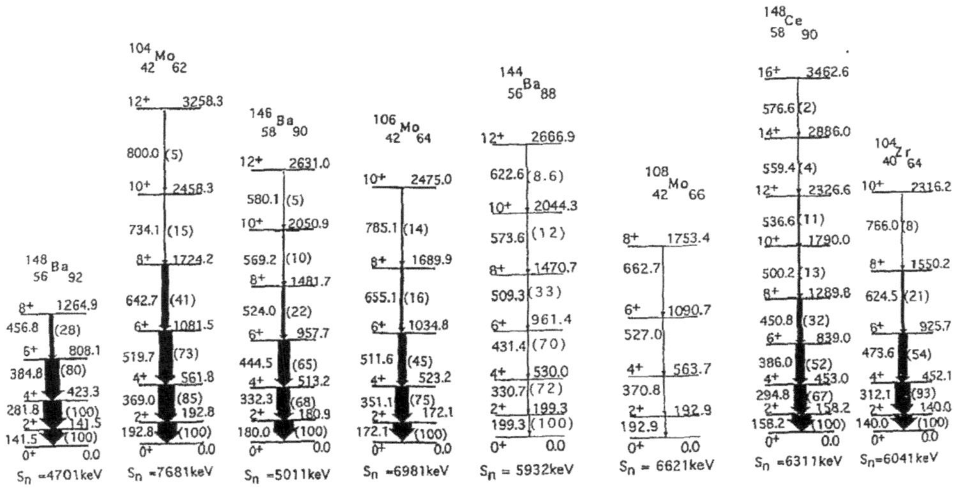

Fig. 5. Zero neutron emission in spontaneous fission of ^{252}Cf: A form of cluster radioactivity.

with a theoretical paper including Greiner on such cluster radioactivity.[21] We also discovered zero neutron emission cold ternary spontaneous fission with the third particle being an α particle another form of cluster radioactivity, co-authored with Greiner along with several theoretical studies.[22]

6. Super Heavy Elements

In 1960, macroscopic nuclear theory predicted that elements with Z $>$ 100 should not exist since they would have spontaneous fission lifetimes of $\leq 10^{-14}$ s. The introduction of microscopic shell corrections to explain the fission barriers of the actinides led to the prediction of new deformed and spherical shell gaps for Z $>$ 100 and N $>$ 150 as shown in Fig. 6. Several groups, including Prof. Greiner,[23–26] in the second half of the 1960s predicted that elements beyond Z $>$ 100 should be observed leading up to an Island of Stability around a spherical shell gap at N $=$ 184 with different theories predicting a spherical shell gap for Z at 114 or 120 or 126. The isotopes in and around the Island of Stability were predicted to have lifetimes possibly even as long as the age of the Earth. These 1960s predictions led to various laboratories seeking to push the discoveries of new elements with Z $>$ 102 by different techniques. A review of all approaches can be found in Ref. 27.

The GSI laboratory used what is called cold fission that involves bombarding stable targets of ^{208}Pb and ^{209}Bi with projectiles of ^{48}Ca to ^{64}Ni. The reactions are called cold fusion because after fusion the nucleus of the new element had only enough internal energy to boil off one neutron. By using cold fusion, the new elements with Z $=$ 107–112 were discovered at GSI and an isotope of 113 was found in RIKEN in Japan.

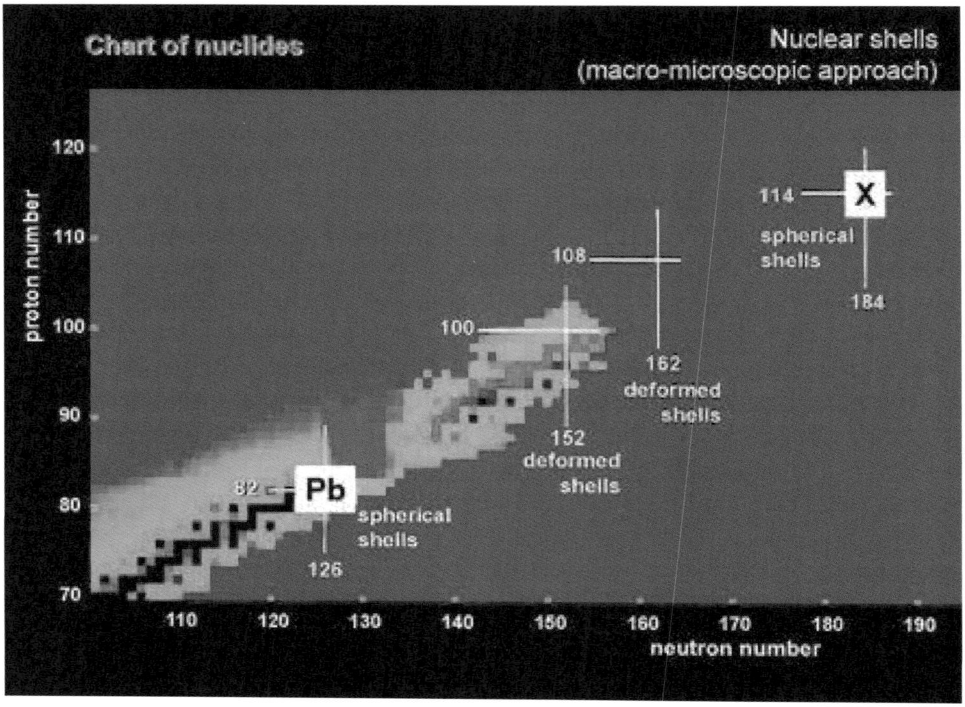

Fig. 6. Chart of nuclides showing deformed and spherical shell gaps.

The Flerov Laboratory for Nuclear Reactions pioneered hot fusion reactions to discover new elements with Z = 113–118. In hot fusion, the internal energy is sufficient to boil all 3–5 neutrons. New elements 114 and 116 were first identified and certified. The discovery of these two elements were approved by IUPAC in June, 2011. They were given the names Flerovium for 114 and Livermorium for 116.

In 2005 Oganessian, who had been cooperating in nuclear structure research with Vanderbilt for over 10 years, asked Vanderbilt's help to obtain ^{249}Bk target material from the High Flux Reactor in Oak Ridge National Laboratory in order to discover the unknown element 117. Oganessian and Hamilton met with the reactor scientists. The cost of making just a ^{249}Bk target was prohibitive but were told that from time to time commercial companies pay for a campaign to make ^{252}Cf. The ^{249}Bk is a by-product of making ^{252}Cf and for \$600,000 they could chemically separate out the ^{249}Bk. However, they had no such campaign planned. So for the next three years, Hamilton called the reactor scientists every three months to see if a campaign was under way. Three years later, in August 2008, the answer was yes. In September 2008, Vanderbilt held a small symposium in honor of Hamilton's 50 Years of Teaching and Research at Vanderbilt. At the meeting, Hamilton arranged a luncheon with Oganessian and J. Roberto, Deputy

Director of ORNL, to have Oganessian explain the importance of the ^{249}Bk target. Roberto was excited by the presentation and subsequently asked K. Rykaczewski (ORNL) and Hamilton to write him a proposal for \$500,000 which they did and it was funded. Then Hamilton called his collaborator in structure research, M. Stoyer at Lawrence Livermore National Laboratory, to see if LLNL would like to join the experiment and provide the additional \$100,000 for the cost of the chemical separation to which they agreed.

The target material came out of the reactor in December 2008 and was allowed to cool for three months to allow the short lived radioactinide to decay. In March 2009 Oganessian and Hamilton went to ORNL to see the start of the chemical separation which took three months (Fig. 7). The ^{249}Bk is in the long aluminum tube inside the first hot cell (Fig. 7). The target material was sent to Dubna in June and the experiment began in late July with the Vanderbilt group there for the starting week.

Fig. 7. Hamilton and Oganessian observing the radiated target with Berkelium as it came out the ORNL reactor.

The experiment was successful and was published in Physical Review Letters in April 2010 with only three days between submission and acceptance. The data presented were featured on the cover (Fig. 8).[28] This was the first new element to be reported in over five years and was reported in over 250 news reports in newspapers and TV worldwide. The results were presented at the International Nuclear Physics Conference in Vancouver on June 20, 2010[29] and also highlighted in another talk

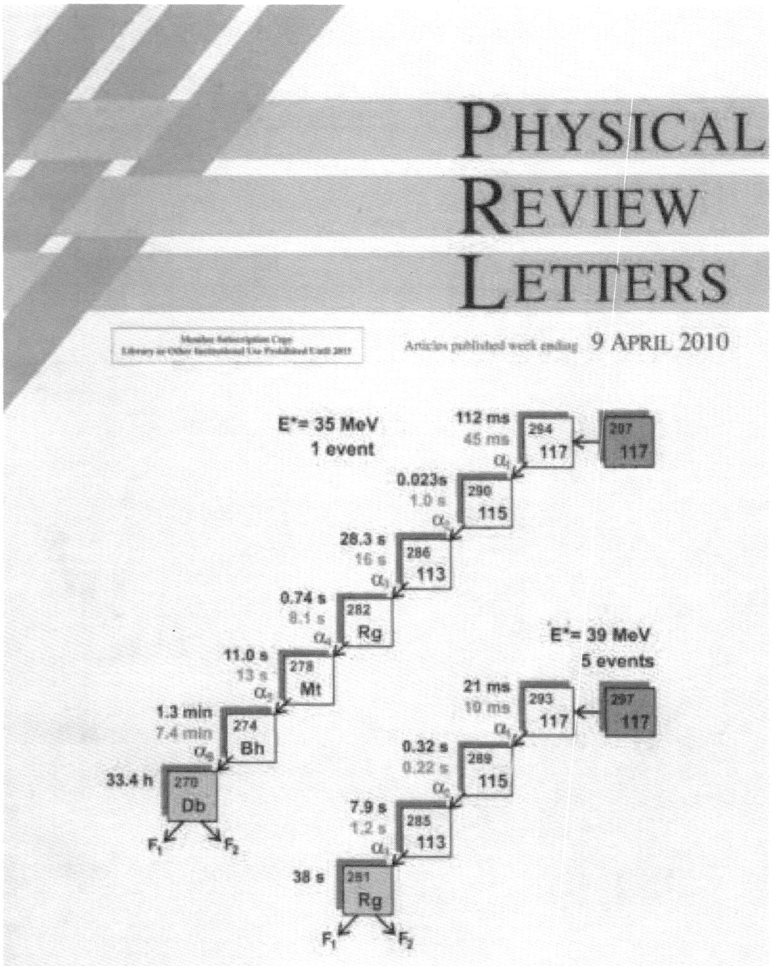

Fig. 8. Evidence for the discovery of the new element 117.

there as an example of the US Department of Energy's response to the President's request to carry out important basic research.

To provide definitive evidence for the new elements 113 and 115, the FLNR, ORNL, VU, LLNL collaboration carried out two sets of experiments from 11/2010–2/2012 of the reaction ^{243}Am + ^{48}Ca at five beam energies.[30,31] The results were first reported at the Nucleus–Nucleus Collisions Conference in May 2012.[29] We obtained 28 new events of ^{289}115 and very important four events of ^{289}115 for the first time in this reaction. These later ^{289}115 events provided cross bombardment check to give definitive evidence for the discoveries of 113, 115, 117 as shown in Fig. 9.[30,31]

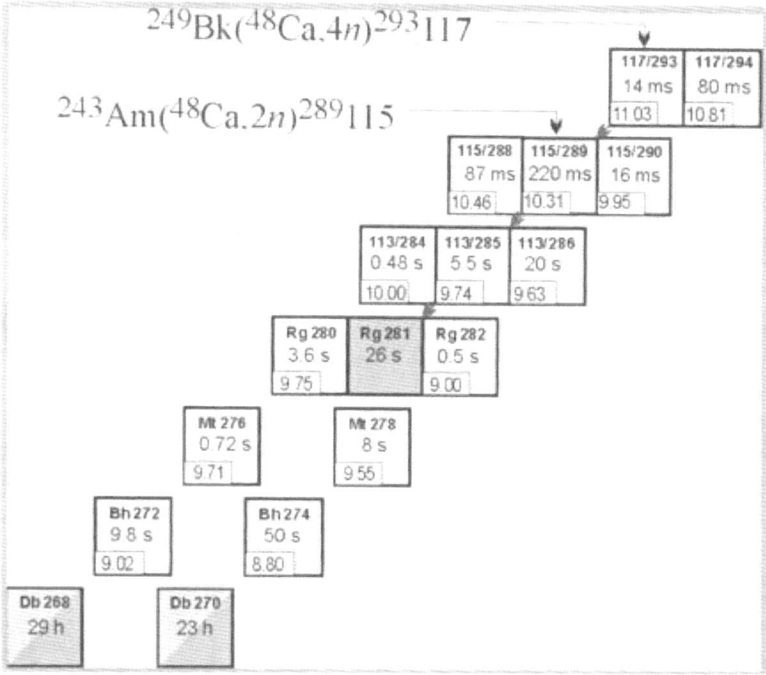

Fig. 9. Cross bombardment checks give definitive evidence for 113, 115, 117.

With new ^{249}Bk target material divided between FLNR and TASCA at GSI, 11 new events of ^{293}117 and three new events of ^{294}117 were found at FLNR along with one new ^{249}Cf + ^{48}Cf $\rightarrow$ ^{294}118 + 3n reaction toward the end of the run where ^{249}Bk had beta decayed to ^{249}Cf which was then 30% of the target.[32] At TASCA at GSI, the reaction ^{249}Bk + ^{48}Ca was studied and our discovery[28] of the new element 117 was confirmed.[33]

December 30, 2015, International Union of Pure and Applied Chemistry (IUPAC) approved our discoveries and asked our collaboration to submit their names. Spring, 2016, in a video conference call between Russian, ORNL, VU, LLNL scientists the names moscovium, tennessine, oganesson were submitted. June 8, 2016, these proposed names were accepted as provisional names with 5 months of public review. In November 2016, the IUPAC approved the names of elements 113, 115, 117, 118 including of the new element 117 tennessine after our Tennessee state, with symbol Ts as proposed by Vanderbilt group.

In 2010, while serving on a Review Committee for the High Flux Reactor, Hamilton found out that there were old ^{252}Cf sources which had been returned with mixed ^{249,250,251}Cf isotopes that included tens of milligrams of ^{251}Cf. On discussing this discovery with ORNL and FLNR, it was decided to seek to make a mixed target to study the ^{251}Cf + ^{48}Ca $\rightarrow$ ^{295,296}118 + 4, 3n reactions. With two neutrons closer to the center of the Island of Stability at N = 184, $^{296}118$ should have a factor of 100 or more longer lifetime and its decay products too. These longer lifetimes should open up studies of the chemistry of elements with $Z \geq 113$ to test theoretical predictions that these nuclei should have different chemical projections than elements above them in the Periodic Table. Oak Ridge successfully made a target wheel of this very radioactive material and this experiment is now in progress in Dubna.

7. Our Collaborations

In our collaborations, Professor Greiner and Vanderbilt scientists co-authored 20 regular journal articles plus 8 letter articles, 36 conference proceedings, 1 book article and 5 articles in books edited by Professor Greiner. In addition, 6 articles were co-authored with his former Ph.D. students. From 1979–93, Greiner gave a full course every spring in theoretical physics at Vanderbilt for graduate students and brought 2 of his young graduate students with him. Hamilton received a Humboldt Prize in 1979–80 to work in Frankfurt and GSI and later spent 2 months in the summer on 3 occasions. Ramayya received a Humboldt Award for a year's research at Frankfurt.

We have especially enjoyed our close friendship and collaboration with him for over 45 years. We will sincerely miss his inciteful discussions in our collaborations and pointing us to new physics.

Acknowledgments

The work at Vanderbilt University was supported by the US Department of Energy through grant DE-FG-05-88ER40407.

References

1. A. Faessler and W. Greiner, *Z. Phys.* **177**, 190 (1964); A. Faessler, W. Greiner and R. K. Sheline, *Nucl. Phys.* **70**, 33 (1965).
2. A. Bohr and B. R. Mottelson, *Kgl. Danske Videnskab. Selskab, Mat.-Fys. Medd.* **27**, N016 (1953).
3. L. L. Riedinger, N. R. Johnson and J. H. Hamilton, *Phys. Rev. Lett.* **19**, 1243 (1967).
4. L. L. Riedinger, Ph.D. Thesis, Vanderbilt University, 1969.
5. Bohr, A. and Mottelson, B. R., Nuclear Structure. Number v. 1,2 in Nuclear Structure. (1969).
6. D. L. Hill and J. A. Wheeler, *Phys. Rev.* **89**, 1102 (1953).
7. H. Morinaga, *Phys. Rev.* **101**, 254 (1956).
8. W. Greiner, in *Proceedings of the International Conference on Neutron Capture*, Jülich, Germany (1965).
9. V. G. Soloviev, *Phys. Lett.* **21**, 311 (1966).
10. K. Kumar, in the *Electromagnetic Interaction in Nuclear Spectroscopy*, W. D. Hamilton, ed., (North-Holland, Amsterdam, 1975), p. 119.
11. J. H. Hamilton *et al.*, *Phys. Rev. Lett.* **32**, 239 (1974).

12. J. H. Hamilton *et al.*, *Phys. Rev. Lett.* **35**, 562 (1975).
13. J. M. Eisenberg and W. Greiner, *Nuclear Models* (North-Holland, Amsterdam, 1987), pp. 278–285.
14. G. Gneuss, U. Mosel and W. Greiner, *Phys. Lett. B* **30**, 397 (1969).
15. H. B. Ding *et al.*, *Phys. Rev. C* **74**, 054301 (2006).
16. J. G. Wang, S. J. Zhu, J. H. Hamilton *et al.*, *Phys. Lett. B* **675**, 430 (2009); H. H. Liu *et al.*, *Phys. Rev. C* **88**, 054311 (2013).
17. A. Sandulescu, D. N. Poenaru and W. Greiner, *Sov. J. Part. Nucl.* **11**, 528 (1980).
18. A. Sandulescu and W. Greiner, *Rep. Prog. Phys.* **55**, 1423 (1992).
19. A. Sandulescu, D. N. Poenaru, W. Greiner and J. H. Hamilton, *Phys. Rev. Lett.* **54**, 490 (1985).
20. J. H. Hamilton *et al.*, *J. Phys. G: Nucl. & Part. Phys.* **20**, L85 (1994).
21. A. Sandulescu *et al.*, *Int. J. Mod. Phys. E* **7**, 625 (1998).
22. A. V. Ramayya *et al.*, *Phys. Rev. C* **57**, 2370 (1998).
23. A. Sobiczewski *et al.*, *Phys. Lett.* **22**, 500 (1966).
24. H. Meldner, *Ark. Fys.* **36**, 593 (1967).
25. S. G. Nilsson *et al.*, *Nucl. Phys. A* **115**, 545 (1968).
26. U. Mosel and W. Greiner, *Z. Phys.* **228**, 371 (1969).
27. J. H. Hamilton, S. Hofmann, and Y. T. Oganessian, *Annu. Rev. Nucl. Part. Sci.* **63**, 383–405 (2013).
28. Yu. Ts. Oganessian *et al.*, *Phys. Rev. Lett.* **104**, 142502 (2010).
29. J. H. Hamilton *et al.*, *J. Phys. Conf. Series* **312**, 082026 (2011).
30. Yu. Ts. Oganessian *et al.*, *Phys. Rev. Lett.* **108**, 022502 (2012).
31. J. H. Hamilton *et al.*, 11th Int. Conf. on NN Collisions, *J. Phys. Conf. Series* **420**, 012011 (2013).
32. Yu. Ts. Oganessian *et al.*, *Phys. Rev. C* **87**, 054621 (2013).
33. J. Khayagbaatar *et al.*, *Phys. Rev. Lett.* **112**, 172501 (2014).

The Power of the Geometrical Model of the Nucleus

Peter O. Hess

Instituto de Ciencias Nucleares, UNAM, Circuito Exterior,
C.U., A.P. 70-543, 04510 México, D.F., Mexico
Frankfurt Institute for Advanced Studies,
D-60438 Frankfurt am Main, Germany
hess@nucleares.unam.mx

The first and very fruitful contribution of Walter Greiner was in the area of the Geometrical Model of the nucleus, namely on the *Rotation–Vibration Model* (RVM), also extended later on to the *General Collective Model* (GCM). Further developments were on the microscopic relation of this model. A particular application of the RVM was its extension to *Nuclear Molecules*. In this contribution, we give a resumé on all these developments and recent advances. A primary on group theory is presented, which is important to understand the relation to the *Interactive Boson Approximation* and it will be shown that they are intimately connected.

1. Introduction

In 1976 I entered the *Institute of Theoretical Physics* at the *Johann-Wolfgang von Goethe University* (Frankfurt am Main, then West-Germany). Walter proposed as a topic the *Geometrical Model of the Nucleus* (GMN),[1,2] with a large weight on group theory. The reason for that was that just before M. Moshinsky and collaborators[3,4] have developed a basis for that model, using heavy group theory. I was very interested in these techniques and accepted. Since then, Walter was one of my principle advisers, in science and privately. Through this collaboration, the GMN was extended to non-polynomial terms in the collective potential.[5,6]

Walter was then already known as one of the founders of the *Rotation–Vibration Model* (RVM),[7] which has been extended in Ref. 1 to more general potentials. My homework consisted in extending it to non-polynomial terms and to use the new found basis.[3,4]

But our collaboration did not end there. Later we extended the RVM to nuclear molecules of two and three clusters,[8,9] showing how powerful geometric considerations can be.

In the mid 1970s an alternative model came on the market, namely the *Interaction Boson Approximation* (IBA),[10] which introduces very powerful methods, based on group theory (or a more catchy word: *symmetry*), into nuclear physics. In this contribution, some remarks will be given on the relation to the geometric model.

The chapter is organized as follows: In Section 2 the RVM is briefly discussed and the extension to nuclear molecules. In Section 3, recent and not so recent developments on the relation to microscopic models are explained. We will show that the *Symplectic Model* of the nucleus[11,12] and its contracted version[13] can be used to derive a *Potential Energy Surface* (PES) of the geometric quadrupole model of the nucleus, after applying a geometrical mapping using coherent states.[14] We report on some recent developments which enable us to even discuss the quadrupole structure of super-heavy elements, at least approximately. It will be shown that the driving force for the structure of the potential is the observation of the *Pauli Exclusion Principle* (PEP). In Section 4 we present a brief *primary course* on group theory, as used to construct algebraic models. We show the marvelous power of this technique but also that there is a danger to it. It will be applied in Section 5 to the geometric model, which will be reformulated in terms of creation and annihilation operators and also to the IBA. It will be clear how both models are related and complementary. This will close the circle of contributions and important consequences of the work of Walter on the field of geometric models. Other chapters in this book will discuss various important contributions, initiated and/or realized by Walter.

2. The Rotation–Vibrational Model, its Extension and Nuclear Molecules

With J. M. Eisenberg, W. Greiner published three books on Nuclear Physics, which still have an important impact in the community. In the first one,[2] his first mayor contribution in nuclear physics is presented in all its details, namely the *Rotation–Vibration Model* (RVM).[7]

The point of partition is the definition of the collective variables α_{2m} ($m = \pm 2, \pm 1, 0$) which describe the expansion of the nuclear surface in the quadrupole degree of freedom. These variables are transformed into the *intrinsic system* where the quadrupole operator is diagonal. Denoting in this system the deformation variables as a_m, they are related to the deformation β and the triaxiality γ via

$$a_0 = \beta\cos\gamma \ , \ a_{+1} = a_{-1} = 0 \ , \ a_2 = a_{-2} = \frac{\beta}{\sqrt{2}}\sin\gamma \ . \tag{1}$$

The collective potential is expanded around the minimum located at ($a_0 = \beta_0 + \xi$, $a_2 = \eta$), for prolate nuclei.

The collective potential is a function in these variables and in the RVM an expansion around a deformed position is made, namely

$$V = \frac{1}{2}C_0\xi^2 + C_2\eta^2 \ . \tag{2}$$

The kinetic energy needs a more involved treatment. Starting from the Bohr-Mottelson kinetic energy[15] it is a sum of a vibrational kinetic energy (in ξ and η) and a mixed term which describes the *Rotation–Vibrational* coupling. This coupling leads to a moment of inertia which varies within a band toward higher spin.

The RVM was one of the first to describe the collective spectra of deformed nuclei (besides the Davidov Model[2] and the microscopic $SU(3)$ Model by Elliott[16]). It was also the starting point to more involved models, based on the quadrupole degree of freedom, which are all described in Ref. 2. Today, this group of models is generally referred as the *Frankfurt* model and is still in use.

It also inspired the extension to nuclear molecules[2,8] with two clusters and later to three clusters.[9] New modes appeared as the butterfly motion of a two cluster molecules (see Fig. 1). This work also found its way into text books.[17]

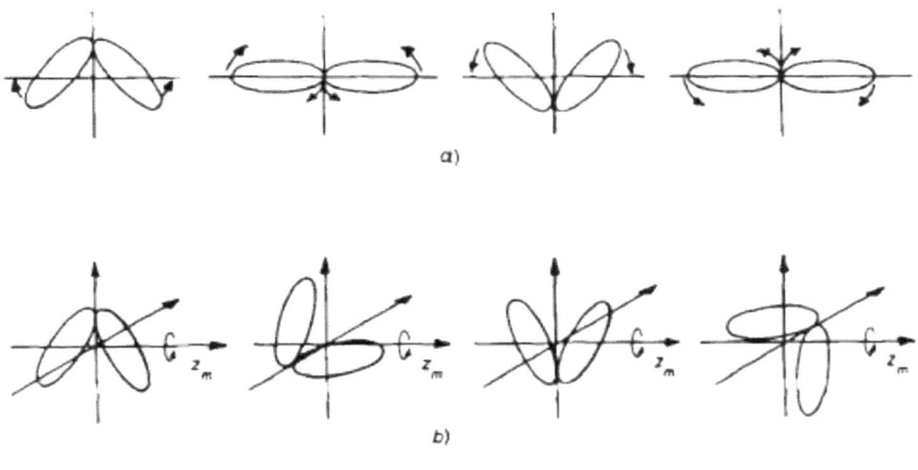

Fig. 1. One of the main modes for a two-cluster molecule, where both clusters are well separated, are the *butterfly motion* (upper part of the figure) and the *Belly-Dancer mode* (lower part of the figure).

As in the RVM, analytic solutions for the two- and three-cluster molecules were obtained. For the two cluster molecules, the analytical formula is

$$E = \frac{\hbar^2}{2\mu r_0^2}\left[I(I+1) - K^2\right] + \hbar\omega_r\left(n_r + \frac{1}{2}\right) + \hbar\omega_K\left(\mid K \mid + 2n_\epsilon + 1\right), \tag{3}$$

where $\hbar\omega_r$ and $\hbar\omega_K$ are the frequency of the relative and butterfly motion, n_r is the oscillation quanta of the relative motion and n_ϵ the one of the butterfly motion. The I is the spin and K its projection to the intrinsic z-axis. The μ is the reduced mass of the system and r_0 the distance of the two centers of the two clusters. The *Belly-Dancer mode* is obtained through a rotation around the z-axis.

3. Microscopic Relation of the GCM to the Shell Model

In Refs. 11 and 12 the symplectic model of the nucleus is introduced. It is based on the shell model and the Hamiltonian is a function of the $SU(3)$ generators of the shell model and further generators which connect states with a difference of $2\hbar\omega$ in the oscillator shells. Important here to know is that the band-head of each symplectic irrep is given by a $SU(3)$ irrep of the shell model in the $0\hbar\omega$ shell. The model was constructed such hat the similarities to the geometric quadrupole model becomes obvious. This is also the reason why we used it as a starting point to map from the microscopic Hamiltonian a quadrupole PES. Instead of the full symplectic model its contracted version[13] is used, which means nothing but to approximate the generators, which connect shells with $2\hbar\omega$ difference, by simple bosons. In Ref. 18 the mapping to the collective potential is explained and goes as follows:

The details can be deduced from Ref. 18. One corner stone is the contracted Symplectic Model (PSM).[13] This Hamiltonian provides a measure for the relative distances in energy of the $SU(3)$ irreducible representations (irreps). Here we outline briefly the procedure exposed in a more detailed form in Refs. 18–20.

The PSM is an extension of the pseudo-$SU(3)$ model[21,22] taking into account $2n\hbar\omega$ shell excitations. Within this model the orbitals are divided into intruders, called *unique orbitals* and the *normal orbitals*, which are presented by the rest. The intruders are the orbitals with maximal spin in each oscillator shell. By inspecting the normal orbitals and redefining the orbital angular momentum into pseudo-orbital angular momentum and pseudo-spin, a new degeneration is observed, with the same content of an oscillator shell with one quantum less. This allows us to introduce a pseudo-$SU(3)$ (denoted from here on as $\widetilde{SU}(3)$).[21,22] The PSM is equivalent to the extension of the Elliott model of $SU(3)$[16] to the symplectic model,[11,12] including an approximation called contraction.[13] In the PSM the nucleons in the intruder orbitals are treated as spectators. In order to take into account the nucleons in the intruder orbitals, an effective charge with a definite scaling is used in terms of Z and A, the number of charge and nucleons, and Z_n and A_n the same for the particles in the normal orbitals.

In order to determine Z_n and the number of neutrons N_n in the normal orbitals ($A_n = Z_n + N_n$), one needs the external information of the deformation of the nucleus and the Nilsson diagram for either $\beta < 0$ (or equivalently $\beta > 0$ with $\gamma = 60°$) or $\beta > 0$. The deformation can be deduced experimentally[23] or by a microscopic model, which is indispensable for obtaining a Nilsson diagram. In some cases experimental deformation values are used. Here, we consult the macroscopic-microscopic calculations of Möller *et al.*, for both the quadrupole deformation[24] and the Nilsson diagram.[25] In this sense, our result will contain the effects of relevant microscopic interactions. Once the numbers β, Z_n and N_n are determined, the content of the shell model space in the valence shells (protons and neutrons) is determined using the tables of Ref. 26 or the $SU(3)$-package routines of Refs. 27–29. This gives a list of $SU(3)$ irreps (λ_π, μ_π) for protons and (λ_ν, μ_ν) for neutrons. The pseudo-symplectic space is then obtained by allowing $2n\hbar\omega$ excitations on top of $(\lambda, \mu) = (\lambda_\pi + \lambda_\nu, \mu_\pi + \mu_\nu)$, i.e. the stretched $SU(3)$ irreps of the proton and neutron part. Only stretched irreps contribute because no-stretched irreps represent protons moving against neutrons, i.e., giant resonances corresponding to states at higher energies.

In order to estimate the relative positions in energy of the different symplectic irreps, the following Hamiltonian is used

$$H = \hbar\omega\tilde{N} - \frac{1}{2}\chi\left[Q^c \cdot Q^c - (Q^c \cdot Q^c)_{\mathrm{shell}}\right], \tag{4}$$

where $\tilde{N}$ is the phonon number operator of the protons plus the neutrons, $\hbar\omega$ is $45A^{1/3}$-$25A^{2/3}$ MeV for light and $41A^{1/3}$ MeV for heavy nuclei. The Q^c_m is the physical quadrupole operator and $(Q^c \cdot Q^c)_{\mathrm{shell}}$ is the *trace equivalent part* of the quadrupole-quadrupole interaction, which guarantees that even for large deformation in average the shell structure is conserved.[30]

Following Ref. 18, the PES is defined as the expectation value of the microscopic Hamiltonian H with respect to a trial state[18] $V = \langle\Phi(\alpha_{2\mu}) \mid H \mid \Phi(\alpha_{2\mu})$. The $\alpha_{2\mu}$ are related to the geometrical quadrupole variables.[2] The trial state is the product of a vector coherent state,[14] describing inter-shell mixtures, and the Elliott $\widetilde{SU}(3)$ state,[16] describing the irreps within the valence shell. The vector coherent state $\mid \Phi(\alpha_{2\mu})\rangle$ is given by an operator, which is an exponential function in the symplectic operators which raise the number of phonons by two, applied to a lowest weight state of an $\widetilde{SU}(3)$ irrep within the valence shell.

Restricting to *only one* symplectic irrep, in lowest order, the potential is given as follows:

$$\begin{aligned} V \approx\ & \frac{65}{16\pi}(\chi N_s^2)\tilde{\beta}_s^2 + \frac{5}{4\pi}N_s\left(\hbar\omega - \frac{7}{6}\chi N_s\right)(\tilde{\beta} - \tilde{\beta}_s)^2 \\ & - (\chi N_s^2)\frac{5}{\pi}\tilde{\beta}_s\tilde{\beta} + \frac{5}{2\pi}N_s\left(\hbar\omega + \frac{5}{6}\chi N_s\right)\tilde{\beta}_s\tilde{\beta}\left(1 - \cos(\gamma - \gamma_s)\right), \end{aligned} \tag{5}$$

with N_s the total number of phonons in the lowest $SU(3)$-state plus $3\frac{(\tilde{A}-1)}{2}$ and $\tilde{\beta} = \frac{\beta}{\kappa}$ $(\tilde{\beta}_s = \frac{\beta_s}{\kappa})$. The value κ is approximately given by $\frac{1}{2}(\frac{\tilde{\eta}_p+1.5}{\tilde{\eta}_p+0.5} + \frac{\tilde{\eta}_n+1.5}{\tilde{\eta}_n+0.5})$ with $\tilde{\eta}_p$ and $\tilde{\eta}_n$ being the pseudo valence shell number for protons and neutrons respectively. $(\tilde{\beta}_s, \gamma_s)$ in (5) is related to the deformation of the symplectic band head in the $0\hbar\omega$,[13] with

$$\tilde{\beta}_s = \left[\left(\frac{4\pi}{5N_s^2}\right)\left(\widetilde{\lambda}^2 + \widetilde{\lambda}\widetilde{\mu} + \widetilde{\mu}^2 + 3\widetilde{\lambda} + 3\widetilde{\mu}\right)\right]^{\frac{1}{2}}. \tag{6}$$

The total PES is then obtained by the envelope of all potentials related to the sum of all symplectic irreps.

The expression in (5) depends on the parameters χ, N_s and $\hbar\omega$. The $\hbar\omega$ and the N_s are fixed. The χ-value is obtained by requiring that the total potential has its minimum either at the experimentally deduced position β_{exp} or one can use the predictions of the macroscopic-microscopic model,[24] or any other source of information and theoretically calculated values of the deformation from a microscopic model.

The only parameter, which cannot be deduced is the collective mass B_2 of the geometrical model.[2] This parameter has to be adjusted to, e.g., a particular state in the ground state band. Also assuming for neighboring nuclei the same value of B_2 is in general far more accurate than using the Cranking Model. Other microscopic contributions to the mass parameter come from nucleons in the intruder orbitals. Using the B_2 as a parameters, implies that it contains this information as a black box.

The microscopic shell model space in the Pseudo-Symplectic Model can be obtained by first filling the Nilsson orbitals from below at a given deformation, counting the number of nucleons in the normal orbitals, and finally use the tables of the reduction $U(n) \times U(2) \supset SU(3)$,[26] with $U(2)$ as the spin group. An alternative is to use the $SU(3)$ package of the Louisiana group.[27–29] In the reduction of $U(n) \times U(2) \supset SU(3)$ only spin $S = 0$ states were considered, because states with larger S are lying higher in energy (they involve the breaking of nucleon pairs).

We emphasize that the method presented in Ref. 18 gives a procedure to obtain the PES for a given nucleus, *if* the deformation of a nucleus is known (or proposed by a theoretical model) and *if* the distribution of nucleons in the normal and unique orbitals can be obtained from a Nilsson diagram, deduced by a microscopic model as the one in Ref. 25.

In Ref. 19 the spectrum of a nucleus could be determined by solving numerically the Schrödinger equation using the finite element method. Unfortunately, the program is not available any more. We wrote our own program to fit the GCM potential to the microscopically derived one. In general, we got a very good agreement, smoothing out the wiggles in the microscopically derived PES due to the geometrical mapping.

The GCM potential, defined in Ref. 2, has the form

$$V = \frac{C_2}{\sqrt{5}}\beta^2 - \sqrt{\frac{2}{35}}C_3\cos(3\gamma) + \frac{C_4}{5}\beta^4 - \sqrt{\frac{2}{175}}C_5\beta^5\cos(3\gamma) + \left(\frac{2}{35}\right)C_6\beta^6\cos^2(3\gamma) + \frac{D_6}{5\sqrt{5}}\beta^6 \ . \tag{7}$$

This potential is adjusted to the microscopically deduced PES. The kinetic energy consists in the sum of a quadratic term in the momentum, whose parameter is B_2, and a term with a deformation-dependent collective mass, whose parameter is called P_3.[5]

The final calculation, in order to obtain a spectrum and B(E2) transition values, is performed using the program published in Ref. 31.

We applied our method to nuclei with a known collective structure.[32] The motivation was to show that our proposal is able to reproduce the collective quadrupole properties of nuclei.

In a first step, the distribution of nucleons in normal and unique orbitals is determined, using the Nilsson scheme and filling at a given deformation β_0 the nucleons from below into the orbitals. Because the Nilsson scheme is calculated only for prolate and oblate nuclei, the question is when to use the prolate (positive β_0) and when to use the oblate side (negative β_0) of the Nilsson scheme. We apply the following procedure: (a) when both proton and neutron pseudo-shells are less than half filled we use the positive sign of the physical deformation β_0, (b) when the valence pseudo-shells of the protons is less than half filled and the one for neutrons more than half (or vice versa), we consult the tables of Möller–Nix,[24] (c) when both proton and neutron valence pseudo-shells are more than half filled we use the negative sign of the physical deformation. This criterion will not work well when the nucleus is triaxial, i.e., when it exhibits a deformation with a γ-value near 30°, though it serves still as an estimate.

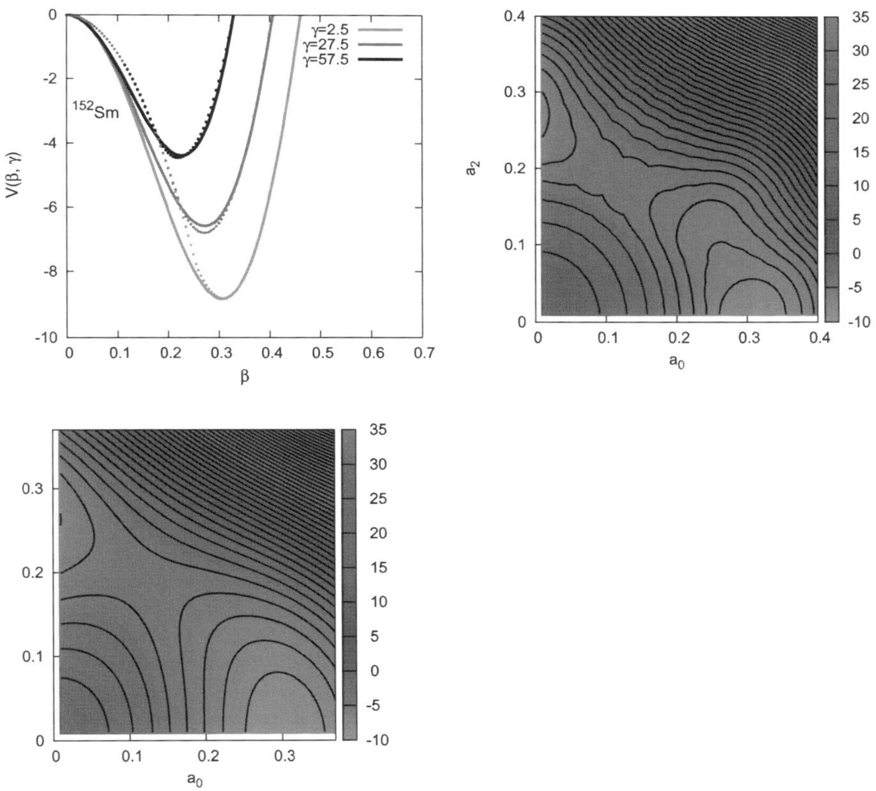

Fig. 2. Comparison of microscopic and phenomenological potential energy surfaces of ^{152}Sm.

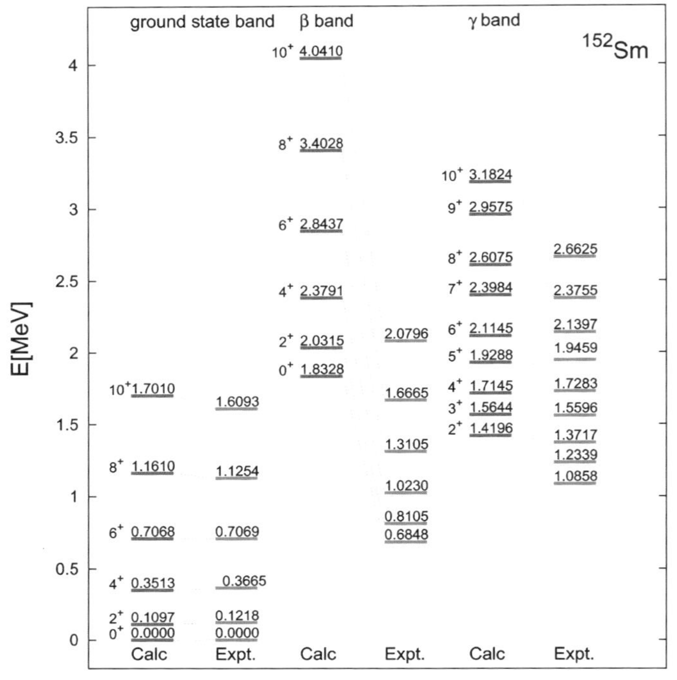

Fig. 3. The spectrum of ^{152}Sm, as obtained by the mapping procedure. Experimental data were taken from Ref. 33.

In Fig. 2 we just present a typical example from Ref. 32, for ^{152}Sm. The left figure in the first row depicts the comparison of the microscopically derived PES to the GCM-PES. Only cuts are shown. The right figure in the first row is the microscopically mapped PES. The figure in the second row gives the GCM-PES adjusted to the microscopic derived PES and Fig. 3 shows the spectrum. As noted, the agreement to experiment is not bad at all, considering that practically no parameters were used. The $B(E2)$-transition values are also in very good agreement to experiment (please, consult[32]). In the spectrum one notes a too strong staggering in the β- and γ-bands. This is part due to the fitting of the GCM-PES to the microscopic PES and also due to the simplicity of the derivation. Other interactions have to be includes in order to get a better fit to the microscopic PES.

Because the content in terms of symplectic irreps determines the main structure of the PES and this content is a pure result of the PEP, we can claim that this principle is determinant for the structure of the PES.

The final remark is that the geometric quadrupole model of the nucleus is not dead and still provides new insights and applications to modern topics in the collective structure of the nucleus.

4. A Primary on Group Theory

This section is of importance to the understanding of the relation of the IBA to the geometric model.

The section serves also to show that once the number of basic degrees of freedom are known, group theory leads directly to possible Hamiltonians, transition operators and more. The lesson to learn is that *when two theories exhibit the same degrees of freedom, one has to be careful to implement the basic assumptions of each theory in order to get a difference. Not doing so, leads to jumping from one theory to another*, often without noticing it.

We consider only bosonic theories, though the *same* can be done for fermions with the same results!

Let us assume a system with n degrees of freedom. We will not specify which degrees of freedom, it is of no importance here. Defining the operators $(i, j, ... = 1, ..., n)$

$$\boldsymbol{C}_i^{\ j} = \boldsymbol{b}_i^\dagger \boldsymbol{b}^j, \tag{8}$$

where $\boldsymbol{b}_i^\dagger$ and $\boldsymbol{b}^j$ are boson creation and annihilation operators. We distinguish lower and upper indices because a creation operator transforms co- and an annihilation operator contra-variantly. Raising or lowering indices depends on the definition of phases, but are of no important here. The operators $\boldsymbol{C}_i^{\ j}$ form an algebra:

$$\left[\boldsymbol{C}_i^{\ j}, \boldsymbol{C}_p^{\ q}\right] = \delta_{pj}\mathbf{C}_i^{\ q} - \delta_{iq}\mathbf{C}_p^{\ j}, \tag{9}$$

which can be shown to be the commutation relation of an $U(n)$ group. The *rank r of an algebra* is defined as *the maximal number of commuting generators*, which in this particular case is $r = n$ and the commuting operators are $\boldsymbol{C}_i^{\ i} = \boldsymbol{b}_i^\dagger \boldsymbol{b}^i$.

The lesson from here is that when n degrees of freedom are present, one can *always associate an $U(n)$ group* to this system. Two theories, with complete different initial assumptions but *with the same degrees of freedom*, are described by *the same symmetry group*. Therefore, we stress, in order to distinguish both theories one has to implement the basic assumptions and differences.

Also Casimir operators can be obtained in a straight forward manner. The formula is

$$\begin{aligned}
\boldsymbol{C}_1 &= \sum_{i_1} \boldsymbol{C}_{i_1}^{\ i_1} \\
\boldsymbol{C}_2 &= \sum_{i_1 i_2} \boldsymbol{C}_{i_1}^{\ i_2} \boldsymbol{C}_{i_2}^{\ i_1} \\
\boldsymbol{C}_3 &= \sum_{i_1 i_2 i_3} \boldsymbol{C}_{i_1}^{\ i_2} \boldsymbol{C}_{i_2}^{\ i_3} \boldsymbol{C}_{i_3}^{\ i_1} \\
&\cdots \\
\boldsymbol{C}_m &= \sum_{i_1 i_2 \ldots i_m} \boldsymbol{C}_{i_1}^{\ i_2} \boldsymbol{C}_{i_2}^{\ i_3} \ldots \boldsymbol{C}_{i_m}^{\ i_1} \\
&\cdots \ .
\end{aligned} \tag{10}$$

According to Racah's theorem there are only $r = n$ independent Casimir operators, thus it suffices to use only he first n operators in the list (10). The eigenvalues are readily obtained, applying a Casimir operator to the *maximal weight state* which can be defined by $\boldsymbol{C}_i^{\ j} \mid \max\rangle = 0$ for $i < i$ and defining the eigenvalues for the weight operators $\boldsymbol{C}_i^{\ i}$ to be n_i (the same way as in the angular momentum group the representation is defined by the eigenvalue of J_z on the state $\mid JM_J = +J\rangle$).

The $SU(n)$ group is obtained defining

$$\widetilde{\boldsymbol{C}}_i^{\ j} = \boldsymbol{C}_i^{\ j} - \frac{\delta_{ij}}{n} \sum_k \boldsymbol{C}_k^{\ k}, \tag{11}$$

where $\sum_k \boldsymbol{C}_k^k$ is nothing but the number operator $\boldsymbol{N}$. The $\widetilde{\boldsymbol{C}}_i^{\ j}$ form the same algebra as the $\boldsymbol{C}_i^{\ j}$. Now, the weight operators $\widetilde{\boldsymbol{C}}_i^{\ i}$ are not linear independent because $\sum_k \widetilde{\boldsymbol{C}}_k^{\ k} = 0$, as can be proven easily. Thus the rank of the $SU(n)$ algebra is $r = n - 1$. The Casimir operators are defined in the same way as in $U(n)$ (see (10)) only substituting $\boldsymbol{C}_i^{\ j}$ by $\widetilde{\boldsymbol{C}}_i^{\ j}$ and where $\boldsymbol{C}_1$ is zero by construction.

Having identified the larger group, the next step is to identify a basis, which normally corresponds to a given, convenient group chain. It is required that the angular momentum group $SO(3)$ is contained. This group represents the total spin and a Hamiltonian in nuclear physics conserves angular momentum. Thus, the rough structure of such a chain is

$$U(n) \supset G_1 \supset G_2 \supset \ldots \supset SO(3), \tag{12}$$

which are the groups G_k depend on the system considered and the number of possible chains depends also on it. In general, more than one group chains are possible, all containing the rotation group.

To each group belongs one or more Casimir operators. In a bosonic theory, where the irreducible representation (irrep) of $U(n)$ has to be the complete symmetric irrep $[N]$, there is only one linear independent Casimir operator for each G_k.

When the Hamiltonian is a function of only those Casimir operators belonging to one chain, one speaks of a *dynamical symmetry*. The reason is that a Casimir operator of a higher group in (12) commutes with all of the lower groups, i.e., all commute to each other. Thus, one can define states which are simultaneous eigenstates of all Casimir operators of this group chain. The eigenvalues are known and, therefore, for the eigenvalue of the Hamiltonian *an analytic solution is obtained!*

One can use the generators also to provide *transition operators*. Matrix elements are obtained in the standard way, using the algebra of the group. At the end, one has a lot of freedom to construct Hamiltonians, calculate their eigenvalue, determine transition values, etc. *It is a very powerful method!*

The downside is that here no information is given on what the degrees of freedom are! If, for example, two *different* theories are taken, with the *same number of degrees of freedom, both give the same results*, iff no additional information is taken into account which distinguishes the two theories. For example, if a bosonic theory is compared to a second one which assumes that the basic degrees of freedom are also boson, but are composed by a certain number of fermions, one jumps from the theory with the composed bosons to the pure bosonic theory when the PEP is not taken in to account in the second theory!

The primary of group theory is "only" mathematics, though very powerful and elegant. Nevertheless, one should not forget about the basic assumptions, i.e., *Physics*.

5. Relation to the IBA

Let us first propose to use instead of the standard form of the geometric quadrupole Hamilton, in terms of the collective coordinates α_{2m} and its conjugate momentum π^{2m}, another form in terms of the boson creation $\boldsymbol{d}^\dagger_{2m}$ and annihilation $\boldsymbol{d}^{2m}$ operators. The index "2" refers to the spin of this degree of freedom, corresponding to the quadrupole surface motion. The $\boldsymbol{d}^\dagger_{2m}$ operators represent five degrees of freedom, thus, a $U(5)$ group appears. As done in the GCM a finite cutoff is introduced and the collective Hamiltonian is diagonalized. The final result is checked for convergence, demanding that the low energy results do not change within a small error. This demands a numerical diagonalization of matrices, usually with a finite cut-off N large enough to obtain the convergence of the energy and transition values at low energy. But there is a trick which taught us the IBA: One can introduce a *cut-off* in the theory, defining a new boson, called $\boldsymbol{\sigma}^\dagger$ and $\boldsymbol{\sigma}$, and requiring that the total number of bosons is fixed:

$$N = n_d + n_\sigma, \tag{13}$$

with $\boldsymbol{n}_d = \sum_m \boldsymbol{d}^\dagger_m \boldsymbol{d}^m$ being the number operator of the d-bosons while $\boldsymbol{n}_\sigma = \boldsymbol{\sigma}^\dagger\boldsymbol{\sigma}$ is the number operator of the σ-boson. Please, note that the σ-boson is completely artificial, it only serves to introduce a cut-off in the d-bosons, namely (13) teaches us that for a *fixed* number of total bosons N the d-bosons vary from zero to maximally N. With the introduction of the σ-degree of freedom, the total number has increased to 6 and, thus, the group $U(6)$ is involved. This implies to rewrite interactions, which change the number of d-bosons such that the total interaction does not change the total number of bosons. For example,

$$\begin{aligned} &\left[\boldsymbol{d}^\dagger \otimes \boldsymbol{d}^\dagger\right]^L_M + \left[\boldsymbol{d} \otimes \boldsymbol{d}\right]^L_M \\ &\rightarrow \\ &\left[\boldsymbol{d}^\dagger \otimes \boldsymbol{d}^\dagger\right]^L_M (\boldsymbol{\sigma})^2 + \left[\boldsymbol{d} \otimes \boldsymbol{d}\right]^L_M \left(\boldsymbol{\sigma}^\dagger\right)^2 . \end{aligned} \tag{14}$$

This interaction is hermitian and does not change the number of quanta. In order to eliminate the contribution from the σ-bosons, artificially introduced, one divides this interaction by N, which is the approximate matrix element of $(\boldsymbol{\sigma}^\dagger)^2$ and $(\boldsymbol{\sigma})^2$ for large N.

Note, that the mathematical structure is identical to the IBA, which uses d- and s bosons, being formed by *fermion pairs* (pairs of nucleons). This relation was found in Ref. 34.

Here, we have an example of two models with the same degrees of freedom. In order to distinguish them, one has to consider the basic assumptions. In the GCM model, the degrees of freedom are surface vibrations which behave exactly as bosons. On the other hand, the bosons in the IBA have a *substructure*, they are pairs of fermions. One restriction is usually taken into account, namely that the maximal number of bosons N is given by *half the number of valence nucleons*. The other property to be taken into account is that these bosons are fermion pairs. As shown in Ref. 35, when the density of bosons is too large the fermions overlap and additional interactions appear including high order and non-polynomial terms (e.g. square root expressions), reflecting the effects of the PEP. This is done, as an example, in models of colliding nuclei by introducing a repulsive potential. In the IBA, however, such higher order terms are not taken into account. However, there is another effect, namely that *the irreps allowed are also affected*! K. T. Hecht showed in Ref. 36 that the density of bosons in the $SU(3)$ dynamical limit is large enough that the lowest $SU(3)$ irrep is not $(2N, 0)$ but has the form of (λ, μ) where both are different from zero. A (λ, μ), with λ and $\mu > 0$, allows also transitions between the ground state and the γ band, which otherwise is forbidden for the case $(2N, 0)$, because the transition operator is a generator of $SU(3)$ and generators do not mix different irreps. This was not taken seriously by the IBA-community and they found a way around it by introducing more degrees of freedom, like g-bosons.[37] It implies, the IBA violates the PEP, one of the most fundamental principles in nature! This was observed in Ref. 38 and denoted as the *fatal flaw* of the IBA. In this reference it is clearly shown that the $(2N, 0)$ irrep is not allowed for mid-shell nuclei, iff the bosons are considered to be *pairs of fermions*. The same finding was reported in Ref. 36 and in Ref. 39 it was also discussed and the need to determine well the $U(6)$ irreps in mid-shell nuclei. Introducing more degrees of freedom[37] does not avoid the question on the violation of the PEP.

However, *the IBA works* and this argument is often used to convince others that the model is correct. As shown in the last section, this does not mean much: If a fundamental principle is violated, it only means that there is a model with the same degrees of freedom, but maybe with a different physical interpretation. And there is one: The geometric quadrupole model of the nucleus, rather this model *with a finite cut-off*.

That for sufficient large N the IBA gives very similar results to a variety of geometric models was proven in Ref. 40 for the case of ^{168}Er. This was a response to the finding in Ref. 41, where ^{168}Er was considered as a test case for the IBA, though the comparisons were only done to oversimplified geometric models.

The conclusion is that the IBA is a very clever way to apply the geometric model to complicated nuclear isotopes. The finite number of the cut-off results in a not too good description of a given nucleus, which was nicely proven for the case of ^{196}Pt.[42] In Ref. 42 is is explicitly stated that the geometrical model works much better than the IBA and that the origin lies in the assumed finite value of the cut-off.

What we have presented here, can be extended to all versions of interacting boson models. That we restricted the consideration to only one type of d- and σ-bosons has practical reasons, it suffices to make the point.

The IBA is not the only algebraic model were the PEP is forgotten. As a further example we mention the *Algebraic Cluster Model.*[43,44] Parting from a triangular shape of the ^{12}C nucleus in term of α-particles, states were predicted and found[44] including the denominated *Hoyle-state*. However, within a cluster model one has to take into account the PEP in order to account for the sub-structure of the α-particle, which is important due to their large overlap. Not doing so means that the model is *not a cluster model*. Here we have another example of a working model, when the claim is imposed that the fundamental particles are composed of fermions. The reason why it works is that the model is equivalent to a geometric model, namely a triangle which is stretched and squeezed. Using the geometric symmetry of this system leads to correct predictions and demonstrates the power of symmetry.

6. Conclusions

A brief introduction to the initial scientific activity of Walter Greiner was presented, related to the collective quadrupole model of the nucleus. As all his contributions, they are still in use and of importance. A brief presentation was given on the *Rotation-Vibration-Model* from the 60s, over the most general description from the 70s and 80s, to nuclear molecules of two and three clusters from the 90s and part of the first decade of the new century. An important effort was made to relate it in the most simplest way to the microscopic structure of the shell model and it was shown that the *Pauli Exclusion Principle* dominates the collective quadrupole structure, besides other microscopic information treated as a black box by the mass and deformation parameter.

A primary on group theory was given, in order to show that another very successfully collective model, the *Interaction Boson Approximation* introducing very elegant concepts of symmetry, is the same as the geometric collective model with a finite cut-off. The reason is that the IBA does not take into account the *Pauli Exclusion Principle*. This chapter serves also to show that many other algebraic models, which are based on the same procedure as explained in this chapter, have the same problem. We urge to include in these models the basic assumptions correctly, as that some bosons consist of two or four fermions.

The list of Walter Greiner's contributions is huge and cannot be covered in this book. However, some topics will be treated.

Acknowledgments

Peter O. Hess acknowledges the financial support from DGAPA-PAPIIT (IN100315) and from CONACyT (Project No.: 251817)

References

1. G. Gneuss and W. Greiner, *Nucl. Phys.* **171** (1971), 449.
2. J. Eisenberg and W. Greiner, *Nuclear Theory I: Nuclear Models* (North-Holland, Amsterdam, 1987).
3. E. Chacón, M. Moshinsky and R. T. Sharp, *J. Math. Phys.* **17** (1976), 668.
4. E. Chacón and M. Moshinsky, *J. Math. Phys.* **18** (1977), 870.
5. P. O. Hess, M. Seiwert, J. Maruhn and W. Greiner, *Z. f. Phys.* **296** (1980), 147.
6. P. O. Hess, J. Maruhn and W. Greiner, *J. Phys. G* **7** (1981), 737.
7. A. Faessler and W. Greiner, *Z. Phys.* **168** (1962), 425.
8. P. O. Hess and W. Greiner, *Il Nuovo Cimento* **83** (1984), 76.
9. P. O. Hess, W. Scheid, W. Greiner and J. H. Hamilton, *J. Phys. G* **25** (1999), L139.
10. Arima and F. Iachello, *Ann. Phys. (N.Y.)* **99** (1976), 253
11. D. J. Rowe, *Rep. Prog. Phys.* **48** (1985), 1419.
12. D. J. Rowe, *Prog. Part. Nucl. Phys.* **17** (1996), 265.
13. O. Castaños and J. P. Draayer, *Nucl. Phys. A* **491** (1989).
14. K. T. Hecht, *The Vector Coherent State Method and its Application to Problems of Higher Symmetries*, Lecture Notes in Physics 290 (Springer, Heidelberg, 1987).
15. A. Bohr and B. R. Mottelson, *Vol. 1: Nuclear Structure* (World Scientific, Singapore, 1999).
16. J. P. Elliott, *Proc. R. Soc. London A* **245** (1958), 128, 562.
17. W. Greiner, J. Y. Park and W. Scheid, *Nuclear Molecules* (World Scientific, Singapore, 1995).
18. O. Castaños, P. O. Hess, J. P. Draayer and P. Rochford, *Phys. Lett. B* **277**, (1992), 27.
19. D. Troltenier, *Das Generaliesierte Kollektivmodell* (Ph.D. thesis, University of Frankfurt, 1992), p. 55.;
   D. Troltenier, J. A. Maruhn, W. Greiner and P. O. Hess, *Zeit. f. Phys. A* **343** (1992), 25.
20. H. van Geel, P. O. Hess, J. A. Maruhn, W. Greiner and D. Troltenier, *Nucl. Phys. A* **577** (1994), 605.
21. K. T. Hecht and A. Adler, *Nucl. Phys. A* **137** (1969), 129.
22. A. Arima, M. Harvey and K. Shimizu, *Phys. Lett. B* **30** (1969), 517.
23. S. Raman, C. H. Malarkey, W. T. Milner, C. W. Netsor and P. H. Stelson, *Atom. Data and Nucl. Data Tables* **36** (1987), 1.
24. P. Möller, J. R. Nix, W. D. Myers and W. J. Swiatetecki, *Atom. Data and Nucl. Data Tab* **59** (1995), 185.
25. P. Möller, J. R. Nix and K. L. Kratz, *Atom. Data and Nucl. Data Tab* **66** (1997), 131.
26. J. P. Draayer and Y. Leschber, *Branching Rules for Unitary Symmetries of the Nuclear Shell Model*, internal report, Louisiana State University, Institute of Physics and Astronomy, USA (1987).
27. J. P. Draayer and Y. Akiyama, *J. Math. Phys.* **14**, 1904 (1973);
   J. Escher and J. P. Draayer, *J. Math. Phys.* **39** (1998), 5123.
28. J. P. Draayer *et al.*, SU(3)-package routines, Louisiana, USA (2002).

29. C. Bahri, D. J. Rowe and J. P. Draayer, *Comput. Phys. Commun.* **159** 121 (2004).
30. G. Rosensteel and J. P. Draayer, *Nucl. Phys. A* **436** (1985), 445.
31. D. Troltenier, J. A. Maruhn and P. O. Hess, *A numerical application of the geometric collective Model*, in *Computational Nuclear Physics 1: Nuclear Structure*, eds. K. Langanke *et al.*, Chapter 6, 105–128 (Springer, Heidelberg, 1991).
32. M. Ermamatov and P. O. Hess, *Ann. Phys. (N.Y.)* **371** (2016), 125.
33. www.nndc.bnl.gov/ensdf.
34. O. astaños, A. Frank, P. O. Hess, M. Moshinsky, *Phys. Rev. C* **24** (1981), 1367.
35. A. Klein and E. R. Marshalek, *Rev. Mod. Phys.* **63** (1991), 375.
36. K. T. Hecht, *Notas de Física* **8** (1985), 165.
37. P. v. Isacker, K. Heyde, M. Waroquier and G. Wenes, *Phys. Lett. B* **104** (1981), 5; P. v. Isacker, K. Heyde, M. Waroquier and G. Wenes, *Nucl. Phys.* **376** (1982), 213.
38. J. P. Draayer, *Algebraic Approaches to Nuclear Structure: Interacting Boson and Fermion Models*, Contemporary Concepts in Physics, Vol. 6, edt. R. F. Casten *et al.*, (Harwood Academic Press, New York, 1993).
39. H. G. Ganev, *Eur. Phys. J. A* **51** (2015), 84.
40. M. Seiwert, J. A. Maruhn and P. O. Hess, *Phys. Rev. C* **30** (1984), 1779.
41. D. D. Warner, R. F. Casten and W. F. Davidson, *Phys. Rev. C* **24** (1981), 1713.
42. A. Mauthofer, K. Stelzer, J. Idzko *et al.*, *Z. Phys. A* **336** (1990), 263.
43. R. Bijker, F. Iachello and A. Leviatan, *Ann. of Phys. (N.Y.)* **236** (1994), 69.
44. D. J. Marín-Lámbarri, R. Bijker, M. Freer, M. Gai, Tz. Kokalova, D. J. Parker, and C. Wheldon, *Phys. Rev. Lett.* **113** (2014), 012502.

Pseudo-complex General Relativity: Theory

Peter O. Hess[1,2] and Thomas Boller[3]

[1]*Instituto de Ciencias Nucleares, UNAM, Circuito Exterior, C.U., A.P. 70-543, 04510, México, D.F., Mexico*

[2]*Frankfurt Institute for Advanced Studies, D-60438 Frankfurt am Main, Germany*

[3]*Max-Planck Institute for Extraterrestial Physics, Giessenbachstrasse, D-85748 Garching, Germany*

[1]*hess@nucleares.unam.mx*

[3]*bol@mpe.mpg.de*

In the last years of his life, one of Walter Greiner's main interests was to extend General Relativity such that a black hole has no event horizon. Here, we will resume his contribution to this topic and show that indeed the event horizon can be eliminated, using the pseudo-complex extension of General Relativity and a phenomenological approach to include quantum effects. Most importantly, Greiner was interested in how the predicted observable consequences of the theory can be verified. We will resume the main properties of the extended theory and some of their applications. The theory contains a minimal length and it will be related to other theories with a minimal length. Observable predictions will be resumed in a separate chapter.

1. Introduction

Since I knew Walter Greiner, he always looked for a possibility to avoid the appearance of the event horizon of a black hole. In the 70s, when I was a student at the Institute of Physics in Frankfurt, he investigated in collaboration with some of his students the internal structure of stars, which could prevent the complete collapse of a star. This was when the first ideas of strange stars was born! He did not succeed in it because there was always a limit, after which the star continues to collapse.

In 2008 Walter approached one of us (P.O.H.) with the idea of how to avoid the event horizon of a black hole. In the two years before, a pseudo-complex extension of the field theory was proposed,[1] which did not suffer from infinities, i.e., the theory is already regularized. He wondered if a pseudo-complex extension of General Relativity (GR) could eliminate the coordinate singularity represented by the event horizon. He believed that "God would not allow to hide part of the space to a nearby observer." His way of arguing and thinking was characteristic for him, a manner which did lead in the past to many new developments and important contributions in physics. Without his insistence and intuition, the pseudo-complex General Relativity (pcGR) would not have been developed.

Another argument is that the gravitational field near a massive object, such as a black hole, is very strong and one has to consider that under such a situation GR may not be complete. One such addition are quantum effects, which should be described by a quantized theory of gravity. However, such a theory does not exist and one option is to proceed along a phenomenological part, making educated guesses of the nature of quantum effects.

Many attempts to extend the GR have been published. In Refs. 2 and 3 a complex metric was introduced; in Refs. 4 and 5 an algebraic extension to complex coordinates was proposed; in Ref. 6 more general metrics were considered; and in Ref. 7 a maximal acceleration was introduced.

Concerning the algebraic extension of the coordinates, in Ref. 8 it is shown that only pseudo-complex coordinates do not produce non-physical states (the pseudo-complex coordinate were denoted in this reference as hyper-complex).

In Ref. 9 the algebraic extension to pseudo-complex coordinates was proposed (called pseudo-complex General Relativity (pcGR)) and the Einstein field equations were obtained, by applying a modified variational principle as proposed in Ref. 10. In Ref. 11 is was finally shown that this is equivalent of using a standard variational principle with a constraint. In Section 2 we will therefore resume the latest results on the formulation of pcGR.

In Section 2 some basic results, obtained up to now, will be discussed. They are important for understanding the predictions of structures near a black hole. Section 3 will be dedicated to the relation with other theories with a minimal length and in Section 4 conclusions will be drawn. In the next contribution by P. O. Hess and T. Boller, observational consequences will be discussed.

Throughout this contribution, the metric signature $(-+++)$ is used and $c = G = 1$.

2. The pseudo-complex General Relativity

In an algebraic extension of GR the real coordinates are proposed to be of another type. They may be complex, pseudo-complex (pc) (other names that also exist in literature are hyperbolic, hypercomplex, para-complex, ...) quaternions, etc. But as shown in Ref. 8 only the extension to pc-coordinates make sense; they do not lead to ghost and/or tachyon solutions in the limit of weak gravitational fields.

Pc-coordinates have the structure

$$X^\mu = x^\mu + I y^\mu \quad , \tag{1}$$

with $I^2 = +1$ and where x^μ is the standard coordinate in space-time and y^μ the pseudo-complex component.

In what follows, some elementary properties of pseudo-complex variables will be discussed:

- Instead of the division in a pseudo-real and a pseudo-complex component, there is an alternative form

$$X^\mu = X_+^\mu \sigma_+ + X_-^\mu \sigma_- \quad ,$$
$$\sigma_\pm = \frac{1}{2}(1 \pm I) \quad . \tag{2}$$

- The $\sigma_\pm$ satisfy the relations

$$\sigma_\pm^2 = \sigma_\pm \ , \ \sigma_+\sigma_- = 0 \quad , \tag{3}$$

which is called the *zero-divisor basis* (see text below).
- The last relation in (3) states that an element proportional to σ_+ multiplied by one proportional to σ_- gives zero, i.e., there is a *zero-divisor* and the pc-variables do not form a field but rather a ring.
- In both zero-divisor component ($\sigma_\pm$) the analysis is very similar to the standard complex analysis.

In pcGR the metric is also pseudo-complex

$$g_{\mu\nu} = g_{\mu\nu}^+ \sigma_+ + g_{\mu\nu}^- \sigma_- \quad . \tag{4}$$

Due to the existence of the zero-divisor, a GR can be constructed independently in the component of σ_+ and σ_-.

This causes the problem of how to connect both components and get real answers.

Here, we will discuss the latest formulation, published in Refs. 11 and 12, using a standard variational principle through the implementation of the constrained of a real line element. We will not use the modified variational principle as in Refs. 9, 13 and 12, though in Ref. 12 the standard variational principle was partially already discussed as an alternative.

The infinitesimal pc length element squared is given by

$$\begin{aligned} d\omega^2 &= g_{\mu\nu} dX^\mu dX^\nu \\ &= g_{\mu\nu}^+ dX_+^\mu dX_+^\nu \sigma_+ + g_{\mu\nu}^- dX_-^\mu dX_-^\nu \sigma_- \quad , \end{aligned} \tag{5}$$

written in terms of the zero-divisor components. In terms of the pseudo-real and pseudo-imaginary components, one has

$$\begin{aligned} d\omega^2 = {} & g_{\mu\nu}^s \left(dx^\mu dx^\nu + dy^\mu dy^\nu\right) + g_{\mu\nu}^a \left(dx^\mu dy^\nu + dy^\mu dx^\nu\right) \\ & + I\left[g_{\mu\nu}^a \left(dx^\mu dx^\nu + dy^\mu dy^\nu\right) + g_{\mu\nu}^s \left(dx^\mu dy^\nu + dy^\mu dx^\nu\right)\right] \quad , \end{aligned} \tag{6}$$

with $g_{\mu\nu}^s = \frac{1}{2}\left(g_{\mu\mu}^+ + g_{\mu\nu}^-\right)$ and $g_{\mu\nu}^a = \frac{1}{2}\left(g_{\mu\mu}^+ - g_{\mu\nu}^-\right)$. The upper indices s and a refer to a *symmetric* and *antisymmetric* combination of the metrics.

The connection between the two zero-divisor components is achieved, requiring that the infinitesimal length element squared in (6) *is real*, i.e., in the zero divisor components

$$(\sigma_+ - \sigma_-)\left(g^+_{\mu\nu} dX^\mu_+ dX^\nu_+ - g^-_{\mu\nu} dX^\mu_- dX^\nu_-\right) = 0 \quad . \tag{7}$$

This constraint has to be taken into account at the moment a *standard* variational procedure is applied. The $X^\mu_\pm$ denote the local pc-position of a fluid element, which are fixed but arbitrary. The 4-velocity components are defined by $u^\mu = \frac{dx^\mu}{ds}$.

The action used is

$$S = \int dx^4 \sqrt{-g}\,(\mathcal{R} + 2\alpha) \quad , \tag{8}$$

where $\mathcal{R}$ is the pc-Riemann scalar. The last term in the action integral allows us to introduce the cosmological constant in cosmological models, where α has to be constant in order not to violate the Lorentz symmetry. This however changes when a system with a uniquely defined center is considered, which has spherical (Schwarzschild) or axial (Kerr) symmetry. In these cases, the α is allowed to be a function in r, for the Schwarzschild solution, and a function in r and ϑ, for the Kerr solution.

The variation of the action with respect to the metric is $\frac{\delta S}{\delta g_{\mu\nu}} \delta g_{\mu\nu} = \frac{\delta S_+}{\delta g^+_{\mu\nu}} \delta g^+_{\mu\nu} \sigma_+ + \frac{\delta S_-}{\delta g^-_{\mu\nu}} \delta g^-_{\mu\nu} \sigma_- = 0$. The variation is performed independently in both zero-divisor components, leading to

$$\int \left[\mathcal{R}^{\mu\nu}_\pm - \frac{1}{2} g^{\mu\nu}_\pm \mathcal{R}_\pm - \alpha g^{\mu\nu}_\pm\right] \delta g^\pm_{\mu\nu} \sqrt{-g} d^4x = 0 \quad , \tag{9}$$

where $\mathcal{R}^{\mu\nu}_\pm$ is the Ricci tensor and $\mathcal{R}_\pm$ the Riemann scalar in each zero-divisor component. The expression in the square bracket cannot be set to zero, due to the constraint which renders the metric variables linear dependent.

Adding the constraint gives

$$\int \left[\mathcal{R}^{\mu\nu}_\pm - \frac{1}{2} g^{\mu\nu}_\pm \mathcal{R}_- - \alpha g^{\mu\nu}_\pm - \lambda\left(\dot{X}^\mu_\pm \dot{X}^\nu_\pm\right)\right] \delta g^\pm_{\mu\nu} \sqrt{-g} d^4x = 0 \quad . \tag{10}$$

The expression in the square bracket is set to zero, resulting the modified Einstein equations

$$\begin{aligned} \mathcal{R}^\pm_{\mu\nu} - \frac{1}{2} g^\pm_{\mu\nu} \mathcal{R}_\pm &= \lambda\left(\dot{X}^\pm_\mu \dot{X}^\pm_\nu\right) + \alpha g^\pm_{\mu\nu} \\ &= \lambda u_\mu u_\nu + \lambda\left(\dot{y}_\mu \dot{y}_\nu \pm u_\mu \dot{y}_\nu \pm u_\nu \dot{y}_\mu\right) + \alpha g^\pm_{\mu\nu} \\ &= 8\pi T^\Lambda_{\pm\;\mu\nu} \quad , \end{aligned} \tag{11}$$

where the derivative of the coordinate ($X^\mu_\pm$) with respect to $s = t$ was expressed in terms of the pseudo-real and pseudo-imaginary component of the $X^\mu_\pm$. The right-hand side of the Einstein equations is identified with an energy-momentum tensor, using the appropriate expression of λ (Ref. 12, see also text below).

When only the pseudo-real part is taken, the right-hand side of (11) can be rewritten in terms of an energy-momentum tensor of an anisotropic fluid: Renaming

$$\begin{aligned} \lambda &= 8\pi\tilde{\lambda} \ , \ \alpha \ = \ 8\pi\tilde{\alpha} \quad , \\ \tilde{\lambda} &= \left(p_\vartheta^\Lambda + \varrho_\Lambda\right) \ , \ \tilde{\alpha} \ = \ p_\vartheta^\Lambda \ , \ \tilde{\lambda} y_\mu y_\nu \ = \ \left(p_r^\Lambda - p_\vartheta^\Lambda\right) k_\mu k_\nu \quad , \end{aligned} \tag{12}$$

the real part of the energy-momentum tensor acquires the structure of an ideal anisotropic fluid

$$T_{\mu\nu,R}^\Lambda = (\varrho_\Lambda + p_\vartheta^\Lambda) u_\mu u_\nu + p_\vartheta^\Lambda g_{\mu\nu} + \left(p_r^\Lambda - p_\vartheta^\Lambda\right) k_\mu k_\nu \quad , \tag{13}$$

where p_ϑ^Λ and radial p_r^Λ are the tangential and pressure respectively. For an isotropic fluid $p_\vartheta^\Lambda = p_r^\Lambda = p^\Lambda$. The u^μ are the components of the 4-velocity of the elements of the fluid and k^μ is a space-like vector ($k_\mu k^\mu = 1$) in the radial direction. It satisfies the relation $u_\mu k^\mu = 0$. Within pcGR, the fluid is anisotropic due to the presence of y_μ.

Up to now, the path followed mainly mathematical arguments. The theory, as it is, cannot tell where the additional energy-momentum tensor comes from, only that *it has to be there*. A fully quantized GR should give the answer. However, in the absence of a such theory, one option is to follow a phenomenological approach. The pcGR is much more general because it contains the information on a minimal length,[12] but for practical reasons we restrict to the pseudo-real part of this theory.

In order to get information about the structure of the energy-momentum tensor, we use the results obtained in Ref. 14, where semi-classical quantum mechanical calculations were performed[15] (obtaining the Casimir effect). The result is that at large distances the density roughly falls off as $1/r^6$. The semi-classical Quantum Mechanics can only be applied when the background metric is not too strong. Near the Schwarzschild radius the field is very strong and in Ref. 14 an explosion of the density is obtained. For strong gravitational fields, this result is not confidential but it gives evidence that the density has to increase significantly toward the center.

We treat the vacuum fluctuations as a classical ideal anisotropic fluid, thus we are free to propose a different fall-off of the negative energy density, which is finite at the Schwarzschild radius. The main result obtained in Ref. 14 is only used as an argument that the density has to increase toward the center, being large near the Schwarzschild radius $R_S = 2m$.

A general conjecture emerges: *mass not only curves the space but also changes the space- (vacuum-) properties, which in turn influences the metric.* As a consequence *the event horizon disappears*!

For the dark energy density we used

$$\varrho_\Lambda = \frac{B}{8\pi r^5} \quad , \tag{14}$$

which is strong enough in order not to contribute to the known observations within the solar system and others with not too strong gravitational fields.[16]

With the assumed density, the metric for the Kerr solution changes to[13]

$$g_{00} = -\frac{r^2 - 2mr + a^2\cos^2\vartheta + \frac{B}{2r}}{r^2 + a^2\cos^2\vartheta} \quad ,$$

$$g_{11} = \frac{r^2 + a^2\cos^2\vartheta}{r^2 - 2mr + a^2 + \frac{B}{2r}} \quad ,$$

$$g_{22} = r^2 + a^2\cos^2\vartheta \quad ,$$

$$g_{33} = (r^2 + a^2)\sin^2\vartheta + \frac{a^2\sin^4\vartheta\left(2mr - \frac{B}{2r}\right)}{r^2 + a^2\cos^2\vartheta} \quad ,$$

$$g_{03} = \frac{-a\sin^2\vartheta\, 2mr + a\frac{B}{2r}\sin^2\vartheta}{r^2 + a^2\cos^2\vartheta} \quad , \tag{15}$$

where $0 \leq a \leq 1$ is the spin parameter of the Kerr solution, in units of m. The Schwarzschild solution is obtained, setting $a = 0$. The parameter $B = bm^3$ measures the coupling of the dark energy to the central mass, where the value of b has to be fixed.

3. Relations to other Theories with a Minimal Length and Further Applications

There were several attempts to extend the GR in the past. One of the well-known ones was proposed by A. Einstein,[2,3] of which the metric was extended to a complex one, where the real part is the standard metric $g_{\mu\nu}$ and the imaginary part is the electromagnetic field tensor $F_{\mu\nu}$: $G_{\mu\nu} = g_{\mu\nu} + iF_{\mu\nu}$.

Einstein did not introduce complex coordinates. This was done in Refs. 4 and 5, introducing a coordinate $X^\mu = x^\mu + iy^\mu$, where x^μ is the usual coordinate and y^μ the complex component. But, as shown in Ref. 8 this theory leads to ghost solutions and thus are not physical.

Another approach was adopted by Born.[17,18] He noticed that while the Quantum Mechanics is symmetric in the coordinates and momenta, while the GR is not. He added a term to the infinitesimal length element squared proportional to the square of infinitesimal momenta:

$$d\omega^2 = g_{\mu\nu}dx^\mu dx^\nu + \left(\frac{l}{m}\right)^2 g_{\mu\nu}dp^\mu dp^\mu \quad . \tag{16}$$

The factor in front of the momentum part is due to dimensional reasons. The disadvantage is the appearance of the mass m, thus the formulation is tight to a specific particle.

A way around it was proposed by Caianiello[7] who introduced the length element

$$d\omega^2 = g_{\mu\nu}\left[dx^\mu dx^\mu + dy^\mu dy^\nu\right] \quad , \tag{17}$$

where $dy^\mu = l\, du^\mu$, with u^μ the four velocity. Again, the factor in l is introduced for dimensional reasons. The l is interpreted as a minimal length.

In pcGR, the real part of the length element squared is

$$d\omega^2 = g^S_{\mu\nu}\left[dx^\mu dx^\mu + dy^\mu dy^\nu\right] + g^a_{\mu\nu}\left[dx^\mu dy^\mu + dy^\mu dx^\nu\right] \quad , \tag{18}$$

where the index s refers to the symmetric and a to the antisymmetric combination of the metric as defined above. Iff $g^-_{\mu\nu} = g^+_{\mu\nu} = g_{\mu\nu}$, then the $g^a_{\mu\nu} = 0$ and $g^s_{\mu\nu} = g_{\mu\nu}$. Further, iff we map y^μ to lu^μ, then the length element squared in pcGR reduces to the one of Caianiello's. However, only in flat space-time (then $g^\pm_{\mu\nu} = \eta_{\mu\nu}$) one can prove that y^μ is proportional to u^μ.[12] Thus, all theories in GR with a minimal length can be deduced from pcGR as special cases.

That pcGR contains a minimal length is for itself very interesting. However, all attempts to get some measurable effects for this minimal length leads nowhere, because l is too small and expected to be of the order of the Planck-length. This is the reason why up to now the minimal length was set to zero, as will be used when the observational effects are discussed. The minimal length may become important when cyclic universes are considered, then all the universe tends to reduce to a point. A minimal length may then avoid a complete collapse. However, this is only a plan for the future.

The pcGR has been applied to the pc-Robertson-Walker and Reissner-Nordström metric.[12,19] Also the influence of the dark energy on the internal structure of a star was considered. In Ref. 20 it was shown that a primitive linear coupling of the matter with the dark energy distribution leads stable starts of up to six solar masses. In Ref. 21 this coupling was modified, using results of calculations within the semi-classical Quantum Mechanics[15] leading already to stable stars of up to 200 solar masses. Unfortunately, we are lacking of theories/models to describe the interior of the star and its coupling to the dark energy. Nevertheless, the results of Refs. 20 and 21 indicate that stars of arbitrary masses may be possible, which was the dream of Walter Greiner.

Another application is the interpretation of the gravitational wave event reported in Ref. 22. Using GR, the source is a system of two black holes, each of approximately 30 solar masses, and the distance of the event is at approximately 1.3 billion light years. Within pcGR, however, the source is a binary system with a mass of several thousand solar masses and a distance of up to a redshift of 12.[23] The event can be interpreted as the merger of two large black holes from centers of two primordial galaxies in the early universe. This shows that, though gravitational waves give a chance to test GR in strong gravitational fields, the interpretation of the source depends still on the theory used. In the next contribution on observational consequences, simulations of accretion disks around black holes will be presented and that a definite difference between GR and pcGR appears, which can discriminate between both theories.

4. Conclusions

We gave a brief introduction to the pseudo-complex General Relativity (pcGR), introduces two metrics belonging to the two zero-divisor components. Both are related by the constraint that the length element squared $d\omega^2$ is real, which leads in form to the Einstein equations with the addition of an energy-momentum tensor on its right-hand side. The difference to the standard Einstein equations is that the Einstein tensor and the energy-momentum tensor are pseudo-complex. Setting the pseudo-imaginary component to zero, leads to the standard Einstein equations, but with the non-vanishing energy-momentum tensor. This tensor is interpreted as the results of a building up of dark energy towards the center of the mass.

We showed that the pcGR contains theories which also introduce a minimal length. This suggests that the structure of space is pseudo-complex, though, the final results are real.

It seems that one dream of Walter Greiner became true: The possibility to avoid the event horizon. The definite answer may come from observations of the *Event Horizon Telescope*, whose results are expected in near future. A complete and consistent description can only be given, once a theory of quantized gravity is available. Until then, phenomenological approached may help to shed some light on the nature of GR in strong gravitational fields.

Acknowledgments

Peter O. Hess acknowledges the financial support from DGAPA-PAPIIT (IN100315).

References

1. P. O. Hess and W. Greiner, *Int. J. Mod. Phys. E* **16**, 1643 (2007).
2. A. Einstein, *The Ann. of Math.* **46**, 578 (1945).
3. A. Einstein, *Rev. Mod. Phys.* **20**, 35 (1948).
4. C. L. M. Mantz, *Hermitian graity and cosmology*, PhD thesis, Univ. Utrecht (2007).
5. C. L. M. Mantz and T. Prokopec, *Found. Phys.* **41**, 1597 (2011).
6. W. Moffat, *Phys. Rev. D* **19**, 3554 (1979).
7. E. R. Caianiello, *Il Nuovo Cim. Lett.* **32**, 65 (1981).
8. P. F. Kelly and R. B. Mann, *Class. Quant. Grav.* **3**, 705 (1986).
9. P. O. Hess and W. Greiner, *Int. J. Mod. Phys. E* **18**, 51 (2009).
10. F. P. Schuller, M. N. R. Wohlfarth and T. W. Grimm, *Class. Quantum Gravity* **20**, 4269 (2003).
11. P. O. Hess and W. Greiner, *Centennial of General Relativity: A Celebration*, ed. C. A. Zen Vasconcellos (World Scientific, Singapure, 2017), p. 97.
12. P. O. Hess, M. Schäfer and W. Greiner, *Pseudo-complex General Relativity* (Springer, Heidelberg, 2015).
13. G. Caspar, T. Schönenbach, P. O. Hess, M. Schäfer and W. Greiner, *Int. J. Mod. Phys. E* **21**, 1250015 (2012).
14. M. Visser, *Phys. Rev. D* **54**, 5116 (1996).

15. N. D. Birrell and P. C. W. Davies, *Quantum Field in Curved Space* (Cambridge University Press, 1986).
16. C. M. Will, *Living Rev. Relativ.* **9**, 3 (2006).
17. M. Born, *Proc. Roy. Soc. A* **165**, 291 (1938).
18. M. Born, *Rev. Mod. Phys.* **21**, 463 (1949).
19. P. O. Hess, L. Maghlaoui and W. Greiner, *Int. J. Mod. Phys. E* **19**, 1315 (2010).
20. I. Rodrguez, P. O. Hess, S. Schramm and W. Greiner, *J. Phys. G* **41**, 105201 (2014).
21. G. Caspar, I. Rodríguez, P. O. Hess and W. Greiner, *Int. J. Mod. Phys. E* **25**, 1650027 (2016).
22. B. P. Abbott *et al.*, (LIGO Scientific Collaboration and Virgo Collaboration), *Phys. Rev. Lett.* **116**, 061102 (2016).
23. P. O. Hess, *MNRAS* **462**, 3026 (2016).

Observational Tests of the Pseudo-complex Theory with Black Hole Imaging

Thomas Boller[1] and Peter O. Hess[2,3]

[1]*Max-Planck Institute for Extraterrestial Physics, Giessenbachstrasse, D-85748 Garching, Germany*

[2]*Instituto de Ciencias Nucleares, UNAM, Circuito Exterior, C.U., A.P. 70-543, 04510, México, D.F., Mexico*

[3]*Frankfurt Institute for Advanced Studies, D-60438 Frankfurt am Main, Germany*

[1]*bol@mpe.mpg.de*

[2]*hess@nucleares.unam.mx*

The pseudo-complex theory of gravitation has been proposed by Hess and Greiner in 2009.[1] The motivation was to remove singularities from the standard theory of gravitation. Besides that it is important to work out observational tests that can be used to discriminate between pseudo-complex General Relativity (pcGR) and standard General Relativity (GR). Here we show that the upcoming Event Horizon Telescope will provide robust tests for the pcGR and standard GR theory.

1. Preface by T. Boller — My Personal Relation with Walter Greiner

Walter Greiner was born on October 29, 1935, in Neuenbau, a small village in the Thuringian Wood, 10 km away from Sonneberg with its observatory containing the second largest collection of photo plates for sky monitoring. Walter was growing up in a house located only 400 m from the house which my grandfather finished building in 1934 (cf. Fig. 1). My father was born in 1935, the same year as Walter, on May 16, and it became natural that both shared their daily live together until 1949, when Walter left Neuenbau for the western part of Germany. When I met Walter first in 1991 at the Institute for Theoretical Physics in Frankfurt, he immediately told me about his friendship with my father and their activities, especially their common most favored activities like fishing, skiing, and enjoying the beautiful environment in Neuenbau.

Figure 1 shows the landscape of Neuenbau as in 1940. Neuenbau is located at a height of about 700 meters, the summers are short and beautiful, but also in winter time there was and is still a lot of show. In this picture, you can see a village with small houses surrounded by wood and grassland. Walter liked to walk and even to lay down on this grassland that is unique in the Thuringian Wood.

Fig. 1. (Color online) Top: Neuenbau, as in 1940, the place where Walter Greiner was born (© O. Boller). Bottom: Red arrow (1) shows the house of my parents, red arrow (2) shows the position of the house where Walter Greiner was living (© Google Maps).

The earliest pictures from Walter together with my father and other friends at Neuenbau are shown Fig. 2. The picture on left shows Walter (standing on left) together with my father Manfred Boller. This picture was taken at around 1947 in front of Walter's house. At that time Walter and my father shared the same school

Fig. 2. Left: Walter Greiner (on the left) and my father Manfred Boller in front of Walter's house in 1947. Right: Walter Greiner on the right in the back, my father (Manfred Boller) on the left, together with other friends in the winter of 1942 close to my father's house in Neuenbau. Photos were taken by Ottomar Boller, my grandfather.

in Neuenbau and I always remember my father's comments on the early interest of Walter when physical and mathematical issues were taught at school. Already at that time Walter was brilliant according to my father's memory. The picture on right is the earliest picture in our private archive, showing Walter on the right side in the back and my father on the most left. This picture was taken in the winter of 1942.

As mentioned above Walter left Neuenbau in 1949, while my father stayed and he and my mother are still living in their house in Neuenbau. I was growing up in Neuenbau from 1958 onwards, went to the same school Walter and my parents had attended. After finishing the high school in Sonneberg I studied physics at the Friedrich Schiller University in Jena. As my wish was always to study astrophysics, I was happy to get a PhD position at the Astronomical Institute in Potsdam, which I finished my study in 1988. At that time Germany was still divided, but when I was visiting Moscow in 1988 for a workshop, I received an invitation from the director of the Max Planck Institute for extraterrestrial physics (MPE) in Garching, Prof. Dr. Joachim Trümper, to visit MPE in 1990. In 1988 I was not allowed to travel to western countries, and there was a long discussion within the Potsdam institute, allowing me at the end to visit MPE for half a year. This decision was made half a year before the wall was finally fallen. However, in April 1990 I started to work at MPE Garching and I am still working there. In 1991 my father mentioned about his friend Walter Greiner and he recommended me to phone Walter. I clearly remember my first telephone conversation with Walter which took about half an hour, instantaneously putting coins to the public phone, as it was state of the art at that time. Walter Greiner invited me to come to Frankfurt and to give a talk about my work. This was the beginning of a yearly visit with Walter and in 1996 promoted me to write my habilitation treatise to become an adjunct professor at the Institute for Theoretical Physics (ITP) in Frankfurt, which I successfully defended in 2004. From 1996 to 2009 I taught weekly basic and advanced courses in astronomy for students at ITP Frankfurt. After the retirement of Prof. Kegel in 2000, I took over the main lecturing courses in astrophysics, with a weekly teaching of 5 hours per day, flying in the morning to Frankfurt and returning late in the night, every Wednesday during the semesters. During that time, up to about 200 students visited my lectures every week and I have worked out completely new courses in astrophysics. This also gave me the chance to talk with Walter mainly every Wednesday during the semesters and we often talked about black holes and the singularity problems in standard GR theory. In 2009 Walter then asked me to work out observational consequences of the pseudo-complex theory and we both agreed that a theory requires tests against other theories. Together with Andreas Müller from the Excellence Cluster Munich we worked out tests on Galactic binaries, having both relativistic Fe K lines and quasiperiodic oscillations. We found the discrepancy between the origin of the QPO signal and Fe K emission vanishes when applying the pcGR theory. This is described in detail in Ref. 12. However, the Event Horizon Telescope will bring the robust test for the pcGR theory. I will describe this in the following Section how the successful campaign between April 5 and 15, 2017 and the expected first black hole images in Spring 2018 will give us a definitive observational proof (or disproof) of the theory of Walter Greiner and Peter Hess.

To finish with the private remarks let us mention that in July 2014 the school in Neuenbau had their 100th birthday. In June Walter and me were attending the Boppard conference and Walter said to me, we definitively must go to Neuenbau and celebrate the school's birthday. In Fig. 3 we show the local press release showing Walter and his brother (right middle figure), my father (top right figure, at the right) and me (bottom down). To my knowledge this was the last time Walter and I met in our birthplace. After that we met at FIAS Frankfurt and the last meeting was in 2015 where we invited Heino Falcke, head of the European Event Horizon Community. In October 2016 I attended the IWARA conference in Gramado. After coming back to home on October 15, my mother phoned me about Walter who died on October 5, shortly before we started to fly to Gramado. It is still a shock to us that our friend Walter is now not anymore with us.

Kleine Schule brachte Große hervor

Seltsame „Methode“

Fig. 3. Press release of the 100th anniversary of the school in Neuenbau. Walter Greiner, my parents and me attended this school. It was my last meeting with Walter in Neuenbau.

2. Observational Tests of the Pseudo-complex Theory

2.1. *Galactic Binaries with quasiperiodic oscillations and relativistic Fe K emission*

A first test proposed by Boller and Müller[12] was applying the pcGR theory to Galactic Binaries with both Quasi-Periodic Oscillations (QPOs) and relativistic Fe K line emission. Only a handful of these objects do have dynamically measured masses and they are listed in Ref. 12. The problem with these sources in standard GR is that the relativistic Fe K lines are not originating from the region where the QPOs occur. The Fe K line originates very close to the black hole within about 2 to 5 gravitational radii. In contrast, the QPOs arise at much larger distances of about 50 to 500 gravitational radii. This is true for all the objects investigated by Boller and Müller.[12] We then applied the pcGR theory to these systems and found that the regions of the relativistic Fe K line and the QPOs are consistent in terms of distance to the black holes and arise at around $2R_G$ (cf. Fig. 4). While this is not a

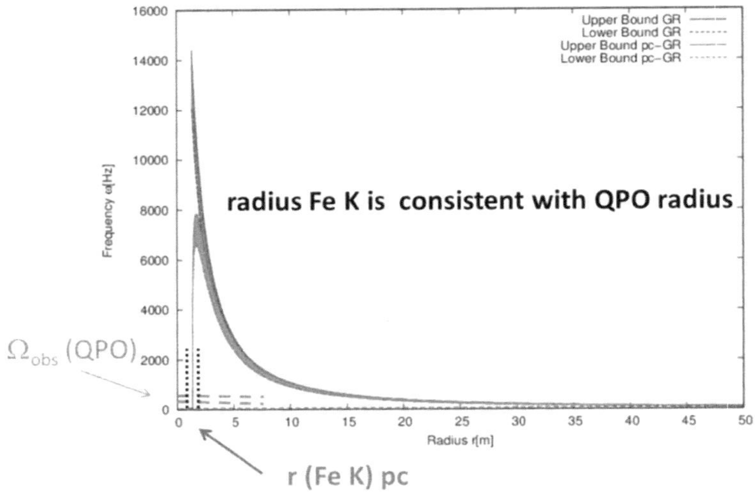

Fig. 4. Applying pcGR theory to GRO J1655-40 indicates that the Fe K emission and QPO emission arises at the same distances to the black hole. This is not the case in the standard GR regime.

definitive proof that pcGR is correct, it was the first hint that pcGR theory can be tested with observations and hat there are significant deviations in the observations between standard and pcGR theory.

The most convincing test will however come from the analysis of the successful EHT observations in April 2017 and we describe our proposed test in the following section.

2.2. *Event Horizon Telescope images: Constraining GR theories*

The Event Horizon Telescope is a long planned experiment by combining the most powerful radio telescopes including most importantly the ALMA telescope. This combination is expected to achieve a spatial resolution of 10 micro arcsec, allowing to resolve the event horizon in Sgr A* and in M87. Only for these two black holes it is possible to resolve the event horizon and produce first images of the black hole and their surroundings. We[12] have contributed to the ALMA White paper writing (cf. Ref. 7) by pointing out that the observational properties resulting from the standard GR and from the pcGR are significantly different and that we most probably can test both theories with very high significance. The pcGR theory developed by Hess and Greiner, and associated papers can be found in Refs. 1–5 and 7–11.

In Fig. 5 we show the radio telescopes which have simultaneously observed SGR A* and M87 between April 5th to April 15th, 2017. The observations have been done at 230 GHz with mmVLBI observing during 5 nights with ALMA, APEX, IRAM 30m, LMT, SMT, SMA, JCMT, and SPT. The observations are reported by C. Goddi (EHT instrument head, private communication) as very successful in terms of the behavior of the instruments, teams, and weather. The data correlation

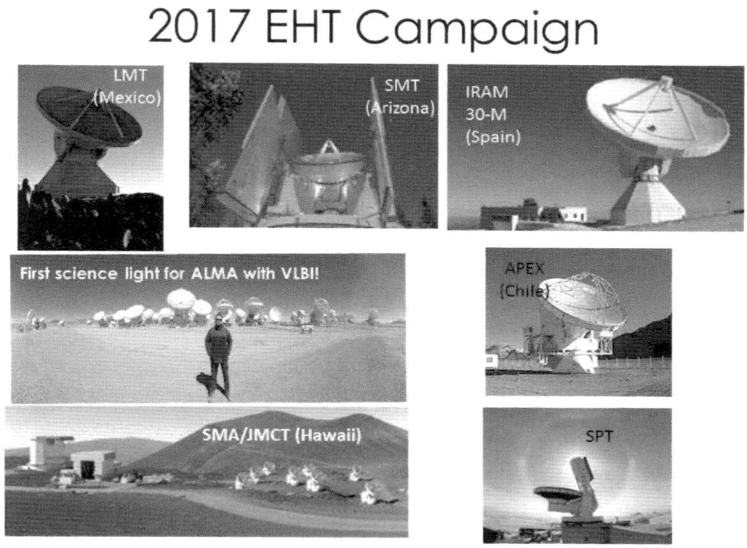

Fig. 5. Radio telescope in the 2017 April 5th to 15th EHT observations. Credit: C. Goddi.

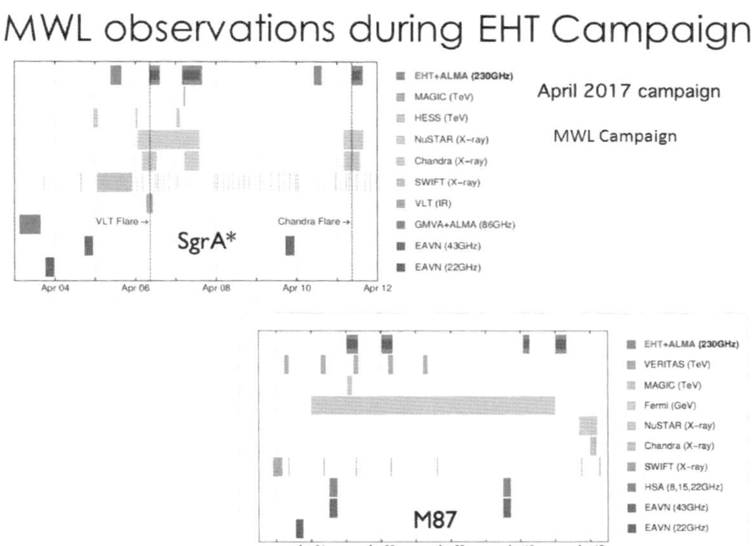

Fig. 6. Multiwavelength observations taken during the EHT campaign. (C. Goddi).

and analysis is presently ongoing and fringes for calibrators have been detected at all baselines. The data from the South-Pole Telescope will be shipped in winter our time out of the Antarctica and will then be added to the other observations.

Predictions are that in spring 2018 first results will become available to the community. In Fig. 6 we show the multi-wavelength observations performed during the EHT observations. The observations range from the TeV range, over the X-ray- Infrared and down to GHz observations. The involved telescopes are listed in Fig. 6, both for Sgr A* and M87. These multi-wavelength observations will be very useful to constrain the in fall and outflow rates, densities and magnetic fields strengths as well as the physical processes at different distances from the black hole and at different frequencies.

We are prepared for the analysis of the EHT data and have worked since 2009 intensively on that, forced by Walter Greiner in order to test his pcGR theory. The pcGR theory has been described by Hess and Greiner[1,2,9,10,12] In the following Section we summarize our observational test to disentangle between pcGR and standard GR.

The existence of a maximum angular velocity (cf. Fig. 7) is a new feature in GR theories and results in Zero flux emission at 1.72R_G. This gives rise to an emission region with Zero intensity, a black ring, while towards closer distances to the black hole there is an onset of emission. This is unique feature of pcGR and below we describe when discussion the ray-tracing simulations, that there are significant differences between standard GR and pcGR.

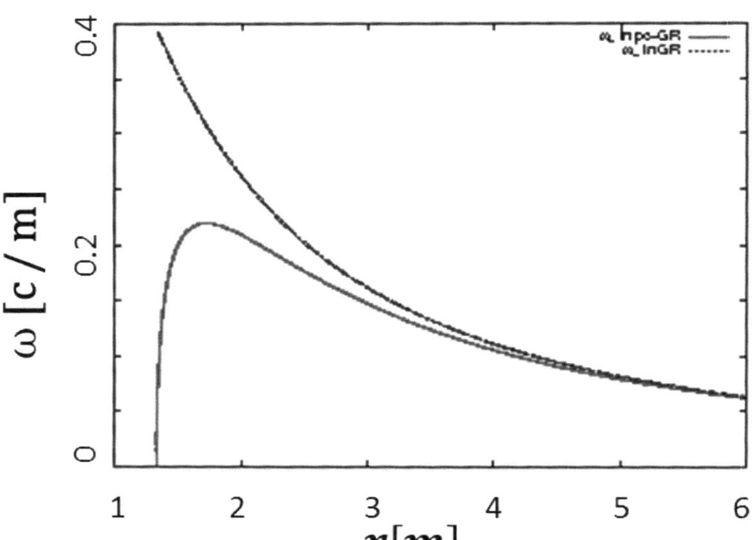

Fig. 7. (Color online) Angular velocity in the standard GR regime (blue curve) and in the pcGR theory (red curve). While in standard GR the velocity increase towards the event horizon, in the pcGR the test particles move slower, reach a maximum at 1.72 m, and then decrease towards smaller distances to the black hole. The maximum orbital velocity is about 9.4 minutes for the mass of the Galactic Center assuming a spin parameter of 0.9.

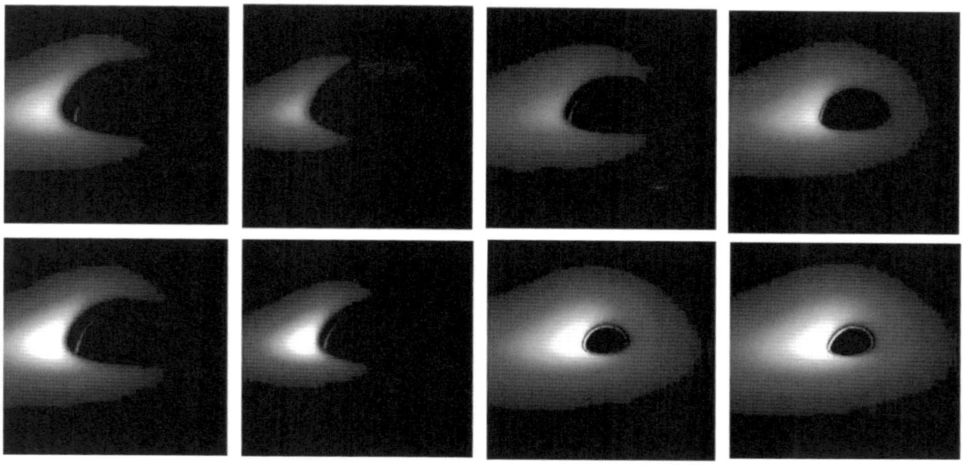

Fig. 8. Upper panel: Ray-tracing simulations performed by Schönenbach *et al.*[5] for the black hole in the Galactic Center for spin values of 0, 0.3, 0.6, and 0.9, respectively. Lower Panel: Same as above but now for the pcGR theory. Note the appearance of a black ring at $1.72\mathrm{R_G}$ and the onset of emission at lower distances to the black hole. This is a new and unique feature in the pcGR theory.

We have performed intense simulations on the black hole in the Galactic Center both for standard GR and pcGR theory. The upper panels of Fig. 8 shows simulations of accretion discs around Sgr A* in standard GR for spin values of 0, 0.3, 0.6, and 0.9. Lower Panel: Same but for pcGR. Note that at spin values above 0.6, a white emission ring appears close to the black hole. At $1.72\mathrm{R_G}$ the emissivity is Zero (dark ring), due to a constant angular velocity of particles orbiting the black.

In the pcGR theory the emission is brighter, due to reduced gravitational redshift (see the contribution from Hess in this book and related references). While for spin values of 0 and 0.3 we do see the same light bending as in standard GR.

The most robust test comes from the intensity distribution in direction towards the black hole predicted by standard GR and pcGR. In standard GR (see the red line in Fig. 9) the intensity is slightly increasing up to about $10\mathrm{R_G}$ and then decreasing to increasing gravitational redshift. At $4\mathrm{R_G}$ the emission goes to Zero. The intensity variations are different in pcGR. First, there is a steeper increase in intensity, reaching a maximum at $4\mathrm{R_G}$, where the emission from standard GR is already Zero. Secondly, below $4\mathrm{R_G}$ the flux is decreasing and reaches at $1.72\mathrm{R_G}$ Zero, giving rise to the zero emissivity seen in Fig. 8. Third, and most important, the flux is increasing again with a very steep slope up to about $1.34\mathrm{R_G}$. The vertical dashed lines will be used to investigate the intensity behavior towards the black hole. As the intensity slopes, relative minimum and maximum values, the appearance of a zone with zero emission, and the onset of the reemission close to the black hole are different to the standard GR theory, we do have a factor up to about 100 between both intensity curves. A factor of 100 is rare in observational astrophysics, and

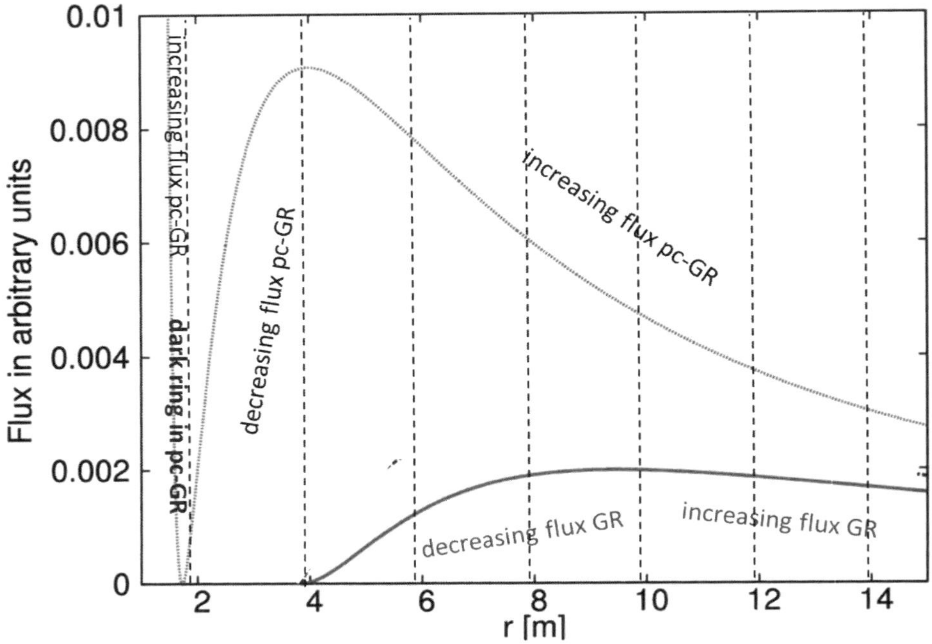

Fig. 9. (Color online) Emission from matter falling into the black hole calculated for the Galactic Center both for the standard GR (red curve) and pcGR (grey curve).

we expect to disentangle between standard GR and pcGR with high significance. Summarizing, the ultimate test of pcGR will be to analyze EHT flux slices into the direction of the center of the black hole and to determine relative flux ratios between GR and pcGR as a function of RG.

If the EHT observations will be successful and if the pcGR theory will be in favor against the standard GR, this would make black holes easier to understand, this would extend Einstein's theory and would be a late honor of our friend Walter Greiner.

Acknowledgments

P.O.H. acknowledges financial support from DGAPA-PAPIIT (IN 100315).

References

1. P. O. Hess and W. Greiner, *Int. J. Mod. Phys. E* **18**, 51 (2009).
2. P. O. Hess, L. Maghlaoui and W. Greiner, *Int. J. Mod. Phys. E* **19**, 1315 (2010).
3. G. Caspar, T. Schönenbach, P. O. Hess, M. Schäfer and W. Greiner, *Int. J. Mod. Phys. E* **21**, 1250015 (2012).

4. T. Schönenbach, G. Caspar, P. O. Hess, T. Boller, A. Müller, M. Schäfer, W. Greiner, *MNRAS* **442**, 121 (2014).
5. T. Schönenbach, G. Caspar, P. O. Hess, T. Boller, A. Müller, M. Schäfer, W. Greiner, *MNRAS* **430**, 2999 (2015).
6. R. Tilanus, T. Kirchbaum, J. Zensus, A. Baudry, and 158 coauthors including T. Boller, A. Müller, P. Hess, *Future mmVLBI Research with ALMA: A European vision*, arXiv:1406.4650.
7. M. Schäfer, P. Hess, W. Greiner, *Astron. Nachr.* **335**, 751 (2014).
8. P. Hess, B. Algalan, T. Schönenbach, W. Greiner, *Astron. Nachr.* **336**, 739 (2015).
9. P. Hess, *MNRAS* **462**, 3026 (2016).
10. P. Hess, I. Rodriguez, W. Greiner, T. Boller, *J. Phys. Conf. Ser.* **578**, 012008 (2015).
11. T. Boller and A. Müller, in *Exciting Interdisciplinary Physics*, FIAS Interdisciplinary Science Series, ed. W. Greiner (Springer International Publishing Switzerland, 2013), p. 293.
12. P. Hess, W. Greiner, T. Schönenbach and G. Capsar, in *Exciting Interdisciplinary Physics*, FIAS Interdisciplinary Science Series, ed. W. Greiner (Springer International Publishing Switzerland, 2013), p. 313.

From Strangeness Enhancement to Quark–Gluon Plasma Discovery

Peter Koch
Bitfabrik GmbH & Co. KG, D-63110 Rodgau, Germany

Berndt Müller
Department of Physics, Duke University, Durham, NC 27708, USA

Johann Rafelski
Department of Physics, The University of Arizona, Tucson, AZ 85721, USA

Dedicated to our mentor Walter Greiner

This is a short survey of signatures and characteristics of the quark–gluon plasma in the light of experimental results that have been obtained over the past three decades. In particular, we present an in-depth discussion of the strangeness observable, including a chronology of the experimental effort to detect QGP at CERN-SPS, BNL-RHIC, and CERN-LHC.

1. Introduction

Just 15 years after the coincident creation in 1964 of two great ideas governing the strong interactions — quarks and the Hagedorn temperature $T_{\rm H}$ — these two concepts merged, giving birth to a new discipline, the physics of the novel fifth state of matter, the quark–gluon plasma (QGP). Today there is consensus that QGP filled the cosmos during the first 20 μs after the Big Bang. For three decades laboratory experiments at the European Center for Particle Physics (CERN) and Brookhaven National Laboratory (BNL) have been exploring this primordial phase of matter colliding nuclei at relativistic energies.

As the ideas about QGP formation in relativistic heavy ion collisions matured a practical challenge emerged: How can the locally color deconfined QGP state be distinguished from a gas of confined hadrons? In the period 1979–86 the strangeness signature of QGP was developed for this purpose, with our 1986 review[1] in essence completing the theoretical foundations. In a review of 1995 one of us (BM) presented a compendium of possible QGP signatures.[2]

During this period and in the following years many scientists wondered whether QGP could be detected as a matter of principle. Could it be that a quark–gluon based description is merely a change of the Hilbert space basis, i.e. a unitary change between quark and hadron bases? If so, maybe there would be nothing to discover!

A globally color-singlet fireball composed of quarks and gluons and several Fermi in diameter is, in principle, a hadron, i.e. a strongly interacting object. Today it is also understood that for an infinite QCD system there is no discontinuity in the equation of state of the baryon symmetric QCD matter. But the key to understanding why the QGP is a physically meaningful and observable concept is to ask the following question: Do hadronic states exist in nature containing many more than three quarks which cannot be factorized into color-singlet components, each containing a few quarks? Even if the complete Fock space of hadrons includes an extended QGP, such a state is distinct from states, such as ordinary atomic nuclei, in which color-singlet hadrons containing few quarks propagate across arbitrary distances while colored quarks and gluons are confined inside such locally color-singlet hadrons.

BM presented these thoughts at the *Quark Matter 1991* conference in Gatlinburg.[3] The ensuing heated discussions reverberate in retrospect, as they lie at the core of the lingering doubts about the observability of QGP formation in relativistic heavy ion collisions. The doubts remained widespread even among experts for many years. As we discuss below, this misunderstanding was, in part, exacerbated by the fact that many observables that were proposed as QGP "signatures" are not sensitive to local color deconfinement, the defining property of the QGP.

The discussions at *Quark Matter 1991* are recalled here as a reminder against which odds the experimental efforts aimed at discovering the QGP had to struggle. They also explain why the discovery of QGP, which had in fact occurred at the time of that conference, (a) was not recognized as valid by one of the discovering experimental groups; (b) needed to wait nine more years to be announced by the CERN Laboratory where the experiments were being performed; (c) had to wait 14 years before a competing laboratory, BNL, concurred after an intense intellectual struggle; and (d) is still a sometimes disputed discovery a quarter century later.

We describe below the pivotal CERN-SPS experiments that, to the apparent disbelief of some of the involved scientists (see Subsec. 4.2), created the QGP phase of matter (Subsec. 4.3). In the following years these results found their confirmation in the Pb+Pb collision program at the CERN-SPS (Subsec. 4.4), leading to the CERN February 2000 QGP announcement based on two campaigns of experiments and many refereed and published articles. The QGP discovery was confirmed five years later by experiments at the BNL Relativistic Heavy Ion Collider (RHIC) employing new, independent probes that helped establish a broad public consensus (Sec. 5), and later by further experiments at CERN, both at SPS and LHC (Subsec. 4.5).

The strangeness signature, which enabled the first clear observation of the QGP, was originally conceived and proposed by one of us (JR) at CERN and later developed to full maturity by us initially in Walter Greiner's Institute. Walter was at that time among the vocal skeptics of our work and of QGP research more generally. However, his broad principled opposition to the subject provided additional inspiration for our work, which continued at the University of Cape Town.

In hindsight, it is puzzling why Walter originally considered all QGP research with such skepticism, because he generally loved innovative, even "exotic" physics. He became, for example, a strong advocate of searches for stable or metastable multistrange cold quark drops called strangelets. Whatever the reasons may have been, in later years Walter Greiner's institute, with his strong support, became the preeminent German center for theoretical QGP physics, and today QGP is a core component of the research program of the Frankfurt Institute for Advanced Study (FIAS), which he co-founded.

2. Strangeness: The Pivotal QGP Signature

The existence and observability of a new phase of elementary matter, the QGP, must be demonstrated by experiment. This requires identification of probes of QGP that are:

(1) operational on the collision time scale of 10^{-23} s;
(2) sensitive to the local color charge deconfinement allowing color charges to diffuse freely throughout the matter;
(3) dependent on the gluon degree of freedom, which is the characteristic new dynamical degree of freedom.

The heaviest of the three light quark flavors, strangeness, emerged during 1980–82 as the pivotal signature of QGP satisfying these three conditions. When color bonds are broken, the chemically equilibrated deconfined state contains an unusually high abundance of strange quark pairs.[4,5] This statistical argument was soon complemented by a study of the dynamics of the strangeness (chemical) equilibration process. We found that predominantly the gluon component in the QGP produces strange quark pairs rapidly, and just on the required time scale.[6] Our work also connected strangeness enhancement to the presence of gluons in the QGP. The high density of strangeness at the time of QGP hadronization was a natural source of multistrange hadrons,[7–9] if hadronization proceeded predominantly by the coalescence of pre-existing quarks and antiquarks.[1]

By Spring 1986 we had developed a detailed model and presented predictions showing how the high density and the mobility of already produced strange and antistrange quarks in the fireball favors the formation of multistrange hadrons during hadronization.[1] We also showed that these particles are produced quite rarely if only individual hadrons collide. We presented a detailed discussion of how a fireball of deconfined quarks turns into strangeness carrying hadrons and showed that multistrange antibaryons are the most characteristic signatures of the QGP nature of the source. By distinguishing the relative chemical equilibrium from the absolute yields of quark pairs, we introduced what today is called the statistical hadronization model which allows us to measure the chemical properties of the hadron source.[10–12]

Though the production of final state hadrons characterizes the conditions in the QGP fireball at the time of its breakup (hadronization). The total strangeness flavor yield provides additional information about conditions arising in the first instants of the reaction in situations when chemical equilibrium in QGP fireball is hard to achieve. In this sense strangeness alone, when studied in depth, can provide a wealth of insights about the formation and evolution of the QGP fireball. As we discuss below, the other are observables available to explore the properties of the early stage QGP.

Considering all produced hadronic particles, it is possible to evaluate the property of the dense matter fireball. A fireball of QGP that expands and breaks apart should do this in a manner that does not remember in great detail the mechanisms that led to the formation of the thermal fireball. Indeed, one of the important findings emerging from studies of the hadronization process is that the hadron chemical freeze-out conditions are universal.[13–15] This universality is further consistent with the sudden hadronization mechanism we first studied 30 years ago.[1]

The strangeness observable was and is experimentally popular since strange hadrons are produced abundantly and can be measured over a large kinematic domain. Therefore, a large body of experimental results is available today. All of these results are consistent with hadronic particle production occurring from a dense source in which the deconfined strange quarks are already created before hadrons are formed. These (anti-)strange quarks are free to move around or diffuse through the QGP and are readily available to form hadrons.

Once one has confirmed that a QGP was formed, other observables can be interpreted on that basis. However, few, if any, other QGP observables probe the characteristic nature of the source in a way that would uniquely pinpoint a QGP with local color deconfinement at the time of hadron formation. Let us discuss a few examples:

- Fluid dynamics: The fireball is recognized to consist of matter described by hydrodynamical simulations,[16,17] which implies the fireball is comprised of a near minimal specific viscosity liquid.[18] It is natural to associate this result with a fluid composed of relativistic, strongly interacting particles, e.g. quarks and gluons, but does not by itself signal that the fluid is a QGP. Indeed, it is still not entirely clear how the QGP acquires its nearly "perfect" liquid nature at thermal length scales.

- Jet quenching is observed in a clear and convincing way in relativistic heavy ion collisions at sufficiently high energy.[19–22] Arguably, this property signals the formation of a fireball endowed with a high density of color fields, which impedes the escape of high energy particles (Subsec. 5.4).
- Quarkonium production can occur in primary collisions[23] and charm recombinant hadronization.[24] Heavy quarkonia, like jet emission, can be suppressed[25] in interaction with dense matter. Therefore the yield is determined by the interplay of at least three different mechanisms with processes contributing needing to be modeled in detail. Today, the increased charmonium yield at the LHC caused by $c\bar{c}$ recombination, similar to the enhanced formation of multistrange baryons observed over a wide energy range, is often considered as the most convincing quarkonium signature of QGP formation (Subsec. 5.5).
- Electromagnetic signals: Photons and dileptons are the most penetrating probes promising insights into the initial dynamics of QGP formation and evolution.[23,26,27] These observables will without doubt come of age in the future high luminosity RHIC and LHC runs. Today we can use them in a semi-quantitative manner to estimate, e.g. the initial temperature of the fireball (Subsec. 3.1).

In the next section we briefly recapitulate the properties and evolution of a thermal QGP fireball formed in relativistic heavy ion collisions, followed by a review of the chronology of the strangeness signature providing evidence for QGP formation in Sec. 4. There we describe the initial CERN-SPS research program, the first series of experiments, and the more mature Pb+Pb collision experiments that reconfirmed the early observations and led to the February 2000 CERN announcement of the "discovery" of the QGP, as well as more recent strangeness developments. The QGP discovery announcement by the RHIC community relied on other observables and is described in Sec. 5.

3. Quark–Gluon Plasma

3.1. *Evolution of fireball in time*

In laboratory experiments involving collisions of large nuclei at relativistic energies, several (nearly) independent reaction steps occur and ultimately lead to hadron production:

(1) Formation of the primary fireball; a momentum equipartitioned partonic phase comprising in a compressed space–time domain most of the final state entropy;
(2) The cooking of the energy content of the hot matter fireball towards chemical (flavor) equilibrium in a hot QGP phase;
(3) Emergence of transient massive quarks due to spontaneous chiral symmetry breaking and disappearance of free gluons; from this point on the fireball cannot be in chemical equilibrium if entropy, energy, baryon number, and strangeness are to be conserved;

(4) Hadronization, i.e. is the coalescence of effective and strongly interacting up, down, and strange quarks and antiquarks into the final state hadrons, with the coalescence probability weighted by accessible phase space. The hadronization process can be subject to detailed experimental study, resulting in determination of the physical properties and statistical parameters that govern the process. We discuss this below.

Experimental information about the maximum temperatures reached in heavy ion collisions can be derived from the spectrum of radiated photons in the energy range $E_T = 1$–3 GeV, where direct photon emission is dominated by thermal radiation. The analysis of the measured spectrum in terms of the thermal properties of the fireball is somewhat model dependent, since the observed yield spans the entire collision history. In practice, the hydrodynamical simulations of the time evolution set a lower limit to the initial temperature. In Au+Au collisions at RHIC ($\sqrt{s_{\rm NN}} = 200$ GeV), this initial temperature exceeds 300 MeV; in Pb+Pb collisions at the LHC ($\sqrt{s_{\rm NN}} = 2.76$ TeV) the initial temperature is at least 450 MeV. This increase is in good agreement with the observed scaling of the particle multiplicity from RHIC to LHC, which indicates a substantial increase in the fireball entropy content.

3.2. *QCD matter in heavy ion collisions*

Lattice gauge theory has made impressive progress on the calculation of static thermodynamic properties of baryon symmetric QCD matter. The equation of state for physical quark masses is now known for $\mu_{\rm B} \lesssim \mu_{\rm B,cr}$ with precision. The quasi-critical temperature where susceptibilities related to chiral symmetry peak has been determined in lattice QCD simulations:[28] $T_c(\mu_{\rm B} = 0) \approx 150$ MeV. A critical point is expected in the $\mu_{\rm B}$ domain that can be explored in experiments carried out at SPS and RHIC-BES, as is illustrated in left-hand part of Fig. 1, which shows the approximate range of T explored in RHIC and LHC experiments with the corresponding energy density versus temperature at $\mu_{\rm B} = 0$ curve calculated by lattice QCD.

On the right panel in Fig. 1 we show how these results stack up against the results obtained across 20 years of hadron chemical freeze-out points; chemical freeze-out is where the yields of produced hadrons will not change and hence these results *must be* below quasicritical temperature. We see two lines describing results of statistical hadronization model final state analysis that span a range of $\mu_{\rm B}$ fitted to the data in SPS, RHIC and LHC experiments. In between these two is the Lattice QCD critical temperature obtained and reconfirmed in past five years. From these results emerges clearly that the chemical nonequilibrium description of hadron production by a QGP is the only model compatible with the current understanding of strongly interacting matter. Note that the curve marked SHARE assumes, as in our 1986 work relative chemical equilibrium, but allows the freedom in the yields of all quark-pairs. For further details see Ref. 15.

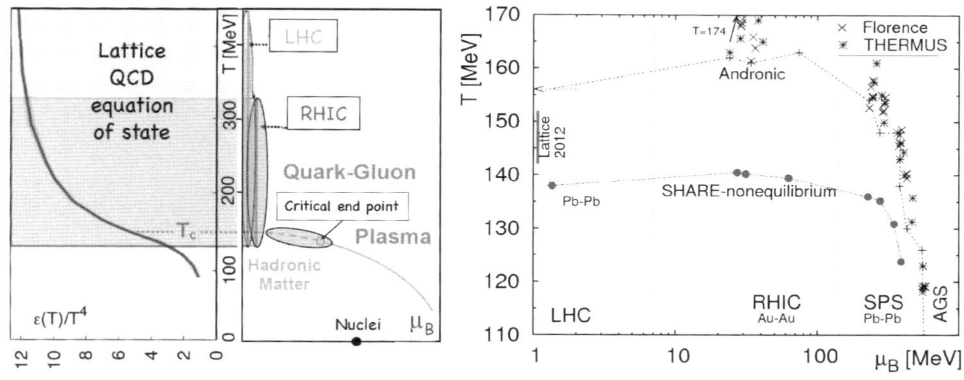

Fig. 1. On the left: Phase diagram of QCD matter (right part of left panel) overlaid with regions covered by LHC and RHIC. The experimentally covered ranges are projected onto the energy density versus temperature at $\mu_B = 0$ curve calculated by lattice QCD (left part of left panel). The current status of statistical hadronization model analysis showing different results obtained for the chemical freeze-out points in the $T - \mu_B$ plane, update of results shown in Ref. 15.

3.3. *Properties of QCD matter*

It is worthwhile asking which intrinsic properties of the quark–gluon plasma medium dynamics we can hope to determine experimentally and from which observables. A not-exhaustive list seen already in the 1992 presentation[3] includes:

- The equation of state of the matter, given by relations among the components of the energy–momentum tensor $T_{\mu\nu}$ at equilibrium and their temperature dependence are reflected in the spectra of emitted particles. Lattice QCD is able to compute these quantities reliably. The analysis of chemical freeze-out conditions provides pressure P, energy density ϵ and entropy density σ.[15]
- Transport coefficients of the quark–gluon plasma, especially the shear viscosity η, the coefficient $\hat{q}$ governing the transverse momentum diffusion of a fast parton (often called the *jet quenching parameter*), the coefficient of linear energy loss $\hat{e}$, and the diffusion coefficient κ of a heavy quark, are related to the final-state flow pattern and the energy loss of fast partons that initiate jets. Lattice gauge theory presently cannot reliably calculate these dynamical quantities.
- The static color screening length λ_D (the inverse Debye mass m_D) governs the dissolution of bound states of heavy quarks in the quark–gluon plasma. This static quantity can be reliably calculated on the lattice.
- The electromagnetic response function of the quark–gluon plasma is reflected in the emission of thermal photons and lepton pairs. This dynamical quantity is difficult to calculate on the lattice, but moderate progress has been made recently.

All but the last of these properties are microscopically related to correlation functions of the gauge field. This implies that the associated experimental observables are mostly sensitive to the gluon structure of the quark–gluon plasma. On the other hand, much more is known theoretically from lattice simulations about the quark structure of hot QCD matter, because it is much easier to construct operators from quark fields that can be reliably calculated. In this respect, lattice calculations and heavy ion experiments are to a certain degree complementary. The presence of jets in heavy ion collisions at LHC and RHIC tell us that at high virtuality Q^2 or high momenta p, the QGP is weakly coupled and has quasi particle structure. On the other hand, the collective flow properties of the matter produced in the collisions tell us that at thermal momentum scales the quark–gluon plasma is strongly coupled. At which Q^2 or p does the transition between strong and weak coupling occur? Does the quark–gluon plasma still contain quasi particles at the thermal scale? Which observables (jets?) can help us pinpoint where the transition occurs?

Leaving aside these theoretical contemplations we now turn to the empirical observable, the strangeness content of QGP fireball. The reason that this is of interest is the undisputed observation that the yield of strange quark pairs is noticeably greater in relativistic heavy ion collisions compared to typical high energy elementary particle reactions. How this happens was addressed in our 1986 review.[1] In the following we focus on what happens next: how this medium heavy flavor hadronizes, what type of particles are produced, and what this tells about the physics of the source, so that our discussion can concentrate on the physics of the fireball source as observed through the "eye" of strangeness.

4. Probing QCD Matter with Strangeness

4.1. *CERN-SPS experiments: Overview*

Figure 2 shows the timeline of those CERN experiments conceived in the early-to-mid 1980s that contributed to the observation of strangeness by means of study of emitted hadrons. Note that CERN accelerated oxygen (^{16}O) ion beams at 200 A GeV in 1986, followed within a year by sulfur (^{32}S) beams at the same energy, and in the mid-1990s by lead (^{208}Pb) beams at 158 A GeV. First results from Pb+Pb collisions were reported in 1996; this is indicated on the left in Fig. 2. We will address below the pivotal results obtained by the two experimental series WA85–NA57 and NA35–NA49.

The WA85 experiment started a bit later than the other Day-1 experiments, and it is instructive to understand the reasons for this delay. The initial experimental proposal was directed at our strange antibaryon signature of QGP; however, the CERN-SPS advisory committee was influenced by arguments by senior theorists that the strange antibaryon enhancement could not be observed since, when QGP converts into hadrons, the multistrange antibaryon ($\bar{\Xi}$ and Ω, $\bar{\Omega}$) signature would be erased by annihilation in the baryon-rich environment produced in collisions at the SPS energy.

As we know today, the assumptions we made in our work are, in fact, realized in nature. The QGP fireball explodes so rapidly after hadronization occurs that many hadrons are produced free-streaming into the vacuum, and chemical reactions

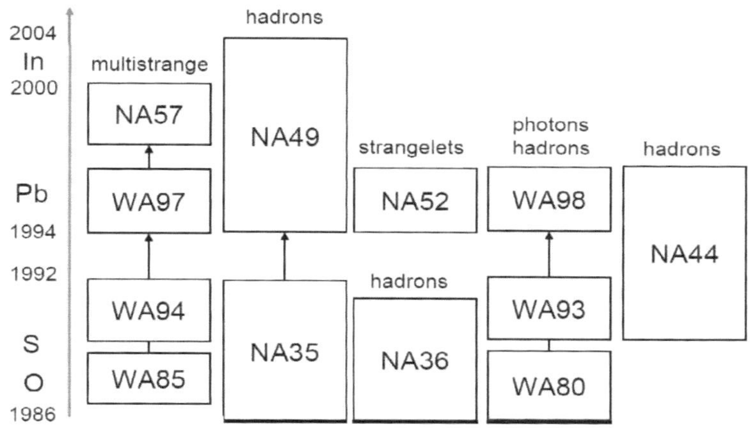

Fig. 2. The multitude of CERN experiments measuring hadron production on line prior to the year 2000 QGP announcement.

involving strangeness quickly fall out of equilibrium during the explosive expansion. Strange antibaryon annihilation does not prevail. This can easily be seen by comparing baryon and antibaryon $p_\perp$-spectra as we show below. Another piece of evidence for the sudden hadronization process, as this became known, is that the yields of hadrons follow the predicted pattern imposed by entropy, baryon number, strangeness content of QGP fireball, which means that the hadron yields follow a scenario today called "chemical nonequilibrium."

The experiment WA85 was eventually approved in Fall 1986, (a) because a way was found to modify the CERN Ω-spectrometer to observe less exotic strange particles including kaons $(\bar{s}q)$, $(\bar{q}s)$ and singly strange antihyperons $\bar{\Lambda}(\bar{q}\bar{q}\bar{s})$; and (b) on the strength of the arguments presented in our 1986 published work.[1]

4.2. *CERN-NA35 experiment*

The CERN-NA35 experiment was an extension of the LBNL-GSI collaboration at the BEVALAC with LBNL's Howell Pugh being the main force for strangeness. Howell was a member of both the NA35 and the NA36 experiments; however, NA36 was highly advanced and had instrumental difficulties, whereas NA35 relied on well established technology. Initially the objective of NA35 was the exploration of the equation of state of dense nuclear matter, a direct continuation of the effort carried out at BEVALAC by most of those involved in the experiment. The fade-out of the strangeness focused NA36 presented the NA35 experimental program with the opportunity to expand into the strangeness signature of QGP.

Writing in 2000, Grażyna Odyniec[29] of LBNL commented: "From the very beginning Howell [Pugh], with firmness and clarity, advocated the study of strange baryon and antibaryon production. He played a leading role in launching two of the major CERN heavy ion experiments: NA35 and NA36, the latter being exclusively dedicated to measurements of hyperons. Strangeness enhancement predicted by theorists was discovered by NA35 and reported at *Quark Matter* 1988." The NA35 results were presented in their extended and final form in 1990. The published article[31] stated in its abstract: "Significant enhancement of the multiplicities of all observed strange particles relative to negative hadrons was observed in central $S+S$ collisions, as compared to $p+p$ and $p+S$ collisions." In the concluding section the authors commented: "Thus our observation... appears to be consistent with a dynamical evolution that passes through a deconfinement stage."

Yet the article refrained from making a discovery claim by continuing: "However... this may not be the only explanation because the possible pre-equilibrium aspects of the early interpenetration stage, or even the conceivable overall off-equilibrium nature of the entire dynamics, may also lead to enhanced strangeness production, even without plasma formation." In retrospect, this statement appears puzzling, because it had already been shown five years earlier[32] that the strangeness production and evolution within the hadronic gas phase could not chemically equilibrate strangeness. In their *Quark Matter* 1990 proceedings[33] report, the NA35 Collaboration distanced itself even further from a possible QGP discovery claim by stating: "We have demonstrated a two-fold increase in the relative $s+\bar{s}$ concentration in central $S-S$ collisions, both as reflected in the K/π ratio and in the hyperon multiplicities. A final explanation in terms of reaction dynamics has not been given as of yet."

Clearly, the NA35 Collaboration missed a unique opportunity to claim the first observation of our QGP strangeness signature by refraining to interpret their experimental results in these terms. In the following, the publications of NA35 strangeness results regularly side-stepped any explicit mention of QGP formation as an explanation of their observations. It is tempting to speculate why this occurred: a motivation may have been that the idea of QGP formation in collisions of mid-sized nuclei in the SPS energy range ran counter to the views of many members of the high energy physics community, some being members of NA35, who thought that only RHIC with its tenfold higher collision energy would be capable of creating a QGP. We leave further discussion of this question to future historians of science.

4.3. *CERN-WA85 experiment*

Against this background the late arriving CERN Ω' spectrometer experiment WA85 under the leadership of Emanuele Quercigh took center stage of QGP search at SPS with published results on Λ and $\bar{\Lambda}$,[35] Ξ^-, $\bar{\Xi}^-$,[36] and a systematic exploration of the parametric dependence of both observables, showing characteristics of a QGP.[37] In contrast to NA35, the WA85 Collaboration takes a much firmer position, citing evidence in favor of QGP formation: "The(se) results indicate that our $\bar{\Xi}^-$ production rate, relative to $\bar{\Lambda}$, is enhanced with respect to pp interactions; this result is difficult to explain in terms of non-QGP models[13] or QGP models with complete hadronization dynamics.[14] We note, however, that sudden hadronization from QGP near equilibrium could reproduce this enhancement."[2]

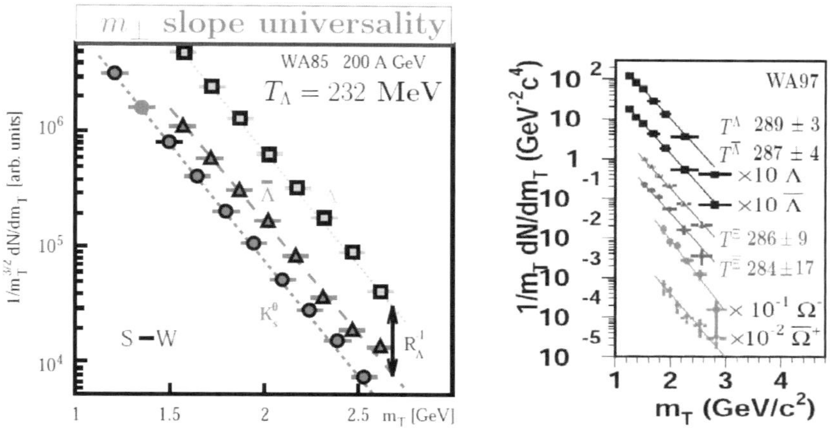

Fig. 3. The universality of baryon–antibaryon $m_\perp$ spectra obtained by WA85 (left), and by WA97 (right) demonstrate that baryons and antibaryons are produced in the same explosive manner by the fireball of dense QGP.

Reference 2 is an analysis[38] of these WA85 results within the nascent Statistical Hadronization Model published in March 1991 by one of us (JR). In this work strange baryon and antibaryon particle production data for S–W collisions were used to determine the "chemical" properties of the particle source, i.e. the chemical potentials and phase space occupancy. In the abstract of this analysis we read: "Experimental results on strange antibaryon production in nuclear S–W collisions at 200 A GeV are described in terms of a simple model of an explosively disintegrating quark–gluon plasma (QGP)." The summary closes with, "We have presented here a method and provided a wealth of detailed predictions, which may be employed to study the evidence for the QGP origin of high $p_\perp$ strange baryons and antibaryons."

The WA85 paper[37] cited above echoed this point of view, leading the WA85 Collaboration in early 1991 to claim the QGP discovery. From today's vantage point, we can say that with this 1990/91 analysis method and the WA85 results and claims of the period, the QGP had been discovered; however, the discovery was not universally accepted by the broader physics community, not even by the majority of the community involved in relativistic heavy ion experiments at the CERN-SPS. For a more extensive discussion of the WA85 results and the arguments behind the discovery claim we refer to the popular review by the spokesman of the WA85 Collaboration, Emanuele Quercigh, prepared with JR.[39]

4.4. *The path to CERN QGP discovery announcement*

Strangeness results continued to be published by these two CERN experimental groups: NA35/NA49 (evolving into NA61) and WA85/WA94/WA97 (evolving into NA57) in the ensuing decade through the early 21st century. When CERN announced in early February 2000 the discovery of a new phase of matter, this event

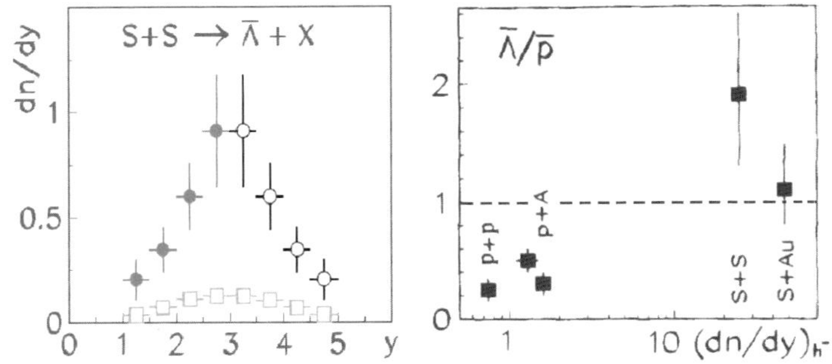

Fig. 4. NA35-antihyperon results for 200 A GeV $S-A$ collisions of July 1995. Left: Distribution in rapidity of $\bar{\Lambda}$ for $S-S$ compared to the yield in pp collisions scaled up with relative abundance of negatives h^-, from Foka thesis;[41] Right: $\bar{\Lambda}/\bar{p}$ ratio as a function of mean h^- multiplicity; results adapted from Ref. 42.

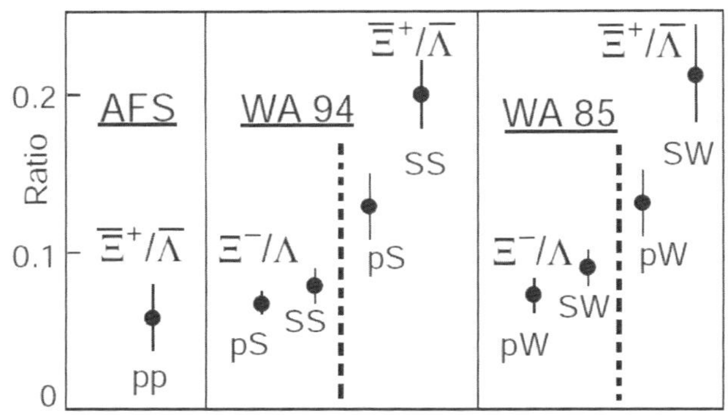

Fig. 5. WA85 and WA94 hyperon and antihyperon results for 200 GeV/A $S-A$ collisions of January 1995, adapted from Ref. 44.

followed a decade of experimental work with dozens of refereed papers published showing agreement of QGP strangeness signature with our QGP based predictions.

The NA35 Collaboration presented the ratio $\bar{\Lambda}/\bar{p} \sim 1.4$ measured near mid-rapidity in Summer 1995,[42] showing a three-to-fivefold enhancement, dependent on the collision system as compared to measurement in more elementary reactions. This was the QGP signature of the first strangeness papers in 1980.[4,5] Due to the shift of the central rapidity for asymmetric collisions the decrease in this ratio as the asymmetry increases is in agreement with theoretical expectations; these results are shown on the right in Fig. 4.

The WA85/94 Collaboration focused on multistrange baryon and antibaryon ratios, for $\bar{\Xi}/\bar{\Lambda}$, see e.g. the 1993 review of David Evans.[43] A full summary of all results is contained in the review of Federico Antinori of 1997[44] and shown in Fig. 5 with data referring to the WA85/94 reports presented at the *Quark Matter 1995* conference.[45,46]

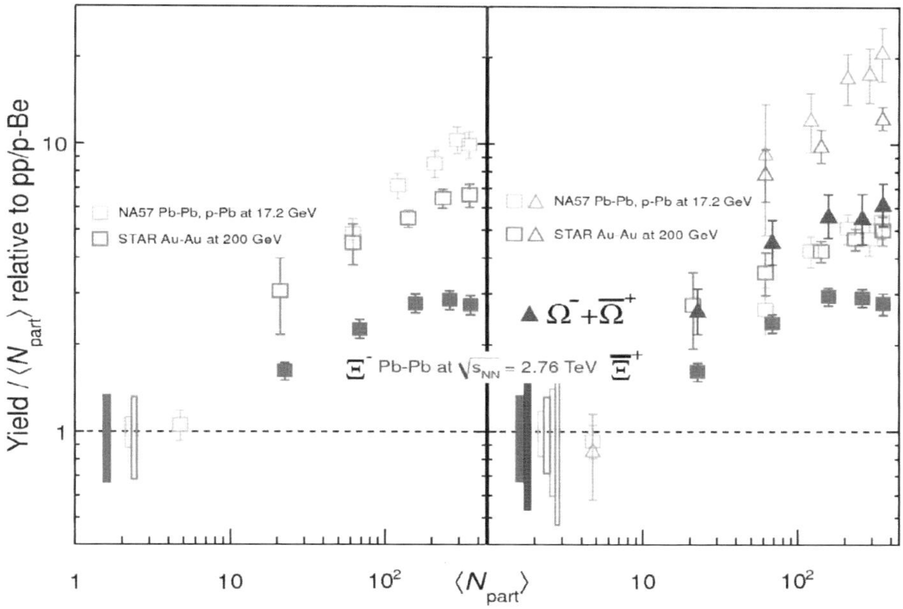

Fig. 6. ALICE-LHC, STAR-RHIC, and NA57-SPS N_{part} normalized hyperon (left), and antihyperon (right) yields as function of N_{part}; yields renormalized arbitrarily to unity for smallest N_{part} available, adopted from Ref. 47.

Seeing these developments one could not help but being convinced that the then forthcoming Pb+Pb experiments at CERN would confirm and cement the strangeness based QGP discovery before the end of the old millennium. This was indeed the case as we see, for example, in the retrospective summary, Fig. 6: in this 2013 strangeness enhancement review by the CERN-ALICE Collaboration we see that the Pb+Pb CERN-SPS antihyperon enhancement at $\sqrt{s_{\text{NN}}} = 17.2$ GeV (equivalent to 158 A GeV) reaches the value 20 for the triply strange $\Omega + \bar{\Omega}$ baryons (open triangles). This is the most dramatic medium modification result ever recorded in relativistic heavy ion collisions.

Despite these impressive (anti-)hyperon results available at the time of the CERN QGP announcement[48] in February 2000, the announcement was based on a "consensus" presentation of all heavy-ion SPS-CERN physics results. There was a strategic problem, since some of these experiments had results that had little to do with the QGP discovery itself or were statistically marginal. With the "consensus" strategy CERN opened its QGP discovery up to criticism, where both the interpretation of the data and the statistical validity of some of the results could be questioned. This problem spilled over to the strangeness signature which should have been uncontested on both grounds. If CERN management was not willing to base its claim of QGP discovery on the impressive observations of strangeness enhancement alone, many believed there was a scientific reason for this lack of confidence!

Would an announcement of the QGP discovery have been more compelling, if it had been based solely on the strangeness signature? While it is tempting to think so, it probably was not practical, because a sizable fraction of the relativistic heavy ion physics community questioned then, and has continued to question, the relevance of strangeness as a signature of the QGP, in spite of the fact that strangeness enhancement is the sole QGP signature for which every single prediction has been quantitatively confirmed experimentally and no comprehensive alternative explanation has been given.

In summary, we firmly believe that CERN had a justifiable case for claiming the discovery of QGP with SPS results addressing strangeness and multistrange antihyperon production. However, the lack of consensus within the community precluded this simple approach and the global "all experiments" claim advanced by CERN was not convincing to outside observers. It took another experimental program, that of the Relativistic Heavy Ion Collider commencing in 2000, with a broader range of accessible observables, to create a wider consensus, leading up to a second announcement of QGP discovery in 2005.

4.5. *Developments at CERN after* 2000

The NA49 Collaboration evolving into NA61 focused its main objective since the CERN QGP discovery announcement on the search of a threshold in beam energy for the previously observed phenomena. In a systematic experimental study of the K^+/π^+ ratio[53] as function of energy the so-called "Marek's Horn" was discovered, named after Marek Gazdzicki, now spokesman of the NA61 Collaboration.

The Statistical Hadronization Model analysis of this feature is seen in Fig. 7 on the left, where the K^+/π^+-ratio for both experiment and theoretical fit adapted from Ref. 54 is shown. The question that we need to answer is, what it is that the "horn" signals? To this end we look at the right-hand panel of Fig. 7 showing the ratio of the thermal fireball energy to the number of strange quark pairs E/s. We clearly see that at the location of the horn-like feature the energy cost of producing a pair of strange quarks is leveling off at a low value, signaling onset of a new, more energy efficient, production mechanism.

This E/s curve on the right panel of Fig. 7 is derived from the fit to the observed particle yields which fit the "horn" but needs further interpretation. Considering our prior extensive work showing that quark based processes are less effective than gluon mediated processes in producing strange quark pairs, we conclude that in the knee of E/s the glue degrees of freedom must have been fully activated. The conclusion is that QGP has been formed at collision energies above $\sqrt{s_{\rm NN}} = 7$ GeV. Below this threshold we see a transition domain where with decreasing energy more and more fireball energy is required to make strange quark pairs. Future experiments (the second RHIC Beam Energy scan will probe down to $\sqrt{s_{\rm NN}} = 3$ GeV) will tell if this rise saturates at lower collision energies.

Another interesting result, combining RHIC and SPS data, is that the ratio $\Xi(ssq)/\phi(s\bar{s})$ of those two different double strange particles is an energy independent

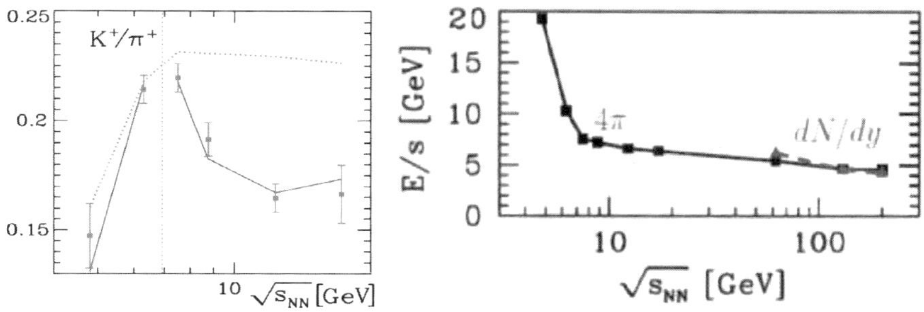

Fig. 7. (Color online) Left: K^+/π^+-ratio adapted from Ref. 54; AGS (lowest $\sqrt{s_{\rm NN}}$) and NA49-SPS energy range results are connected by the (blue) line indicating at the edges the theoretical fit result within chemical nonequilibrium model, the dotted line shows best chemical equilibrium result. Right: Cost in fireball thermal energy of a strangeness pair, E/s as a function of CM collision energy $\sqrt{s_{\rm NN}}$. 4π results (black) are estimates for RHIC, line guides the eye; RHIC domain (blue) shows $(dE/dy)/(ds/dy)$ (update of result of Ref. 13).

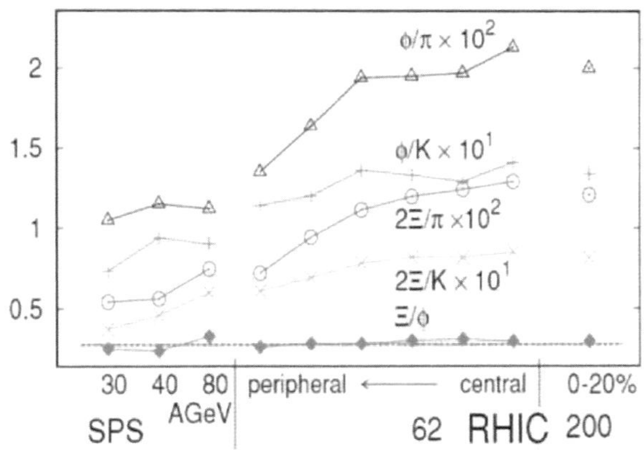

Fig. 8. (Color online) Data points of Ξ/ϕ (bottom red diamonds) and a straight line at 0.281; compared to ratios of these particles with π and K which are highly variable. To eliminate in baryon-rich domain dependence on chemical potential a geometric mean of Ξ with $\bar{\Xi}$ is shown (adapted from Ref. 55).

constant (see Fig. 8). This observation decisively resolves the discussion about canonical strangeness suppression, i.e. the volume dependent suppression caused by overall net strangeness conservation. The data provide for a clear connection of multistrange particle production to total strangeness content, but not to net strangeness content, which is zero for the ϕ. This rules out canonical suppression as a viable explanation of the multiplicity dependence of strangeness enhancement seen in Figs. 5, 6 and 8. Moreover, the universal value of the ratio is signaling that irrespective of how the fireball is formed there is no significant alteration in the final state of the yields of these double strange particles. For further analysis of the implications of these data we refer to Petran's work.[55]

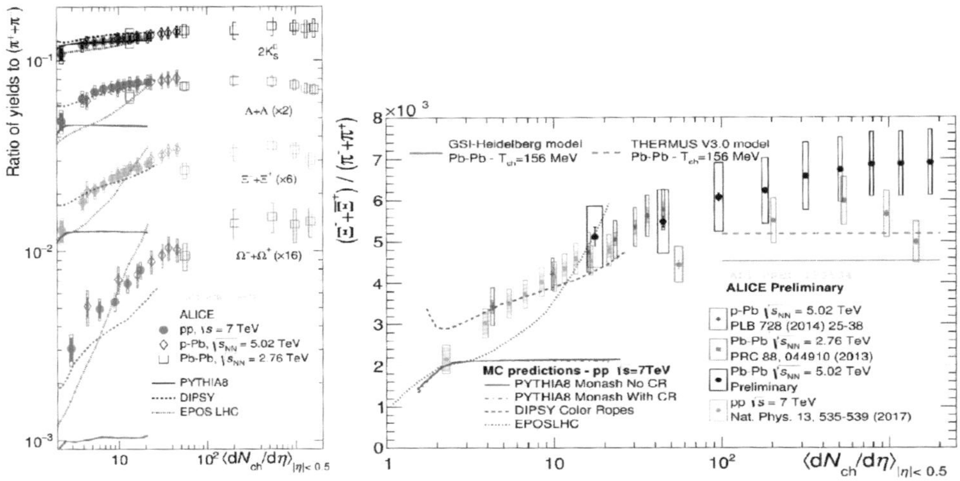

Fig. 9. On left overview of results for strangeness signature of QGP from ALICE published Spring 2017.[57] On right the results of July 2017 SQM meeting showing the need to revisit the old analysis of Ξ-yields. QCD Monte Carlo simulations are not able to reproduce the observed results (curves marked PYTHIA8, DIPSY, EPOS LHC).

Returning now to continue the LHC contribution to QGP physics, the decrease of the enhancement effect with increasing collision energy seen in Fig. 6 signals that there is an increase in strange antihyperon production in the control *pp* or *pA* collisions. This effect has been nicely demonstrated by the ALICE-LHC Collaboration, which recently published[57] the enhancement over minimum bias *pp* yields as a function of the charged hadron multiplicity $dN_{\rm ch}/d\eta$, combining *AA*-results[47] with *pA* (Ref. 56) and new *pp* results as shown in Fig. 9 on the left. The *pp* and *pA* results merge into the *AA* results when compared for the same multiplicity at central rapidity. We see a smooth increase with $dN_{\rm ch}/d\eta$, and in most cases a yield saturation at large hadron multiplicity, indicating that QGP in internal thermal and chemical equilibrium is achieved as the volume of the fireball grows.

Such an equilibrated QGP fireball hadronizes into an out of chemical equilibrium hadron abundance, an important insight we discussed already in our 1986 review.[1] The predominant modification of the fireball at hadronization for collisions in which strangeness is chemically saturated in the QGP is the fireball hadronization volume. A potential spoiler in Fig. 9 is the behavior of Ξ in *AA* which seems to be on the low end of the *pA* results, an effect less than two standard deviations but somewhat disturbing the otherwise consistent pattern. Today this has attracted additional attention since the ALICE Collaboration presented (preliminary) results at the highest LHC energy, $\sqrt{s_{\rm NN}} = 5.02$ TeV, at the *SQM2017* conference.

These new $\sqrt{s_{\rm NN}} = 5.02$ TeV Ξ results differ significantly from those at $\sqrt{s_{\rm NN}} = 2.76$ TeV as shown on the right panel of Fig. 9. The remarkable, yet (for us) expected outcome is that the results at $\sqrt{s_{\rm NN}} = 5.02$ are where we expected the data from $\sqrt{s_{\rm NN}} = 2.76$ TeV to be, in agreement with the findings for other strange hadrons (results available, not shown in Fig. 9). The ALICE Collaboration announced at SQM2017 the intention to review the analysis of the Ξ data from the 2012/13 $\sqrt{s_{\rm NN}} = 2.76$ TeV runs — the trends of pp and pA results suggest that the revised yields will agree with the new $\sqrt{s_{\rm NN}} = 5.02$ TeV results.

Let us now assume that the corrected $\sqrt{s_{\rm NN}} = 2.76$ TeV antihyperons will indeed track the still preliminary $\sqrt{s_{\rm NN}} = 5.02$ TeV data. This would demonstrate that the strangeness signature of QGP is driven by the global properties of the thermal fireball, in particular its volume and/or lifetime, and not by the collision system or the collision energy. This is so since the hadronization condition of QGP fireball is known to be universal across a wide range of collision energies and centralities.[13–15]

We conclude that CERN-SPS results have shown (a) the onset of deconfinement near $\sqrt{s_{\rm NN}} = 7$ GeV, and, that the systematics of multistrange particle production eliminates alternate enhancement mechanisms, as borne out, e.g. by the Ξ/ϕ ratio. The LHC data indicate that for the thousandfold higher LHC energy, QGP can already be formed in high multiplicity pp and pA reactions. Across a wide range of accessible collision energies at SPS, RHIC, and LHC strangeness in the QGP saturates, and the fireball hadronizes in nearly identical fashion at universal physical conditions, with the final geometric size determining the total particle yields. The formation of the thermal QGP fireball depends on entropy content (hadron multiplicity) and not on how this state has been produced, the pp, pA and AA collisions being equivalent.

5. The BNL-RHIC Contribution to the QGP Discovery

5.1. *Flow dynamics of QCD matter*

The "standard model" of the dynamics of a relativistic heavy ion collision begins with a very brief period of kinetic equilibration — most likely less than 1 fm/c. After that, the space–time evolution of QGP can be described by relativistic viscous hydrodynamics. Hydrodynamics is the effective theory of the transport of energy and momentum in matter on long distance and time scales. In order to be applicable to the description of QGP created in relativistic heavy ion collisions, which forms tiny, short-lived droplets of femtometer size, the hydrodynamic equations must be relativistic and include the effects of (shear) viscosity. The causal relativistic theory of viscous fluid has been worked out over the past decade. It is based on the framework of the Müller–Israel–Stewart formulation of second-order hydrodynamics, which includes relaxation effects for the dissipative part of the stress tensor. Schematically, the equations have the form $\partial_\mu T^{\mu\nu} = 0$ with

$$T^{\mu\nu} = (\varepsilon + P)u^\mu u^\nu + \Pi^{\mu\nu}\,, \tag{1}$$

$$\tau_\pi(d\Pi^{\mu\nu}/d\tau) + \Pi^{\mu\nu} = \eta(\partial^\mu u^\nu + \partial^\nu u^\mu - \text{trace})\,. \tag{2}$$

It turns out that the ratio of the shear viscosity η to the entropy density S jointly with the equation of state is the quantity that most directly controls the behavior of the fluid. The quantity η/S is the relativistic generalization of the well known kinematic viscosity. Since in kinetic theory η is proportional to the mean free path of particles in the fluid, which is inversely proportional to the transport cross-section, unitarity limits how small η can become under given conditions. An interesting consequence of this observation is that the quantity η/S has an apparent lower bound of the order of 0.08 (in units of $\hbar$). The existence of such a bound was conjectured already three decades ago, but it was quantitatively derived only recently using the technique of holographic gravity duals. It is now believed that $\eta/S \geq (4\pi)^{-1}$ for most sensible quantum field theories[58] at $\mu_B \approx 0$.

The experimental handle for the determination of η/S is the azimuthal anisotropy of the flow of final-state particles in off-central heavy ion collisions, where the nuclear overlap region is elongated in the direction perpendicular to the reaction plane. Hydrodynamics converts the anisotropy of the pressure gradient into a flow anisotropy, which sensitively depends on the value of η/S. The average geometric shape of the overlap region in symmetric nuclear collisions is dominated by the elliptic eccentricity, resulting an elliptic flow anisotropy characterized by the second Fourier coefficient v_2. Event-by-event fluctuations of the density distribution with the "Standard" model of dn the overlap region generate higher Fourier coefficients for the initial geometry and final flow, encoded in higher Fourier coefficients v_3, v_4, etc. Their measurement is analogous to the mapping of the amplitudes of multipoles in the thermal fluctuations of the cosmic background radiation.

The precise results of such an analysis of event-by-event fluctuations of the flow distribution depends somewhat on the structure of the initial-state density fluctuations, especially their radial profile and spatial scale. The study of this kind to date starts from the fluctuations of the gluon distribution in the colliding nuclei, evolves them for a brief period using classical Yang–Mills equations, and then inserts the fluctuating energy density distribution into viscous hydrodynamics.[59] The conclusion of this study is that the average value of η/S (averaged over the thermal history of the expansion) in Au+Au collisions at the top RHIC energy is 0.12; whereas the value for Pb+Pb collisions at LHC is 0.20 (see Fig. 10). While

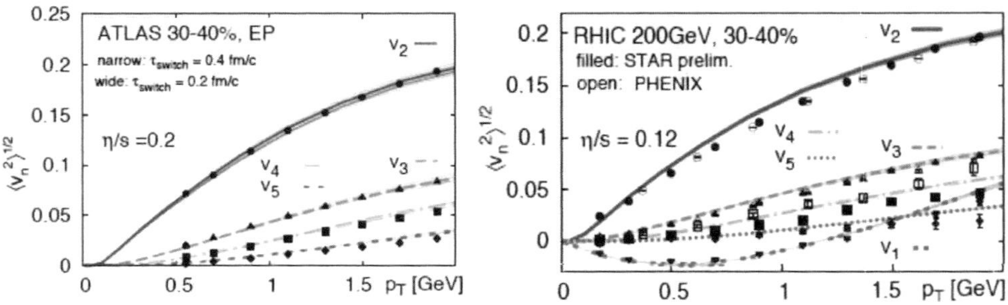

Fig. 10. Fourier components of collective flow, $v_n(p_T)$, for Au+Au collisions at RHIC (left panel) and Pb+Pb collisions at LHC (right panel), in comparison with viscous hydrodynamics calculations. The deduced average value of the ratio η/S is 60% larger at LHC (0.20) than at RHIC (0.12) (adapted from Gale *et al.*[59]).

each of these values has systematic uncertainties of at least 50%, the ratio of these two values is probably rather stable against changes in the assumptions for the initial state.

One interpretation of this result is that the average value of η/S at the 10 times higher LHC energy is somewhat higher than at RHIC, indicating a significant temperature dependence of this quantity. In the frame of this explanation the quark–gluon plasma at the lower temperature reached at RHIC is more strongly coupled and a more "perfect" liquid, making this energy domain especially interesting for the study of this observable.[60]

5.2. *Valence quark recombination*

If the term *quark–gluon plasma* is to truly apply to the hot QCD matter created in heavy ion collisions, it must contain excitations with the quantum numbers of quarks and gluons that are not confined into color singlet objects. As discussed in the context of the strangeness signature, one then expects that hadrons are formed by coalescence of valence quarks[1,61,62] when the matter cools back down below T_c. This predicted the strongly enhanced production of hadrons containing multiple strange quarks as a characteristic signature of the formation of a quark–gluon plasma. This enhancement was clearly observed in Au+Au collisions at RHIC, as it had been before in AA collisions at the CERN-SPS, but its value as a quark–gluon plasma signature had been questioned on the basis of the fact that it is already present at the lower collision energies and also occurs to a lesser degree in proton–nucleus collisions. Recent results from the beam energy scan at RHIC, which clearly indicate the presence of partonic collective behavior at the top SPS energy domain, and from $p+$ Pb collisions at LHC and $d+$ Au collisions at RHIC, which produced strong evidence for the presence of collective flow and showed a continuity of strangeness enhancement with final-state multiplicity, have all but eliminated these doubts.

Support for quark recombination idea came early in the RHIC physics program from particle identified spectra measured by PHENIX.[63] These showed an enhancement in the ratio of protons to pions in the transverse momentum range $p_T =$ 1–3 GeV/c. This finding became known as the "proton anomaly." The data also showed an apparent deviation from the mass hierarchy of the elliptic flow $v_2(p_T)$ of identified hadrons in the same momentum range.[64] Hydrodynamics predicts that heavier hadrons should exhibit a smaller flow anisotropy at the same momentum p_T, but PHENIX data showed that the v_2 of protons and antiprotons exceeds that of pions for $p_T > 2$ GeV/c.

The concept of valence quark recombination explained both experimental findings. If the collective transverse flow is carried by quarks and these quarks recombine at the moment of hadronization, then protons carrying three valence quarks receive a larger transverse momentum boost from the collective expansion than pions, which contain only two valence quarks. The same argument applies of course to all baryons and mesons. The application of the sudden recombination model relies on the insight that valence quarks coalescing into a hadron with a few GeV/c

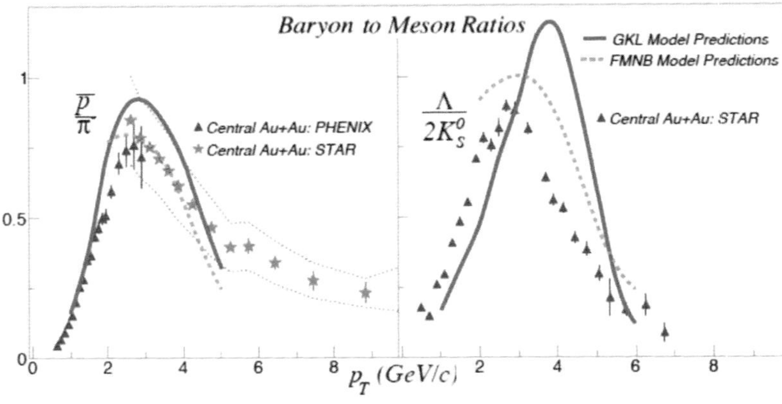

Fig. 11. Antiproton-to-pion ratio and hyperon-to-kaon ratio in Au+Au collisions as a function of transverse momentum in comparison with two quark recombination models (adapted from R. J. Fries[67]).

transverse momentum leave the quark–gluon plasma at nearly the speed of light and thus make a sudden transition from the dense matter into the surrounding vacuum.

Theoretical considerations show that the mechanism of quark recombination from a thermal quark–gluon plasma[65,66] with the transverse flow generated by the expansion at RHIC exceeds the contribution to hadron formation by parton fragmentation for transverse momenta $p_T < 3$–4 GeV/c, precisely the regime where the proton puzzle was observed. Using reasonable values for the expansion velocity at the moment when the cooling matter hadronizes led to quantitative predictions for the transverse momentum dependence of the $\bar{p}/\pi$ and Λ/K^0_s ratios as well as the elliptic flow of protons and pions, which reproduced the essential features of the PHENIX and STAR data (see Fig. 11).

An interesting relationship is obtained for the elliptic flow spectrum of different hadron species containing n valence quarks:[68] $v_2(p_t) \approx n v_2^q(p_T/n)$, which relates the elliptic flow spectrum of mesons ($n = 2$) to that of baryons ($n = 3$). At low transverse momenta, where mass effects are not negligible, it has been suggested that the transverse momentum p_T variable should be replaced by the transverse kinetic energy $m_T = \sqrt{p_T^2 + m^2}$. With this heuristic substitution, the valence quark scaling of elliptic flow was found to hold over the entire range of available data[69] (see Fig. 12).

In a more differential way than the strangeness enhancement signature, the sudden recombination model for hadron emission from the nuclear fireball provided evidence for the formation of a new state that contains collectively flowing matter composed of independently moving quarks and antiquarks. In the summary of their original publication Fries *et al.*[65] concluded: "... we propose a two component behavior of hadronic observables in heavy ion collisions at RHIC. These components include fragmentation of high-p_T partons and recombination from a thermal parton distribution.... Our scenario requires the assumption of a thermalized partonic

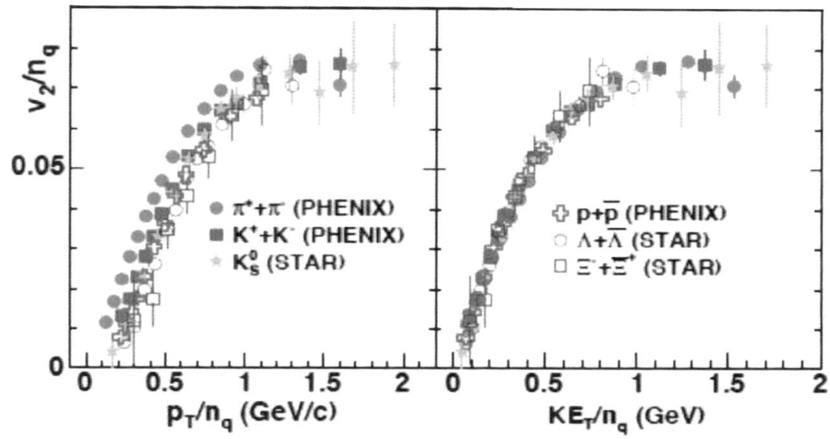

Fig. 12. Valence quark number scaled elliptic flow for various hadron species. Left panel: scaling by transverse momentum p_T; right panel: scaling by transverse mass m_T (adapted from Adare *et al.*[69]).

phase characterized by an exponential momentum spectrum. Such a phase may be appropriately called a quark–gluon plasma."

5.3. *Beam energy scan*

During the years 2010–2011 RHIC conducted a beam energy scan (BES) for Au+Au collisions that covered the energies $\sqrt{s_{\rm NN}} = 7.7$, 11.5, 19.6, 27, 39 GeV. The energy provided an essential link between the results previously obtained at the CERN-SPS and the data measured at the full RHIC collision energy of $\sqrt{s_{\rm NN}} = 200$ GeV. A compilation of the data for observables related to the chemical and kinetic freeze-out parameters of the medium can be found in Ref. 70. The BES confirmed the strong enhancement of strangeness production observed in the CERN-SPS experiments, including detailed features like Marek's "horn," the peak in the K^+/π^+ ratio near $\sqrt{s_{\rm NN}} = 7$ GeV. The strangeness flavor was found to be chemically equilibrated in central collisions over the entire energy range, and the main difference between different energies can be attributed to a strong variation in the baryon chemical potential from $\mu_B \approx 100$ MeV at the high-energy end to $\mu_B \approx 400$ MeV at the low-energy end of the BES. In summary, the BES established the continuity of chemical and thermal bulk properties of the medium from the CERN-SPS to the BNL-RHIC energy range and provided compelling evidence that a quark–gluon plasma is temporarily created across this range of collision energies.

5.4. *Jet quenching*

Energetic partons, the precursors of later emerging hadronic jets, lose energy while traversing the quark–gluon plasma either by elastic collisions with the medium constituents or by gluon radiation. At high energies, radiation should dominate;

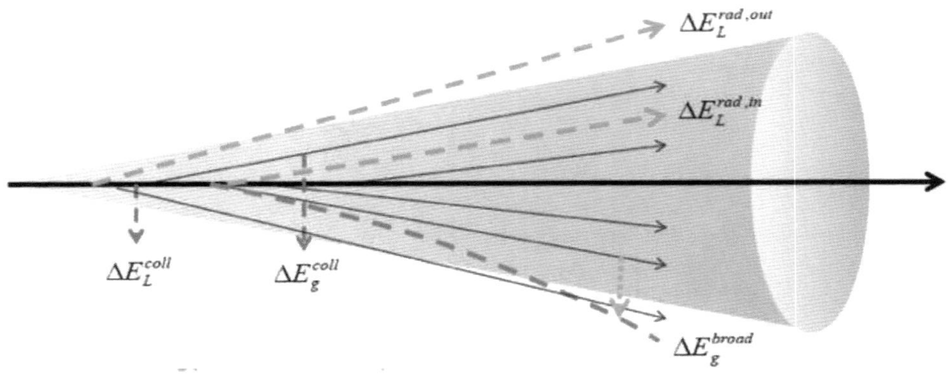

Fig. 13. Schematic representation of matter induced processes that contribute to jet quenching, i.e. the loss of energy out of the jet cone.

collisional energy loss is expected to be important for intermediate energy partons and for heavy quarks. Each mechanism is encoded in a transport coefficient, $\hat{e}$ for collisional energy loss and $\hat{q}$ for radiative energy loss:[71]

$$(dE/dx)_{\rm coll} = -C_2 \hat{e}\,, \quad (dE/dx)_{\rm rad} = -C_2 \hat{q} L\,, \tag{3}$$

where L denotes the path length traversed in matter and C_2 is the quadratic Casimir of the fast parton. The value of $\hat{q}$ is given by the transverse momentum broadening of a fast light parton per unit path length.

The evolution of a jet in the medium, shown schematically in Fig. 13, is characterized by several scales: the initial virtuality $Q_{\rm in}$ associated with the hard scattering process; the transverse scale at which the medium appears opaque, also called the saturation scale Q_s; and the transverse geometric extension of the jet, $r_\perp$. Those components of the jet, for which $r_\perp > Q_s^{-1}$, will be strongly modified by the medium. This means that the core of the jet will remain rather inert, except for an overall energy attenuation of the primary parton, but strong modifications are expected at larger angles and soft components of the jet. These features of jet modification can be encoded in a transport equation for the accompanying gluon radiation:[72]

$$\frac{d}{dt} f(\omega, k_\perp^2, t) = \hat{e}\frac{\partial f}{\partial \omega} + \frac{\hat{q}}{2}\frac{\partial f}{\partial k_\perp^2} + \frac{dN_{\rm rad}}{d\omega\, dk_\perp^2\, dt}\,, \tag{4}$$

where the last term denotes the gluon radiation induced by the medium.

The "jet quenching parameter" $\hat{q}$ can be determined by analyzing the suppression of leading hadrons in $A+A$ collisions, compared with the scaled $p+p$ data, usually given by a suppression factor R_{AA}, which is of the order of 0.2 for hadrons of transverse momenta in the range of 10 MeV/c in Au+Au at RHIC and Pb+Pb at LHC. A systematic analysis of available data from RHIC and LHC was published by the JET Collaboration[73] (see Fig. 14). It suggests that the temperature averaged value of $\hat{q}$ grows slightly less than linear with the matter density between RHIC

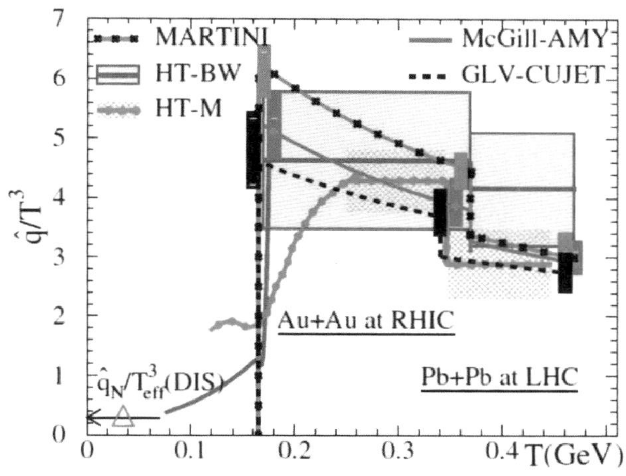

Fig. 14. Temperature scaled jet quenching parameter $\hat{q}$ deduced by the JET Collaboration from measurements of inclusive hadron suppression at RHIC and LHC (adapted from Burke *et al.*[73]).

and LHC. This confirms the notion that the quark–gluon plasma formed at higher temperatures is somewhat less strongly coupled.

Using the values of $\hat{q}$ and $\hat{e}$ determined by comparison with R_{AA} data, one can also explain the strong increase of the di-jet asymmetry observed in central Pb+Pb collisions at LHC. This gives confidence that the basic mechanisms of jet modification and parton energy loss are reasonably well understood. The phenomenology of jet quenching at the LHC, exemplified by the CMS data from Pb+Pb collisions[22] shown in Fig. 15, agrees qualitatively well with the expectation that modifications are concentrated at large cone angles and soft momentum fractions within the jet.

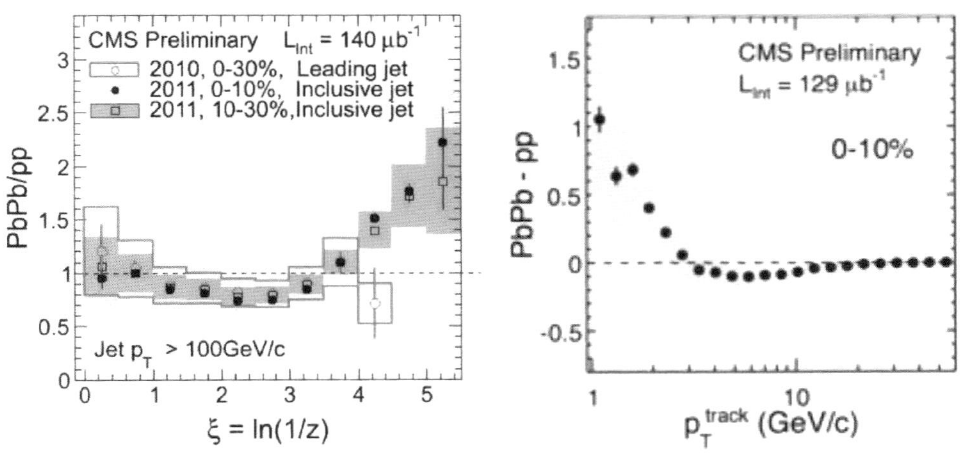

Fig. 15. Jet modifications observed in Pb+Pb collisions at LHC by the CMS Collaboration (adapted from Chatrchyan *et al.*,[22] CMS Collaboration).

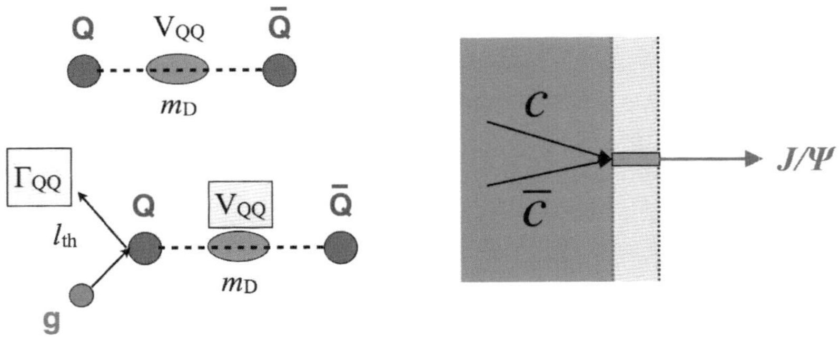

Fig. 16. Mechanisms contributing to matter induced changes in the yield of quarkonia: Color screening (upper left); ionization by thermal gluons (lower left); and recombination (right).

5.5. *Quarkonium melting*

Bound states of heavy quarks, especially quarkonia (J/ψ, ψ', the Υ states), are sensitive to the distance at which the color force is screened in the quark–gluon plasma. Several mechanisms contribute to nuclear modification of the quarkonium yield as is illustrated in Fig. 16. At sufficiently high temperatures the screening length becomes shorter than the size of the quarkonium radius and the $Q\bar{Q}$ bound state "melts." Since the radii of the quarkonium states vary widely — from approximately 0.1 fm for the Υ ground state to almost 1 fm for ψ' — the sequential melting of these states could enable at least a semiquantitative determination of the color screening length.

The static screening length, which is relevant for heavy quarks, can be calculated within the context of lattice QCD. However, it has become well understood in recent years that static color screening is only part of the picture of quarkonium melting, and that quarkonium yields cannot only be suppressed by the action of the medium, but also enhanced by recombination, if the density of heavy quarks and antiquarks is large enough. An important loss mechanism is ionization by absorption of thermal gluons. This mechanism gains in importance as the binding energy of a quarkonium state is lowered by color screening. The absorption channel can be included in the dynamical evolution of the amplitude as an imaginary part of the potential with a corresponding noise term ensuring ultimate approach to the equilibrium distribution:

$$i\hbar\frac{\partial}{\partial t}\Psi_{Q\bar{Q}} = \left[\frac{p_Q^2 + p_{\bar{Q}}^2}{2M_Q} + V_{Q\bar{Q}} - \frac{i}{2}\Gamma_{Q\bar{Q}} + \xi_{Q\bar{Q}}\right]\Psi_{Q\bar{Q}}\,. \tag{5}$$

Recombination of a heavy $Q\bar{Q}$-pair can occur at or near hadronization, similar to the sudden recombination mechanism that is thought to be responsible for the valence quark scaling of the identified particle elliptic flow. The yield of quarkonia formed in this manner grows quadratically with the heavy quark yield. Recombination of charm quark pairs into J/ψ and ψ' is thus expected to be much more frequent at LHC than at RHIC. This expectation is borne out by a comparison

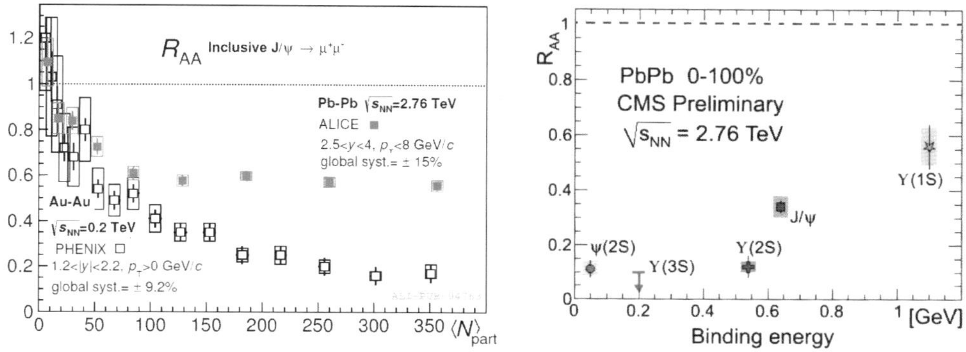

Fig. 17. (Color online) Left panel: Nuclear suppression factor R_{AA} for J/ψ as function of collision centrality in Au+Au at RHIC (red) and Pb+Pb at LHC (blue). Right panel: R_{AA} in Pb+Pb at LHC for different quarkonium states. The surprisingly small amount of suppression for J/ψ at LHC is evidence for recombination. (Adapted from Andronic[74] (ALICE) and Calderon de la Barca Sanchez[75] (CMS).)

of the centrality dependence of J/ψ suppression observed by PHENIX in Au+Au at RHIC and by ALICE in Pb+Pb at LHC (see Fig. 17). The LHC data show *less* suppression in central collisions than the RHIC data, although the significantly hotter matter at LHC energy must surely be more effective in melting the J/ψ state. Whether it is possible to measure enough observables in order to not only disentangle the action of these different mechanisms, but also to determine the color screening length, will need to be seen. On the positive side, the theory of quarkonium transport in hot QCD matter has now reached a state of sophistication where this seems possible.

In summary of this section we note that the RHIC-BNL contributed several additional and convincing experimental observables that further evolved comparing to the contemporary LHC results (which we did not discuss beyond strangeness). Without further discussion let us say that there is reason for hope that details of the initial state structure can be separated from viscous effects, and both can be separately extracted from the data. Jet physics opens new avenues of probing the quark–gluon plasma at different scales. The quarkonium data from the LHC suggest that recombination dominates in central Pb+Pb collisions for the $c\bar{c}$ states.

6. Summary

Since we have offered sub-summaries at the end of each section, here we can be brief: We have described in some detail the experimental developments that followed on the publication of our review of the strangeness signature of the QGP[1] in 1986. While over the past 30 years much has been learned in terms of experimental data accumulation and theoretical analysis, the basic insights described in our 1986 review have withstood the test of time. No other viable interpretation of all soft hadron production data measured in relativistic heavy ion collisions exists than the formation of a thermally and chemically equilibrated QGP fireball that explosively disintegrates preserving entropy, strangeness and baryon number content established at the boundary between QGP and hadron gas. The fleeting presence of the QGP is most clearly witnesses by the large overabundance of strange antibaryons.

Our chronology of the CERN-SPS strangeness research shows that the key experimental results were available as early as 1992 and were several times confirmed and published before CERN finally announced the QGP discovery in February 2000. We have made an attempt to explain why this announcement was less broadly accepted than probably would have been the case if it had been primarily based on the observation of the predicted and quantitatively confirmed strangeness enhancement. Given how it was presented in 2000, many doubts remained, and a whole new set of experiments at BNL-RHIC was required to establish a broad consensus with respect to the discovery of a new state of matter on the basis of additional phenomena not accessible at the SPS energies.

We did not dwell on the numerous attempts made over the past 30 years to question the usefulness of strangeness as signature of QGP, as none of the related arguments has withstood the test of time. However, as an example of such proposals we discussed the fact that the Ξ/ϕ ratio is constant over the whole range of SPS and RHIC energies while the degree of strangeness enhancement varies substantially, discrediting the family of models called "canonical enhancement."

With new results for strange baryon and antibaryon enhancement over a wide range of system sizes now emerging from experiments at LHC, it becomes possible to explore in detail how the strangeness flavor is chemically equilibrated as function of QGP lifetime and size. As charm becomes an abundant flavor in the LHC energy domain, flavor probes of the QGP are further expanding their reach as primary bulk signals and probes of QGP formation. While the basic picture is now well established, much awaits discovery.

References

1. P. Koch, B. Müller and J. Rafelski, *Phys. Rep.* **142**, 167 (1986).
2. B. Müller, *Rep. Prog. Phys.* **58**, 611 (1995).
3. B. Müller, *Nucl. Phys. A* **544**, 95 (1992).
4. J. Rafelski and R. Hagedorn, From hadron gas to quark matter. 2, in *Statistical Mechanics of Quarks and Hadrons*, ed. H. Satz (North-Holland, 1980), pp. 253–272, also: Preprint CERN-TH-2969, October 1980.
5. J. Rafelski, Extreme states of nuclear matter, in *Proc. "Workshop on Future Relativistic Heavy Ion Experiments"*, GSI, Darmstadt, Germany, 7–10 October 1980, pp. 282–324 [*Eur. Phys. J. A* **51**, 115 (2015)].
6. J. Rafelski and B. Müller, *Phys. Rev. Lett.* **48**, 1066 (1982) [Erratum: *ibid.* **56**, 2334 (1986)].
7. J. Rafelski, Formation and observables of the quark–gluon plasma, preprint UFTP-80-1982.
8. M. Jacob and J. Tran Thanh Van, *Phys. Rep.* **88**, 321 (1982); see pp. 331–347.
9. J. Rafelski, *Nucl. Phys. A* **374**, 489C (1982).

10. M. Petran, J. Letessier, J. Rafelski and G. Torrieri, *Comput. Phys. Commun.* **185**, 2056 (2014).
11. G. Torrieri, S. Jeon, J. Letessier and J. Rafelski, *Comput. Phys. Commun.* **175**, 635 (2006).
12. G. Torrieri, S. Steinke, W. Broniowski, W. Florkowski, J. Letessier and J. Rafelski, *Comput. Phys. Commun.* **167**, 229 (2005).
13. J. Letessier and J. Rafelski, *Eur. Phys. J. A* **35**, 221 (2008).
14. M. Petran and J. Rafelski, *Phys. Rev. C* **88**, 021901 (2013).
15. J. Rafelski, *Eur. Phys. J. A* **51**, 114 (2015).
16. J. D. Bjorken, *Phys. Rev. D* **27**, 140 (1983).
17. J. Y. Ollitrault, *Phys. Rev. D* **46**, 229 (1992).
18. H. Song, S. A. Bass, U. Heinz, T. Hirano and C. Shen, *Phys. Rev. Lett.* **106**, 192301 (2011) [Erratum: *ibid.* **109**, 139904 (2012)].
19. J. D. Bjorken, Energy loss of energetic partons in quark–gluon plasma: Possible extinction of high $p(t)$ jets in hadron–hadron collisions, FERMILAB-PUB-82-059-T.
20. X. N. Wang and M. Gyulassy, *Phys. Rev. Lett.* **68**, 1480 (1992).
21. PHENIX Collab. (K. Adcox *et al.*), *Phys. Rev. Lett.* **88**, 022301 (2002).
22. CMS Collab. (S. Chatrchyan *et al.*), *Phys. Rev. C* **84**, 024906 (2011).
23. E. V. Shuryak, *Phys. Lett. B* **78**, 150 (1978) [*Sov. J. Nucl. Phys.* **28**, 408 (1978); *Yad. Fiz.* **28**, 796 (1978)].
24. R. L. Thews, M. Schroedter and J. Rafelski, *Phys. Rev. C* **63**, 054905 (2001).
25. T. Matsui and H. Satz, *Phys. Lett. B* **178**, 416 (1986).
26. E. L. Feinberg, *Nuovo Cimento A* **34**, 391 (1976).
27. J. I. Kapusta, P. Lichard and D. Seibert, *Phys. Rev. D* **44**, 2774 (1991) [Erratum: *ibid.* **47**, 4171 (1993)].
28. S. Borsanyi, *Nucl. Phys. A* **904-905**, 270c (2013).
29. G. Odyniec, *J. Phys. G: Nucl. Part. Phys.* **27**, 255 (2001).
30. NA35 Collab. (M. Gazdzicki *et al.*), *Nucl. Phys. A* **498**, 375C (1989).
31. NA35 Collab. (J. Bartke *et al.*), *Z. Phys. C* **48**, 191 (1990).
32. P. Koch and J. Rafelski, *Nucl. Phys. A* **444**, 678 (1985).
33. NA35 Collab. (R. Stock *et al.*), *Nucl. Phys. A* **525**, 221C (1991) [presented at QM90: *VIIIth International Conference on Ultrarelativistic Nucleus-Nucleus Collisions*, 7–11 May 1990, Menton, France].
34. NA35 Collab. (J. Bachler *et al.*), *Z. Phys. C* **58**, 367 (1993).
35. WA85 Collab. (S. Abatzis *et al.*), *Phys. Lett. B* **244**, 130 (1990).
36. WA85 Collab. (S. Abatzis *et al.*), *Phys. Lett. B* **259**, 508 (1991).
37. WA85 Collab. (S. Abatzis *et al.*), *Phys. Lett. B* **270**, 123 (1991).
38. J. Rafelski, *Phys. Lett. B* **262**, 333 (1991).
39. E. Quercigh and J. Rafelski, *Phys. World* **13**, 37 (2000).
40. NA35 Collab. (T. Alber *et al.*), *Z. Phys. C* **64**, 195 (1994).
41. P. Foka, Study of strangeness production in central nucleus-nucleus collisions at 200 GeV/nucleon by developing a new analysis method for the NA35 streamer chamber pictures, University Geneva, July 1995.
42. NA35 Collab. (T. Alber *et al.*), *Phys. Lett. B* **366**, 56 (1996).
43. WA85 Collab. (D. Evans *et al.*), *Nucl. Phys. A* **566**, 225C (1994).
44. F. Antinori, The heavy-ion physics programme at the CERN OMEGA spectrometer, eds. M. Jacob and E. Quercigh, The CERN OMEGA spectrometer, pp. 43–49, Yellow Report CERN-97-02.
45. WA85 Collab. (D. Di Bari *et al.*), *Nucl. Phys. A* **590**, 307C (1995).
46. WA94 Collab. (J. B. Kinson *et al.*), *Nucl. Phys. A* **590**, 317C (1995).

47. ALICE Collab. (B. B. Abelev *et al.*), *Phys. Lett. B* **728**, 216 (2014) [Corrigendum: *ibid.* **734**, 409 (2014)].
48. U. W. Heinz and M. Jacob, Evidence for a new state of matter: An Assessment of the results from the CERN lead beam program, arXiv:nucl-th/0002042.
49. BRAHMS Collab. (I. Arsene *et al.*), *Nucl. Phys. A* **757**, 1 (2005).
50. PHENIX Collab. (K. Adcox *et al.*), *Nucl. Phys. A* **757**, 184 (2005).
51. PHOBOS Collab. (B. B. Back *et al.*), *Nucl. Phys. A* **757**, 28 (2005).
52. STAR Collab. (J. Adams *et al.*), *Nucl. Phys. A* **757**, 102 (2005).
53. M. Gazdzicki, M. Gorenstein and P. Seyboth, *Acta Phys. Pol. B* **42**, 307 (2011).
54. J. Rafelski and J. Letessier, *PoS* **CONFINEMENT 8**, 111 (2008).
55. M. Petran and J. Rafelski, *Phys. Rev. C* **82**, 011901 (2010).
56. ALICE Collab. (J. Adam *et al.*), *Phys. Lett. B* **758**, 389 (2016).
57. ALICE Collab. (J. Adam *et al.*), *Nature Phys.* **13**, 535 (2017).
58. P. Kovtun, D. T. Son and A. O. Starinets, *Phys. Rev. Lett.* **94**, 111601 (2005).
59. C. Gale, S. Jeon, B. Schenke, P. Tribedy and R. Venugopalan, *Phys. Rev. Lett.* **110**, 012302 (2013).
60. S. A. Bass, J. E. Bernhard and J. S. Moreland, *Nucl. Phys. A* **967**, 67 (2017).
61. T. S. Biro and J. Zimanyi, *Nucl. Phys. A* **395**, 525 (1983).
62. J. Rafelski and M. Danos, *Phys. Lett. B* **192**, 432 (1987).
63. PHENIX Collab. (S. S. Adler *et al.*), *Phys. Rev. Lett.* **91**, 172301 (2003).
64. PHENIX Collab. (S. S. Adler *et al.*), *Phys. Rev. Lett.* **91**, 182301 (2003).
65. R. J. Fries, B. Müller, C. Nonaka and S. A. Bass, *Phys. Rev. Lett.* **90**, 202303 (2003).
66. V. Greco, C. M. Ko and P. Levai, *Phys. Rev. Lett.* **90**, 202302 (2003).
67. R. J. Fries, *PoS* **CERP 2010**, 008 (2010).
68. R. J. Fries, B. Müller, C. Nonaka and S. A. Bass, *Phys. Rev. C* **68**, 044902 (2003).
69. PHENIX Collab. (A. Adare *et al.*), *Phys. Rev. Lett.* **98**, 162301 (2007).
70. STAR Collab. (L. Adamczyk *et al.*), *Phys. Rev. C* **96**, 044904 (2017).
71. A. Majumder and M. Van Leeuwen, *Prog. Part. Nucl. Phys.* **66**, 41 (2011).
72. G. Y. Qin and B. Müller, *Phys. Rev. Lett.* **106**, 162302 (2011) [Erratum: *ibid.* **108**, 189904 (2012)].
73. JET Collab. (K. M. Burke *et al.*), *Phys. Rev. C* **90**, 014909 (2014).
74. ALICE Collab. (A. Andronic), *J. Phys. Conf. Ser.* **455**, 012002 (2013), arXiv:1304.6332 [nucl-ex].
75. CMS Collab. (M. C. de la Barca Sánchez), *J. Phys. Conf. Ser.* **458**, 012011 (2013), arXiv:1305.5560 [nucl-ex].

Time-Dependent Perturbation Theory as a Basis for Combined Many-Body-Perturbation and QED

Ingvar Lindgren
Physics Department, University of Gothenburg, Sweden
ingvar.lindgren@physics.gu.se

A general procedure has been developed for time-dependent perturbative calculations, based upon the previously developed covariant-evolution operator (CEO) procedure. This will make it possible to include time- or energy-dependent perturbations, like the QED ones, into the expansion of the standard atomic or molecular wave function.

The QED perturbations in atomic theory are of two kinds, non-radiative effects (electron correlation and virtual electron-positron pairs) and radiative ones (self energy, vacuum polarization and vertex correction). Both types have been included in our calculations, and in the radiative case it was found necessary to perform the calculations in the Coulomb gauge in order to reach sufficient numerical stability. The Coulomb gauge is more complicated to use than the commonly used Feynman gauge and has previously not been used in this kind of calculation. We have developed a workable procedure for dimensional regularization in Coulomb gauge and applied that to highly-charged helium-like ions, where very accurate experimental data are available.

The procedure can also be applied to *dynamical problems*, such as scattering, also when bound states are involved. This leads to an analogue for bound states of the well-known *optical theorem* for free-particle states. Applications are under way.

1. Introduction

The standard procedure for taking care of QED effects in atomic calculations is to calculate the atomic and QED parts separately and then add the energy contributions together.[1] This is not always satisfactory, however, since the two parts are not strictly additive. There are combined effects that in some cases could be significant. The correct way is instead to evaluate the two effects simultaneously in a coherent fashion, which requires that the QED effects are included in the atomic wave function. This can be achieved by means of the covariant-evolution operator method (CEO) that we have previously developed.[2,3] Here, there is a fundamental problem in that the QM and QED procedures have different structures. In QM we have the same time for all particles, while in field theory there are individual times for each particle, which is a consequence of relativistic covariance. This requires some approximation.

2. Standard Many-Body Perturbation Theory

2.1. *The Bloch equation*

The procedure mostly used for many-body perturbative calculations (MBPT) is the Rayleigh–Schrödinger (RS) procedure, which has the definite advantage over the alternative Brillouin–Wigner (BW) procedure of being *size extensive*.[4,5] In 1955 Claude Bloch[6] formulated the first form of the so-called Bloch equation

$$(E_0 - H_0)|\Psi\rangle = V|\Psi\rangle - |\Psi\rangle|\Psi_0\rangle\langle\Psi_0|V|\Psi\rangle, \tag{1}$$

which can be used to generate the RS expansion to all orders. Here, $|\Psi\rangle$ is the target function under investigation and $|\Psi_0\rangle$ is an intermediately normalized model function that is an approximation of the target function and an eigenfunction of a model Hamiltonian, H_0,

$$H_0\langle\Psi_0| = E_0\langle\Psi_0|; \qquad H = H_0 + V; \qquad \langle\Psi_0|\psi\rangle = 1, \tag{2}$$

V being the perturbation.

If we have several target functions, $|\Psi^\alpha\rangle$, forming a target space, we introduce for each of them a model function, $|\Psi_0^\alpha\rangle$, which together form a model space with the projection operator P. We shall assume that the model functions are the projections of the target functions onto the model space

$$\langle\Psi_0^\alpha| = P\langle\Psi^\alpha|, \tag{3}$$

and we introduce a wave operator, Ω, that transforms the model functions back to the original target functions

$$\langle\Psi^\alpha| = \Omega\langle\Psi_0^\alpha|. \tag{4}$$

We can now rewrite the Bloch equation (1) as

$$(E_0 - H_0)\Omega = V\Omega - \Omega PV\Omega, \tag{5}$$

which holds for a degenerate model space with a single energy level, E_0. The Bloch equation is frequently used as the starting point in calculations in atomic physics and quantum chemistry.

The Bloch equation was generalized to a model space with several energy levels ("quasi-degeneracy") by Baird Brandow in 1977 by using a double perturbative expansion.[7] A more direct derivation was given by Lindgren in 1974, leading to the commonly used commutator form,[5,8]

$$\boxed{\big[\Omega, H_0\big]P = \big(V\Omega - \Omega PV\Omega\big)P.} \tag{6}$$

This equation holds for an arbitrary model space that might be degenerate, non-degenerate or quasi-degenerate, i.e., with very closely spaced energy levels.

It is convenient to introduce an effective Hamiltonian, $H_{\rm eff}$, which generates the exact energies by operating on the model functions

$$H_{\rm eff}\Psi_0^\alpha = E^\alpha\Psi_0^\alpha. \tag{7}$$

The perturbative part of this operator is referred to as the effective interaction, W,

$$H_{\rm eff}P = PH_0P + W. \tag{8}$$

The Bloch equation (6) can now be expressed

$$\boxed{\big[\Omega, H_0\big]P = \big(V\Omega - \Omega W\big)P.} \tag{9}$$

In the case of a degenerate model space the Bloch equation can be expressed

$$\boxed{\Omega P = \Gamma_Q\big(V\Omega - \Omega PV\Omega\big)P,} \tag{10}$$

where

$$\Gamma_Q = \frac{Q}{E_0 - H_0} \qquad (Q = 1 - P) \tag{11}$$

is the *reduced resolvent.*

Using second quantization and Wick's theorem,[5] the perturbation can be transformed into a diagrammatic expansion in terms of Goldstone diagrams. This leads to so-called linked and unlinked diagrams, and it turns out that all unlinked diagrams cancel, if the model space is complete, i.e., contains all configurations that can be formed from the open shells (see e.g., Ref. 5). This is the Brueckner–Goldstone linked-diagram theorem.[9,10]

2.2. *Relativistic effects*

Relativistic MBPT is normally based upon the *projected Dirac–Coulomb–Breit Hamiltonian*[11]

$$H = \Lambda_+\Big[\sum_{i=1}^{N} h_D(i) + V_{\rm C} + V_{\rm B}\Big]\Lambda_+, \tag{12}$$

using the Coulomb gauge. Here, the first term represents the sum of single-particle Dirac Hamiltonians, the second term the Coulomb interaction and the third term the *instantaneous Breit interaction* between the electrons,

$$V_{\rm B} = -\frac{e^2}{8\pi\epsilon_0}\sum_{i<1}\Big[\frac{\boldsymbol{\alpha}_i\cdot\boldsymbol{\alpha}_j}{\boldsymbol{r}_{ij}} + \frac{(\boldsymbol{\alpha}_i\cdot\boldsymbol{r}_{ij})(\boldsymbol{\alpha}_j\cdot\boldsymbol{r}_{ij})}{r_{ij}^2}\Big], \tag{13}$$

where $\boldsymbol{\alpha}_i$ is the Dirac alpha vector matrix for particle i. The projection operators, Λ_+, exclude negative-energy states that can lead to singularities (Breit–Ravenhall disease[12]). The Breit interaction represents the magnetic interaction as well as the leading part of the *retardation*, caused by the fact that the electromagnetic radiation propagates with the finite speed of light.

3. QED effects

The Hamiltonian (12) is correct to order α^2 Rydbergs (or $\alpha^4 mc^2$), where α is the fine-structure constant. This is referred to as the *No-Virtual-Pair Approximation* (NVPA). Effects beyond this approximation — of order α^3 Rydbergs and higher — are conventionally referred to as *QED effects.*

The QED effects are of two kinds, non-radiative (electron correlation and virtual pairs) and radiative ones (self energy and vertex correction). Some low-order effects are illustrated in Fig. 1.

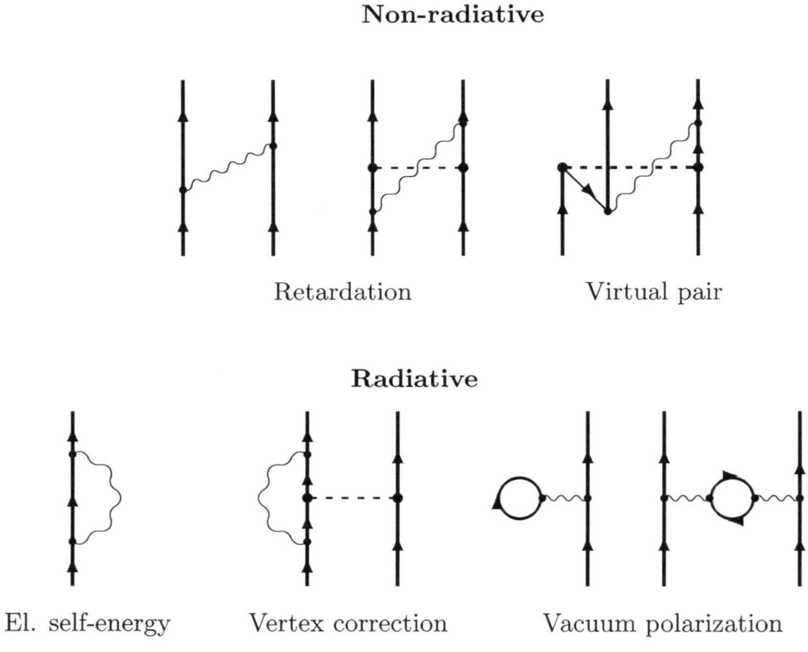

Fig. 1. Diagrams representing the first-order QED effects. The heavy vertical lines represent bound-state orbitals, generated in an external (nuclear) potential. The dotted line represent the instantaneous Coulomb interaction and the wavy line the covariant photon exchange.

4. The S-Matrix and the Covariant-Evolution Operator

4.1. *The time-evolution operator*

The state vector evolves in time according to ($\hbar = 1$)

$$|\Psi(t)\rangle = \mathrm{e}^{-\mathrm{i}H(t-t_0)}\,|\Psi(t_0)\rangle, \tag{14}$$

known as the *Schrödinger picture.* In the *interaction picture*(IP) the state vectors and the operators are transformed according to

$$|\Psi_{\mathrm{I}}(t\rangle = \mathrm{e}^{\mathrm{i}H_0 t}\,|\Psi_{\mathrm{S}}(t)\rangle; \qquad \mathcal{O}_{\mathrm{I}}(t) = \mathrm{e}^{\mathrm{i}H_0 t}\,\mathcal{O}_{\mathrm{S}}\,\mathrm{e}^{-\mathrm{i}H_0 t}. \tag{15}$$

We shall later also need an alternative picture, the Heisenberg picture,

$$|\Psi_{\rm H}(t\rangle = {\rm e}^{{\rm i}Ht}\,|\Psi_{\rm S}(t)\rangle; \qquad \mathcal{O}_{\rm H}(t) = {\rm e}^{{\rm i}Ht}\,\mathcal{O}_{\rm S}\,{\rm e}^{-{\rm i}Ht}. \tag{16}$$

The Schrödinger equation is in IP transformed to

$${\rm i}\hbar\frac{\partial}{\partial t}|\Psi_{\rm I}(t)\rangle = V_{\rm I}(t)\,|\Psi(t)\rangle \tag{17}$$

with the solution

$$|\Psi_{\rm I}(t)\rangle = |\Psi_{\rm I}(t_0)\rangle - {\rm i}\int_{t_0}^{t} {\rm d}t_1\; V_{\rm I}(t_1)\,|\Psi_{\rm I}(t_1)\rangle. \tag{18}$$

$V_{\rm I}(t_1)$ is the perturbation in the interaction picture, which is assumed to be *time independent in the Schrödinger picture.*[a] For a *stationary state* of energy E the time dependence is ${\rm e}^{-{\rm i}Et}$, and it then follows that the state in the IP is of the form

$$|\Psi(t)\rangle = {\rm e}^{-{\rm i}t(E-H_0)}\,|\Psi(t=0)\rangle. \tag{19}$$

The *time-evolution operator* in IP, $U(t,t_0)$, is defined by

$$\boxed{|\Psi(t)\rangle = U(t,t_0)\,|\Psi(t_0)\rangle \qquad (t>t_0).} \tag{20}$$

This operator satisfies the differential equation

$${\rm i}\hbar\frac{\partial}{\partial t}U(t,t_0) = V(t)\,U(t,t_0), \tag{21}$$

which leads to the expansion

$$\begin{aligned} U(t,t_0) &= 1 - {\rm i}\int_{t_0}^{t}{\rm d}t_1\,V(t_1)\,U(t_1,t_0) \\ &= 1 - {\rm i}\int_{t_0}^{t}{\rm d}t_1\;V(t_1) + {\rm i}^2\int_{t_0}^{t}{\rm d}t_1\;V(t_1)\int_{t_0}^{t_1}{\rm d}t_2\,V(t_2)\,U(t_2,t_0), \end{aligned} \tag{22}$$

etc., where T is the *Wick time-ordering* operator. By extending all integrations from t_0 to t, this can be expressed

$$U(t,t_0) = \sum_{n=0}^{\infty}\frac{(-{\rm i})^n}{n!}\int_{t_0}^{t}{\rm d}t_1\ldots\int_{t_0}^{t}{\rm d}t_n\,T\big[V(t_1)\ldots V(t_n)\big]. \tag{23}$$

The interaction can be expressed by means of the *Hamiltonian density* by

$$V(t) = \int {\rm d}^3\boldsymbol{x}\,\mathcal{H}(t,\boldsymbol{x}), \tag{24}$$

which gives

$$\boxed{U(t,t_0) = \sum_{n=0}^{\infty}\frac{(-{\rm i}/c)^n}{n!}\int_{t_0}^{t}{\rm d}x_1^4\ldots\int_{t_0}^{t}{\rm d}x_n^4\,T\big[\mathcal{H}(x_1)\ldots\mathcal{H}(x_n)\big].} \tag{25}$$

(The factor of c in the denominator is due to the fact that we now use the integration variable $x_0 = ct$.)

[a]In the following we shall work in the IP, if not stated otherwise, and then leave out the subscript "I".

A *damping factor* $\mathrm{e}^{-\gamma|t|}$ is applied to the perturbation, where γ is a small, positive number, which eventually $\to 0$. This implies that the evolution operator becomes unperturbed, when $t \to \pm\infty$.

We assume here that the interaction is that between the electrons (of charge $-e$) and the electromagnetic radiation field

$$\boxed{\mathcal{H}(x) = -\hat{\psi}^{\dagger}(x) ec\alpha^{\mu} A_{\mu}(x)\hat{\psi}(x).} \tag{26}$$

Here, α^{μ} is the Dirac operator and A_{μ} is the covariant radiation field operator

$$A_{\mu}(x) = \sqrt{\frac{\hbar}{2\epsilon_0 \omega V}} \sum_{\boldsymbol{k}r} \varepsilon_{\mu r} \left[a^{\dagger}_{\boldsymbol{k}r}\, \mathrm{e}^{ikx} + a_{\boldsymbol{k}r}\, \mathrm{e}^{-ikx}\right]. \tag{27}$$

4.2. *The S-matrix*

The *S-matrix* is defined

$$S = U(\infty, -\infty) \tag{28}$$

or with (25)

$$S = \sum_{n=0}^{\infty} \frac{1}{n!} \left(\frac{-\mathrm{i}}{c}\right)^{n} \int_{-\infty}^{\infty} \mathrm{d}x_1^4 \ldots \int_{-\infty}^{\infty} \mathrm{d}x_n^4\, T\big[\mathcal{H}(x_1) \ldots \mathcal{H}(x_n)\big]. \tag{29}$$

This is frequently used for QED calculations on bound systems, but restricted to cases where the model space contains only a single energy level.[13] The more general case can be treated by means of the covariant-evolution operator, to be discussed later.

4.3. *The Green's function*

Next, we consider the single-particle *Green's function*, which for a one-dimensional model space can be defined[14]

$$G(x, x_0) = \frac{\langle 0_{\mathrm{H}} | T[\hat{\psi}_{\mathrm{H}}(x) \hat{\psi}^{\dagger}_{\mathrm{H}}(x_0)] | 0_{\mathrm{H}} \rangle}{\langle 0_{\mathrm{H}} | 0_{\mathrm{H}} \rangle}, \tag{30}$$

where $\hat{\psi}_{\mathrm{H}}$, $\hat{\psi}^{\dagger}_{\mathrm{H}}$ are the electron-field operators in the Heisenberg representation (see above). The state $|0_{\mathrm{H}}\rangle$ represents the vacuum in this representation, i.e., the state with no particles or holes. Both the numerator and the denominator of the definition can be singular in the limit of vanishing damping, but the *ratio* is always regular. The expression above can also be expressed

$$G(x, x_0) = \frac{\langle 0 | T\big[\hat{\psi}(x)\, S\, \hat{\psi}^{\dagger}(x_0)\big] | 0 \rangle}{\langle 0 | S | 0 \rangle} = \frac{1}{\langle 0 | S | 0 \rangle} \sum_{n=0}^{\infty} \frac{1}{n!} \left(\frac{-\mathrm{i}}{c}\right)^{n}$$
$$\times \int_{-\infty}^{\infty} \mathrm{d}x_1^4 \ldots \int_{-\infty}^{\infty} \mathrm{d}x_n^4\, \langle 0 | T\big[\hat{\psi}(x) \mathcal{H}(x_1) \ldots \mathcal{H}(x_n) \hat{\psi}^{\dagger}(x_0)\big] | 0 \rangle, \tag{31}$$

using (29).

The additional electron-field operators in the expression are contracted to the free ends of the S-matrix, forming electron propagators (see Fig. 2(b)). Hence, time can flow in both directions, and the Green's function is relativistically covariant. The denominator in (31) eliminates the disconnected parts that are singular. Then the Green's function is completely connected and regular.

In second order the one-body Green's function becomes

$$G(x,x_0)=\frac{1}{\langle 0|S|0\rangle}\Big(\frac{-\mathrm{i}}{c}\Big)^2\iint \mathrm{d}^4x_1\mathrm{d}^4x_2\langle 0|T\big[\hat{\psi}(x)\mathcal{H}(x_1)\,\mathcal{H}(x_2)\,\hat{\psi}^\dagger(x_0)\big]|0\rangle. \tag{32}$$

With a perturbation $V(x_1,x_2)$

$$\overbrace{\mathcal{H}(x_1)\mathcal{H}_2}=\hat{\psi}^\dagger(x_1)\,\mathrm{i}V(x_1,x_2)\hat{\psi}(x_2), \tag{33}$$

where the "hook" represents a contraction, this gives, after further contractions, (see Fig. 2(b))

$$G(x,x_0)=\frac{1}{\langle 0|S|0\rangle}\frac{1}{c^2}\bigg\{\iint \mathrm{d}^4x_1\mathrm{d}^4x_2\,\mathrm{i}S_\mathrm{F}(x,x_2)\,(-\mathrm{i})V(x_2,x_1)\,\mathrm{i}S_\mathrm{F}(x_1,x_0)\bigg\}. \tag{34}$$

S_F is the electron propagator. (Restricting the integrations to $x_2>x_1$, the factor of 1/2 can be left out.)

4.4. *The covariant-evolution operator*

If we add incoming and outgoing electron-field operators to the Green's function, represented in Fig. 2(b), then we obtain an evolution operator, where time can flow in both directions. This is relativistically covariant and we refer to it as the *covariant-evolution operator* (Fig. 2(c)).

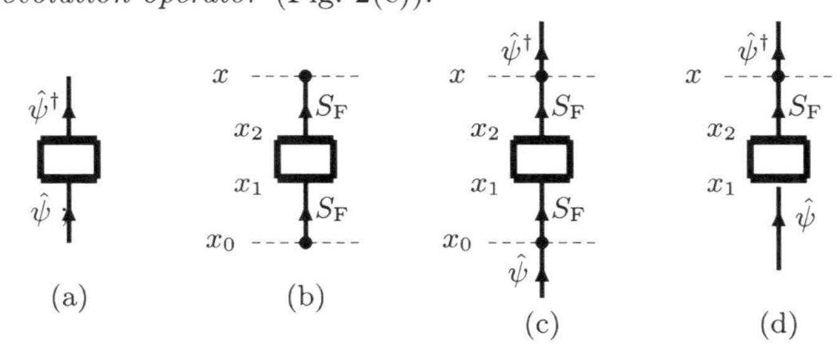

Fig. 2. One-particle evolution operator or S-matrix (a), Green's function (b) and the covariant evolution operator (CEO) (c). (d) is the CEO with initial time $t_0=-\infty$.

In Fig. 3 we show the corresponding two-particle diagrams with a single photon exchange.

The standard, non-relativistic evolution operator represents the time evolution of the non-relativistic wave function or state vector (20)

$$\Psi(t)=NU(t,t_0)\,\Psi(t_0), \tag{35}$$

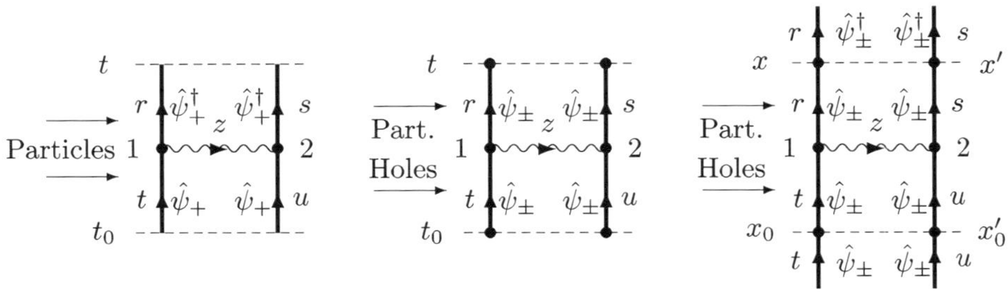

Fig. 3. Comparison between the standard evolution operator, the Green's function and the covariant evolution operator for single-photon exchange in the equal-time approximation.

where N is a normalization constant. This is illustrated for single-photon exchange in the first diagram of Fig. 3. Since here time flows only in the positive direction, this operator is *not* relativistically covariant. By replacing the evolution operator by the corresponding covariant one, we obtain the corresponding formula for the relativistic evolution

$$\Psi_{\rm Rel}(t) = N_{\rm Rel} U_{\rm Cov}(t, t_0)\, \Psi_{\rm Rel}(t_0). \tag{36}$$

Generally, the single-particle covariant-evolution operator can be defined

$$U^1_{\rm Cov}(t, t_0) = \iint \mathrm{d}^3\boldsymbol{x}\, \mathrm{d}^3\boldsymbol{x}_0\, \hat\psi^\dagger(x) \langle 0_{\rm H}|T[\hat\psi_{\rm H}(x)\hat\psi^\dagger_{\rm H}(x_0)]|0_{\rm H}\rangle\, \hat\psi(x_0), \tag{37}$$

using the same vacuum expectation as in the definition of the Green's function (30). In the interaction picture (15) this leads to the expansion

$$U^1_{\rm Cov}(t, t_0) = \sum_{n=0}^{\infty} \frac{1}{n!} \iint \mathrm{d}^3\boldsymbol{x}\, \mathrm{d}^3\boldsymbol{x}_0 \left(\frac{-\mathrm{i}}{c}\right)^n \int \mathrm{d}^4x_1 \cdots \int \mathrm{d}^4x_n$$
$$\times\, \hat\psi^\dagger(x)\, T\Big[\hat\psi(x)\, \mathcal{H}(x_1) \cdots \mathcal{H}(x_n)\, \hat\psi^\dagger(x_0)\Big] \hat\psi(x_0)\, \mathrm{e}^{-\gamma(|t_1|+|t_2|\cdots)}, \tag{38}$$

coupled to a one-body operator. Here, γ is the parameter of the adiabatic damping. Note that a contraction of TWO interactions of $\mathcal{H}$ is needed to form a single-photon exchange (see (34)). Generally, when the evolution operator contains unpaired interactions of the type (2), it will fall in the *photonic Fock space*, where the number of photons is no longer conserved.

In the standard evolution operator (20) time evolves only in the positive direction, and this operator is therefore *not covariant* and not useful in a relativistic treatment. By associating electron propagators to the operator, time can run in both directions, and the operator will then be relativistically covariant, the *Covariant Evolution Operator* (CEO).

4.5. *The Green's operator, the time-dependent wave operator*

The CEO (37) becomes singular (or quasi-singular) when a state degenerate (or quasi-degenerate) with the initial state is involved in the expansion. In order to remedy the situation, a *Green's operator*, $\mathcal{G}(t, t_0)$, is introduced by the relation[2,3]

$$\boxed{U(t, t_0)P = \mathcal{G}(t, t_0) \cdot PU(0, t_0)P.} \tag{39}$$

Here, the Green's operator does not operate beyond the heavy dot.

When the relation above is expanded perturbatively, negative *counter-terms* appear that are singular and cancel the corresponding singularities of the time-evolution operator, U. This leads to a finite remainder, referred to as *model-space contribution.* The resulting Green's operator is then completely regular.

It follows from the definitions (20) and (39) that the state vector at arbitrary time, t, is given by

$$\big|\chi^\alpha(t)\big\rangle = N^\alpha U(t, -\infty)\big|\Phi^\alpha\big\rangle = N^\alpha \mathcal{G}(t, -\infty) \cdot PU(0, -\infty)\big|\Phi^\alpha\big\rangle = \mathcal{G}(t, -\infty)|\Psi_0^\alpha\rangle, \tag{40}$$

where

$$N^\alpha PU(0, -\infty)\big|\Phi^\alpha\big\rangle = P|\Psi^\alpha\rangle = |\Psi_0^\alpha\rangle \tag{41}$$

is the model state (3). This implies that *the Green's operator is a generalization into the time-independent domain of the wave-operator concept, Ω, in* (4).

For $t = 0$ the first-order Green's operator is the same as the first-order wave-operator (6)

$$\mathcal{G}^{(1)} = U^{(1)} = \Gamma_Q V = \Omega^{(1)}. \tag{42}$$

In second order we have

$$\mathcal{G}^{(2)} = \Gamma_Q V \mathcal{G}^{(1)} + \frac{\delta \mathcal{G}^{(1)}}{\delta \mathcal{E}} PVP = \Gamma_Q V \mathcal{G}^{(1)} - \Gamma_Q\, \mathcal{G}^{(1)}\, PVP + \Gamma_Q\, \frac{\delta V}{\delta \mathcal{E}} PVP. \tag{43}$$

The last two terms form the model-state contribution that is left over after eliminating the singularities. This can be compared with the standard second-order wave operator (6)

$$\Omega^{(2)} = \Gamma_Q V \Omega^{(1)} - \Gamma_Q \Omega^{(1)} PVP. \tag{44}$$

Then we see that the Green's operator (43) contains an additional term, containing the energy derivative of the perturbation, V. This illustrates the fact that the Green's operator can handle also time- or energy-dependent perturbations, such as the QED perturbations, and mix them into the wave function together with time- or energy-independent perturbations that form the electron correlation. This is the basic principle for the unified QED-MBPT procedure we have developed.

A procedure that is feasible and at the same time quite accurate is to combine the first-order QED effects in Fig. 1 with all-order electron correlation, using the Coulomb gauge.[15] In that gauge most of the transverse-photon effect (Breit interaction) can be included in the electron correlation by means of the instantaneous Breit interaction (13). This implies that almost all two-photon effects are included in the procedure, as well as the dominating higher-order effects. This has the consequence that first-order QED with electron correlation can be considerably more accurate than complete second-order QED.

The leading radiative corrections are represented in Fig. 1. Calculating the radiative QED interaction, involving a self-energy loop, leads to singularities and have to be regularized and renormalized. The most versatile regularization procedure is the dimensional regularization, where the number of dimensions is set to 4-η, where η is a small positive number that eventually tends to zero.

5. Numerical Results

Essentially all previous calculations of radiative QED effects are performed using the Feynman gauge, but for our purpose of combining with electron correlation it would be advantageous to use the Coulomb gauge, where the electronic interaction is represented by the instantaneous Coulomb interaction. A workable procedure for dimensional regularization of QED interactions in Coulomb gauge was developed by Johan Holmberg and Daniel Hedendahl[16,17] in our research group, based upon fundamental works of Adkins.[18,19] The results obtained for some highly charged hydrogen-like ions are exhibited in Table 1 in terms of the $F(Z\alpha)$ function,

$$\Delta E^{SE} = \frac{\alpha}{\pi}(Z\alpha)^4 mc^3 F(Z\alpha), \tag{45}$$

Z being the nuclear charge and α the fine-structure constant.

Table 1. Values of the function $F(Z\alpha)$ in Eq. (45) for the self-energy of the ground state of hydrogenlike ions (from Hedendahl and Holmberg[17]).

Z	Coulomb gauge	Feynman gauge
18	3.444 043(9)	3.444 04(3)
26	2.783 762(3)	2.783 77(1)
36	2.279 314(2)	2.279 316(7)
54	1.181 866 2(6)	1.781 868(3)
66	1.604 461 5(4)	1.604 462(2)
82	1.487 258 4(4)	1.487 259(1)
92	1.472 424 1(4)	1.472 425(1)

It is interesting to note that the numerical accuracy is actually higher, when the Coulomb gauge is used. This is a general observation.

Our study of the radiative corrections, primarily the self-energy and the vertex correction, on helium-like ions has yielded interesting new results.[15] The total result is gauge independent, but there are enormous cancellations between various contributions in the Feynman gauge, which is not the case in the Coulomb gauge. This will make the results numerically more unstable in the Feynman gauge.

In going beyond second order it is not feasible to evaluate all diagrams, and the series has to be truncated. We have found that sensible results can only be achieved in the Coulomb gauge.

6. Comparison with Experiments

Extensive comparison between theoretical and experimental X-ray energies have been performed for helium like ions by Artemyev *et al.*,[20] and the agreement is generally quite good. Discrepancies have been found,[21,22] and the question has been whether these could be due to higher-order QED effects, not included in the theoretical analysis, or have other causes. Our calculations have shown that these effects are too small to explain the claimed discrepancies, some of which have furthermore been eliminated by new measurements.

7. Dynamic Processes

The standard procedure for dealing with *dynamical* processes for free or unbound particles is the S-matrix formalism. When bound states are involved, however, this procedure may fail due to possible intermediate model-space states, which might make the S-matrix singular. These singularities can be eliminated by considering the Green's operator, in a similar way as in static problems, considered above.

In energy calculations for static systems we are mainly concerned with the real part of the effective Hamiltonian. It follows from the so-called *optical theorem* for scattering processes that the scattering cross section is proportional to the *imaginary* part of this operator, which explains why essentially the same procedure can be used in the two — seemingly quite different — cases.

7.1. *Optical theorem for free particles*

The amplitude τ for a scattering process for free particles from an initial state $|p\rangle$ to a final state $\langle q|$ is related to the S-matrix by

$$\langle q|S|p\rangle = 2\pi \mathrm{i}\delta(E_p - E_q)\,\tau(p \to q). \tag{46}$$

Considering the forward scattering, leads to the **optical theorem for free particles**

$$-2\mathrm{Im}\langle p|\mathrm{i}S|p\rangle = \sum_q \Big|2\pi\delta(E_p - E_q)\tau(p \to q)\Big]^2. \tag{47}$$

This implies that ***the imaginary part of the forward scattering amplitude is proportional to the total cross section.***

7.2. *Optical theorem for bound particles*

When bound states are involved in the process, discrete intermediate model-space states may appear in the S-matrix, which leads to singularities, so-called *model-space contributions* (MSC). By replacing the S-matrix by the Green's operator, $\mathcal{G}(\infty, -\infty)$, these singularities are eliminated,

$$-2\mathrm{Im}\langle p|\mathrm{i}\,\mathcal{G}(\infty, -\infty)|p\rangle = \sum_q \Big|2\pi\delta(E_p - E_q)\tau(P \to q)\Big]^2, \tag{48}$$

which is the **optical theorem for bound particles**.

It can be shown that the S-matrix is closely related to the effective Hamiltonian (7),

$$2\mathrm{Im}\langle p|\mathrm{i}S|p\rangle = 2\pi\delta(E_p - E_q)\langle p|\mathrm{Im}(H_{\mathrm{eff}})|p\rangle, \tag{49}$$

and a similar relation holds for the Green's operator in the bound case,

$$2\mathrm{Im}\langle p|\mathrm{i}\mathcal{G}(\infty,-\infty)|p\rangle = 2\pi\delta(E_p - E_q)\langle p|\mathrm{Im}(H_{\mathrm{eff}})|p\rangle. \tag{50}$$

This leads to a formulation of the optical theorem

$$\boxed{-2\mathrm{Im}\langle p|(H_{\mathrm{eff}})|p\rangle = \sum_q 2\pi\delta(E_p - E_q)\Big|\tau(p\to q)\Big]^2} \tag{51}$$

that is **valid for free as well bound particles**.

The optical theorem can be applied to scattering processes as well as to other dynamical processes. We have applied it to the process of radiative recombination and to atomic transitions.[23,24]

Acknowledgments

In 1993 I received a research award from the Humboldt Foundation, and Walter Greiner was my host during my stays in Germany, financed by this award. I benefitted greatly from these stays, and I want to express my gratitude to the Foundation for this award and to the late Walter Greiner for his great support. I also want to express my gratitude to my coworkers, Sten Salomonson and Johan Holmberg, together with whom most of this work has been performed.

References

1. D. R. Plante, W. R. Johnson, and J. Sapirstein, *Phys. Rev. A* **49**, 3519 (1994).
2. I. Lindgren, S. Salomonson, and B. ÅAsén, *Physics Reports* **389**, 161 (2004).
3. I. Lindgren, *Relativistic Many-Body Theory: A New Field-Theoretical Approach, Second edition* (Springer-Verlag, New York, 2016).
4. R. J. Bartlett and G. D. Purvis, *Int. J. Quantum Chem.* **14**, 561 (1978).
5. I. Lindgren and J. Morrison, *Atomic Many-Body Theory* (Second edition, Springer-Verlag, Berlin, 1986, reprinted 2009).
6. C. Bloch, *Nucl. Phys.* **6/7**, 329/451 (1958).
7. B. H. Brandow, *Adv. Quantum Chem.* **10**, 187 (1977).
8. I. Lindgren, *J. Phys. B* **7**, 2441 (1974).
9. K. A. Brueckner, *Phys. Rev.* **100**, 36 (1955).
10. J. Goldstone, *Proc. R. Soc. London, Ser. A* **239**, 267 (1957).
11. J. Sucher, *Phys. Rev. A* **22**, 348 (1980).
12. G. E. Brown and D. G. Ravenhall, *Proc. R. Soc. London, Ser. A* **208**, 552 (1951).
13. P. J. Mohr, G. Plunien, and G. Soff, *Physics Reports* **293**, 227 (1998).
14. C. Itzykson and J. B. Zuber, *Quantum Field Theory* (McGraw-Hill, New York, 1980).
15. J. Holmberg, S. Salomonson, and I. Lindgren, *Phys. Rev. A* **92**, 012509 (2015).
16. J. Holmberg, *Phys. Rev. A* **84**, 062 504 (2011).
17. D. Hedendahl and J. Holmberg, *Phys. Rev. A* **85**, 012 514 (2012).

18. G. Adkins, *Phys. Rev. D* **27**, 1814 (1983).
19. G. Adkins, *Phys. Rev. D* **34**, 2489 (1986).
20. A. N. Artemyev, V. M. Shabaev, V. A. Yerokhin, G. Plunien, and G. Soff, *Phys. Rev. A* **71**, 062104 (2005).
21. C. T. Chantler *et al.*, arXiv-ph 0988193 (2012).
22. C. T. Chantler, *Phys. Rev. Lett.* **109**, 153001 (2012).
23. I. Lindgren, S. Salomonson, and J. Holmberg, *Phys. Rev. A* **89**, 062 504 (2014).
24. A. N. Artemyev, J. Holmberg, and A. Surzhykov, *Phys. Rev. A* **92**, 042510 (2015).
25. V. M. Shabaev, *Physics Reports* **356**, 119 (2002).

The Fullerene-like Structure of Superheavy Element $Z = 120$ (Greinerium) — A Tribute to Walter Greiner

Ş. Mişicu
Department of Theoretical Physics,
National Institute for Nuclear Physics and Engineering,
P.O. Box MG-6, RO-077125 Bucharest-Magurele, Romania
misicu@theory.nipne.ro

I. N. Mishustin
Frankfurt Institute for Advanced Studies,
D-60438, Frankfurt am Main, Germany
Russian Research Center, Kurchatov Institute,
123182 Moscow, Russia
mishustin@fias.uni-frankfurt.de

We review the basic ideas and some theoretical models behind the concept of fullerene-like structures made of α-particles. The possibility of such a peculiar nuclear shape developing in a double magic superheavy nucleus with $Z = 120$ was mentioned for the first time in the literature by Walter Greiner and then constantly defended by him. We provide estimates of the energy of such metastable states within the liquid-drop model. In the second part of this paper we discuss a simple model rooted in the non-topological soliton model consisting of a complex scalar field describing a finite system of α bosons with non-linear self-interactions coupled to the electromagnetic field. We demonstrate that this model predicts density depletion in the central region of a soliton-like structure for a range of model parameters.

1. Introduction

More than two decades ago, Walter Greiner conjectured that some superheavy elements assume a fullerene-like structure formed of α-clusters. This speculation was born out from relativistic mean-field (RMF) calculations for nuclei with charges around $Z = 120$ which are pointing to a pronounced central density depletion.[1] In the following years he constantly advocated this exotic picture in superheavies.[2,3]

The α-fullerene is a nuclear aggregate, hollow inside, with 60 ^{4}He nuclei distributed on the vertices of 20 hexagons and 12 pentagons in a manner analogous to the buckminsterfullerene (C_{60})[4] (see Fig. 1(a)). Taking as an example the nucleus $Z = 120$, $N = 184$, Greiner suggested that apart of the protons and neutrons

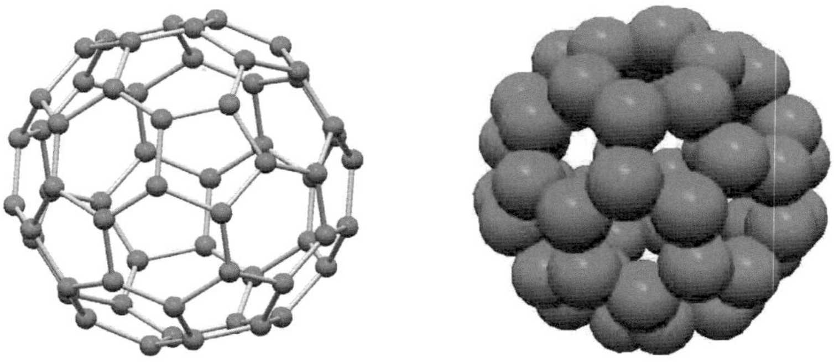

(a) (b)

Fig. 1. (a) The structure of the fullerene (C_{60}) which served as a model of a hollow superheavy nucleus with $Z = 120$ as proposed by Walter Greiner. The edges between two adjacent hexagons are double bonding. (b) Following the analogy devised by Greiner, the Carbon atoms of the buckminsterfullerene are replaced by α clusters.

distributed over 60 α-particles, the rest of 60 neutrons (neglecting the last four neutrons) are insuring the additional bonding, i.e. one neutron per alpha.[2] He speculated that the two bond lengths that are manifest in C_{60} have their nuclear counterpart in the presence of neutrons that are not associated in α particles. As Greiner asserted: "such a structure would immediately explain the semi-hollowness of that superheavy nucleus."[2] An artist view of the Greiner fullerene is given in Fig. 1(b). The RMF framework applied by Greiner and collaborators in 2002 to the nucleus $^{292}120$ suggests the existence of a pronounced depletion of matter in the interior of this superheavy nucleus[3] (see Fig. 2). We compare this old case with a new calculation for the $N = Z$ superheavy $^{240}120$ where the depletion is also visible.

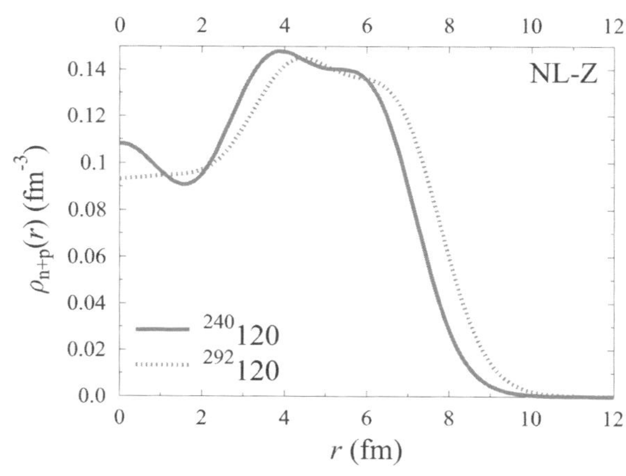

Fig. 2. Total density profiles of the superheavy isotopes of $Z = 120$: $N = 120$ (solid line) and $N = 172$ (dashed line). Calculations are made within the spherical RMF model with the NLZ parametrization.[5]

The α-particle model of nuclei is a ubiquitous presence in the literature since the early days of nuclear physics.[6–8] In this picture nuclei are viewed as crystaline structures formed of closely packed structureless spherical α-particles, viz. a classical assumption is made on the nuclear wave function which exhibits strong four-particle correlations (quartet correlations). Each structure is characterized by a number of bonds or pairs of adjacent particles. As concluded in a study on α-conjugate $N = Z$ nuclei,[9] the systematics of the binding energy is satisfactorly described for light nuclei in the α-particle model.

On the other hand, over the years the possible existence of bubble and semi-bubble shapes of heavy and superheavy nuclei was also frequently discussed in the literature.[10–14] Very recently RMF calculations[15] have shown the appearance of low nucleonic density in the central regions of lighter nuclei: ^{22}O and ^{34,36}Si were emphasized as good candidates of being spherical bubble nuclei, whereas ^{24}Ne, ^{32}Si and ^{34}Ar were classified as deformed bubble nuclei.

2. Stability of Fullerene-like Nuclei

It was argued that since the α-particle model provides an explanation of the binding energy systematics for $N = Z$ nuclei and the quartetting properties in the ground state, the liquid-drop model qualifies for an approximate description of the binding energies in terms of the mass number (semi-empirical mass formula for α-nuclei).[9] The liquid-drop model framework applied to nuclei with a strongly non-uniform distribution of matter inside, as it is the case for bubble-like nuclear shapes, requires some modifications compared to the standard case of "normal" nuclei with highly uniform distribution of nuclear matter. Below we present a rough estimation of the LDM energy for the same superheavy nucleus taking into account only the Coulomb and surface energy.

For a charge distribution $\rho_c(r)$, the Coulomb potential energy is

$$E_C = \frac{1}{2}\int dr \int dr' \frac{\rho_c(\boldsymbol{r})\rho_c(\boldsymbol{r}')}{|\boldsymbol{r}-\boldsymbol{r}'|}, \tag{1}$$

where the charge density of the semi-bubble reads density

$$\rho_c(\boldsymbol{r}) = \rho_{c1}\Theta(R_1 - r) + \rho_{c2}[\Theta(R_2 - r) - \Theta(R_1 - r)]. \tag{2}$$

The charge density of the core, containing Z_1 protons is

$$\rho_{c1} = \frac{3}{4\pi}\frac{Z_1}{R_1^3}$$

and of the outer shell, containing Z_2 protons is

$$\rho_{c2} = \frac{3}{4\pi}\frac{Z_2 e}{R_2^3 - R_1^3}$$

such that

$$Z_1 = \int_0^{R_1} \rho_{c1} d\boldsymbol{r}, \quad Z_2 = \int_{R_1}^{R_2} \rho_{c2} d\boldsymbol{r}. \tag{3}$$

We use the RMF input for charge densities from Ref. 3. After some lengthy calculations we arrive at the following expression of the Coulomb energy

$$E_C = \frac{3}{5}\frac{(Ze)^2}{R_2}\frac{p^3(1-q)[p^2(3-2q)-5]+2}{2[1-p^3(1-q)]^2}, \quad (4)$$

where by $p = R_1/R_2$ we denote the "breathing deformation" and by $q = \rho_1/\rho_2$ the core density to outer shell density ratio.

Assuming a similar dependence for the total density, the surface energy can be calculated using the Yukawa-plus-exponential interaction,[16] which for the particular case of a semi-bubble nucleus splits into the contribution from the external surface, internal surface and mixed terms[17]

$$E_S = E_{\mathrm{Y+E}\ 1} + E_{\mathrm{Y+E}\ 2} + E_{\mathrm{Y+E}\ 12} + E_{\mathrm{Y+E}\ 21}. \quad (5)$$

The diagonal terms are casted in the form

$$E_{\mathrm{Y+E}\ i=1,2} = c_S(1-\kappa_S I^2)\left(\frac{R_i}{r_0}\right)^2(1-q\delta_{i,1})^2 \times\left[1-3\left(\frac{a}{R_i}\right)^2+\left(1+\frac{R_i}{a}\right)\left(2+3\frac{a}{R_i}+3\left(\frac{a}{R_i}\right)^2\right)e^{-2R_i/a}\right], \quad (6)$$

whereas for the mixed term we get

$$E_{\mathrm{Y+E}\ 12} = \frac{2c_S(1-\kappa_S I^2)}{ar_0^2}(1-q)(R_1R_2)^{3/2} \times\left[\frac{R_1}{a}I_{1/2}\left(\frac{R_1}{a}\right)K_{3/2}\left(\frac{R_2}{a}\right)-\frac{R_2}{a}I_{3/2}\left(\frac{R_1}{a}\right)K_{\frac{1}{2}}\left(\frac{R_2}{a}\right) - 3I_{\frac{3}{2}}\left(\frac{R_1}{a}\right)K_{\frac{3}{2}}\left(\frac{R_2}{a}\right)\right]. \quad (7)$$

The last term plays a crucial role in the stability against coupled quadrupole oscillations of the inner and outer surfaces.[17] The physical constants and the meaning of the special functions entering the above formulas are given in Ref. 18.

We employ the above macroscopic formalism to the case illustrated in Fig. 3. The entire charge Z is assumed to be distributed in the outer layer together with an equal number of neutrons, i.e. $N_{\mathrm{shell}} = Z$, both species being associated in $Z/2$ α-particles. To this shell $\Delta N_{\mathrm{shell}}$ neutrons are added ($N_2 = Z + \Delta N_{\mathrm{shell}}$), which according to Greiner's scenario insure the supplementary bonding between the α-particles. The rest of neutrons, i.e. $N_1 = N - N_{\mathrm{shell}} - \Delta N_{\mathrm{shell}}$ are distributed in the "hollow" region of the semi-bubble nucleus.

We plot in Fig. 4 the sum of the Coulomb (4) and surface (5) energies as a function of the ratio $p = R_1/R_2$ relative to the equivalent spherical values $E_S^{(0)} + E_C^{(0)}$. For an outer shell composed of equal number of protons and neutrons ($\Delta N_{\mathrm{shell}} = 0$) a bubble-like state with $R_1 \approx 0.6R_2$ is unstable. The addition of neutrons in the outer shell and the dilution of the neutron fluid in the core region produces a metastable state with the breathing deformation p decreasing with increasing $\Delta N_{\mathrm{shell}}$. Therefore, a perfect bubble ($N_1 = N$, $N_2 = 0$) is the most favourable configuration for the nucleus $^{292}120$ in this version of the liquid-drop model, with a completely hollow central region of small radius ($R_1 < 0.2R_2$).

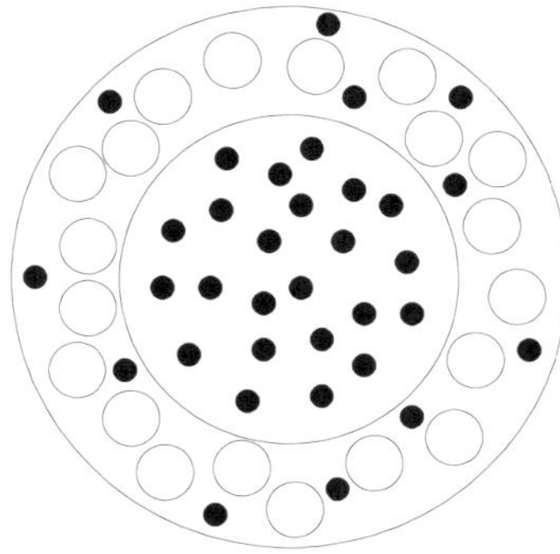

Fig. 3. Artist view of the semi-bubble nucleus resembling the endohedral cluster fullerene. Open circles are α-particles while small black circles represent neutrons. The core region has radius R_1 and contains only neutrons. The shell extends from R_1 to R_2 and contains all protons and an equal number of neutrons bound in α clusters, plus an additional number of neutrons that might contribute to the above mentioned double bondings.

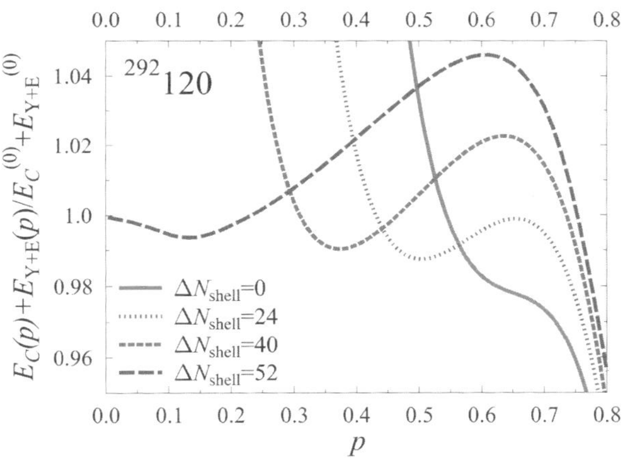

Fig. 4. Dependence of the total LDM energy surface (surface + Coulomb) to the equivalent spherical value on the breathing deformation for the superheavy nucleus $^{292}120$. Along with the 60 "α-particles" the outer shell is doped with $\Delta N_{\rm shell} = 0$ (solid-curve), 24 (dotted-curve), 40 (green dashed-curve) and 52 (long-dashed-curve) neutrons.

3. Q-Balls of α-Clusters as Non-Topological Solitons

Very recently we proposed with Walter Greiner a RMF model which includes, besides the Dirac–Fermi fields describing standard baryonic matter composed of protons and neutrons, also α-particles described by a scalar complex field such that the corresponding Lagrangian is allowed to contain quartic and sextic self-interaction terms.[19,20] The non-relativistic counterpart of these interactions is represented by two- and three-body zero-range interactions between α-particles with strengths fixed by the α–α scattering lengths extracted via the Calogero equation (see Ref. 19 for more details). In the present work we consider the simplified version of the model (without scalar σ and vector ω meson fields), and take into account the electromagnetic nature of α-particles, i.e. we consider the case of the charged Q-ball.[21,22] These physical objects are extensions of the simple Q-balls.[23,24] The stability of such an aggregate of α-particles results from the balance between the non-linear self-interactions of the complex scalar field and the Coulomb repulsion between the α-particles. In our recent work[19] we reported some interesting conclusions regarding the stability of neutral Q-balls if the meson fields are removed: (1) the tendency for over-binding is weakened, due to the absence of the σ-field attraction; (2) the equation-of-state (EOS) becomes much softer, due to the absence of the ω-field repulsion, making thus the Q-ball more easy to compress; (3) attractive quartic self-interactions, supplemented by repulsive sextic self-interactions lead to stable configurations.

Below we consider a complex scalar field ϕ, which describes α-particles of charge $q = +2e$ coupled to a U(1) gauged (electromagnetic) field A_μ. The Lagrangian density is

$$\mathcal{L} = (\partial^\mu \phi + iqA^\mu \phi)^* (\partial_\mu \phi + iqA_\mu \phi) - U(|\phi|^2) - \frac{1}{4} F_{\mu\nu} F^{\mu\nu}, \tag{8}$$

where $F_{\mu\nu} = \partial_\mu A_\nu - \partial_\nu A_\mu$ is the electromagnetic field strength and the scalar potential function is chosen in the form

$$U(|\phi|^2) = M_\alpha{}^2 |\phi|^2 + g_2 |\phi|^4 + \frac{1}{3} g_3 |\phi|^6, \tag{9}$$

where M is the α-particle mass and non-linear terms describe the self-interaction of the scalar field.

Assuming for the complex scalar field a time-dependence of the form

$$\phi(r,t) = \frac{1}{\sqrt{2}} \varphi(r) e^{-i\omega t} \tag{10}$$

and defining the radial function

$$\gamma(r) = \omega - qA_0(r), \tag{11}$$

one can represent the Lagrangian density in the form

$$\mathcal{L} = \frac{1}{2} \left[\frac{1}{q^2} \left(\frac{d\gamma}{dr} \right)^2 + \varphi^2 \gamma^2 - \left(\frac{d\varphi}{dr} \right)^2 - 2U(\varphi) \right]. \tag{12}$$

Introducing the canonical variables (ϕ, π) and (A_0, Π), where

$$\pi = \frac{\partial \mathcal{L}}{\partial \dot{\phi}}, \quad \Pi = \frac{\partial \mathcal{L}}{\partial \dot{A}_0}, \tag{13}$$

the "differential Hamilton function"[25]

$$\mathcal{H} = \pi \dot{\phi} + \pi^* \dot{\phi}^* + \Pi \dot{A}_0 - \mathcal{L} \tag{14}$$

is eventually casted in the form

$$\mathcal{H} = \frac{1}{2}\left[\frac{1}{q^2}\left(\frac{d\gamma}{dr}\right)^2 + \varphi^2\gamma^2 + \left(\frac{df}{dr}\right)^2 + 2U(\varphi)\right]. \tag{15}$$

The α-particle density reads[19]

$$\rho = j_0 = i\left(\frac{\partial\mathcal{L}}{\partial(\partial_\mu\varphi^*)}\varphi^* - \frac{\partial\mathcal{L}}{\partial(\partial_\mu\varphi)}\varphi\right) = g\varphi^2. \tag{16}$$

The corresponding equations of motion are obtained by varying the action integral[25]

$$I = \int dt \int d\boldsymbol{r}\mathcal{L}(\varphi, \varphi', \gamma, \gamma') \tag{17}$$

with respect to φ and γ at fixed ω, i.e.

$$\varphi'' + \frac{2}{r}\varphi' + g^2\varphi - \frac{\partial U(\varphi)}{\partial\varphi} = 0, \tag{18}$$

$$\gamma'' + \frac{2}{r}\gamma' - q^2\varphi^2\gamma = 0. \tag{19}$$

The solutions of the above set of second-order non-linear differential equations were discussed in the literature (see for example Ref. 21) and concluded that for a range of values of the coupling constants (g_2, g_3 in our case) the charge is pushed to the surface, the topological configuration of the gauged Q-ball resembling thus a semi-bubble. In the case of a neutral Q-ball ($q = 0$), the complex scalar field is constant inside the ball, i.e. $\varphi = \varphi_0$. We take for the squared charge, $q^2 = 4 \times 1.44/(\hbar c) = 0.029$.

In Ref. 21 it was established that for a Q-ball with total number of α-particles Q the radius can be expressed in the compact form

$$R = \left[\frac{3Q}{4\pi\varphi_0\sqrt{2U(\varphi_0)}}\right]^{1/3}\left(1 + \frac{C^{2/3}}{45}\right); \quad C = 3e^3Q\sqrt{\frac{2\pi\varphi_0^4}{U(\varphi_0)}}. \tag{20}$$

Since the strengths g_2 and g_3 assume large values we rescale the variables and the physical constants according to

$$\varphi = \frac{M_\alpha}{\sqrt{|g_2|}}f, \quad \gamma = M_\alpha g, \quad q^2 = |g_2|e^2 \tag{21}$$

such that the scalar potential (9) assume the simple form used in the literature[21,22]

$$U(f) = \frac{1}{2}f^2 - \frac{1}{4}f^4 + \frac{\lambda^2}{6}f^6, \tag{22}$$

where the strength λ is related to g_2 and g_3 via

$$\lambda^2 = \frac{g_3 M_\alpha^2}{4g_2^2}. \tag{23}$$

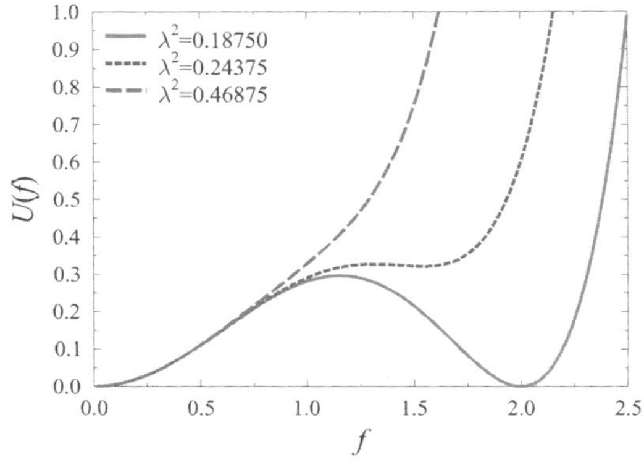

Fig. 5. Scalar potential $U(f)$ for three different values of λ^2.

As argued by Coleman[23] the scalar potential fulfills $U(f) \geq 0$, a condition which sets an inferior bound on the dimensionless constant λ, i.e. $\lambda^2 > 3/16$. In Fig. 5 we displayed the scalar potential (22) first for this limiting value (solid-curve). We observe that in this case $U(f)$ has two degenerate minima at $f = 0$ and at $f \approx 2$. By increasing λ^2, this second, non-trivial minimum, becomes shallower (dashed-curve) and eventually disappears (long-dashed-curve). For the last case, when $\lambda^2 \approx 0.47$, the corresponding quartic g_2 and sextic g_3 strengths are in the range of values established by the elastic scattering phase-shift analysis carried out in our previous and last work in collaboration with Walter Greiner.[19] More precisely the term that mimicks the two-body interaction is attractive and the one corresponding to the three-body interactions is repulsive with a value very close to the one used by us in the paper on neutral Q-balls.

We solved the system (18) and (19) in the new representation with dimensionless quantities:

$$f'' + \frac{2}{r}f' + (g^2 - 1)f - g_2 f^3 - \lambda^2 f^5 = 0,$$
$$g'' + \frac{2}{r}g' - e^2 f^2 g = 0. \tag{24}$$

The system (24) was treated as a non-linear coupled-channel problem with interior and exterior boundary conditions. In order to obtain a reasonable guess for $f(0)$, we make use of the above mentioned positiveness condition on $U(f)$ for all r. In our numerical experiments we take $f(0) < \sqrt{3}/2\lambda$. We obtained a guess for $g(0)$ by solving the second equation of the system (24) in the thin-wall approximation and applied the l'Hopital rule for $r \to 0$. In order to find finite-energy solutions of the above equations of motion, the following initial values are assumed for the derivatives

$$f'(0) = g'(0) = 0. \tag{25}$$

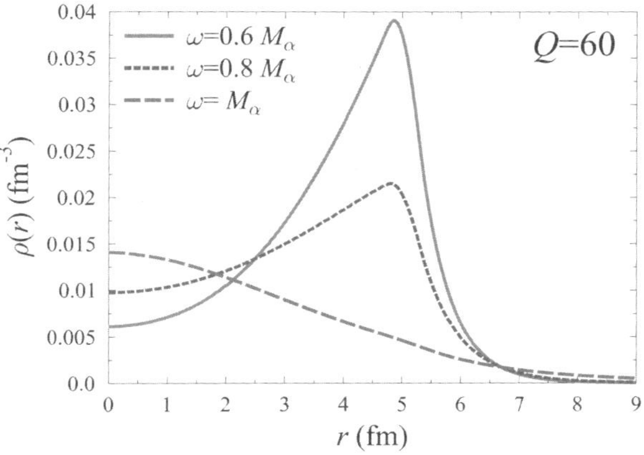

Fig. 6. Profile of the α-particle condensate with $Q = 60$ for three values of the frequency ω. We set λ^2 to the value corresponding to the green long-dashed-curve in Fig. 5.

The exterior boundary condition was specified at $r = R$ where we assume that the nonlinearities cease to be important. The radius R was taken to be the one provided by the estimation given in Eq. (20). Consequently the two equations from (24) are amenable to analytic solutions:

$$f(r) \sim e^{-kr} U(1+\kappa, 2, 2kr), \quad \omega < M_\alpha \tag{26}$$

$$f(r) \sim \frac{K_1(\sqrt{\beta r})}{\sqrt{r}}, \quad \omega = M_\alpha \tag{27}$$

$$g(r) = \omega - \frac{e^2 Q}{r}. \tag{28}$$

Above, U is the confluent hypergeometric function, K_1 the first-order modified Bessel function[26] and

$$k = \sqrt{M_\alpha^2 - \omega^2}, \quad \kappa = \frac{\omega e^2 Q}{kr}, \quad \beta = 8e^2 Q M_\alpha. \tag{29}$$

The numerical implementation consists in integrating a system of four ordinary differential equations, as results from (24), with the help of the Runge–Kutta method of orders 5 and 6.[27]

The profile (16) is plotted in Fig. 6 for values up to the maximum $\omega = M_\alpha$. We chose the following values for the cuartic and sextic strengths: $g_2 = -30000$, $g_3 = 4786193.6$ fm^2. For large ω ($\omega \to M_\alpha$) the bubble disappears. Thus, a bubble structure of the α condensate is favored by small ω. According to Ref. 28 this circumstance happens for large Q when the energy of the non-topological soliton realize a minimum in the ω coordinate. However, it is the repulsive sextic interaction that it is mainly responsible for the depletion of the scalar field inside the Q-ball.

4. Summary and Outlook

It is natural to ask how can one validate such an atypic nuclear shape. Due to the hollow spherical shape and of the large number of α-particles, this aggregate might easily develope particular rotational and vibrational modes of excitation with a different manifestation compared to normal nuclei. The investigation of such collective modes was carried out for the C_{60} fullerene within a classical model of elastic continua for a homogenous, spherical shell of atoms of zero thickness.[29] A similar approach can easily be extended to α-like fullerenes with non-vanishing thickness once the elastic constants of nuclear matter are derived from an energy density functional. Such exotic structures in nuclei could also be discriminated by considering high-lying collective excitations: giant resonances are expected to manifest themselves in a different manner in fullerene-like nuclei compared to normal fluid nuclei. One of us (Ş. M.) and Walter Greiner discussed this circumstance more than 15 years ago.[3] If one takes as an example the giant quadrupole resonance, we expect that the isovector axial symmetric mode at its amplitude produces a further depletion of proton matter inside the nucleus and a corresponding increase at the polar tips (see Ref. 30, Chap. 15, Fig. 2). Consequently, if α particles are already present at the nuclear periphery and the oscillations push them in regions of low Coulomb barriers, i.e. at the two poles of the dynamically prolate deformed nucleus, the α-decay of the superheavy nucleus is a highly probable doorway channel. The decay process could take place by multiple emission of α-particles! If on the other hand we consider the excitation of the giant dipole resonance, one should take into account that the role of different terms in the energy density in building such collective modes in bubble nuclei can be fundamentally different compared to normal nuclei. Indeed, if the pure neutron core of low density oscillates in anti-phase to the higher density outer-shell, the elastic restoring constants of the hydrodynamical Steinwedel-Jensen model[30] will receive a consistent contribution from the compression energy.

In this paper we discussed an ideal situation, i.e. an $N = Z$ nucleus. Since in real situations we deal with a large excess of neutrons then, if we stick to the fullerene picture of the superheavy $Z = 120$, one may expect that the neutrons that are not associated with the α-particle are instead confined inside the α-fullerene cage. One should remind the reader that an outstanding property of atomic fullerenes is their ability to trap atoms, ions, clusters or small molecules.[4] This association of fullerenes with other species leads to the so-called endohedral cluster fullerene. In an analogous manner we expect that the excess of neutrons wanders inside the α-cage instead of sharing the same peripheral region with the α-clusters. One can push the analogy even further and imagine a core-like structure (with proton and neutron numbers close to the magic numbers) in the center of the α-fullerene and a gas o neutrons filling the space between the core and the outer shell.

Although the possibility of a fullerene-type structure consisting of α-clusters is highly speculative, as stated in a very recent review dedicated to the state-of-art on element $Z = 120$,[31] "the highly advanced experimental technology should be used also for some experiments to search for such really exotic phenomena in the region of SHN and beyond, which is accesible using the heaviest beams and targets."

Acknowledgments

One of the authors (Ş. M.) is grateful to Prof. P.-G. Reinhard for elucidating him some aspects related to the numerical implementation of the relativistic mean-field model and to Mrs. C. Matei for assistance with the artwork. Ş. M. acknowledges the financial support received from the Institute of Atomic Physics-IFA, through the national programme PN III 5/5.1/ELI-RO, Project 04-ELI/2016 ("QLASNUC") and the Ministry of Research and Innovation of Romania, through the Project PN 16 42 01 05/2016. I. N. Mishustin acknowledges financial support from the Helmholz International Center for FAIR (Germany).

References

1. M. Bender, K. Rutz, P.-G. Reinhard, J. A. Maruhn and W. Greiner, Shell structure of superheavy nuclei in self-consistent mean-field models, *Phys. Rev. C* **60**, 034304 (1999).
2. W. Greiner, Nuclear Cluster Structure: Superheavies, cluster-radioactivity and exotic fission processes, *Heavy Ion Phys.* **13**, 61 (2001).
3. Ş. Mişicu, T. Bürvenich, T. Cornelius and W. Greiner, Properties of some collective excitations in spherical nuclei from the superheavy island, *J. Phys. G* **28**, 1441 (2002).
4. K. D. Sattler (ed.), *Carbon Nanomaterials Sourcebook: Graphene, Fullerene, Nanotubes and Nanodiamonds, Vol. 1* (CRC Press, Taylor & Francis Group, Boca Raton, 2016).
5. P.-G. Reinhard, M. Rufa, J. Maruhn and W. Greiner, The ground-state properties in a relativistic meson-field theory, *Z. Phys. A, At. & Nucl.* **323**, 13 (1986).
6. W. Wefelmeier, Ein geometrisches Modell des Atomkerns, *Z. Phys.* **107**, 332 (1937).
7. L. Rosenfeld, *Nuclear Forces*, section II (North-Holland, Amsterdam, 1949).
8. L. Pauling, The closed-packed-spheron model of atomic nuclei and its relation to the shell model, *Proc. Nat. Acad. Sci.* **54**, no. 4, 989 (1965).
9. W. von Oertzen, Dynamics of α-clusters in $N = Z$ nuclei, *Eur. Phys. J. A* **29**, 133 (2006).
10. H. A. Wilson, A spherical shell nuclear model, *Phys. Rev.* **69**, 538 (1946).
11. Ph. J. Siemens and H. A. Bethe, Shape of heavy nuclei, *Phys. Rev. Lett.* **18**, 704 (1967).
12. C. Y. Wong, Toroidal and spherical bubble nuclei, *Ann. Phys. (N.Y.)* **77**, 279 (279).
13. L. G. Moretto, K. Tso and G. J. Wozniak, Stable Coulomb bubbles, *Phys. Rev. Lett.* **78**, 824 (1997).
14. J. Dechargé, J.-F. Berger, M. Girod and K. Dietrich, Bubbles and semi-bubbles as a new kind of superheavy nuclei, *Nucl. Phys. A* **716**, 55 (2003).
15. A. Shukla, S. Åberg and A. Bajpeyi, Systematic study of bubble nuclei in relativistic mean-field model, *Phys. Part. Nucl.* **79**(1), 11 (2016).

16. H. J. Krappe, J. R. Nix and A. J. Sierk, Unified nuclear potential for heavy-ion elastic scattering, fusion, fission, and ground-state masses and deformation, *Phys. Rev. C* **20**, 992 (1979).
17. Ş. Mişicu and J. A. Maruhn, Macroscopic deformation energy and oscillations of bubble and semi-bubble nuclei, unpublished.
18. P. Moeller, J. R. Nix, W. D. Myers and W. J. Swiatecki, Nuclear ground-state masses and deformations, *Atomic Data and Nuclear Data Tables* **59**(2), 189 (1995).
19. Ş. Mişicu, I. N. Mishustin and W. Greiner, Q-balls of clusterized baryonic matter, *Mod. Phys. Lett. A* **32**, 1750010 (2017).
20. Ş. Mişicu, I. N. Mishustin and W. Greiner, Instability of α boson vacuum in highly compressed baryonic matter, *J. Phys. G* **42**, 075104 (2015).
21. K. Lee, J. A. Stein-Schabes, R. Watkins and L. M. Widrow, Gauged Q-balls, *Phys. Rev. D* **39**, 1665 (1989).
22. I. E. Gulamov, E. Ya. Nugaev and M. N. Smolyakov, Theory of $U(1)$ gauged Q-balls revisited, *Phys. Rev. C* **89**, 085006 (2014).
23. S. Coleman, Q-balls, *Nucl. Phys. B* **262**, 263 (1985).
24. I. N. Mishustin and W. Greiner, Multipion droplets, *J. Phys. G* **19**, L101 (1993).
25. G. Wentzel, *Quantum Theory of Fields* (Dover Publication Inc., New York, 2003).
26. M. Abramowitz and I. A. Stegun, *Handbook of Mathematical functions with Formulas, Graphs and mathematical Tables*, (National Bureau of Standards, Washington, 1964).
27. J. C. Butcher, *Numerical Methods for Ordinary Differential Equations*, 2nd edn. (John Willey & Sons, 2008)
28. A. Kusenko, Small Q-balls, *Phys. Lett. B* **404**, 285 (1997).
29. M. Apostol, Deformation of a spherical molecule, *Acta Phys. Pol. A* **88**, 315 (1995).
30. J. Eisenberg and W. Greiner, *Nuclear Theory. Volume I: Nuclear Models* (North-Holland, Amsterdam, 1987).
31. S. Hoffman *et al.*, Review of even element super-heavy nuclei and search for element 120, *Eur. Phys. J. A* **52**, 180 (2016).

Cluster Radioactivity — Past and Future

D. N. Poenaru[1,2] and R. A. Gherghescu[1,2,*]

[1]*Horia Hulubei National Institute of Physics and Nuclear Engineering (IFIN-HH), P.O. Box MG-6, RO-077125 Bucharest-Magurele, Romania*

[2]*Frankfurt Institute for Advanced Studies, Johann Wolfgang Goethe University, Ruth-Moufang-Str. 1, D-60438 Frankfurt, Germany*

[*]*poenaru@nipne.ro*

We start by reminding few corner stones during the development of the new branch of nuclear physics: cluster radioactivity (CR), in which our friend Walter Greiner played a very important role. In the perspective of the future production of superheavy nuclei with atomic numbers $Z > 118$ we present some calculations, for 7 isotopes of $Z = 119$ superheavy, $^{297-303}119$, and for 12 isotopes of $Z = 120$ superheavy, $^{295-306}120$. We found that in two cases, $^{298,302}120$, it makes sense in experiments to look for CR, since the decimal logarithm of the branching ratio is $B_\alpha = 0.25$, and 0.49, respectively. When we plot the half-lives, T_α and $T_{\rm CR}$ plus the branching ratio, B_α, against the mass number A, an even–odd staggering is clearly observed. When $Z = 119$, B_α of odd A isotopes is higher than that of the neighbors, T_α is longer for even A, and $T_{\rm CR}$ is also longer for even A. When $Z = 120$, B_α of even A isotopes is higher than that of the neighbors, T_α is longer for odd A, and $T_{\rm CR}$ is also longer for odd A.

1. Introduction

The most cited article in the field of cluster radioactivity (CR) was published in Dubna in 1980 in Russian. The journal is translated in English, hence it was available to everybody.[1] It was published in Dubna because at that time A. Săndulescu was vice-director of the Joint Institute for Nuclear Research. Four theories have been used to predict CR: fragmentation theory developed in Frankfurt;[2,3] extension of α decay (αD) to smaller mass asymmetry,[4] the analytical (ASAF) and numerical (NuSAF) superasymmetric fission model, developed in Bucharest;[5,6] see also the semi-empirical model based on fission theory.[7,8] Until 1984 the paper[1] was ignored, even by the first two experimental publications[9,10] reporting the ^{14}C radioactivity of ^{223}Ra. Then, the comment[11] helped the researchers to find out the truth, and shortly after, the majority of experimentalists and theorists cited.[1] Moreover, Encyclopedia Britannica, both the mutli-volume book and the online version[12] mentioned: "In 1980 A. Săndulescu, D. N. Poenaru, and W. Greiner described calculations indicating the possibility of a new type of decay intermediate between alpha decay and spontaneous fission. The first observation of heavy ion radioactivity was that of a 30 MeV Carbon-14 emission from Radium-223 by H. J. Rose and G. A. Jones in 1984."

One of us (DNP) spent a total of almost 11 years of research stages with Walter Greiner in Frankfurt am Main (Institut für Theoretische Physik and Frankfurt Institute for Advanced Studies, Johann Wolfgang Goethe University) and rarely in Darmstadt at GSI, which explains the productivity of the group: three books;[13–15] 11 chapters in books edited by others; 17 chapters in books edited by us; 100 articles; 64 invited talks and oral contributions at International Conferences; 14 E-prints and preprints. I also wrote two papers of popularization in Romanian about Walter Greiner[16] and the Romanian-German fruitful scientific collaboration,[17] in which he played a very important role. By using the criteria of SPIRES (Stanford Linear Accelerator Center Database) we may select from our common publications with Walter Greiner one famous publication (250–499 citations)[1] and 8 very well-known publications (100–249 citations).[13,18–24]

Until now the following cluster radioactivities have been experimentally confirmed: ^{14}C, ^{20}O, ^{23}F, $^{22,24-26}$Ne, ^{28,30}Mg, ^{32,34}Si with half-lives in good agreement with predicted values within our ASAF model. Particularly useful to guide the experiments have been the comprehensive tables.[20,22] The review paper[25] gave the results of the most important experiments. Besides Rose and Jones, important contributions have been made by P. B. Price from Berkeley, S. P. Tretyakova from JINR Dubna, by E. Hourany and M. Hussonnois from IPN Orsay, by R. Bonetti and A. Guglielmetti from Milano, by A. A. Ogloblin from Moscow, etc. The experiments have been performed in Universities and Research Institutes from: Oxford, Moscow, Orsay, Berkeley, Dubna, Argonne, Livermore, Geneva, Milano, Vienna, and Beijing. Unfortunately both Roberto Bonetti (killed in a car accident) and Svetlana Tretyakova are no longer with us. This is the main reason why only one experiment has been made after 2007.[26]

The heaviest superheavy (SH) nucleus[27,28] produced until now is an isotope of $Z = 118$ (Oganesson). In general there are two kinds of fusion reactions employed on that purpose: (1) almost cold fusion (with one evaporated neutron) at GSI Germany[29,30] and RIKEN Japan[31] based on the doubly magic target ^{208}Pb or its neighbor ^{209}Bi, and (2) hot fusion (with three or four evaporated neutrons) at JINR Dubna, Russia and Livermore National Laboratory, USA[32,33] with the, quite expensive, ^{48}Ca projectile. The most important decay modes of these nuclides are αD and spontaneous fission (SF). When the most important (shortest half-life) decay mode is αD it is much easier to identify the synthesized nucleus by a chain of decays leading to a well-known final product.

Nobody attempted to determine a cluster decay mode (CR) of a SH nucleus until now. Based on our calculations[24,34] we shall show in the present work that there are some possibilities to investigate experimentally CR. The candidates for the forthcoming experiments taken by us into consideration are neutron-deficient nuclides near the proton drip line, which is limited, according to the WS4 atomic mass model[35,36] by $A = 296$ for $Z = 119$ and $A = 294$ for $Z = 120$.

It was shown[37] that among 20 models of atomic masses, the best result is given in Refs. 35 and 36. Also, among the 18 semi-empirical relationships in the same article[37] it is mentioned: "SemFIS2 formula is the best one to predict the alpha-decay half-lives...In addition, the UNIV2 formula with fewest parameters and...with fewer parameters work well in prediction on the SHN alpha-decay half-lives." They cited[38] our semFIS (semi-empirical based on fission theory) and UNIV (universal curve).

We shall employ semFIS, UNIV, ASAF, and AKRA[39] to calculate the αD half-lives and ASAF for CR.

2. Models

In a decay mode we are interested, a parent nucleus AZ decays into an emitted particle A_eZ_e and a daughter A_dZ_d:

$$^{A}Z \rightarrow^{A_e} Z_e +^{A_d} Z_d. \tag{1}$$

The corresponding half-life is given by

$$T = [(h \ln 2)/(2E_v)]\exp(K_{ov} + K_s), \tag{2}$$

where we use the WKB quasiclassical approximation, according to which the action integral is

$$K = \frac{2}{\hbar} \int_{R_a}^{R_b} \sqrt{2B(R)[E(R) - Q]}dR \tag{3}$$

with $B = \mu$ — the reduced mass, $K = K_{ov} + K_s$ (overlapping and separated fragments), and $E(R)$ is the total deformation energy. R_a and R_b are the turning points, defined by $E(R_a) - Q = E(R_b) - Q = 0$.

2.1. *AKRA*

The new formula[39] was introduced by adding two parameters (d, e) and I to the one developed by G. Royer:[40]

$$T_{1/2} = a + bA^{1/6}\sqrt{Z} + \frac{cZ}{\sqrt{Q_\alpha}} + dI + eI^2, \tag{4}$$

where $I = (N - Z)/A$ and the parameters a, b, c, d and e are obtained by fit with experimental data.

2.2. *ASAF*

We replace in Eq. (3) $E(R) - Q$ by $[E(R) - E_{\text{corr}}] - Q$. E_{corr} is a correction energy similar to the Strutinsky shell correction, also taking into account the fact that Myers–Swiatecki's liquid drop model (LDM) overestimates fission barrier heights, and the effective inertia in the overlapping region is different from the reduced mass. The interaction energy at the top of the barrier, in the presence of a non-negligible angular momentum, $l\hbar$, is given by:

$$E_i = E_c + E_l = e^2 Z_e Z_d / R_t + \hbar^2 l(l+1)/(2\mu R_t^2). \tag{5}$$

The two terms of the action integral K, corresponding to the overlapping (K_{ov}) and separated (K_s) fragments, are calculated by analytical formulas (approximated for K_{ov} and exact for K_s in case of separated spherical shapes within the LDM).

The choice $E_v = E_{\text{corr}}$ allows to get a smaller number of parameters. For a given cluster radioactivity there are four values of the coefficients a_i, the largest for even–even parent and the smallest for the odd–odd one (see Fig. 1 of Ref. 22). The shell effects for every cluster radioactivity is implicitly contained in the correction energy due to its proportionality with the Q value.

2.3. *UNIV (universal formula)*

In cluster radioactivity and α decay the (measurable) decay constant $\lambda = \ln 2/T$, can be expressed as a product of three (model dependent) quantities

$$\lambda = \nu S P_s, \tag{6}$$

where ν is the frequency of assaults on the barrier per second, S is the preformation probability of the cluster at the nuclear surface, and P_s is the quantum penetrability of the external potential barrier. The frequency ν remains practically constant, the preformation differs from one decay mode to another but it is not changed very much for a given radioactivity, while the general trend of penetrability follows closely that of the half-life.

One may assume that preformation probability only depends on the mass number of the emitted cluster, $S = S(A_e)$ and that $\nu(A_e, Z_e, A_d, Z_d) = \text{constant}$. In this way it was obtained a single straight line *universal curve* on a double logarithmic scale

$$\log T = -\log P_s - 22.169 + 0.598(A_e - 1), \tag{7}$$

where

$$-\log P_s = c_{AZ}\left[\arccos\sqrt{r} - \sqrt{r(1-r)}\right] \tag{8}$$

with $c_{AZ} = 0.22873(\mu_A Z_d Z_e R_b)^{1/2}$, $r = R_t/R_b$, $R_t = 1.2249(A_d^{1/3} + A_e^{1/3})$, $R_b = 1.43998 Z_d Z_e / Q$, and $\mu_A = A_d A_e / A$.

2.4. *SemFIS (semiempirical relationship based on fission theory of α decay)*

The neighborhood of the magic numbers of nucleons is badly described by all these relationships. The semFIS formula based on the fission theory of α-decay gives

$$\log T = 0.43429 K_s \chi - 20.446, \tag{9}$$

where

$$K_s = 2.52956 Z_{da} [A_{da}/(A Q_\alpha)]^{1/2} [\arccos \sqrt{x} - \sqrt{x(1-x)}] \ ;$$
$$x = 0.423 Q_\alpha (1.5874 + A_{da}^{1/3})/Z_{da} \tag{10}$$

and the numerical coefficient χ, close to unity, is a second-order polynomial

$$\chi = B_1 + B_2 y + B_3 z + B_4 y^2 + B_5 yz + B_6 z^2 \tag{11}$$

in the reduced variables y and z, expressing the distance from the closest magic-plus-one neutron and proton numbers N_i and Z_i:

$$y \equiv (N - N_i)/(N_{i+1} - N_i) \ ; \ N_i < N \leq N_{i+1} \tag{12}$$

$$z \equiv (Z - Z_i)/(Z_{i+1} - Z_i) \ ; \ Z_i < Z \leq Z_{i+1} \tag{13}$$

with $N_i = \ldots, 51, 83, 127, 185, 229, \ldots$, $Z_i = \ldots, 29, 51, 83, 115, \ldots$, and $Z_{da} = Z-2$, $A_{da} = A - 4$. The coefficients B_i are obtained by fit with experimental data, using a computer program making automatically the best fit.[8]

3. Results

We present in Fig. 1 the most important results of our study. One can see an even–odd staggering of the three quantities we plotted. When $Z = 119$, B_α of odd A isotopes is higher than that of the neighbors, T_α is longer for even A, and $T_{\rm CR}$ is also longer for even A. When $Z = 120$, B_α of even A isotopes is higher than that of the neighbors, T_α is longer for odd A, and $T_{\rm CR}$ is also longer for odd A.

In Table 1 we present the half-lives, $\log_{10} T_\alpha(s)$, obtained with different models. One can see that sometimes the difference in half-life from model to model may be larger than 5 orders of magnitude. The result obtained by UNIV for $^{297}120$ is certainly wrong. When the half-life is shorter than one μs, it is not possible to be measured.

In conclusion, the most important result of the present work is the possibility of a very strong cluster radioactivity of $^{298,302}120$ for which we report branching ratio with respect to α decay larger than one, $B_\alpha = 0.25$, and 0.49, respectively.

Acknowledgments

This work was supported within the IDEI Programme under Contracts No. 43/05.10.2011 and 42/05.10.2011 with UEFISCDI, and NUCLEU Programme PN16420101/2016 Bucharest.

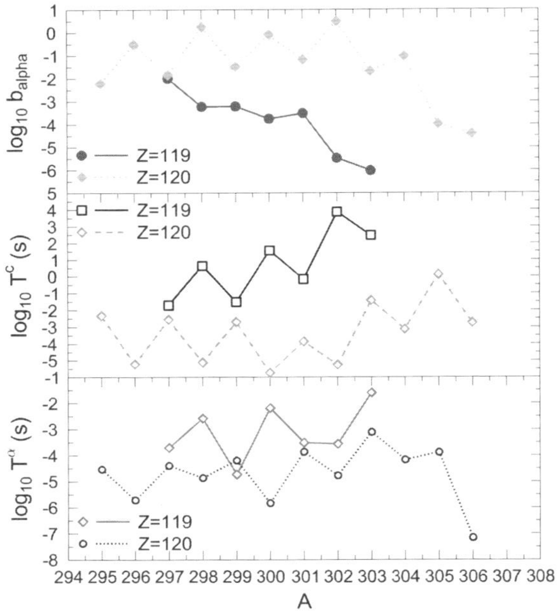

Fig. 1. Decimal logarithms of α decay (bottom), cluster radioactivities (center) and branching ratio against α decay (top) versus mass number for 7 isotopes of $Z = 119$ and 12 isotopes of $Z = 120$ superheavy nuclei.

Table 1. Half-lives, $\log_{10} T_\alpha(s)$, obtained with different models: AKRA, ASAF, UNIV and semFIS.

Parent nucleus	AKRA	ASAF	UNIV	semFIS
$^{297}119$	−0.450	−3.695	−4.124	−3.717
$^{298}119$	−3.057	−2.576	−3.945	−2.084
$^{299}119$	−1.357	−4.730	−5.142	−4.739
$^{300}119$	−2.879	−2.190	−3.762	−1.615
$^{301}119$	−0.184	−3.675	−4.157	−3.672
$^{302}119$	−2.551	−1.604	−3.421	−0.937
$^{303}119$	0.051	−3.559	−4.074	−3.545
$^{295}120$	−6.208	−4.531	−4.905	−4.021
$^{296}120$	−2.166	−5.706	−3.885	−3.290
$^{297}120$	−6.330	−4.386	13.104	−3.827
$^{298}120$	−3.429	−4.860	−5.139	−4.802
$^{299}120$	−6.409	−4.190	−4.813	−3.579
$^{300}120$	−4.415	−5.833	−6.113	−5.765
$^{301}120$	−6.365	−3.856	−4.628	−3.188
$^{302}120$	−3.367	−4.769	−5.117	−4.682
$^{303}120$	−5.952	−3.109	−4.086	−2.365
$^{304}120$	−2.792	−4.174	−4.577	−4.068
$^{305}120$	−6.922	−3.875	−4.886	−3.121
$^{306}120$	−5.150	−7.158	−6.881	−7.190

References

1. A. Sandulescu, D. N. Poenaru and W. Greiner, *Soviet Journal Particles and Nuclei* **11**, 528 (1980).
2. H. J. Fink, J. A. Maruhn, W. Scheid and W. Greiner, *Zeitschrift für Physik* **268**, 321 (1974).
3. J. A. Maruhn, W. Greiner and W. Scheid, Theory of fragmentation, in *Heavy Ion Collisions*, ed. R. Bock (North Holland, Amsterdam, 1980) p. 399.
4. R. Blendowske, T. Fliessbach and H. Walliser, Many-body approach to alpha and cluster radioactivity, in *Nuclear Decay Modes*, ed. D. N. Poenaru (IOP Publishing, Bristol, 1996) pp. 337–349.
5. D. N. Poenaru and M. Ivaşcu, *Journal de Physique (Paris)* **45**, 1099 (1984).
6. D. N. Poenaru, M. Ivaşcu and W. Greiner, Unified approach of alpha-decay, heavy ion emission and cold fission, in *Particle Emission from Nuclei, Vol. III: Fission and Beta-Delayed Decay Modes*, eds. D. N. Poenaru and M. Ivaşcu (CRC Press, Boca Raton, Florida, 1989) pp. 203–235.
7. D. N. Poenaru, M. Ivaşcu and D. Mazilu, *Journal de Physique Lettres* **41**, L589 (1980).
8. D. N. Poenaru, M. Ivaşcu and D. Mazilu, *Computer Physics Communications* **25**, 297 (1982).
9. H. J. Rose and G. A. Jones, *Nature* **307**, 245 (1984).
10. S. Gales, E. Hourani, M. Hussonnois, J. P. Schapira, L. Stab and M. Vergnes, *Physical Review Letters* **53**, 759 (1984).
11. A. Săndulescu, D. N. Poenaru, W. Greiner and J. H. Hamilton, *Physical Review Letters* **54**, 490 (1985).
12. The New Encyclopaedia Britannica, Online http://www.britannica.com/EBchecked/topic/465998/.
13. D. N. Poenaru (ed.), *Nuclear Decay Modes* (Institute of Physics Publishing, Bristol, 1996).
14. D. Poenaru and W. Greiner (eds.), *Handbook of Nuclear Properties* (Clarendon Press, Oxford, 1996).
15. D. N. Poenaru and W. Greiner (eds.), *Experimental Techniques in Nuclear Physics* (Walter de Gruyter, Berlin, 1997).
16. D. Poenaru, *Academica*, 20 (1993).
17. D. N. Poenaru, *Revista de Politica Stiintei si Scientometrie Serie Noua* **5**, 293 (2016).
18. D. N. Poenaru, M. Ivaşcu, A. Săndulescu and W. Greiner, *Physical Review, C* **32**, 572 (1985).
19. D. N. Poenaru and M. Ivaşcu (eds.), *Particle Emission from Nuclei* 3 volumes (CRC Press, Boca Raton, 1989).
20. D. N. Poenaru, W. Greiner, K. Depta, M. Ivaşcu, D. Mazilu and A. Săndulescu, *Atomic Data and Nuclear Data Tables* **34**, 423 (1986).
21. D. N. Poenaru, I. H. Plonski and W. Greiner, *Physical Review C* **74**, 014312 (2006).
22. D. N. Poenaru, D. Schnabel, W. Greiner, D. Mazilu and R. Gherghescu, *Atomic Data and Nuclear Data Tables* **48**, 231 (1991).
23. D. N. Poenaru, M. Ivaşcu, A. Săndulescu and W. Greiner, *Journal of Physics G: Nuclear Physics* **10**, L183 (1984).
24. D. N. Poenaru, R. A. Gherghescu and W. Greiner, *Phys. Rev. Lett.* **107**, 062503 (2011).
25. R. Bonetti and A. Guglielmetti, *Rom. Rep. Phys.* **59**, 301 (2007), ISSN 1221-1451.
26. A. Guglielmetti, D. Faccio, R. Bonetti, S. Shishkin, S. Tretyakova, S. Dmitiev, A. Ogloblin, G. Pik-Pichak, N. van der Meulen, G. Steyn, T. van der Walt,

C. Vermeulen and D. McGee, *Journal of Physics: Conference Series (JPCS)* **111**, 012050 (2008).
27. J. H. Hamilton, S. Hofmann and Y. Oganessian, *Ann. Rev. Nucl. Part. Sci.* **63**, 383 (2013).
28. J. Khuyagbaatar *et al.*, *Phys. Rev. Lett.* **112**, 172501 (2014).
29. S. Hofmann and G. Münzenberg, *Review of Modern Physics* **72**, 733 (2000).
30. S. Hofmann, *Radiochim. Acta* **99**, 405 (2011).
31. K. Morita *et al.*, *J. Phys. Soc. Jpn.* **76**, 045001 (2007).
32. Y. T. Oganessian, F. S. Abdullin, P. D. Bailey, D. E. Benker, M. E. Bennett, S. N. Dmitriev, J. G. Ezold, J. H. Hamilton, R. A. Henderson, M. G. Itkis, Y. V. Lobanov, A. N. Mezentsev, K. J. Moody, S. L. Nelson, A. N. Polyakov, C. E. Porter, A. V. Ramayya, F. D. Riley, J. B. Roberto, M. A. Ryabinin, K. P. Rykaczewski, R. N. Sagaidak, D. A. Shaughnessy, I. V. Shirokovsky, M. A. Stoyer, V. G. Subbotin and R. Sudowe, *Phys. Rev. Lett.* **104**, 142502 (2010).
33. Y. T. Oganessian and V. K. Utyonkov, *Reports on Progress in Physics* **78**, 036301 (2015).
34. D. N. Poenaru, R. A. Gherghescu and W. Greiner, *Phys. Rev. C* **85**, 034615 (2012).
35. M. Liu, N. Wang, Y. Deng and X. Wu, *Phys. Rev. C* **84**, 014333 (2011).
36. N. Wang, M. Liu, X. Z. Wu and J. Meng, *Phys. Lett. B* **734**, 215 (2014).
37. Y. Wang, S. Wang, Z. Hou and J. Gu, *Phys. Rev. C* **92**, 064301 (2015).
38. D. N. Poenaru, R. A. Gherghescu and N. Carjan, *Europhysics Letters* **77**, 62001 (2007).
39. D. Akrawy and D. Poenaru, *Alpha decay calculations with a new formula*, arXiv:1702.05598v1 [nucl-th] 2017.
40. G. Royer, *Nuclear Physics, A* **848**, 279 (2010).

Nuclear Mean-Field Models and Super-Heavy Elements

P.-G. Reinhard
Institut für Theoretische Physik, Universität Erlangen/Nürnberg,
Staustst r. 7, D-90158 Erlangen, Germany
paul-gerhard.reinhard@fau.de

This article reviews the description of super-heavy elements (SHE) in terms of self-consistent mean-field models. The Skyrme–Hartree–Fock (SHF) approach is taken as one typical example for such models in the family of nuclear density functional theories. The tour starts with the search for new magic numbers, continues with estimates of stabilization of SHE through the shell-correction energy, and rounds up with a discussion fission barriers and lifetimes as the finally decisive quantities for the existence of SHE.

1. Introduction

The ability to control nuclear reactions opened the door to the production of new elements. This led mid-20th century to intense research on transactinides.[1] Steady progress in accelerator design paved the way to ever heavier nuclei. This inspired since the 1960s a new wave of research on super-heavy elements (SHE).[2,3] The topic remains of high interest and new elements are added every year to the list of SHE, for a few examples from the rich list see Refs. 4–8. SHE are also produced during the r-process[9–11] and their properties are important in determining the upper end of the nucleosynthesis flow. The Frankfurt theory group under W. Greiner joined these efforts from the theoretical side[12] and continued to participate since then. The theoretical methods were steadily refined over the time. The early studies were based on the nuclear shell model.[13] The particular contribution at this stage was the two-center shell model which is sufficiently versatile to cover nuclear shapes up to separate fragments.[14,15] A self-consistent mean-field models in terms of nuclear density functional theory came up in 1970 and developed with the time a high descriptive power for nuclear structure and dynamics, for reviews see, e.g. Refs. 11, 16–19. Naturally, such nuclear density functional theory was also much used to study SHE. In this review, we go along the development of theoretical investigations on SHE with self-consistent models. We do this with a bias on the contributions of the Frankfurt group which is appropriate in this memorial book for Walter Greiner.

Early theoretical explorations of SHE were much concerned with shell structure, in particular the question of new magic numbers possibly showing up. These topics will be discussed in Subsec. 3.1. The key question in the study of SHE is their stability, particularly against spontaneous fission. From a purely macroscopic point of view, all SHE are massively unstable. It is only by shell effects that extra binding is produced which leads to stabilization of SHE. First estimates of stability were thus obtained from the shell-correction energy. This aspect will be discussed in Subsec. 3.2. The ultimate test, of course, is a direct analysis of fission barriers and lifetimes. Self-consistent calculations of fission lifetimes are extremely demanding and thus have come up only recently, see e.g. Refs. 20–22 which are still using approximate masses or Refs. 23 and 24 for fully self-consistent calculations and Ref. 25 for a recent review. Fission of SHE will be discussed in Subsec. 3.3. Competing decay channels are α- and β-decay which we will not address in this article for reason of space. A survey of the balance between the various decay channels is found in Ref. 24.

Before coming to results, the theoretical framework has to be recalled. Subsection 2.1 presents the Skyrme–Hartree–Fock method as a typical example of nuclear density functional theory. A center issue in theoretical extrapolations to SHE is to assess the predictive value of nuclear density functional theory. Techniques for calibration of the energy functionals as well as estimate of extrapolation errors and model dependence are discussed in Subsec. 2.3. Results are presented and discussed in Sec. 3.

2. Formal Framework

2.1. *Self-consistent mean-field theory: Example of Skyrme–Hartree–Fock (SHF)*

2.1.1. *The mean-field state*

In a mean-field theory, one describes a many-body system by a set of single-particle (s.p.) wavefunctions $\{\varphi_\alpha, \alpha = 1, ..., A\}$. To each s.p. state, one associates a pairing occupation amplitude v_α which can take values in the interval $[0, 1]$. The complementary non-occupation amplitude is $u_\alpha = \sqrt{1 - v_\alpha^2}$. States are grouped in conjugate pairs $\alpha \leftrightarrow \overline{\alpha}$. The typical mean-field state then is a BCS state $|\Phi\rangle = \prod_{\alpha>0} \left(u_\alpha + v_\alpha \hat{a}_\alpha^+ \hat{a}_{\overline{\alpha}}^+\right)|0\rangle$ where $|0\rangle$ is the vacuum state and $\hat{a}_\alpha^+$ are Fermion operators generating a particle in state α. The mean-field equations are obtained by variation of the total energy with respect to the single-particle wavefunctions φ_α and variation with respect to the v_α. The starting point is an expression for the total energy from which everything else follows systematically.

2.1.2. *Basic densities and currents*

The energy-density functionals require the knowledge of only a few local densities and currents:

$$\begin{aligned}
\rho_q &= \sum_{\alpha\in q} f_\alpha v_\alpha^2 |\varphi_\alpha|^2, && \text{density,} && \text{even} \\
\vec{s}_q &= \sum_{\alpha\in q} f_\alpha v_\alpha^2 \varphi_\alpha^+ \hat{\vec{\sigma}} \varphi_\alpha, && \text{spin density,} && \text{odd} \\
\vec{j}_q &= \Im m \left\{ \sum_{\alpha\in q} f_\alpha v_\alpha^2 \varphi_\alpha^+ \nabla \varphi_\alpha \right\}, && \text{current,} && \text{odd} \\
\vec{J}_q &= -\mathrm{i} \sum_{\alpha\in q} f_\alpha v_\alpha^2 \varphi_\alpha^+ \nabla \times \hat{\vec{\sigma}} \varphi_\alpha, && \text{spin–orbit density,} && \text{even} \\
\tau_q &= \sum_{\alpha\in q} f_\alpha v_\alpha^2 |\nabla \varphi_\alpha|^2, && \text{kinetic density,} && \text{even} \\
\vec{\tau}_q &= -\mathrm{i} \sum_{\alpha\in q} f_\alpha v_\alpha^2 \nabla \varphi_\alpha^+ \cdot \nabla \hat{\vec{\sigma}} \varphi_\alpha, && \text{kinetic spin-density,} && \text{odd} \\
\xi_q &= \textstyle\sum_{\alpha\in q} f_\alpha u_\alpha v_\alpha \varphi_{\overline{\alpha}} \varphi_\alpha, && \text{pairing density,} && \text{undef.}
\end{aligned} \tag{1}$$

where q labels the nucleon species with $q = p$ for protons and $q = n$ for neutrons and where the last column indicates the time parity. The f_α is a phase-space weight which serves to provide a smooth cutoff of the space of single-particle states included in pairing.[26] All above expressions are local quantities depending on $\vec{r}$ composed from local wavefunctions $\varphi_\alpha = \varphi_\alpha(\vec{r})$. It is advantageous to handle the densities in terms of isospin in the recoupled forms

$$\rho_{T=0} \equiv \rho = \rho_p + \rho_n, \quad \rho_{T=1} \equiv \tilde{\rho} = \rho_p - \rho_n, \tag{2}$$

and similarly for the other densities and currents. The isoscalar density $\rho_{T=0} \equiv \rho$ is equivalent to the total density and the difference $\tilde{\rho} \equiv \rho_{T=1}$ corresponds to the isovector density. The pairing density ξ_q has indefinite time-parity if the state of a system has no definite time parity (e.g. odd nuclei or excited states).

2.1.3. *The energy functional*

The starting point in any density functional theory (DFT) is an expression for the total energy which is composed for SHF as

$$E = E_{\text{kin}} + \int d^3r\, \mathcal{E}_{\text{Sk}} + E_{\text{Coul}} + E_{\text{pair}} - E_{\text{corr}}. \tag{3a}$$

The one-body kinetic energy is given by the exact quantum mechanical expression as

$$E_{\text{kin}} = \int d^3r \left(\frac{\hbar^2}{2m_p} \tau_p + \frac{\hbar^2}{2m_n} \tau_n \right). \tag{3b}$$

Using the exact E_{kin} implies that the full quantum mechanical shell structure is maintained which means that we deal with DFT at the level of the Kohn–Sham approach.[27,28]

The key piece is the SHF energy density $\mathcal{E}_{\rm Sk}$:

$$\mathcal{E}_{\rm Sk} = \sum_{T=0,1} \left(\mathcal{E}_T^{\rm even} + \mathcal{E}_T^{\rm odd}\right), \tag{3c}$$

$$\begin{aligned}\mathcal{E}_T^{\rm even} &= C_T^{\rho}\,\rho_T^2 + C_T^{\rho,\alpha}\,\rho_T^2\rho_0^\alpha + C_T^{\Delta\rho}\,\rho_T\Delta\rho_T \\ &\quad + C_T^{\tau}\,\rho_T\tau_T + C_T^{\nabla J}\rho_T\,\nabla\cdot\vec{J}_T + C_T^{J}\vec{J}_T^2,\end{aligned} \tag{3d}$$

$$\begin{aligned}\mathcal{E}_T^{\rm odd} &= C_T^{s}\mathcal{F}_T^{s}\,\vec{s}_T^2 + C_T^{s,\alpha}\,\vec{s}_T^2\rho_0^\alpha + C_T^{\Delta s}\,\vec{s}_T\cdot\Delta\vec{s}_T + C_T^{j}\,\vec{j}_T^2 + C_T^{sT}\,\vec{s}_T\cdot\vec{\tau}_T \\ &\quad + C_T^{\nabla j}\,\vec{s}_T\cdot\nabla\times\vec{j}_T.\end{aligned} \tag{3e}$$

It is built by the straightforward principle that one includes all conceivable bi-linear forms in the densities (1) up to second order in derivatives and allows only combination that are together time reversal invariant. The $\mathcal{E}^{\rm odd}$ contains all dependencies on time-odd currents while $\mathcal{E}^{\rm even}$ contains only time-even densities. Only the latter part counts in stationary calculations. The $T = 1$ terms appear in bi-linear form which makes the functional also isospin invariant. The tensor spin–orbit term $\propto C_T^J$ is, in fact, more involved.[29] But its details have little impact on basic bulk properties. Thus it is often ignored at all with the choice of $C_T^J = 0$. Even then, the functional (3c) seems to cover a large number of free parameters. However, the number of parameters is restricted substantially by basic principles. Galilean invariance, i.e. invariance under local gauge transformations, requires that currents appear only in the combinations $\rho\tau - \vec{j}^2$, $\vec{s}\cdot\vec{\tau} - \mathbb{J}^2$, and $\rho\nabla\vec{J} + \vec{s}_T\cdot\nabla\times\vec{j}_T$.[30,31] This amounts to six fixed relations between parameters of time-even terms and time-odd ones. The Galilean invariance leaves undetermined the genuine spin terms C_T^s, $C_T^{\Delta s}$. These, however, are not active in time-even systems and in nuclear excitations with natural parity (e.g. giant resonances) which for these applications leaves effectively 11 parameters. Odd nuclei (not appearing in this article) require additional decisions concerning the spin terms, for examples see Ref. 32. For a detailed discussion of parameter restriction and options of the SHF functional, see Ref. 19.

2.1.4. *Coulomb energy*

The Skyrme interaction is complemented by an energy functional accounting for the Coulomb interaction of the nuclear charges:

$$E_{\rm Coul} = E_{\rm Coul,dir} - E_{\rm Coul,ex}, \tag{4a}$$

$$E_{\rm Coul,dir} = \frac{e^2}{2}\int d^3r\, d^3r'\,\frac{\rho_p(\vec{r})\rho_p(\vec{r}')}{|\vec{r}-\vec{r}'|}, \tag{4b}$$

$$E_{\rm Coul,ex} = \frac{3e^2}{4}\left(\frac{3}{\pi}\right)^{1/3}\int d^3r[\rho_p(\vec{r})]^{4/3}, \tag{4c}$$

where $e^2 = 1.44$ MeV fm. The Coulomb exchange is done at the level of the local-density approximation,[28] and is ignored, however, in a few cases. The direct Coulomb energy should employ, in principle, the charge density ρ_C. However, the mere proton density ρ_p is used in most applications.

2.1.5. *Center-of-mass correction and other correlation energies*

The correlation energy may contain several contributions as

$$E_{\text{corr}} = E_{\text{cm}} + E_{\text{rot}} + E_{\text{vib}}, \tag{5}$$

which are a correction for the spurious center-of-mass energy, another one for rotational projection (non-zero only in deformed nuclei), and a term accounting for soft surface vibrations. The leading part which is used in practically all applications is E_{cm}, which exist in two variants:

$$E_{\text{cm}} = E_{\text{cm}}^{(\text{full})} = \frac{\langle \hat{\vec{P}}^2_{\text{cm}} \rangle}{2M}, \quad \hat{\vec{P}}_{\text{cm}} = \sum_{n=1}^{A} \hat{\vec{p}}_n, \tag{6a}$$

$$E_{\text{cm}} = E_{\text{cm}}^{(\text{diag})} = \frac{\langle \sum_{n=1}^{A} \hat{\vec{p}}^2_n \rangle}{2M}, \tag{6b}$$

where $M = Zm_p + Nm_n$ is the total mass. The more complete form is $E_{\text{cm}}^{(\text{full})}$ and it accounts for the effect of center-of-mass projection evaluated in second-order Gaussian overlap approximation.[33] It employs, in fact, a two-body operator which makes its application in variational equations very cumbersome. It is thus only evaluated *a posteriori*, i.e. for the given solution of the SHF mean-field equations. A simplification is $E_{\text{cm}}^{(\text{diag})}$ where only the diagonal terms of $\hat{\vec{P}}^2_{\text{cm}}$ are taken into account. This correction amounts to a modification of the nucleon mass and it can easily be included in the variational equations. However, the modification of the mass has side-effects in nuclear dynamics which are not well understood. For the rotational and vibrational corrections see Ref. 34.

2.1.6. *Pairing functional*

The pairing functional reads

$$\mathcal{E}_{\text{pair}} = \frac{1}{4} \sum_{q\in\{p,n\}} V_{\text{pair,q}} \int d^3r \xi_q^2 \left[1 - \frac{\rho}{\rho_{0,\text{pair}}}\right]. \tag{7a}$$

Letting $\rho_{0,\text{pair}} \to \infty$ suppresses the density-dependent part and recovers what is called volume pairing. Surface pairing is obtained by setting $\rho_{0,\text{pair}} = 0.16\ \text{fm}^{-3}$. Most generally, one allows $\rho_{0,\text{pair}}$ to be a free parameter. By optimizing the pairing functional in this way, one obtains as optimum a mix of volume and surface pairing.[35]

The functional (7a) is incomplete for practical purposes. One must introduce a cutoff in s.p. space.[36] There are several cutoff recipes in the literature. In the following calculations, we use a soft cutoff factor

$$f_\alpha = \left[1 + \exp\left((\varepsilon_\alpha - (\epsilon_F + \epsilon_{\text{cut}}))/\Delta\epsilon\right)\right]^{-1}, \quad \Delta\epsilon = \epsilon_{\text{cut}}/10, \tag{7b}$$

where ϵ_α is the single particle energy of the state α and ϵ_F is the chemical potential. The original recipe used a fixed cutoff energy $\epsilon_{\text{cut}} = 5\text{–}10$ MeV.[37,38] Here, we determine ϵ_{cut} such that the active s.p. space has a fixed size $\propto N^{2/3}$, i.e. relative to the surface in configuration space.[26] This yields a cutoff which depends on the actual nucleon number N_q:

$$\epsilon_{\mathrm{cut}} \quad \longleftrightarrow \quad \sum_{\alpha} f_\alpha = N_q + \eta_{\mathrm{cut}} N_q^{2/3}. \tag{7c}$$

Actually, we use a rather small width parameter $\eta_{\mathrm{cut}} = 1.65$ to keep computational expense low.

2.2. *Observables and nuclear matter parameters (NMP)*

A great variety of observables can be computed from SHF. The prominent ones are energy and local density distributions. The latter yields as global characterizations radius, surface thickness, central depression and more detailed measures.[39,40] There are several other, more subtle, observables. In the present survey, we are mainly discussing energies (total energy, and single-particle energies) that are trivially obtained during the SHF calculations and need no special explanation here. More involved is the computation of fission barriers and lifetimes, which will be explained at the beginning of Subsec. 3.3.

Although not really observable in a laboratory, an intuitive and instructive class of observables are nuclear matter properties (NMP). They are defined at the equilibrium point ($\rho_0 = \rho_{\mathrm{eq}}$, $\rho_1 = 0$) of symmetric nuclear matter. They are summarized in Table 1. Most of them are obvious. Two NMP deserve words of explanation: The slope of symmetry energy L characterizes the density dependence of the symmetry energy which allows us to estimate asymmetry for densities other than equilibrium as they are relevant, e.g., at the surface of finite nuclei. The enhancement factor κ_{TRK} for the Thomas–Reiche–Kuhn (TRK) sum rule[36] is a widely-used way to characterize the isovector effective mass which is obvious from the given expression involving derivative with respect to τ_1.

The NMP in Table 1 are grouped in three classes: first, the (isoscalar) ground state properties ρ_{eq} and $E/A\big|_{\mathrm{eq}}$; second, isoscalar response properties K and m^*/m; third, isovector response properties J, L, κ_{TRK}. The response properties determine zero sound in matter[41] and subsequently they are closely related to giant resonance modes in finite nuclei as we will see later.

The seven NMP in Table 1 are fully equivalent to the seven volume parameters in the SHF functional (3c), namely C_T^ρ, $C_T^{\rho,\alpha}$, α, and C_T^τ. Thus the NMP are often used as the model parameters. This simplifies direct comparison with other mean-field models which are often also fully mappable to NMP, see for example the comparison of symmetry energies in Ref. 42.

In the following, we will use a couple of published parametrizations from different groups: SkP, an attempt to optimize pairing force within the SHF functional;[43] SkI3, a χ^2-fit to energies and formfactor for an SHF functional extended to allow for isovector spin–orbit terms;[44] SLy4 and SLy6, fits to nuclei and nuclear matter properties with a bias on neutron rich systems where SLy4 and SLy6 differ in

Table 1. Definition of the nuclear matter properties (NMP). All derivatives are to be taken at the equilibrium point of symmetric nuclear matter $\rho_0 = \rho_{\rm eq}$ and $\rho_1 = 0$.

Equilibrium density:	$\rho_{\rm eq}$	$\leftrightarrow$	$\left.\frac{d}{d\rho_0}\frac{E}{A}\right\|_{\rm eq} = 0$	isoscalar g.s.
Equilibrium energy:	$\left.\frac{E}{A}\right\|_{\rm eq}$	$=$	$\frac{E}{A}(\rho_0 = \rho_{\rm eq}, \rho_1 = 0)$	isoscalar g.s.
Incompressibility:	K_∞	$=$	$9\,\rho_0^2 \left.\frac{d^2}{d\rho_0^2}\frac{E}{A}\right\|_{\rm eq}$	isoscalar response
Effective mass:	$\frac{\hbar^2}{2m*}$	$=$	$\frac{\hbar^2}{2m} + \left.\frac{\partial}{\partial\tau_0}\frac{E}{A}\right\|_{\rm eq}$	isoscalar response
Symmetry energy:	J	$=$	$\frac{1}{2}\rho_0^2 \left.\frac{d^2}{d\rho_1^2}\frac{E}{A}\right\|_{\rm eq}$	isovector response
Slope of J:	L	$=$	$3\rho_0 \frac{d}{d\rho_0} J$	isovector response
TRK sum-rule enh.:	$\kappa_{\rm TRK}$	$=$	$\frac{2m}{\hbar^2}\rho_0 \left.\frac{\partial}{\partial\rho_1}\frac{\partial}{\partial\tau_1}\frac{E}{A}\right\|_{\rm eq}$	isovector response

their handling of the c.m. correction;[45] BSk1, a fit to binding energies of nuclei throughout the table of isotopes;[46] UNEDF2, a fit also including information on s.p. spectra;[47] SV-min and SV-bas, fits to a large of spherical nuclei scrutinized to have negligible impact from correlations whereby SV-bas tunes additionally NMP;[35] NL-Z(2) and NL3, RMF functionals with non-linear meson self-couplings fitted to nuclear ground-state data Ref. 48 or Ref. 49, DD-ME, fit of an RMF functional with density dependent meson couplings.[50] This broad selection allows us to demonstrate the model dependence of the results. For more systematic exploration, we employ statistical analysis and the sets of systematically varied SHF functionals from Ref. 35, see Subsec. 2.3.

2.3. *Least-squares fits and statistical analysis*

A fully microscopic description of nuclei in terms of many-body theory (see Ref. 51 for a review) or effective QCD models (see e.g. Refs. 52–57) is a task of great current interest. Although remarkable progress has been made, the precision of these ab-initio models does not yet suffice for a high-quality description of nuclear properties. Thus it is still state of the art to adjust the parameters of the functional to empirical data. The method of choice are least-squares (χ^2) fits[58] used for nuclear DFT first in Ref. 59 and being meanwhile a standard technique for calibration of self-consistent mean-field models. One finds in the literature a great variety of functional development using χ^2-fits to empirical data. For example, some concentrate on spherical nuclei with negligible correlation effects,[35] others are particularly concerned with deformed nuclei,[60–62] still others try to adjust also spectra of s.p. energies.[47,63] Although differences occur in details due to different choice of fit

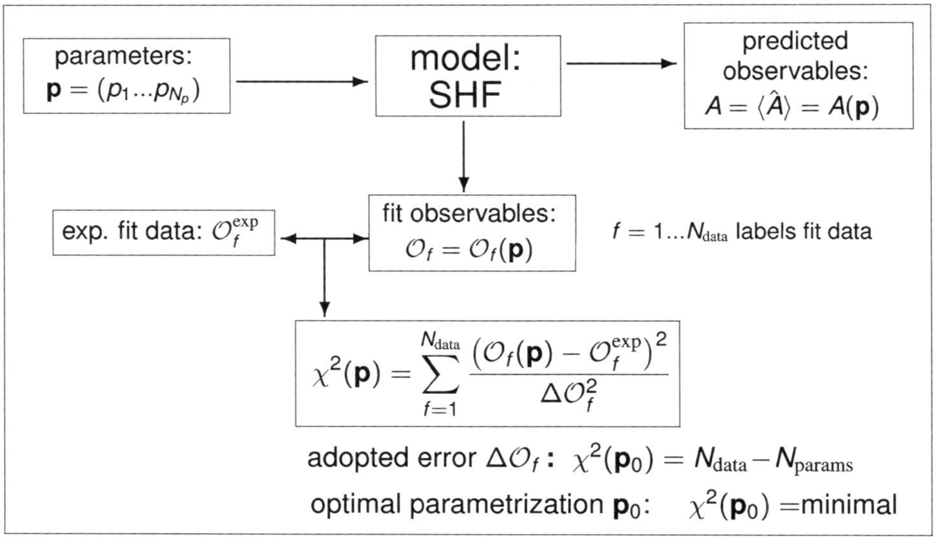

Fig. 1. Schematic representation of χ^2-fits for calibrating effective nuclear DFT.

data, it is reassuring that all these fits provide similar and high-quality results for the reproduction of global nuclear properties.

The principles of χ^2-fits are summarized schematically in Fig. 1. The model given here is a case of SHF, which is considered for the purpose of the fits as a black box which produces for given model parameters $\mathbf{p}$ a great manifold of expectation values of observables A. We select a subgroup of observables $\hat{\mathcal{O}}_f$ for which we have reliable experimental data and from which we assume that they can be described reliably well by DFT. The quality of the model is quantified by χ^2, the sum of squared deviations. Each term is weighted by the associated error $\Delta\mathcal{O}_f$. This quantity has dimension $\mathcal{O}_f$ and represents an "adopted error" that accounts for the experimental uncertainty as well as for the expected theoretical reliability for this observable. There is an automatic feedback built in into the scheme: The adopted errors are chosen correctly if the final χ^2 of the best fit if of order of the number of data points (for a detailed discussion see Refs. 58 and 64). Trivially, the best fit is achieved if χ^2 becomes minimal. What remains is to find the absolute minimum of the function $\chi^2(\mathbf{p})$. This is a non-trivial task for which, however, a great deal of know-how is around.

There is more in the χ^2-scheme than just a minimum. The vicinity of minimum carries also worthwhile information. Parameters which lie not exactly at the minimum can still provide a pretty good description. How good depends, again, on the actual χ^2, which is quantified in terms of a "probability of a parametrization" $W(\mathbf{p})$. This leads to the statistical interpretation of χ^2 sketched in Fig. 2. The maximum probability W is found, of course, for minimal χ^2 and the curvature of χ^2 around the best fit $\mathbf{p}_0$ determines the uncertainty of the parameters. A range of

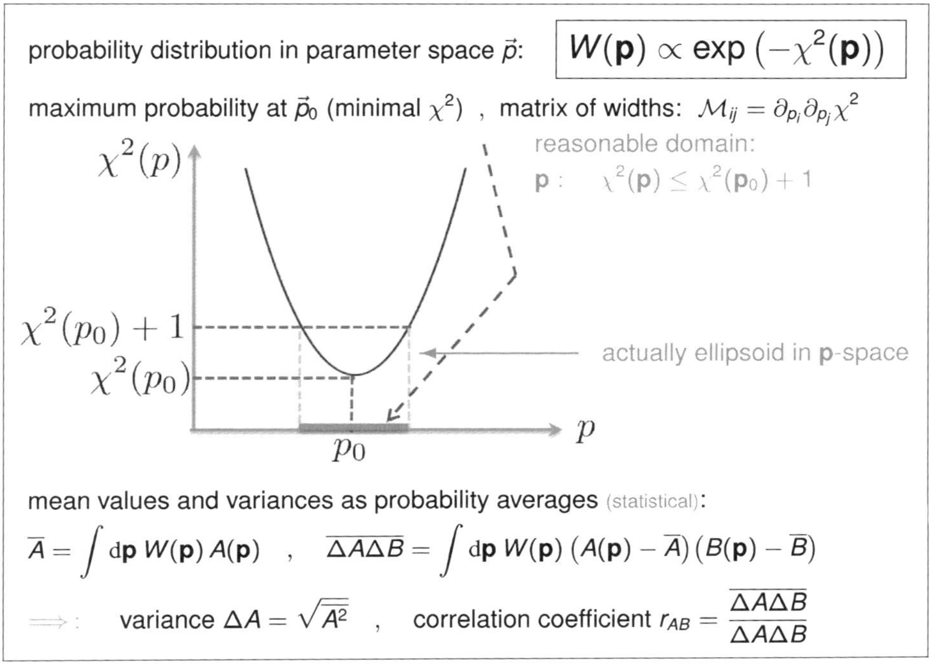

Fig. 2. Illustration of the statistical interpretation of χ^2.

reasonable parameters is found in the space of $\mathbf{p}$ for which $\chi^2 \leq \chi^2(\mathbf{p}_0) + 1$.[58] As $\chi^2(\mathbf{p})$ near the minimum is close to a parabolic function, this range of reasonable parameters covers an ellipsoid in $\mathbf{p}$-space. The probability distribution $W(\mathbf{p})$ allows us to compute statistical averages of observables $\overline{A}$ and (mixed) uncertainties $\overline{\Delta A \Delta B}$, see Fig. 2. For the average, we have usually $\overline{A} \approx A(\mathbf{p}_0)$. The uncertainty of an observable is simply $\Delta A = \sqrt{\overline{\Delta A \Delta A}}$. Mixed uncertainties allow us to deduce the amount of statistical correlation, also called alignment between two observables, which is defined as

$$r_{AB} = \frac{\overline{\Delta A \Delta B}}{\sqrt{\overline{\Delta^2 A}}\,\sqrt{\overline{\Delta^2 B}}}. \tag{8}$$

A value $r_{AB} = \pm 1$ means that the two observables are fully (anti-)correlated. Within the given model, knowledge of B does not add any new information as compared to knowledge of A. In contrast, a value $r_{AB} = 0$ means that A and B are uncorrelated, thus fully independent within the model. Often, one is not interested in the sign of r_{AB} and better considers the coefficient of determination r_{AB}^2.[65] This is what we will present in the results later on.

Correlation analysis is compact and instructive, but often somewhat abstract. Parameter correlations can be visualized by trend analysis. To this end, one defines a starting point with a couple of model parameters (mostly NMP) fixed at chosen optimal values during the χ^2 fit. This base point is in our studies of the parametrization SV-bas.[35] Then one takes one of these frozen model parameters and refits for a deliberately changed value which is then kept frozen during the fit while optimizing all other model parameters. This is done for a couple of different values thus delivering a set of parametrizations for dedicated variation of this one model parameter. Drawing observables as functions of this parameter then yields an impression of the impact on this observable. An example of this strategy will be given in Fig. 12.

3. Stability of Super-Heavy Elements (SHE)

3.1. *Spectral gaps*

Although nuclear binding is dominated by bulk properties of nuclear droplets,[66] the existence of magic numbers (particularly well bound nuclei) indicates that shell structure plays a role for nuclei. The proposal of a working shell model for nuclei was a great breakthrough in theoretical nuclear physics[67] which allowed one to sort a great manifold of nuclear data. In particular, one could explain the magic numbers, $Z = 8, 20, 28, 50, 82$ for protons and $N = 8, 20, 28, 50, 82, 126$ for neutrons as well as the existence of long lived elements beyond the critical charge for Coulomb instability predicted in the droplet model. Extrapolation far beyond the regime of stable nuclei has predicted possible new shell closures in this new area of super-heavy elements (SHE).[12,13] One of the motivation of studying SHE was the aim to find new magic shells. Early theoretical explorations were thus much concerned with prediction of shell structure, in particular shell closures. Besides that, shell structure provides a rather inexpensive first indicator for stability against spontaneous fission. Thus we start with a brief review of the investigations of shell structure.

To recall the situation in well-established doubly magic nuclei, we show in Fig. 3

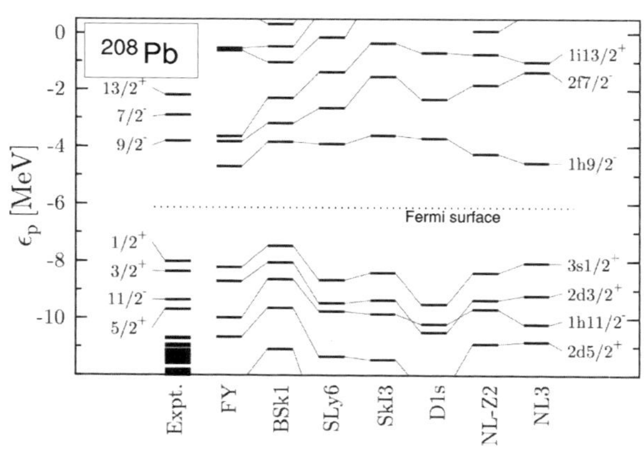

Fig. 3. Single-particle (s.p.) proton level near the Fermi surface for ^{208}Pb computed with a variety of very different models: folded Yukawa shell model (FY), SHF parametrizations BSk1,[46] SLy6,[45] SkI3,[44] Gogny force D1s,[68] and RMF parametrizations NL-Z2[48] and NL3.[49]

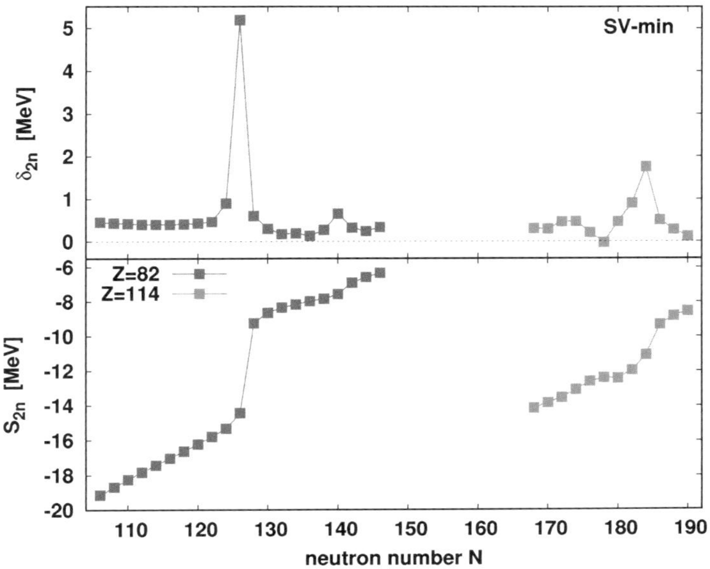

Fig. 4. Two-neutron separation energy S_{2n} (lower) and two-neutron shell gap δ_{2n} along the isotopic chains of lead ($Z = 82$) and flerovium ($Z = 114$) computed with the SHF parametrization SV-min.[35]

the proton s.p. spectra near the Fermi energy for ^{208}Pb. The large spectral gap between HOMO ($3s_{1/2}$) and LUMO ($1h_{9/2}$) is immediately obvious. This gap exists in very similar fashion for all the models shown in the figure. Although the models are very different in formal structure, they are all calibrated to nuclear binding energies and the excess binding for magic nuclei is intimately related to the occurrence of a spectral gap. We see, so to say, the impact of empirical data independent of the model structure. The dream was to find such clearly developed shell closure in SHE again. Thus, Fig. 3 should be kept in mind as a benchmark.

The shell closure can be deduced from the systematics of binding energies. For example, a neutron magic shell is related to a jump in the two-neutron separation energy

$$S_{2n}(Z,N) = E_B(Z,N+2) - E_B(Z,N). \tag{9a}$$

This happens because the first two neutrons added after the magic number have to fill the much less bound LUMO states (see Fig. 3) which makes them much easier to remove than the well-bound nucleons in the HOMO. Such a jump in S_{2n} is clearly observed for $Z = 82$ at between $N = 126$ and 128. A glimpse of a jump could be spotted in the $Z = 114$ chain. A more conclusive signal may be obtained by looking at the second energy difference, the two-neutron shell gap

$$\delta_{2n}(Z,N) = E_B(Z,N+2) - 2E_B(Z,N) + E_B(Z,N-2), \tag{9b}$$

which is shown in the upper panel of Fig. 3. The $\delta_{2n}(Z\!=\!82,N)$ is indeed dramatic, a well isolated and steep peak. The $\delta_{2n}(Z\!=\!14,N)$ does also deliver a better signal

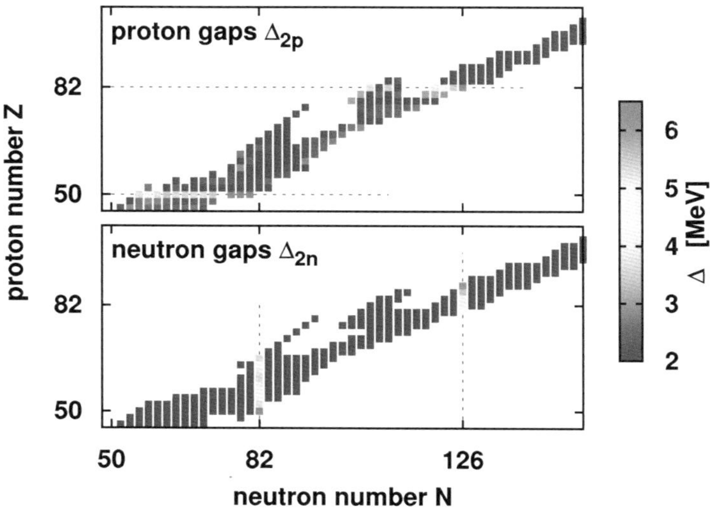

Fig. 5. Two nucleon shell gaps from experimental binding energies in the landscape of heavy nuclei. Lower: two-neutron shell gaps δ_{2n}. Upper: two-neutron shell gaps δ_{2p}.

than the $S_{2n}(Z, N)$ indicating a preference of $N = 184$ as neutron shell closure in this region of SHE. However, the distinction from the neighborhood is not so well developed.

The two-nucleon shell gap is a robust signal in the regime of stable nuclei. Figure 5 shows experimental shell gaps, the two-neutron shell gap (9b) and analogously the two-proton shell gap δ_{2p}, over a whole section of the chart of isotopes. The magic proton numbers 50, 82 and the magic neutron numbers 82, 126 appear consistently for all isotopes, or isotones respectively. However, it is interesting to note that the height of the gap, so to say the "magicitiy", depends on the position in the chain. Take, for example, the $Z = 50$ chain (upper panel). The δ_{2p} increases with the neutron number approaching the magic $N = 82$ and similar trends are observed in the other chains. We conclude from this that shell closures are mutually supporting. This effect may become more important in SHE where shell gaps are weaker anyway.

Figure 6 shows the systematics of two-nucleon shell gaps in the landscape of SHE. The neutron gaps δ_{2n} are low but stay rather robust along the isotopic chains. The proton gaps δ_{2p}, on the other hand, are somewhat larger, but show significant dependence on neutron number even to the extend that the (weak) magic number hops from $Z = 120$ to $Z = 124$ for $N = 180$–190. Moreover, they are not as sharply distinguished from the neighboring nuclei. One gets the impression that these "magic" numbers are not that exclusive as they were for normal nuclei. The whole vicinity experiences some shell stabilization. The notion of "islands of stability" is very appropriate for such a scenario.

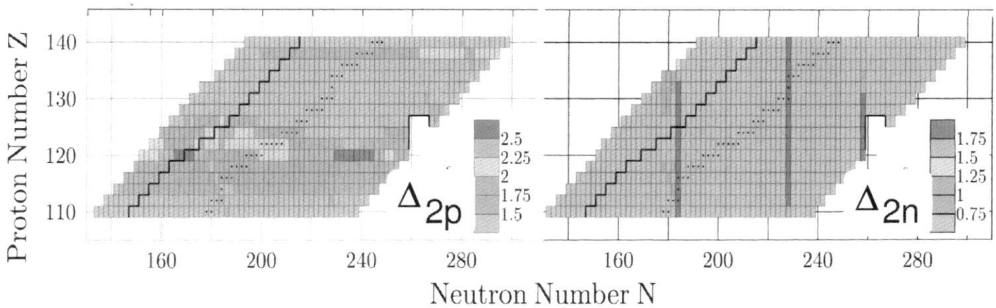

Fig. 6. Two-proton shell gaps δ_{2p} and two-neutron shell gaps δ_{2n} in the landscape of SHE computed with the SHF parametrization SLy6.[45]

3.2. *Shell correction energy*

The search for new magic numbers in the realm of SHE was only partially successful. Particularly for the proton spectrum, we come to an ambiguous situation. This means, however, very little for the existence of SHE. What counts here is the stabilization through shell effects. The binding of nuclei follows in the average a smooth trend which is well described by the droplet or liquid drop model (LDM).[69] Around the LDM average, the energy fluctuates due to varying level density near the Fermi energy. The difference to the average is called the shell-correction energy. SHE exist only by virtue of extra binding by shell corrections. Thus the stability of SHE can be estimated by looking at the shell-correction energy. It can be computed from the s.p. spectrum according to the Strutinsky prescription

$$E_{\text{shell}} = E_{\text{sp}} - \tilde{E}_{\text{sp}}, \quad E_{\text{sp}} = \sum_{\alpha} \varepsilon_\alpha, \quad \tilde{E}_{\text{sp}} = \int d\varepsilon\, \varepsilon \sum_{\alpha} \mathcal{G}(\varepsilon - \varepsilon_\alpha), \tag{10}$$

where $\mathcal{G}$ is a folding function, typically a normalized Gaussian $\mathcal{G}(\epsilon) \propto \exp -\epsilon^2/\Delta^2$ with a folding width Δ of order of twice the shell spacing.[70,71]

The right panel of Fig. 7 shows the shell-correction energy E_{shell} for two the isotonic chains $N = 172$ and $N = 184$ which both are neutron numbers associated with some shell closure for this parametrization SLy4. The left panel shows the two-proton shell gap δ_{2p} for comparison. The patterns are very similar. The spikes in δ_{2p} correspond to the dips in E_{shell} and both signals come along here with rather broad peaks (valleys) indicating that not only the magic numbers, but also its vicinity, take advantage of shell stabilization such that we can expected spacious islands of stability.

Figure 8 shows shell-correction energies all over the landscape of SHE for two different mean-field models, SHF and RMF, and therein a couple of parametrizations. At first glance, the gross patterns are the same for all models. The islands of shell stabilization reside in the same regions for all cases and they are all found to present a broad blob. However, there are great differences in details of average

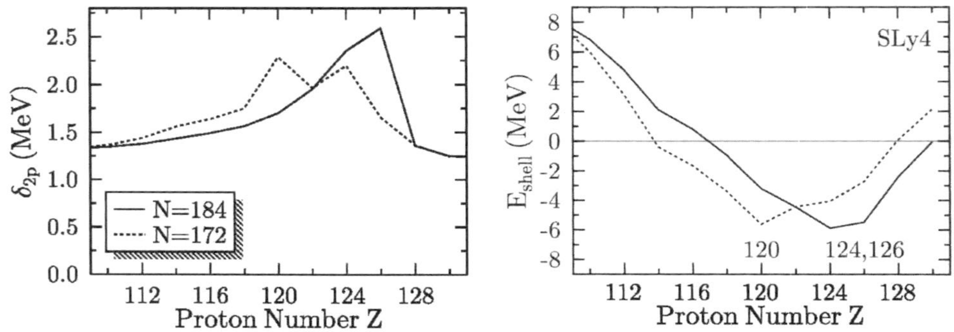

Fig. 7. Indicators of shell stabilization for the isotones at $N = 172$ and $N = 184$ computed with the SHF paramtrization SLy4. Left panel: Two-proton shell gap δ_{2p}. Right panel: Shell correction energy E_{shell}.

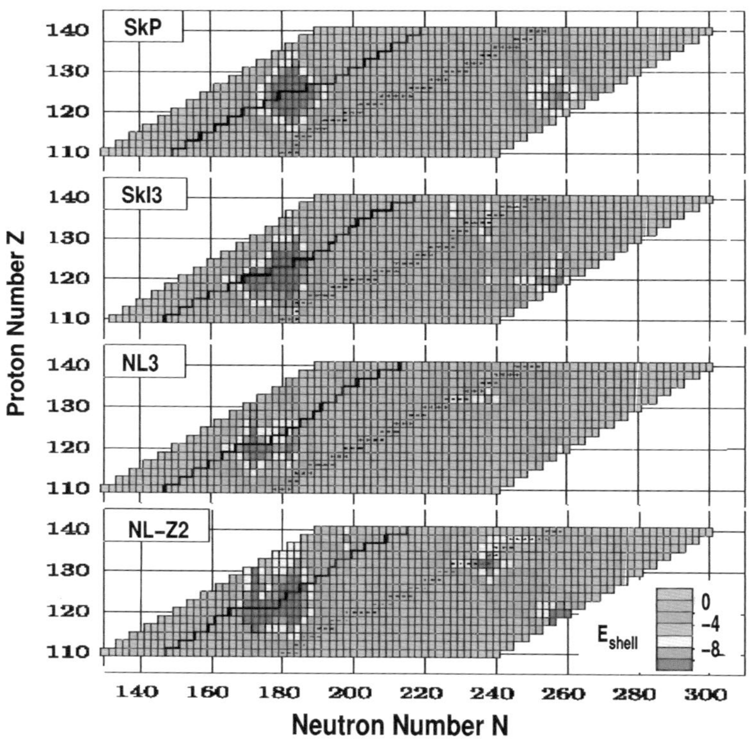

Fig. 8. Shell correction energy in the landscape of SHE for a variety of mean-field models: from SHF SkP[43] and SkI3,[44] and from RMF NL3[49] and NL-Z2.[48]

position and width of the blob. This large variance in details differs very much from what we had seen in stable nuclei. Recall, e.g., Fig. 3 for ^{208}Pb where all models produce practically the same shell gap.

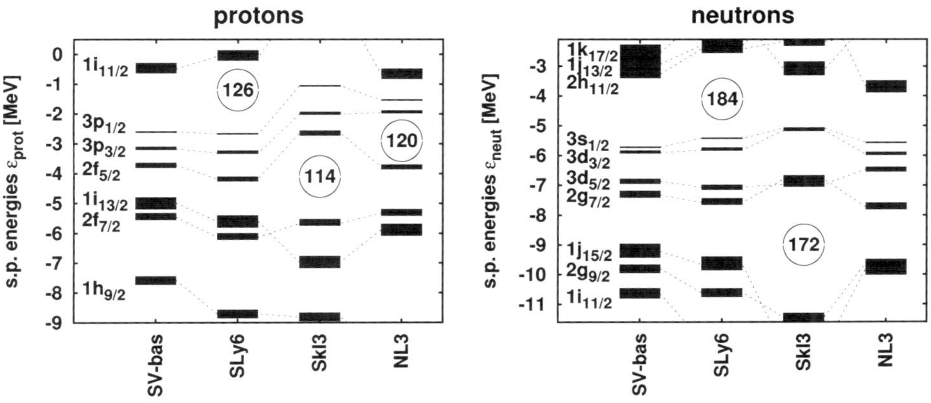

Fig. 9. Proton and neutron s.p. spectra near the Fermi energy for Fl^{298} (having $Z = 114/N = 184$) and a couple of different mean-field models. The multiplicity of each s.p. state is indicated by the width of the corresponding line.

In order to understand what changes with SHE, we show in Fig. 9 proton and neutron spectra near Fermi energy for a typical SHE, here Fl^{298}. The shell gaps Fl^{298} at $N = 172$ and $N = 184$ are robust for all models. However their size differs substantially. The situations is more diffuse for protons where not only the size but also position of shell gaps varies. This explains what we had seen previously, namely that the δ_{2p} produce rather broad bumps rather than sharp peaks, and similarly for the proton shell correction (not shown explicitly here). The reason for this diffuseness is the small band of s.p. states between $Z = 114$ and $Z = 126$. This is floating around with the parametrizations and so delivers floating predictions. However, this band covers states with low multiplicity while the edges for the bands above and below carry heavy weight. Thus this light-weighted intermediate band contributes little to the density of states which, in turn, leaves a large effective shell gap between $Z = 114$ and 126 and this produces conveniently large shell correction energies all over this region as we had seen in the previous figures.

3.3. *Fission*

Shell corrections, as discussed in the previous section, are still a rather convenient means to explore the stability of SHE. But the conclusive answer comes only from computing the decay properties of SHE. The most critical and most cumbersome to evaluate is fission lifetime. This will be addressed in this section.

We summarize here briefly the computation of the fission path and the ingredients of the corresponding collective Hamiltonian, for details of the fission calculation see Refs. 23 and 24. The mean-field equations are complemented by a quadrupole constraint to generate the deformation path. This reads

$$\big[\hat{h} - \lambda\hat{Q}_{20}\big]|\Phi_q\rangle = \mathcal{E}|\Phi_q\rangle, \qquad q = \langle\Phi_q|\hat{Q}_{20}|\Phi_q\rangle, \tag{11a}$$

$$\mathcal{V}(q) = E(\rho_q, \tau_q, \mathbf{J}_q, \chi_q), \tag{11b}$$

where the quadrupole deformation q labels the collective path, $\hat{h}$ is the mean-field Hamiltonian and ρ_q is the local density for the state $|\Phi_q\rangle$, similarly for τ_q, $\mathbf{J}_q$, and χ_q. The collective mass along the fission path is obtained by exploring the dynamical response of the system to changing deformation, commonly called self-consistent or ATDHF cranking.[72] To that end we use the collective-momentum operator $\hat{P}_{20}$ as additional constraint. The $\hat{P}_{20}$ is deduced as generator of deformation through the response equation $\hat{P}_{20}|\Phi_q\rangle \propto \mathrm{i}\partial_q|\Phi_q\rangle$. The doubly constrained mean-field equations $\big[\hat{h} - \lambda\hat{Q}_{20} - \mu\hat{P}_{20}\big]|\Phi_{qp}\rangle = \mathcal{E}|\Phi_{qp}\rangle$ extend $|\Phi_q\rangle$ to a momentum-dependent path $|\Phi_{qp}\rangle$ (with p quantifying the collective momentum) and subsequently total energy, from which the mass is finally obtained as second-order term in collective momentum, $B_{20} = \frac{1}{2}\partial^2_{p_{20}}E(\rho_{qp},...)\big|_{p=0}$. Analogously one computes the momentum of inertia Θ_x for rotations about the x- and y-axis by ATDHF cranking. The raw potential $\mathcal{V}$ as given in Eq. (11b) has to be augmented by quantum corrections to account for angular-momentum projection and spurious zero-point quadrupole motion[72] yielding the net potential

$$\mathcal{V} \quad \longrightarrow \quad V = \mathcal{V} - B_{20}\langle\hat{P}^2_{20}\rangle - \frac{1}{8\langle\hat{P}^2_{20}\rangle}\partial^2_q\mathcal{V} - \frac{\langle\hat{J}^2_x + \hat{J}^2_y\rangle}{2\Theta_x}. \tag{12}$$

These ingredients for fission dynamics are shown in Fig. 10 for the example of ^{260}Rf. The collective mass shows substantial fluctuations along the path. These are caused by level crossings which happen rather frequently in SHE with their high density of s.p. states. The zero-point energy (ZPE) produces basically a global down-shift of the potential which has a slight trend along deformation, although being small, playing quantitatively a role.[73–75]

Having determined the properly corrected collective potential $V(q)$ and inverse collective mass $B(q)$, we compute the fission lifetime in the standard fashion using semi-classical WKB for tunneling probability W and repetition time T in the ground-state potential well:

$$W = \exp\Big(-2\int_b^c dq\sqrt{\frac{V(q)-E}{B_{20}}}\Big), \quad T = \hbar\int_a^b dq\frac{1}{\sqrt{B_{20}(E-V(q))}}. \tag{13}$$

The integrals are evaluated by trapezoidal rule with separate handling at the divergent $\sqrt{E-V}$ at the classical turning points. A quantity which enters very critically these expressions is the tunneling energy E, which is also the ground state energy of the nucleus before fission. That is computed with great care in a fully quantum-mechanical treatment of the collective quadrupole oscillations using all five quadrupole degrees-of-freedom, for details of that part of the calculations see Refs. 34 and 76.

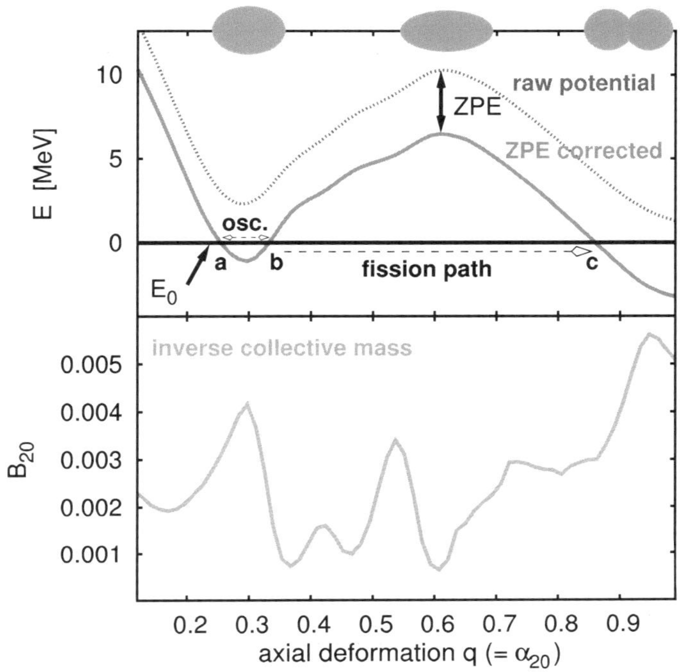

Fig. 10. Potential energy surface (upper panel) and inverse collective mass (lower panel) along the fission path for ^{260}Rf computed with the SHF parametrization SLy6.[45] For the potential energy, one distinguishes the raw energy and the fission potential after correcting for the spurious zero-point energy (ZPE), see text for details. On top of the figure, three typical nuclear shapes along the path are indicated.

It is to be noted that the fission path used here employs only axially symmetric shapes. One knows from actinides that triaxial shapes can produce barriers which are 0–2 MeV lower. The present results are thus to be understood as an upper limit on barriers and lifetimes. Actinides show the famous double-humped barrier. For SHE, however, the outer barrier disappears leaving only one barrier.[77] This barrier is related to still symmetric shapes while asymmetry develops quickly when stepping down further towards fission. The fact that we have only one barrier simplifies the survey of fission properties of SHE.

The strongest hindrance for fission is the fission barrier B_f, the energy difference between potential minimum (in Fig. 10 at $q = 0.3$) and the maximum at the barrier (in Fig. 10 at $\alpha_{20} = 0.6$). The actual barrier is, in fact, the difference between maximum and ground state energy in the ZPE corrected potential. For a quick estimate, one takes just the difference between maximum and minimum in the raw potential. This will be considered in the following as the measure for B_f. Figure 11 shows the variety of B_f for ^{266}Hs as predicted by different mean-field models. The span of values is large. One has to keep in mind, however, that the figure covers different generations of parametrizations and that the calibration of models has

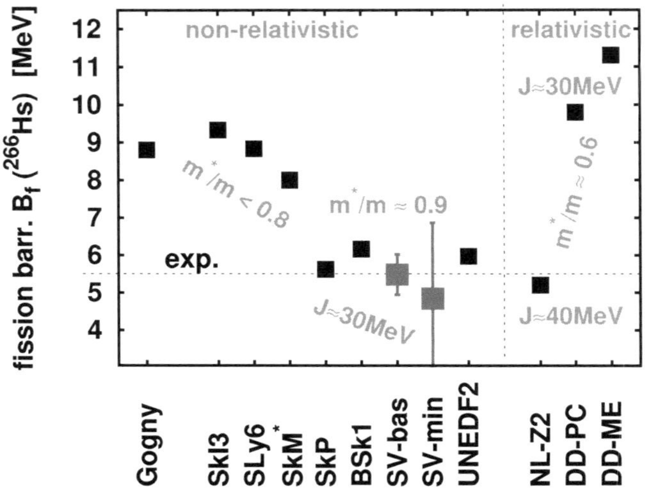

Fig. 11. Fission barrier B_f of ^{266}Hs ($Z = 108$) for a variety of mean-field models and parametrizations. The values of two NMP (symmetry energy J and effective mass m^*/m) are indicated.

much improved over the years because more and more data on exotic nuclei became available. The figure indicates the two NMP which were found to be most crucial for B_f. And indeed, parametrizations with $m^*/m \approx 0.9$ and $J \approx 30$ MeV deliver best B_f. Lower m^*/m enhance shell effect, thus delivering larger shell-correction energies at minimum and consequently larger barriers. There occurs a curious coincidence for the RMF parametrization NL-Z2. The effective mass is extremely low, but the symmetry energy is extremely high. Both unusual NMP together happen to produce a correct B_f.

It is a bit hazardous to pick a couple of parametrizations from literature and to deduce trends from that because there are too many differences in ingredients and bias from the different groups. To find out relations to dedicated nuclear properties, one needs parametrizations where only one chosen property is varied systematically, see Subsec. 2.3. With the pool of systematically varied parametrization from Ref. 35, we have scanned the dependencies for B_f. Figure 12 shows the trends for the three most influential properties together with the statistical uncertainty of each varied parameter. The variation of B_f within the uncertainty band indicates the strength of the correlation of B_f with that parameter. The strongest effects come from variation of pairing strength. This is plausible because pairing tends to wash out the shell fluctuations and just the fission barrier is produced by the fluctuations of shell-correction energy with changing deformation. Effective mass is the next important player, again a quantity strongly related to shell structure. Almost equally important is the symmetry energy J. This is a true bulk property. However, its influence on B_f is mediated through its impact on neutron skin[42,78] which, in turn, plays a role in neck formation.

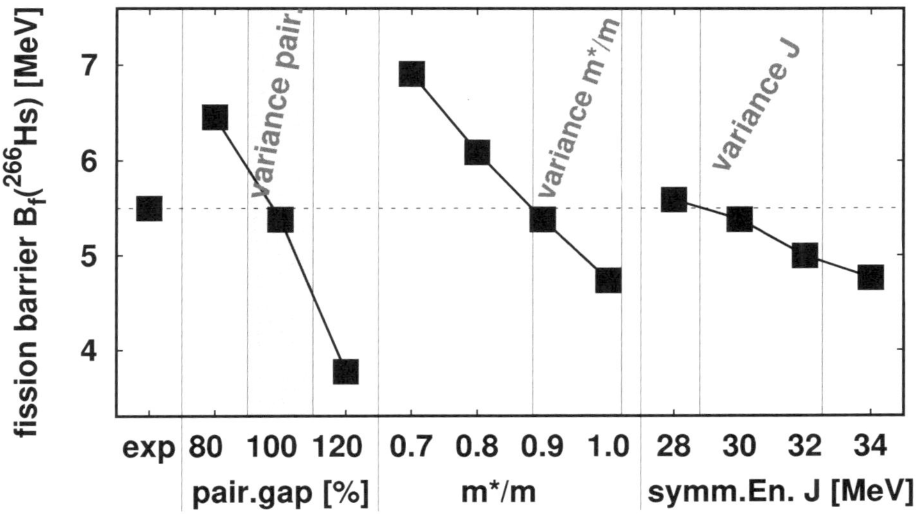

Fig. 12. Fission barrier B_f of ^{266}Hs ($Z = 108$) for three sets of SHF parametrizations with systematically varied properties, according to Ref. 35. The shaded areas indicate the uncertainty of each parameter.

The error bars on the entries for SV-min and SV-bas in Fig. 12 indicate the extrapolation uncertainty ΔB_f according to statistical analysis, see Subsec. 2.3. The uncertainty of SV-min is about ± 2 MeV while SV-bas produces a significantly smaller $\Delta B_f = \pm 0.5$ MeV. Note that SV-min is a fit to ground-state data only. Although the pool of fit data is rather large, the nuclear landscape is narrow in isovector direction (along difference $N - Z$) and thus isovector properties are not so tightly fixed. Moreover, dynamical response as quantified by effective mass m^*/m is naturally not well contained in pure ground-state information either. These loosely determined aspects from ground state fits become apparent in the large ΔB_f of SV-min. SV-bas, on the other hand, was fitted by additionally freezing the four response properties K, m^*/m, J, and κ_{TRK} (see Subsec. 2.2) at values which guarantee a good reproduction of giant resonance and dipole polarizability in ^{208}Pb.[35] This fixes the bulk properties of the functional and leaves only the uncertainty from shell effects (spin–orbit interaction, pairing) which is conveniently small. Not surprisingly, including more information in the calibration helps to confine the predictions of a functional.

Trend analysis as exercised in Fig. 12 is illustrative, but grows cumbersome if many influences are to be compared. Correlation analysis as outline in Subsec. 2.3, provides a more compact measure which allows to visualize many ingredients at once. Figure 13 shows the correlations between the fission barrier in ^{264}Fm and key parameters of the SHF functional in terms of the coefficient of determination (see Subsec. 2.3). There are parameters which are tightly fixed by the fit (in fact, L as such is not tightly fixed, but so strongly correlated with J that only one of these

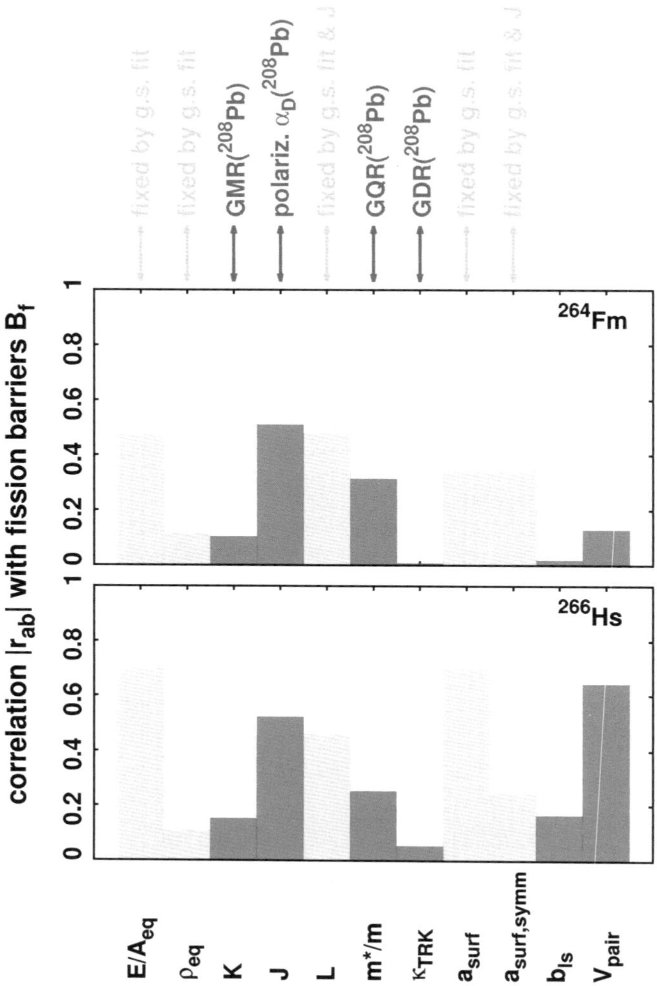

Fig. 13. Coefficient of determination r_{ab}^2 between $B_f(^{264}\text{Fm})$ and key parameters of the SHF functional for the parametrization SV-min.[35] Grey boxes indicate parameters which are tightly fixed by the fit of SV-min such that their correlations have little impact on the variance of B_f.

two entries can be considered as free). Their uncertainty thus is small and their variations do not contribute much to the variation of B_f and we better ignore these correlations. For the remaining parameters, we see indeed that pairing strength, J, and m^*/m have strongest impact. One may be surprised that the importance of J and m^*/m is reverted as compared to Fig. 12. Note that the two figures deal with two different nuclei and the actual strength of correlations is slightly element dependent across the landscape of SHE.

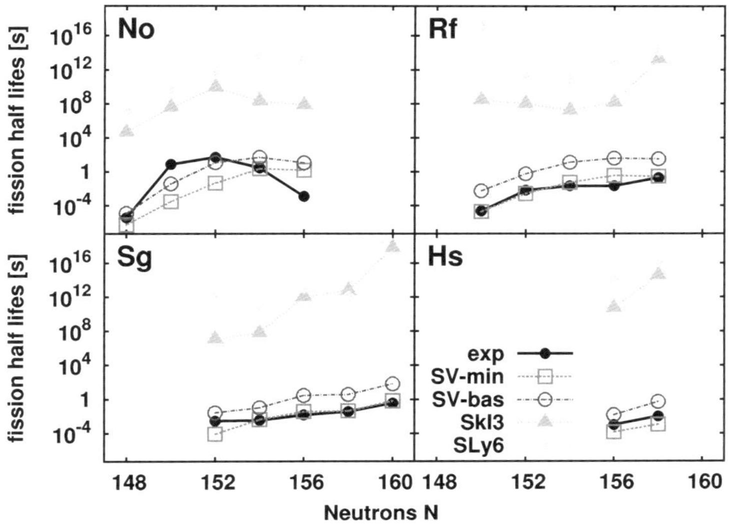

Fig. 14. Fission halflifes τ_f along isotopic chains of four transactinides. Results from four different SHF parametrizations are compared with experimental data (for complete references to experimental fission data see Ref. 24).

Having explored the model dependence of fission barriers, we look now at the system dependence and do this for the finally decisive number, the fission lifetime τ_f. Figure 14 shows trends lifetimes along four isotopic chains at the lower end of SHE for which experimental data exist. The data are compared with results from four SHF parametrizations out of the selection of parametrizations in Fig. 11 where two of them (SV-min and SV-bas) delivered barriers close to data and the other two had too high barriers. The consequence on lifetimes is dramatic as we see in Fig. 14. SLy6 and SkI3 which overestimate barriers in ^{266}Hs by 3–4 MeV produce lifetimes more than 10 orders of magnitude too large. SV-min and SV-bas whose barriers reach data within less than 1 MeV reproduce lifetimes within 2 orders of magnitude. At first glance, one thinks to see a monotonous relation between barrier and lifetime where higher barrier means longer lifetime. This holds for the large differences between the group SV-min with SV-bas versus SLy6 with SkI3 as well as within the group SV-min with SV-bas. But the order is reverted between SLy6 and SkI3. Recall that the computation of τ_f requires many ingredients. Not only the barrier height, but also barrier width and collective mass along the tunneling region are important. The example thus demonstrates in every respect the extreme sensitivity of fission lifetimes. A reproduction of data within two order of magnitude can be considered as great success. This goal can be reached with careful calibration of the mean-field model. We thus dispose now of a reliable description of fission properties of SHE. This can then be used for large-scale surveys in the whole landscape of SHE establishing systematics in the available region and exploring also extremes as, e.g.,

SHE in the r-process region.[24,79] Reviewing these extensive surveys goes beyond the limitations of the present tour through SHE research. For further reading, we refer the reader to the cited articles. Reference 24 addresses also a question of competing decay channels, α- and β-decay, which we had not touched at all in the present article.

4. Conclusions

This article tried to provide a quick tour through studies of super-heavy elements (SHE) using self-consistent mean-field models with a bias on results from the nuclear theory group of Frankfurt university. Three stages of development were considered, the search for shell gaps and new magic numbers, estimates of stability in terms of shell correction energy, and spontaneous fission of SHE.

New magic numbers are hard to assess in the realm of SHE. The density of s.p. states increases with increasing system size such that the spectral shell gaps shrink and render shell closures less dramatic than for normal nuclei. Nonetheless, one may identify robust appearance of shell gaps for neutrons at $N = 172$ and $N = 184$. The situation is more diffuse for protons where shell gaps or shell closures are found to be extremely volatile, depending sensitively on neutron number and model parameters.

Shell correction energies show a similar feature. They do not display sharp and well isolated peaks at particular magic nucleon numbers, but rather broad islands of shell stabilization. The appearance of those islands and their region in the landscape of SHE is a robust feature predicted by all models and parametrizations considered. However, the actual depth of the extra binding and the width of the islands is model dependent. Both features, the weakness of shell closures together with the robustness of islands of stabilization, can be understood from the proton s.p. spectra. In between two well separated bands of high multiplicity weight floats a small band with a few low-multiplicity states. This spoils spectral gaps while shell-correction energy depending on s.p. energies together with multiplicities is still fully at work.

Fission is a most elaborate quantity. In fact, it goes beyond a strict mean-field picture as it involves a whole collective deformation path. As a consequence, many ingredients of the model contribute to the final result. Nonetheless, careful disentangling of the different influences by means of statistical analysis allows us to identify the three force parameters with most impact on fission: pairing strength, effective mass, and symmetry energy. Controlling these features by careful calibration of the mean-field model delivers a prediction of fission barriers within 1 MeV and lifetimes within 2 orders of magnitude. Thus properly designed self-consistent mean-field models provide a reliable tool for systematic surveys of fission in a broad range of SHE and isotopes of astro-physical interest.

Acknowledgments

First of all, my thank goes to Walter Greiner who inspired and accompanied the studies reported here all along the way. There were many contributors over the years which would produce a huge list if accounted for completely. But amongst this large team, I thank particularly my colleagues J.A. Maruhn and W. Nazarewicz. Their active involvement together with their groups was the necessary condition to work out the large body of work on SHE and nuclear energy density functionals.

References

1. E. K. Hyde and G. T. Seaborg, The transuranium elements, in *Handbuch der Physik, Vol. XLII*, (Springer, Berlin, 1957) p. 205.
2. J. R. Nix, *Annual Review of Nuclear Science* **22**, 66 (1972).
3. S. Hofmann and G. Münzenberg, *Rev. Mod. Phys.* 72, 733 (2000).
4. S. Hofmann, F. Heßberger, D. Ackermann, S. Antalic, P. Cagarda, S. Wiok, B. Kindler, J. Kojouharova, B. Lommel, R. Mann, G. Münzenberg, A. Popeko, S. Saro, H. Schtt and A. Yeremin, *Eur. Phys. J. A* **10**, 5 (2001).
5. Y. T. Oganessian, V. K. Utyonkov, Y. V. Lobanov, F. S. Abdullin, A. N. Polyakov, I. V. Shirokovsky, Y. S. Tsyganov, G. G. Gulbekian, S. L. Bogomolov, B. N. Gikal, A. N. Mezentsev, S. Iliev, V. G. Subbotin, A. M. Sukhov, A. A. Voinov, G. V. Buklanov, K. Subotic, V. I. Zagrebaev, M. G. Itkis, J. B. Patin, K. J. Moody, J. F. Wild, M. A. Stoyer, N. J. Stoyer, D. A. Shaughnessy, J. M. Kenneally, P. A. Wilk, R. W. Lougheed, R. I. Il'kaev and S. P. Vesnovskii, *Phys. Rev. C* **70**, 064609 (2004).
6. Y. T. Oganessian, V. K. Utyonkov, Y. V. Lobanov, F. S. Abdullin, A. N. Polyakov, R. N. Sagaidak, I. V. Shirokovsky, Y. S. Tsyganov, A. A. Voinov, G. G. Gulbekian, S. L. Bogomolov, B. N. Gikal, A. N. Mezentsev, S. Iliev, V. G. Subbotin, A. M. Sukhov, K. Subotic, V. I. Zagrebaev, G. K. Vostokin, M. G. Itkis, K. J. Moody, J. B. Patin, D. A. Shaughnessy, M. A. Stoyer, N. J. Stoyer, P. A. Wilk, J. M. Kenneally, J. H. Landrum, J. F. Wild and R. W. Lougheed, *Phys. Rev. C* **74**, 044602 (2006).
7. K. E. Gregorich, J. M. Gates, C. E. Düllmann, R. Sudowe, S. L. Nelson, M. A. Garcia, I. Dragojević, C. M. Folden III, S. H. Neumann, D. C. Hoffman and H. Nitsche, *Phys. Rev. C* **74**, 044611 (2006).
8. J. Dvorak, W. Brüchle, M. Chelnokov, C. E. Düllmann, Z. Dvorakova, K. Eberhardt, E. Jäger, R. Krücken, A. Kuznetsov, Y. Nagame, F. Nebel, K. Nishio, R. Perego, Z. Qin, M. Schädel, B. Schausten, E. Schimpf, R. Schuber, A. Semchenkov, P. Thörle, A. Türler, M. Wegrzecki, B. Wierczinski, A. Yakushev and A. Yeremin, *Phys. Rev. Lett.* **100**, 132503 (2008).
9. G. Martínez-Pinedo, D. Mocelj, N. Zinner, A. Keli, K. Langanke, I. Panov, B. Pfeiffer, T. Rauscher, K.-H. Schmidt and F.-K. Thielemann, *Prog. Part. Nucl. Phys.* **59**, 199 (2007).
10. M. Arnould, S. Goriely and K. Takahashi, *Physics Reports* **450**, 97 (2007).
11. J. Stone and P.-G. Reinhard, *Prog. Part. Nucl. Phys.* **58**, 587 (2007).
12. U. Mosel and W. Greiner, *Z. Phys.* **222**, 1969 (261).
13. A. Sobiczewski, T. Krogulski, J. Błocki and Z. Szymański, *Nucl. Phys. A* **168**, 519 (1971).
14. D. Scharnweber and W. Greiner, *Nuc. Phys. A* **164**, 257 (1971).
15. J. A. Maruhn and W. Greiner, *Z. Phys. A* **251**, 431 (1972).
16. P. Ring, *Prog. Part. Nucl. Phys.* **37**, 193 (1996).

17. M. Bender, P.-H. Heenen and P.-G. Reinhard, *Rev. Mod. Phys.* **75**, 121 (2003).
18. D. Vretenar, A. Afanasjev, G. Lalazissis and P. Ring, *Phys. Rep.* **409**, 101 (2005).
19. J. Erler, P. Klüpfel and P.-G. Reinhard, *J. Phys. G* **38**, 033101 (2011).
20. J. F. Berger, L. Bitaud, J. Decharg, M. Girod and K. Dietrich, *Nuclear Physics A* **685**, 1 (2001).
21. M. Warda, J. Egido and L. Robledo, *Phys. Scr. T* **125**, 226 (2006).
22. A. Staszczak, A. Baran, J. Dobaczewski and W. Nazarewicz, *Phys. Rev. C* **80**, 014309 (2009).
23. N. Schindzielorz, J. Erler, P. Klüpfel, P.-G. Reinhard and G. Hager, *International Journal of Modern Physics E* **18**, 773 (2009).
24. J. Erler, K. Langanke, H. P. Loens, G. Martinez-Pinedo and P.-G. Reinhard, *Phys. Rev. C* **85**, 025802 (2012).
25. A. Baran, M. Kowal, P.-G. Reinhard, L. Robledo, A. Staszczak and M. Warda, *Nucl. Phys. A* **944**, 442 (2015).
26. M. Bender, K. Rutz, P.-G. Reinhard and J. A. Maruhn, *Eur. Phys. J. A* **8**, 59 (2000).
27. W. Kohn and L. J. Sham, *Phys. Rev.* **140**, 1133 (1965).
28. R. M. Dreizler and E. K. U. Gross, *Density Functional Theory: An Approach to the Quantum Many-Body Problem* (Springer-Verlag, Berlin, 1990).
29. T. Lesinski, M. Bender, K. Bennaceur, T. Duguet and J. Meyer, *Phys. Rev. C* **76**, 014312 (2007).
30. Y. M. Engel, D. M. Brink, K. Goeke, S. J. Krieger and D. Vautherin, *Nucl. Phys. A* **249**, 215 (1975).
31. J. Dobaczewski and J. Dudek, *Phys. Rev. C* **52**, 1827 (1995), Erratum in *Phys. Rev. C* **55**, 3177 (1997).
32. K. J. Pototzky, J. Erler, P.-G. Reinhard and V. O. Nesterenko, *subm. Eur. Phys. A* (2010).
33. K. W. Schmid and P.-G. Reinhard, *Nucl. Phys. A* **530**, 283 (1991).
34. P. Klüpfel, J. Erler, P.-G. Reinhard and J. A. Maruhn, *Eur. Phys. J A* **37**, 343 (2008).
35. P. Klüpfel, P.-G. Reinhard, T. J. Bürvenich and J. A. Maruhn, *Phys.Rev. C* **79**, 034310 (2009).
36. P. Ring and P. Schuck, *The Nuclear Many-Body Problem* (Springer–Verl., New York, Heidelberg, Berlin, 1980).
37. P. Bonche, H. Flocard, P.-H. Heenen, S. J. Krieger and M. S. Weiss, *Nucl. Phys. A* **443**, 39 (1985).
38. S. J. Krieger, P. Bonche, H. Flocard, P. Quentin and M. S. Weiss, *Nucl. Phys. A* **517**, 275 (1990).
39. J. Friedrich and N. Vögler, *Nucl. Phys. A* **373**, 192 (1982).
40. B. Schuetrumpf, W. Nazarewicz and P.-G. Reinhard, *subm. Phys. Rev. C* (2017).
41. D. J. Thouless, *The Quantum Mechanics of Many–Body Systems* (Academic Press, New York, 1961).
42. W. Nazarewicz, P.-G. Reinhard, W. Satu la and D. Vretenar, *Eur. Phys. J. A* **50**, 20 (2014).
43. J. Dobaczewski, H. Flocard and J. Treiner, *Nucl. Phys. A* **422**, 103 (1984).
44. P.-G. Reinhard and H. Flocard, *Nucl. Phys. A* **584**, 467 (1995).
45. E. Chabanat, P. Bonche, P. Haensel, J. Meyer and R. Schaeffer, *Nucl. Phys. A* **635**, 231 (1998), *Nucl. Phys. A* **643**, 441(E).
46. M. Samyn, S. Goriely, P.-H. Heenen, J. M. Pearson and F. Tondeur, *Nucl. Phys. A* **700**, 142 (2002).

47. M. Kortelainen, J. McDonnell, W. Nazarewicz, E. Olsen, P.-G. Reinhard, J. Sarich, N. Schunck, S. M. Wild, D. Davesne, J. Erler and A. Pastore, *Phys. Rev. C* **89**, 054314 (2014).
48. M. Rufa, P.-G. Reinhard, J. A. Maruhn, W. Greiner and M. R. Strayer, *Phys. Rev. C* **38**, 390 (1988).
49. M. M. Sharma, G. A. Lalazissis, J. König and P. Ring, *Phys. Rev. Lett.* **74**, 3744 (1995).
50. D. Vretenar, G. Lalazissis, T. Niksic and P. Ring, *Eur. Phys. J. A* **25**, 555 (2005).
51. R. Machleidt and I. Slaus, *J. Phys. G* **27**, R69 (2001).
52. M. Lutz, B. Friman and C. Appel, *Phys. Lett. B* **474**, 7 (2000).
53. N. Kaiser, S. Fritsch and W. Weise, *Nucl. Phys. A* **700**, 343 (2002).
54. W. Weise, *Nucl. Phys. A* **751**, 565 (2005).
55. H. Bohr, C. Providencia and J. da Providencia, *Phys. Rev. C* **71**, 055203 (2005).
56. G. Hagen, T. Papenbrock, D. J. Dean and M. Hjorth-Jensen, *Phys. Rev. Lett.* **101**, 092502 (2008).
57. R. F. G. Ruiz, M. L. Bissell, K. Blaum, A. Ekstrm, N. Frmmgen, G. Hagen, M. Hammen, K. Hebeler, J. D. Holt, G. R. Jansen, M. Kowalska, K. Kreim, W. Nazarewicz, R. Neugart, G. Neyens, W. Nrtershuser, T. Papenbrock, J. Papuga, A. Schwenk, J. Simonis, K. A. Wendt and D. T. Yordanov, *Nature Physics* **12**, 594 (2016).
58. P. R. Bevington and D. K. Robinson, *Data Reduction and Error Analysis for the Physical Sciences* (McGraw-Hill, 2003).
59. J. Friedrich and P.-G. Reinhard, *Phys. Rev. C* **33**, 335 (1986).
60. T. Niksic, D. Vretenar, G. Lalazissis and P. Ring, *Phys. Rev. C* **69**, 047301 (2004).
61. T. Niksic, D. Vretenar and P. Ring, *Phys. Rev. C* **78**, 034318 (2008).
62. M. Kortelainen, J. McDonnell, W. Nazarewicz, P.-G. Reinhard, J. Sarich, N. Schunck, M. V. Stoitsov and S. M. Wild, *Phys. Rev. C* **85**, 024304 (2012).
63. B. A. Brown, *Phys. Rev. C* **58**, 220 (1998).
64. J. Dobaczewski, W. Nazarewicz and P.-G. Reinhard, *Nucl. Phys. A* **693**, 361 (2001).
65. S. A. Glantz, B. K. Slinker and T. B. Neilands, *Primer of Applied Regression & Analysis of Variance* (McGraw Hill, 1990).
66. C. F. von Weizsäcker, *Z. Phys.* **96**, 431 (1935).
67. O. Haxel, J. H. D. Jensen and H. E. Suess, *Phys. Rev.* **75**, 1766 (1949).
68. J. Dechargé and D. Gogny, *Phys. Rev. C* **21**, 1568 (1980).
69. W. D. Myers and W. J. Swiatecki, *Annual Review of Nuclear and Particle Science* **32**, 309 (1982).
70. V. M. Strutinsky, *Nucl. Phys. A* **95**, 420 (1967).
71. M. Brack and R. K. Bhaduri, *Semiclassical Physics* (Addision-Wesley, Reading, 1997).
72. P.-G. Reinhard and K. Goeke, *Rep. Prog. Phys.* **50**, 1 (1987).
73. P.-G. Reinhard, *Nucl. Phys. A* **252**, 120 (1975).
74. P.-G. Reinhard, *Nucl. Phys. A* **306**, 19 (1978).
75. P.-G. Reinhard, J. Maruhn and K. Goeke, *Phys. Rev. Lett.* **44**, 1740 (1980).
76. P. Fleischer, P. Klüpfel, P.-G. Reinhard and J. A. Maruhn, *Phys. Rev. C* **70**, 054321 (2004).
77. T. Bürvenich, M. Bender, J. Maruhn and P.-G. Reinhard, *Phys. Rev. C* **69**, 014307 (2004).
78. P.-G. Reinhard and W. Nazarewicz, *Phys. Rev. C* **81**, 051303 (2010).
79. I. Petermann, K. Langanke, G. Martinez-Pinedo, I. V. Panov, P.-G. Reinhard, and F.-K. Thielemann, *Eur. Phys. J. A* **48**, 122 (2012).

The Octupole Collective Hamiltonian. Does It Follow the Example of the Quadrupole Case?

Stanisław G. Rohoziński
Institute of Theoretical Physics, Faculty of Physics, University of Warsaw, Pasteura 5, 02-093 Warsaw, Poland
Stanislaw-G.Rohozinski@fuw.edu.pl

Leszek Próchniak
Heavy Ion Laboratory, University of Warsaw, Pasteura 5A, 02-093 Warsaw, Poland
prochniak@slcj.uw.edu.pl

Dedicated to the Memory of the late Professor Walter Greiner

A general form of the octupole collective Hamiltonian is introduced and analyzed based on fundamental tensors in the seven-dimensional tensor space. Possible definitions of intrinsic frames of reference possessing cubic symmetry for the octupole tensor are considered. Cubic intrinsic octupole coordinates or deformations are introduced. Shapes of the octupoloid are investigated. The octupole collective Hamiltonian is expressed in intrinsic coordinates. An intrinsic angular momentum carried by the octupole vibrations is discovered. Small oscillations about an axially-symmetric pear shape are analyzed. Formulation of a unified quadrupole-octupole collective model is discussed.

1. Introduction

The idea of attributing a definite multipolarity to nuclear collective excitations came from the nuclear liquid-drop model. It was known before the discovery of atomic nuclei that the normal modes of the surface vibrations of a spherical drop of incompressible liquid have definite multipolarities.[1] Many years later, when atomic nuclei were already known, Siegfried Flügge[2] connected the eigenfrequencies of the surface vibrations of a nuclear liquid drop with the excitation energies of low-lying states of even–even nuclei. These were the beginnings of the collective model.

The most important collective mode in nuclear structure physics is that with multipolarity $\lambda = 2$, because it concerns the lowest excited states in even–even nuclei. These states are well known experimentally and theoretical methods describing them are well developed. One such method, applicable to even–even nuclei only, is a description by means of the Schrödinger equation in a collective space. Initiated by Aage Bohr[3] a long time ago, it is very effective and is still in use. The Bohr Hamiltonian became the label of the model. When applying the method one usually starts with a general classical collective Hamiltonian.[4,5] Its form has been extracted from various microscopic many-body models through methods of the "Adiabatic Time-Dependent Hartree–Fock–Bogolyubov" (ATDHFB) type (see e.g. Chapter 12 in Ref. 6) and then quantized by means of the Podolsky–Pauli prescription.[7–9] Such a quasi-classical approach is still used even today. It would be good to use purely quantum methods to describe the collective states. Therefore, the Generator Coordinate Method (GCM) has been recently employed.[10,11] It leads to integral equations instead of differential ones (see e.g. Chapter 10 in Ref. 6). However, in an approximation to the GCM called the Gaussian Overlap Approximation (GOA) the Hill–Wheeler integral equations for collective motion can be reduced back to differential equations with a collective Hamiltonian.[12]

Apart from the positive-parity quadrupole states, negative-parity levels have been observed for a long time[13] in the low-energy spectra of even–even nuclei. Multipolarity $\lambda = 3$ has been attributed to such states. Information on the octupole states has been collected over many years, see Refs. 14 and 15 for reviews. Recently, measurements of a static octupole deformation in radium (^{224}Ra) and barium (^{144}Ba) isotopes have been reported.[16,17] However, data on the octupole states are not so rich as in the case of quadrupole excitations. This is perhaps connected with experimental and interpretational difficulties. A common opinion seems to be that the quadrupole collective Hamiltonian stands for a pattern for higher multipolarities. However, it is not so easy. The theory of the octupole degrees of freedom is much more complicated than that of the quadrupole ones, or rather the quadrupole case is exceptionally simple. This will be demonstrated further in subsequent Sections. In Sec. 2 a general form of the octupole collective Hamiltonian is introduced and discussed in comparison with that of the quadrupole one. Possible definitions of intrinsic frames of reference for the octupole tensor in analogy to that for the quadrupole tensor are considered in Sec. 3.1. The octupole deformation parameters are introduced and illustrated. The octupole Hamiltonian is expressed in intrinsic coordinates in Sec. 3.2. Its approximate form for small oscillations around a pear shape is given in Sec. 3.3. In Conclusion, in Sec. 4, a draft of a unified quadrupole-octupole collective model is recapitulated.

2. General Form of Collective Hamiltonians

The idea of collective models consists generally in describing some complex (collective) states of a many-body system through the substitution of the coordinates of many particles by a relatively small number of collective variables. An appropriate choice of these collective variables decides the success of the model. Constructing the collective models in question here for the description of low-lying collective states of even–even nuclei one takes the spherical tensors $\boldsymbol{\alpha}_\lambda$ as collective variables.

These tensors are transformed according to irreducible $(2\lambda+1)$-dimensional representations D^λ of the O(3) group of orthogonal transformations in the physical space. The spherical components $\alpha_{\lambda\mu}$ $(\mu=-\lambda,\ldots,\lambda)$ of $\boldsymbol{\alpha}_\lambda$ in the laboratory frame, $\mathrm{U_{lab}}$, are the collective laboratory coordinates. It is assumed that the tensor $\boldsymbol{\alpha}_\lambda$ is electric i.e. has parity $(-1)^\lambda$ and real, which means that its components fulfill the relation $\alpha^*_{\lambda\mu}=(-1)^\mu\alpha_{\lambda-\mu}$. The differential operators[a] $-i\partial/\partial\alpha^*_{\lambda\mu}$ play the role of the momenta canonically conjugate to the coordinates $\alpha_{\lambda\mu}$. The angular momentum operators or the O(3) generators in the collective space fulfill characteristic commutation relations with the coordinates and momenta (see e.g. Appendix A.1 in Ref. 18). The following vector operator:

$$L^{(\lambda)}_{1\mu}(\boldsymbol{\alpha}_\lambda)=(-1)^\lambda\sqrt{\frac{\lambda(\lambda+1)(2\lambda+1)}{3}}\sum_{\kappa\nu}(\lambda\kappa\lambda\nu|1\mu)\alpha_{\lambda\kappa}\frac{\partial}{\partial\alpha^*_{\lambda\nu}} \tag{1}$$

fulfills such commutation relations and thus plays the role of the collective angular momentum.[18,19]

The nuclear collective system is defined by a collective Hamiltonian $H_\lambda(\boldsymbol{\alpha}_\lambda)$. It is assumed that $H_\lambda(\boldsymbol{\alpha}_\lambda)$ has the following properties:

(1) It is a second-order differential operator (like the Schrödinger operator),
(2) it is real, i.e. $H_\lambda(\boldsymbol{\alpha}_\lambda)=H^*_\lambda(\boldsymbol{\alpha}_\lambda)$ (it describes even-even nuclei only),
(3) it is a scalar with respect to the orthogonal group O(3) (rotations and inversion in the physical space) or commutes with the angular momentum operators of Eq. (1),
(4) it is Hermitian with a scalar weight $W(\boldsymbol{\alpha}_\lambda)\geq 0$,
(5) it possesses the lowest eigenvalue (positive kinetic energy),
(6) it is an isotropic function of the coordinates, i.e. it does not depend on any other tensor quantities.

No other assumptions are needed at this stage (cf. Ref. 20). The most general form of $H_\lambda(\boldsymbol{\alpha}_\lambda)$ with properties (1) – (6) given above reads:[12,18]

$$H_\lambda(\boldsymbol{\alpha}_\lambda)=-\frac{1}{2W(\boldsymbol{\alpha}_\lambda)}\sum_{\mu,\nu}\frac{\partial}{\partial\alpha_{\lambda\mu}}W(\boldsymbol{\alpha}_\lambda)B^{-1}_{\lambda\mu\lambda\nu}(\boldsymbol{\alpha}_\lambda)\frac{\partial}{\partial\alpha_{\lambda\nu}}+V(\boldsymbol{\alpha}_\lambda), \tag{2}$$

The Hamiltonian is determined by three quantities: two real scalar functions — weight $W(\boldsymbol{\alpha}_\lambda)$ and potential $V(\boldsymbol{\alpha}_\lambda)$, and a symmetric $(2\lambda+1)\times(2\lambda+1)$ matrix of real so-called inverse inertial functions $B^{-1}_{\lambda\mu\lambda\nu}(\boldsymbol{\alpha}_\lambda)$. Since the matrix is transformed under the O(3) group like a product of two tensors $\boldsymbol{\alpha}_\lambda$ it is called a symmetric bitensor. All these three quantities depend on no other tensors but $\boldsymbol{\alpha}_\lambda$ and therefore they are isotropic tensor fields in the $(2\lambda+1)$-dimensional collective space. They can either be calculated from microscopic many-body models or fitted to experimental data. When the Hamiltonian is obtained by quantization of its classical counterpart the weight is $W(\boldsymbol{\alpha}_\lambda)=\sqrt{\det B_{\lambda\mu\lambda\nu}(\boldsymbol{\alpha}_\lambda)}$.

[a]Units $\hbar=c=1$ are used here.

In order to investigate possible structures of the inverse inertial bitensor it is convenient to express it as a set of (single) tensors $\boldsymbol{T}_{2l}(\boldsymbol{\alpha}_\lambda)$ $(l = 0, \dots, \lambda)$, namely

$$B^{-1}_{\lambda\mu\lambda\nu}(\boldsymbol{\alpha}_\lambda) = \sum_{l=0}^{\lambda} (\lambda\mu\lambda\nu|2l\ m) T_{2l\ m}(\boldsymbol{\alpha}_\lambda). \tag{3}$$

Assumption (6), that the $\boldsymbol{T}_{2l}(\boldsymbol{\alpha}_\lambda)$ are isotropic functions of $\boldsymbol{\alpha}_\lambda$ is essential here. Then, an arbitrary tensor $\boldsymbol{T}_\Lambda(\boldsymbol{\alpha}_\lambda)$ can be expressed in the following form:

$$\boldsymbol{T}_\Lambda(\boldsymbol{\alpha}_\lambda) = \sum_{k=1}^{k_{\lambda\Lambda}} \sigma_k^{(\Lambda)}(\boldsymbol{\alpha}_\lambda) \boldsymbol{\tau}_{k\Lambda}(\boldsymbol{\alpha}_\lambda) \tag{4}$$

by a number of definite fundamental tensors $\boldsymbol{\tau}_{k\Lambda}(\boldsymbol{\alpha}_\lambda)$ for given λ and Λ, and arbitrary scalar coefficients $\sigma_k^{(\Lambda)}(\boldsymbol{\alpha}_\lambda)$ (see Appendix A in Ref. 18). The original Bohr inverse inertial bitensor[3] has the following form:

$$B^{-1}_{\lambda\mu\lambda\nu}(\boldsymbol{\alpha}_\lambda) = (\lambda\mu\lambda\nu|00)(-1)^\lambda \sqrt{2\lambda+1}\frac{1}{B_\lambda} = \frac{(-1)^\mu \delta_{\mu-\nu}}{B_\lambda}, \tag{5}$$

where B_λ is a constant mass parameter.

In the case of a general quadrupole ($\lambda = 2$) collective Hamiltonian it is well known that the inverse inertial bitensor can have at most six independent components out of fifteen possible ones.[5,18] It is interesting to know whether there are similar restrictions for the components of the inverse inertial bitensor in the case of an octupole ($\lambda = 3$) collective Hamiltonian. Unfortunately, the things are much more complicated in that case. The symmetric bitensor $B^{-1}_{3\mu3\nu}(\boldsymbol{\alpha}_3)$ has twenty-eight components. Can they all be independent? To answer this question one should analyze Eqs. (3) and (4) for $\lambda = 3$.

According to Eq. (3) the even (positive-parity) symmetric octupole bitensor, $B^{-1}_{3\mu3\nu}(\boldsymbol{\alpha}_3)$, can be replaced with four tensors $\boldsymbol{T}_{2l}(\boldsymbol{\alpha}_3)$ $(l = 0, \dots, 3)$. These tensors should be even isotropic functions of $\boldsymbol{\alpha}_3$. The tensor fields in the space of the octupole coordinates are built out of some twenty-six elementary tensors

$$\boldsymbol{t}_l^{(n)} = [\underbrace{\boldsymbol{\alpha}_3 \times \ \dots \ \times \boldsymbol{\alpha}_3}_{n}]_l \tag{6}$$

(square brackets $[\dots]_l$ stand for vector coupling to rank l) for $l < 3n$ and related to each other by about two hundred syzygies or relations in the form of rational integral functions.[21] All the elementary tensors split into two groups with positive and negative spin-parity $(-1)^{l+n}$, respectively. Independent fundamental tensors $\boldsymbol{\tau}_{k\Lambda}(\boldsymbol{\alpha}_3)$ from Eq. (4) are constructed by alignment of the elementary tensors of Eq. (6). The positive spin-parity elementary and even (positive parity) fundamental tensors needed to construct the inverse inertial bitensor in question are listed in Sec. A.1. There are twenty-eight relevant fundamental tensors altogether and thus, all twenty-eight components of the bitensor $B^{-1}_{3\mu3\nu}(\boldsymbol{\alpha}_3)$ can be arbitrary tensor fields in the octupole collective space. No additional relations between the components need appear. Twenty-eight scalars $\sigma_k^{(\Lambda)}(\boldsymbol{\alpha}_3)$ for $k = 1, \dots, k_{3\Lambda}$ and $\Lambda = 0,\ 2,\ 4,\ 6$ are functions of the four elementary scalars, Eq. (A.1).

3. Intrinsic Frame and Intrinsic Coordinates

3.1. *Intrinsic frames of reference*

Obviously, the tensor $\boldsymbol{\alpha}_\lambda$ can be represented by different sets of coordinates in different frames of reference. As already stated in Sec. 2, the $\alpha_{\lambda\mu}$ are the components of $\boldsymbol{\alpha}_\lambda$ in the frame $\mathrm{U_{lab}}$. Let us take another frame, say $\mathrm{U_{in}}$, the orientation of which with respect to the laboratory frame $\mathrm{U_{lab}}$ is given by the Euler angles $\omega = (\varphi, \vartheta, \psi)$. We will not use the spherical components of $\boldsymbol{\alpha}_\lambda$ in the frame $\mathrm{U_{in}}$. Instead, we shall use their real and imaginary parts, $a_{\lambda k}$ and $b_{\lambda k}$, respectively, defined in the standard way (cf. e.g. Eq. (A.7) in Ref. 18). Then, the transformation rule between the corresponding components of the tensor $\boldsymbol{\alpha}_\lambda$ with respect to $\mathrm{U_{lab}}$ and $\mathrm{U_{in}}$, respectively, takes the following form (cf. Ref. 19):

$$\alpha_{\lambda\mu} = D^{\lambda(+)}_{\mu 0}(\omega) a_{\lambda 0} + \sum_{k=1,2,3} [D^{\lambda(+)}_{\mu k}(\omega) a_{\lambda k} + D^{\lambda(-)}_{\mu k}(\omega) b_{\lambda k}], \tag{7}$$

where

$$D^{\lambda(+)}_{\mu k}(\omega) = \frac{1}{\sqrt{2(1+\delta_{k0})}}[\mathcal{D}^{\lambda}_{\mu k}(\omega) + (-1)^k \mathcal{D}^{\lambda}_{\mu -k}(\omega)],$$
$$D^{\lambda(-)}_{\mu k}(\omega) = \frac{i}{\sqrt{2}}[\mathcal{D}^{\lambda}_{\mu k}(\omega) - (-1)^k \mathcal{D}^{\lambda}_{\mu -k}(\omega)] \tag{8}$$

are the semi-Cartesian Wigner functions. The Bohr–Mottelson definition of the Wigner functions, $\mathcal{D}^{\lambda}_{\mu k}(\omega)$, is used.[22]

For the frame $\mathrm{U_{in}}$ to be the intrinsic frame, three properly chosen conditions for the coordinates $a_{\lambda k}$ and $b_{\lambda k}$ should be given, namely

$$\Omega_i(a_{\lambda k}(\omega, \alpha_{\lambda\mu}), b_{\lambda k}(\omega, \alpha_{\lambda\mu})) = \Omega_i(\omega, \alpha_{\lambda\mu}) = 0 \tag{9}$$

for $i = 1,\ 2,\ 3$, which determine the three Euler angles, $\omega = (\varphi, \vartheta, \psi)$ as functions of the laboratory coordinates and, in this way, impart the status of intrinsic coordinates to them. The remaining $2\lambda - 2$ independent coordinates are usually called deformations.

The concept of an intrinsic (body-fixed) frame of reference is connected with the descriptions of collective states from the very beginning.[3] The principal axes of the tensor $\boldsymbol{\alpha}_2$ are taken as the intrinsic axes in the quadrupole, $\lambda = 2$, case. According to Eq. (B.1) from Appendix B.2 the well-known definition of the intrinsic frame is $g_{2s} = 0$ for $s = x, y, z$. It is seen from Appendix B that the frame $\mathrm{U_{in}}$ is then $\mathrm{O_h}$-symmetric. The two remaining intrinsic coordinates, $a_{20} = \beta\cos\gamma$ and $a_{22} = \beta\sin\gamma$, are usually parametrized by the well-known deformation parameters β and γ.

The problem of the intrinsic frame for the octupole tensor, $\boldsymbol{\alpha}_3$, has appeared to be less transparent than that for $\lambda = 2$. The tensor has no principal axes and thus no obvious intrinsic frame. Early attempts to define an intrinsic frame failed (cf. Ref. 23). This was, perhaps, the reason why the intrinsic frame of reference was for a long time determined with the quadrupole tensor and octupole coordinates were treated as intrinsic coordinates with respect to that frame (cf. Ref. 19). And what shall we do when any quadrupole tensor is not at our disposal?

In order to define an intrinsic frame for $\lambda = 3$, having the natural symmetry $\mathrm{O_h}$, it is convenient to decompose the tensor representation D^3 of the O(3) group onto the $\mathrm{O_h}$ irreps A_2^-, F_1^- and F_2^-, respectively (see Appendix B.1). The corresponding representations will be called the octupole cubic Wigner functions $A_\mu(\omega)$, $F_{\mu s}(\omega)$, $(s = x, y, z)$ and $G_{\mu s}(\omega)$, $(s = x, y, z)$, respectively. They are the following combinations of the semi-Cartesian Wigner functions:

$$\begin{aligned}
A_\mu(\omega) &= D^{3(-)}_{\mu 2}(\omega), \\
F_{\mu x}(\omega) &= \sqrt{\frac{3}{8}} D^{3(+)}_{\mu 1}(\omega) - \sqrt{\frac{5}{8}} D^{3(+)}_{\mu 3}(\omega), \\
F_{\mu y}(\omega) &= \sqrt{\frac{3}{8}} D^{3(-)}_{\mu 1}(\omega) + \sqrt{\frac{5}{8}} D^{3(-)}_{\mu 3}(\omega), \\
F_{\mu z}(\omega) &= D^{3(+)}_{\mu 0}(\omega), \\
G_{\mu x}(\omega) &= \sqrt{\frac{5}{8}} D^{3(+)}_{\mu 1}(\omega) + \sqrt{\frac{3}{8}} D^{3(+)}_{\mu 3}(\omega), \\
G_{\mu y}(\omega) &= -\sqrt{\frac{5}{8}} D^{3(-)}_{\mu 1}(\omega) + \sqrt{\frac{3}{8}} D^{3(-)}_{\mu 3}(\omega), \\
G_{\mu z}(\omega) &= D^{3(+)}_{\mu 2}(\omega).
\end{aligned} \tag{10}$$

The cubic Wigner functions form a unitary set. When using them the transformation rule of Eq. (7) takes the following form:

$$\alpha_{3\mu}(\omega, b, f) = A_\mu(\omega) b + \sum_{s=x,y,z} [F_{\mu s}(\omega) f_s + G_{\mu s}(\omega) g_s], \tag{11}$$

where b, f_s and g_s are the octupole cubic coordinates in the frame $\mathrm{U_{in}}$, proposed in Ref. 24 and defined in Appendix B.2.

It is seen that we have two possible definitions of the $\mathrm{O_h}$-symmetric frame of reference. We can take Eq. (9) in the two following alternative forms: either $g_s = 0$ or $f_s = 0$ for $s = x, y, z$. Here, we shall explore the former definition. Both of them are briefly discussed in Ref. 25. We see from Eq. (11) that in the former case the following relation between the spherical laboratory coordinates and the Euler angles and the octupole deformations holds:

$$\alpha_{3\mu}(\omega, b, f) = A_\mu(\omega) b + \sum_{s=x,y,z} F_{\mu s}(\omega) f_s. \tag{12}$$

The Jacobian of the transformation $\alpha_\mu \to \omega, b, f_s$ is equal to

$$\begin{aligned}
D_f(\vartheta, b, f_x, f_y, f_z) &= 8 \sin\vartheta \left[b \left(b^2 - \frac{15}{16} \left(f_x^2 + f_y^2 + f_z^2 \right) \right) + \frac{15}{8} \sqrt{\frac{15}{16}} f_x f_y f_z \right] \\
&= 8 \sin\vartheta \, d_f(b, f_s).
\end{aligned} \tag{13}$$

The transformation (12) is reversible for the deformations contained aside from the hyper-surface $D_f(\vartheta, b, f_x, f_y, f_z) = 0$, where ambiguities appear arising from the

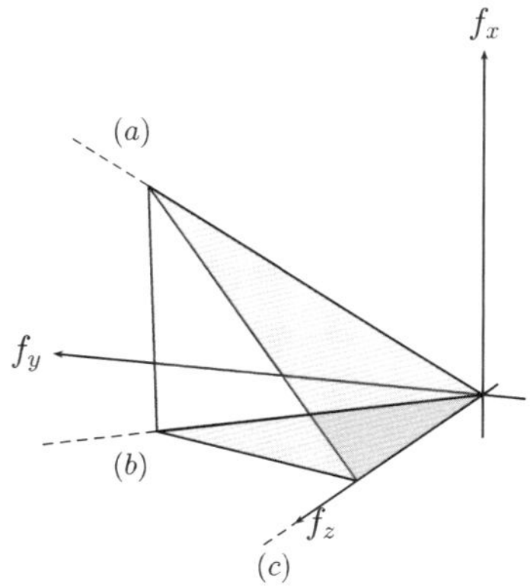

Fig. 1. Infinite triangular pyramid mapping out the three-dimensional region of octupole vector deformations (f_x, f_y, f_z) with the vertex at point (0,0,0). The three pyramid edges are: (a) principal diagonal of octant $f_x, f_y, f_z \geq 0$, (b) diagonal of plane $f_y f_z$, (c) axis f_z.

symmetries of the shape. The F_1^--covariant (vector) deformations (f_x, f_y, f_z) are transformed under the O_h transformations of the intrinsic frame like the Cartesian coordinates x, y, z of a position vector. It follows directly from the O_h symmetry that it is sufficient to consider, for instance, the region $0 \leq f_x \leq f_y \leq f_z$ of vector deformations forming an infinite triangular pyramid in three-dimensional space (see Fig. 1). The forty-eight pyramids obtained by all the O_h transformations fill up the entire space. Deformation b supplements the space of vector deformations to the four-dimensional space. It is invariant under rotations R_1 and R_3 and changes sign under R_2 and inversion P together with the corresponding transformations of the vector deformations. Therefore, there are no additional restrictions on values of b (see Fig. 4 below). The case when $f_x^2 + f_y^2 + f_z^2 = 0$ is an exception: it is sufficient then to consider values $b \geq 0$.

To learn the geometrical meaning of the deformation parameters, b and f_s, one can investigate the shapes of the "octupoloid" given by the following equation in the spherical coordinates R, θ, ϕ:

$$R(\theta, \phi) = R_0 \left[1 - A_0(\phi, \theta, 0)b + \sum_{s=x,y,z} (1 - 2\delta_{sx}) F_{0s}(\phi, \theta, 0) f_s\right] \tag{14}$$

in accordance with the approach of Ref. 26, bearing in mind that $\mathcal{D}^{\lambda}_{0\mu}(\phi, \theta, 0) = (-1)^{\mu}\sqrt{4\pi/(2\lambda+1)}Y_{\lambda\mu}(\theta, \phi)$. The shape of the octupoloid is defined by the set of deformations b, f_x, f_y and f_z up to the cubic group of transformations i.e. the Bohr rotations and mirror reflections (see Appendix B.1). Two octupoloids with identical axial-symmetric (pear) shapes but oriented differently are shown in Fig. 2. Obviously, the two sets of vector deformations then lie inside different pyramids of Fig. 1. Changing the sign of the deformation gives inverted octupoloids with the same shape as that in the figure. Examples of octupoloids with asymmetric shapes are shown in Fig. 3. Their orientations and/or handedness can be changed by cubic

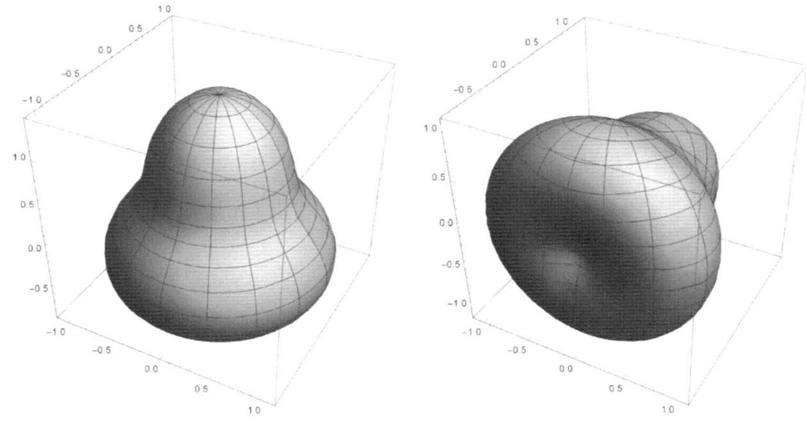

(a) $f_z = 0.5, b = f_x = f_y = 0$ (b) $f_y = 0.5, b = f_x = f_z = 0$

Fig. 2. Two octupoloids with identical axial-symmetric (pear) shapes oriented differently.

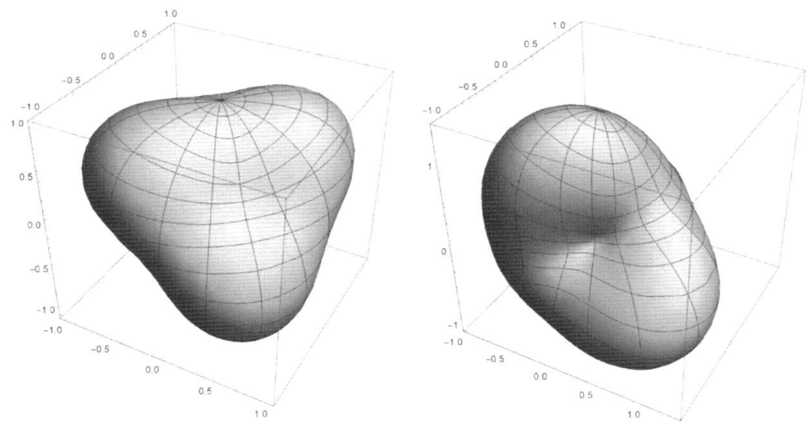

(a) $b = 0.5, f_x = f_y = f_z = 0$ (b) $b = 0.5, f_z = 0.5, f_y = 0.4, f_x = 0.3$

Fig. 3. Examples of octupoloids with asymmetric shapes.

group transormations of the deformations. For instance, changing the signs of all the deformations gives inverted octupoloids. However, when deformations belonging to one out of the two irreducible representations of the cubic group are transformed the shape of the octupoloid is changed, as shown in Fig. 4. Changing the sign of b without changing the vector deformations gives a change of shape.

3.2. *The Hamiltonian in intrinsic coordinates*

A reversible transformation between the laboratory and intrinsic coordinates allows us to use interchangeably one or another set of variables. The use of intrinsic

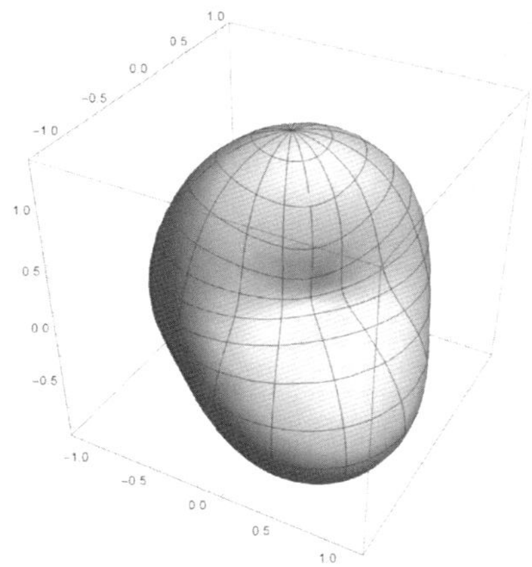

(a) $f_z = 0.5, b = 0.5, f_x = f_y = 0$

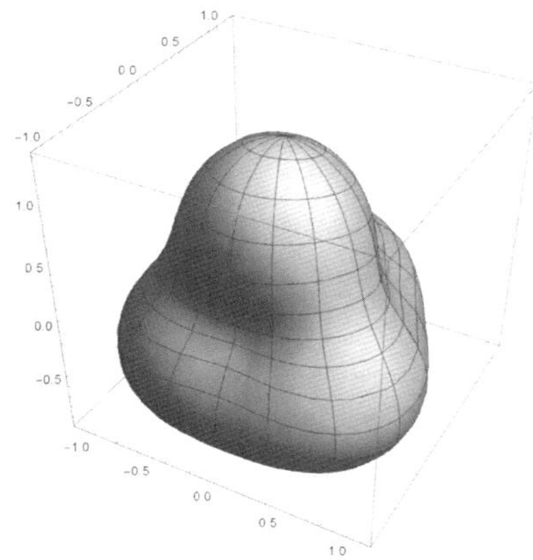

(b) $f_z = 0.5, b = -0.5, f_x = f_y = 0$

Fig. 4. Two octupoloids with opposite signs of deformation b having different shapes.

coordinates is usually more convenient because it gives the possibility to separate variables, especially the Euler angles. It is evident for potentials depending on the laboratory coordinates through the elementary scalars, the number of which coincides with the number of deformations. Hence, the potential in Eq. (2) is a function of the number of deformation parameters characteristic for a given multipolarity: two for $\lambda = 2$ and four for $\lambda = 3$.

To express the kinetic part of the Hamiltonian and angular momenta in intrinsic coordinates we have to convert derivatives with respect to the laboratory coordinates into derivatives with respect to the intrinsic variables. As might be expected (cf. Ref. 27), independently of the multipolarity of the collective space the components of the angular momentum of Eq. (1) can be expressed as

$$L_{1\kappa}^{(\lambda)} = D_{\kappa 0}^{1(+)}(\omega) L_z(\omega) - D_{\kappa 1}^{1(+)}(\omega) L_x(\omega) + D_{\kappa 1}^{1(-)}(\omega) L_y(\omega), \tag{15}$$

where L_x, L_y, L_z are the Cartesian components of the intrinsic angular momentum depending on the Euler angles and their derivatives only and are given by the following standard formulae: (cf. e.g. Eq. (2.15) in Ref. 18).

$$\begin{aligned} L_x(\varphi, \vartheta, \psi) &= -i\left(-\frac{\cos\psi}{\sin\vartheta}\frac{\partial}{\partial\varphi} + \sin\psi\frac{\partial}{\partial\vartheta} + \cot\vartheta\cos\psi\frac{\partial}{\partial\psi}\right), \\ L_y(\varphi, \vartheta, \psi) &= -i\left(\frac{\sin\psi}{\sin\vartheta}\frac{\partial}{\partial\varphi} + \cos\psi\frac{\partial}{\partial\vartheta} - \cot\vartheta\sin\psi\frac{\partial}{\partial\psi}\right), \\ L_z(\varphi, \vartheta, \psi) &= -i\frac{\partial}{\partial\psi}. \end{aligned} \tag{16}$$

For $\lambda = 2$ the procedure for converting the derivatives is well known and is presented in detail, e.g., in Ref. 18. A general quadrupole Hamiltonian expressed in intrinsic variables is divided into the vibrational part depending only on two deformations and the rotational Hamiltonian which contains the angular momenta

L_x, L_y, L_z and the deformation dependent moments of inertia. The intrinsic axes are always the principal axes of the tensor of inertia. When we take the kinetic energy with the Bohr inversed inertial bitensor (5) the structure of the Hamiltonian will not change much. Namely, the mixed term in the vibrational kinetic energy will vanish and the remaining five kinetic-energy terms — vibrational and rotational — contain one common constant mass parameter B_2 instead of different deformation-dependent inertial functions.

The procedure for transforming the octupole collective Hamiltonian to intrinsic coordinates is more involved than that for the case of $\lambda = 2$. The first step of the procedure is conversion of the derivatives with respect to the laboratory coordinates into those with respect to the intrinsic variables. To do this one should calculate the derivatives of the intrinsic with respect to the laboratory coordinates. The transformation the reverse of that of Eq. (12) can be presented in the following entangled form:

$$\begin{aligned} b &= \sum_{\mu} (A_{\mu}(\omega))^* \alpha_{3\mu}, \\ f_s &= \sum_{\mu} (F_{\mu s}(\omega))^* \alpha_{3\mu} \\ 0 &= \sum_{\mu} (G_{\mu s}(\omega))^* \alpha_{3\mu} \end{aligned} \tag{17}$$

for $s = x, y, z$. Derivatives of deformations b and f_s with respect to the laboratory variables $\alpha_{3\mu}$ are equal to

$$\begin{aligned} \frac{\partial b}{\partial \alpha_{3\mu}} &= (A_{\mu}(\omega))^* + \sum_{\nu,\omega} \frac{\partial}{\partial \omega}(A_{\nu}(\omega))^* \alpha_{3\nu} \frac{\partial \omega}{\partial \alpha_{3\mu}}, \\ \frac{\partial f_s}{\partial \alpha_{3\mu}} &= (F_{\mu s}(\omega))^* + \sum_{\nu,\omega} \frac{\partial}{\partial \omega}(F_{\nu s}(\omega))^* \alpha_{3\nu} \frac{\partial \omega}{\partial \alpha_{3\mu}}. \end{aligned} \tag{18}$$

Derivatives of the Euler angles with respect to $\alpha_{3\mu}$ are obtained by solving the following set of linear equations:

$$\begin{aligned} &\sum_{\nu} \left[\left(\frac{\partial}{\partial \varphi}(G_{\nu s}(\omega))^* \right) \frac{\partial \varphi}{\partial \alpha_{3\mu}} + \left(\frac{\partial}{\partial \vartheta}(G_{\nu s}(\omega))^* \right) \frac{\partial \vartheta}{\partial \alpha_{3\mu}} + \left(\frac{\partial}{\partial \psi}(G_{\nu s}(\omega))^* \right) \frac{\partial \psi}{\partial \alpha_{3\mu}} \right] \alpha_{3\nu} \\ &\quad + (G_{\mu s}(\omega))^* = 0 \qquad \text{for} \quad s = x, y, z. \end{aligned} \tag{19}$$

To calculate all the derivatives from Eqs. (18) and(19) one should calculate the derivatives of the octupole cubic Wigner functions with respect to the Euler angles. Using Eqs. (8) and (10) the derivatives of the cubic functions can be expressed by derivatives of the Wigner functions themselves. For example, the handbook Ref. 28 contains formulae for the derivatives in question and other relevant properties of the Wigner functions.[b] Finally, solutions of Eq. (19) for derivatives of the Euler

[b]Note that the Wigner functions from Ref. 28 are the complex conjugate of those used here.

angles with respect to $\alpha_{3\mu}$ are:

$$\begin{aligned}
\frac{\partial\varphi}{\partial\alpha_{3\mu}} &= \frac{1}{2d_f(b,f_s)}\frac{1}{\sin\vartheta}\left[-\cos\psi(\Gamma_{\mu x}(b,f_s,\omega))^* + \sin\psi(\Gamma_y(b,f_s,\omega))^*\right],\\
\frac{\partial\vartheta}{\partial\alpha_{3\mu}} &= \frac{1}{2d_f(b,f_s)}\left[\sin\psi(\Gamma_{\mu x}(b,f_s,\omega))^* + \cos\psi(\Gamma_y(b,f_s,\omega))^*\right],\\
\frac{\partial\psi}{\partial\alpha_{3\mu}} &= \frac{1}{2d_f(b,f_s)}(\Gamma_{\mu z}(b,f_s,\omega))^* - \cos\vartheta\frac{\partial\varphi}{\partial\alpha_{3\mu}},
\end{aligned}\tag{20}$$

where

$$\begin{aligned}
\Gamma_{\mu s}(b,f_x,f_y,f_z,\omega))^* = \frac{15}{16}f_s\left((G_{\mu s}(\omega))^* f_s - (G_{\mu t}(\omega))^* f_t - (G_{\mu u}(\omega))^* f_u\right)\\
- b^2(G_{\mu s}(\omega))^* + \frac{\sqrt{15}}{4}b\left(f_u(G_{\mu t}(\omega))^* + f_t(G_{\mu u}(\omega))^*\right)\omega,
\end{aligned}\tag{21}$$

and s,t,u are circular permutations of x,y,z here and further below. Using Eqs. (16), (18) and (20) we are in a position to express derivatives with respect to the laboratory by derivatives with respect to the intrinsic coordinates in the following two equivalent ways:

$$\begin{aligned}
\frac{\partial}{\partial\alpha_{3\mu}} &= (A_\mu(\omega))^*\frac{\partial}{\partial b} + \sum_{s=x,y,z}\Bigg\{(F_{\mu s}(\omega))^*\frac{\partial}{\partial f_s}\\
&\quad + \frac{i}{2d_f(b,f_x,f_y,f_z)}(\Gamma_{\mu s}(b,f_x.f_y,f_z,\omega))^*\left[L_s(\omega) - J_s^{(f)}(f_t,f_u)\right]\Bigg\}\\
&= \left[\frac{\partial}{\partial b} + \frac{1}{d_f(b,f_x,f_y,f_z)}\frac{\partial d_f(b,f_x,f_y,f_z)}{\partial b}\right](A_\mu(\omega))^*\\
&\quad \sum_{s=x,y,z}\Bigg\{\left[\frac{\partial}{\partial f_s} + \frac{1}{d_f(b,f_x,f_y,f_z)}\frac{\partial d_f(b,f_x,f_y,f_z)}{\partial f_s}\Big)\right](F_{\mu s}(\omega))^*\\
&\quad + \frac{i}{2d_f(b,f_x,f_y,f_z)}\left[L_s(\omega) - J_s^{(f)}(f_t,f_u)\right](\Gamma_{\mu s}(b,f_x,f_y,f_z,\omega))^*\Bigg\},
\end{aligned}\tag{22}$$

where the differential operators

$$J_s^{(f)}(f_t,f_u) = \frac{3}{2}i\left(f_t\frac{\partial}{\partial f_u} - f_u\frac{\partial}{\partial f_t}\right)\tag{23}$$

stand for angular momenta carried by vibrations of the octupole vector deformations f_x, f_y, f_z and will be called the octupole vibrational angular momenta.

Using both versions of the right-hand side of Eq. (22) and taking advantage of the unitarity of the octupole cubic Wigner functions we are able to express the Hamiltonian of Eq. (2) for $\lambda = 3$ in the intrinsic coordinates. To observe the inherent characteristics of the octupole collective motion an octupole Hamiltonian with the simplest inverse inertial bitensor, namely that of Eq. (5), will be presented here.

This Hamiltonian when expressed in the Euler angles and octupole deformations is as follows:

$$
\begin{aligned}
H_3(b, f_x, f_y, f_z, \omega) = &-\frac{1}{2B_3 d_f(b, f_x, f_y, f_z)} \left\{ \frac{\partial}{\partial b} d_f(b, f_x, f_y, f_z) \frac{\partial}{\partial b} \right. \\
&+ \sum_s \frac{\partial}{\partial f_s} d_f(b, f_x, f_y, f_z) \frac{\partial}{\partial f_s} - \sum_{s,s'} (L_s(\omega) - J_s^{(f)}(f_t, f_u)) \\
&\left. \times\ d_f(b, f_x, f_y, f_z) (\hat{I}^{(f)}(b, f_x, f_y, f_z))^{-1}_{ss'} \times (L_{s'}(\omega) - J_{s'}^{(f)}(f_{t'}, f_{u'})) \right\} \\
&+ V(b, f_x, f_y.f_z),
\end{aligned} \tag{24}
$$

where the Cartesian tensor of the moments of inertia is equal to

$$
\hat{I}^{(f)}(b, f_x, f_y, f_z) = \begin{pmatrix} 4b^2 + \frac{15}{4}(f_y^2 + f_z^2) & \frac{15}{4} f_x f_y + 2\sqrt{15} b f_z & \frac{15}{4} f_x f_z + 2\sqrt{15} b f_y \\ \frac{15}{4} f_x f_y + 2\sqrt{15} b f_z & 4b^2 + \frac{15}{4}(f_x^2 + f_z^2) & \frac{15}{4} f_y f_z + 2\sqrt{15} b f_x \\ \frac{15}{4} f_x f_z + 2\sqrt{15} b f_y & \frac{15}{4} f_y f_z + 2\sqrt{15} b f_x & 4b^2 + \frac{15}{4}(f_x^2 + f_y^2) \end{pmatrix}. \tag{25}
$$

It is seen that the intrinsic axes of an octupole system are not the principal axes of the moment of inertia as one would expect. The octupole vibrations, contrary to the quadrupole ones, carry their own angular momentum, which interacts by the Coriolis and centrifugal interactions with the total angular momentum (cf. Ref. 30). This seems to be the most striking feature of the octupole rotations. In conclusion, the kinetic energy part of the Hamiltonian of Eq. (24) consists of ten terms: the four separate vibrations and the six rotational terms. In general, the Hamiltonian of Eq. (2) for $\lambda = 3$, when expressed in the intrinsic variables, can contain additionally six mixed vibrational terms of type $\partial/\partial f_s \ldots \partial/\partial f_{s'}$ $(s \neq s')$ and twelve vibration-rotation terms of type $(\partial/\partial f_s \ldots (L_{s'} - J_{s'}^{(f)}) + \text{h.c.})$.

3.3. *Axially-symmetric deformation*

Many even–even nuclei show a static quadrupole deformation with axial symmetry. In the small-oscillations approximation of the collective Hamiltonian a simple picture of the quadrupole excitations has emerged (consult, e.g. Chapter 6 in Ref. 29). Two separate intrinsic vibrations appear, namely: the β-vibration of deformation $a_{20} \approx \beta$ around the equilibrium deformation $\beta_{\rm eq}$, and the γ-vibration $(a_{22} \approx \beta\gamma)$ strongly coupled to rotations around the symmetry axis. On the other hand, rotations around axes perpendicular to the symmetry axis are weakly coupled to the vibrations and form characteristic rotational bands built on top of every vibrational level.

How is it in the case of a static axially-symmetric octupole deformation? The equilibrium points in the four-dimensional deformation space are then supposed to be $b = f_x = f_y = 0$, $f_z = \pm f_{\rm eq}$. The equilibrium shape is shown in Fig. 2(a). A rough approximation of the small oscillations around the two (because of the mirror symmetry) equilibrium points for the Hamiltonian of Eq. (24) (with the original Bohr kinetic energy) reads as follows:

$$\begin{aligned} H_3(b, f_x, f_y, f_z, \omega) \approx & \sum_{s=x,y} \left[-\frac{1}{2B_3} \frac{\partial^2}{\partial f_s^2} + \frac{1}{2} C_s f_s^2 \right] \\ & - \frac{1}{2B_3} \frac{\partial^2}{\partial f_z^2} + \frac{1}{2} C_z (|f_z| - f_{\text{eq}})^2 \\ & + \frac{1}{2B_3} \left[-\frac{1}{b} \frac{\partial}{\partial b} b \frac{\partial}{\partial b} + \frac{(L_z(\omega) - J_z^{(f)}(f_x, f_y))^2}{4b^2} \right] + \frac{1}{2} C_b b^2 \\ & + \sum_{s=x,y} \frac{2}{15 B_3 f_{\text{eq}}^2} (L_s(\omega) - J_s^{(f)}(f_t, f_u))^2. \end{aligned} \tag{26}$$

From the form of the Hamiltonian given above the following pattern of the small oscillations around the axial-symmetric octupole shape emerges, namely:

- Two harmonic oscillations in coordinates f_x and f_y with stiffnesses C_x and C_y, respectively (x- and y-vibrations),
- The double-oscillator z-vibrations around points $f_z = \pm f_{\text{eq}}$ with stiffness C_z (cf. Ref. 31),
- The b-vibration with stiffness C_b strongly coupled to rotation around the symmetry axis z,
- Rotations around the x- and y-axes perpendicular to the symmetry axis with constant moment of inertia equal to $(15/4) B_3 f_{\text{eq}}^2$.

The rotations around the x and y axes form rotational bands on top of vibrational levels. However, the bands are disturbed by the Coriolis interaction, being a kind of the rotation–vibration interaction. In turn, centrifugal forces affect the four separate vibrations and can mix them with each other.

4. Conclusion

In the previous Sections a formalism for the octupole collective Hamiltonian has been presented and compared to that for the well-known quadrupole one. For a few reasons, like a number of degrees of freedom greater by two, negative parity, additional simplifications in the quadrupole case, the theory of the octupole collective Hamiltonian is essentially more complicated, and therefore less developed than that of the quadrupole collective motion. A substantial feature of the octupole motion, which does not seem to be realized, is that the intrinsic vector x-, y- and z-vibrations carry a non-zero angular momentum. This is obviously not the case for the quadrupole β- and γ-vibrations.

Obviously, a realistic collective model should take into account both modes, the quadrupole and the octupole together.[19] A separate consideration of the $\lambda = 2$ and $\lambda = 3$ cases either serves as a tool for developing a formalism and methods of treatment, or is an approximation. When we take, for instance, the kinetic energies of both modes with the Bohr inverse inertial bitensors of Eq. (5), the total quadrupole–octupole Hamiltonian is the sum of the kinetic energies parametrized by the two mass parameters, B_2 and B_3, respectively, and the potential $V(\boldsymbol{\alpha}_2, \boldsymbol{\alpha}_3)$, which can contain a possible quadrupole–octupole interaction. However, modern collective Hamiltonians are extracted from microscopic theories, which seem to give inverse inertial bitensors $B^{-1}_{\lambda\mu\lambda\mu'}(\boldsymbol{\alpha}_2, \boldsymbol{\alpha}_3)$ dependent on both sets of coordinates for both λ's. Then, for instance, assumption No. (6) from Sec. 2 that the collective Hamiltonians contain isotropic functions of coordinates, is not valid. In consequence, the bitensor $B^{-1}_{2\mu 2\mu'}(\boldsymbol{\alpha}_2, \boldsymbol{\alpha}_3)$ can have more than six independent components. Furthermore, it is natural to allow for the appearance of mixed quadrupole–octupole terms in the total Hamiltonian. These terms would have the following form:

$$H_{23}(\boldsymbol{\alpha}_2, \boldsymbol{\alpha}_3) = -\frac{1}{2W(\boldsymbol{\alpha}_2, \boldsymbol{\alpha}_3)} \left[\sum_{\mu,\nu} \frac{\partial}{\partial \alpha_{2\mu}} W(\boldsymbol{\alpha}_2, \boldsymbol{\alpha}_3) B^{-1}_{2\mu 3\nu}(\boldsymbol{\alpha}_2, \boldsymbol{\alpha}_3) \frac{\partial}{\partial \alpha_{3\nu}} \right.$$
$$\left. + \sum_{\mu,\nu} \frac{\partial}{\partial \alpha_{3\mu}} W(\boldsymbol{\alpha}_2, \boldsymbol{\alpha}_3) B^{-1}_{3\mu 2\nu}(\boldsymbol{\alpha}_2, \boldsymbol{\alpha}_3) \frac{\partial}{\partial \alpha_{2\nu}} \right] + V_{23}(\boldsymbol{\alpha}_2, \boldsymbol{\alpha}_3). \quad (27)$$

Should H_{23} be invariant under space inversion and Hermitian, $B^{-1}_{\lambda\mu\lambda'\mu'}(\boldsymbol{\alpha}_2, \boldsymbol{\alpha}_3)$ is symmetric ($\lambda\mu \rightleftharpoons \lambda'\mu'$) and odd. By analogy to Eq. (3) the mixed bitensor can be presented in the following form:

$$B^{-1}_{\lambda\mu\lambda'\mu'}(\boldsymbol{\alpha}_2, \boldsymbol{\alpha}_3) = \sum_{l=0}^{2} (\lambda\mu\lambda'\mu'|2l+1m) T_{2l+1m}(\boldsymbol{\alpha}_2, \boldsymbol{\alpha}_3) \quad (28)$$

for $\lambda \neq \lambda' = 2, 3$. Tensors T_{2l+1m} should have negative parity. The bitensor of Eq. (28) has 21 components altogether.

When the quadrupole–octupole collective Hamiltonian is considered assumptions Nos. (1)–(6) from Sec. 2 should be extended to the collective space of both tensors, $\boldsymbol{\alpha}_2$ and $\boldsymbol{\alpha}_3$. The practical role of assumption No. (6) is that no material tensors appear for the nuclear medium. Under these extended assumptions the most general form of a quadrupole–octupole Hamiltonian reads as follows:

$$H(\boldsymbol{\alpha}_2, \boldsymbol{\alpha}_3) = -\frac{1}{2W(\boldsymbol{\alpha}_2, \boldsymbol{\alpha}_3)} \left[\sum_{\lambda,\lambda'=2,3} \sum_{\mu,\mu'} \frac{\partial}{\partial \alpha_{\lambda\mu}} W(\boldsymbol{\alpha}_2, \boldsymbol{\alpha}_3) B^{-1}_{\lambda\mu\lambda'\mu'}(\boldsymbol{\alpha}_2, \boldsymbol{\alpha}_3) \frac{\partial}{\partial \alpha_{\lambda'\mu'}} \right]$$
$$+ V(\boldsymbol{\alpha}_2, \boldsymbol{\alpha}_3). \quad (29)$$

It is parametrized by 64 coordinate-dependent inertial functions being components of the three inverse inertial bitensors, scalar weight and potential. The weight can possibly be equal to the square root of the determinant of the 12×12 matrix of components of the inertial bitensors. The potential is a function of the coordinates through nine scalars described as deformations. In order to separate these nine variables from the twelve coordinates, a body-fixed intrinsic frame of reference and intrinsic coordinates have to be introduced. One can do this in different ways. For instance, the principal axes of tensor $\boldsymbol{\alpha}_2$ oriented by the three Euler angles with respect to the laboratory axes can be treated as the intrinsic axes (cf. Ref. 19). Then the two remaining intrinsic components of $\boldsymbol{\alpha}_2$ and all the seven intrinsic components of $\boldsymbol{\alpha}_3$ can be considered as deformations. Another way is to exchange the roles of tensors $\boldsymbol{\alpha}_2$ and $\boldsymbol{\alpha}_3$ and consider one of the frames defined in Sec. 3.1 through the octupole tensor as the intrinsic frame. One can also define two intrinsic frames for tensors $\boldsymbol{\alpha}_2$ and $\boldsymbol{\alpha}_3$ separately and treat the rotation of one frame with respect to the other as an intrinsic motion. Then the relative Euler angles have the status of deformations. Finally, in the case of a weak and well separated interaction between the modes one can treat them separately and then diagonalize the interaction within the product basis.

Only recently an attempt to solve a quadrupole–octupole model, similar to that presented here, however not based on Hamiltonian (29) and with a restricted number of degrees of freedom, has been undertaken.[33] The model has been applied to the positive- and negative-parity collective levels of the ^{156}Gd nucleus. In any case, the problem of the full quadrupole–octupole collective Hamiltonian is apparently complicated enough and still awaits practical applications to the spectroscopy of nuclear collective excitations.

Acknowledgments

The subject of the present paper was initiated some time ago by the first author (S.G.R.), then a Fellow of the Alexander von Humboldt Foundation, in collaboration with Walter Greiner, the then Director of the *Institut für Theoretische Physik der Johann Wolfgang Goethe-Universität.*

The second author (L.P.) gratefully acknowledges support from the Polish National Science Center (NCN), Grant No. 2013/10/M/ST2/00427.

Appendix A. Isotropic Tensor Fields in the Octupole Collective Space

A.1. The positive spin-parity elementary tensors

The twelve positive spin-parity elementary tensors of Eq. (6) in the octupole collective space are as follows:

$$\begin{aligned}
l = 0 \quad & \boldsymbol{t}_0^{(2)},\ \boldsymbol{t}_0^{(4)},\ \boldsymbol{t}_0^{(6)},\ \boldsymbol{t}_0^{(10)}, \\
l = 1 \quad & \boldsymbol{t}_1^{(3)},\ \boldsymbol{t}_1^{(5)},\ \boldsymbol{t}_1^{(7)}, \\
l = 2 \quad & \boldsymbol{t}_2^{(2)},\ \boldsymbol{t}_2^{(4)}, \\
l = 3 \quad & \boldsymbol{t}_3^{(1)},\ \boldsymbol{t}_3^{(3)}, \\
l = 4 \quad & \boldsymbol{t}_4^{(2)}.
\end{aligned} \tag{A.1}$$

A.2. Independent fundamental even tensors

Sets of fundamental even tensors with even ranks from 0 to 6 are listed in Table A.1. The choice of independent tensors need not be unique. This is because all fundamental tensors of a given rank (all possible alignments of the elementary tensors) are related to each other through a number of syzygies which can eliminate this or that tensor.

Table A.1. Fundamental even tensors with ranks $\Lambda = 0,\ 2,\ 4,\ 6$.

Λ	$k_{3\Lambda}$	$\boldsymbol{\tau}_{k\Lambda}(\boldsymbol{\alpha}_3)$ for $k = 1, \ldots, k_{3\Lambda}$
0	1	1
2	5	$\boldsymbol{t}_2^{(2)},\ \boldsymbol{t}_2^{(4)},\ [\boldsymbol{t}_1^{(3)} \times \boldsymbol{t}_1^{(3)}]_2,\ [\boldsymbol{t}_1^{(3)} \times \boldsymbol{t}_1^{(5)}]_2,\ [\boldsymbol{t}_1^{(3)} \times \boldsymbol{t}_1^{(7)}]_2$
4	9	$\boldsymbol{t}_4^{(2)},\ [\boldsymbol{t}_3^{(1)} \times \boldsymbol{t}_1^{(3)}]_4,\ [\boldsymbol{t}_2^{(2)} \times \boldsymbol{t}_2^{(2)}]_4,\ [\boldsymbol{t}_2^{(2)} \times \boldsymbol{t}_2^{(4)}]_4,\ [\boldsymbol{t}_3^{(1)} \times \boldsymbol{t}_1^{(5)}]_4,$
		$[\boldsymbol{t}_2^{(2)} \times \boldsymbol{t}_1^{(3)} \times \boldsymbol{t}_1^{(3)}]_4,\ [\boldsymbol{t}_3^{(1)} \times \boldsymbol{t}_1^{(7)}]_4,\ [\boldsymbol{t}_2^{(2)} \times \boldsymbol{t}_1^{(3)} \times \boldsymbol{t}_1^{(5)}]_4,\ [\boldsymbol{t}_2^{(2)} \times \boldsymbol{t}_3^{(1)} \times \boldsymbol{t}_1^{(7)}]_4$
6	13	$[\boldsymbol{t}_3^{(1)} \times \boldsymbol{t}_3^{(1)}]_6,\ [\boldsymbol{t}_2^{(2)} \times \boldsymbol{t}_4^{(2)}]_6,\ [\boldsymbol{t}_3^{(3)} \times \boldsymbol{t}_3^{(1)}]_6,\ [\boldsymbol{t}_2^{(4)} \times \boldsymbol{t}_4^{(2)}]_6,\ [\boldsymbol{t}_1^{(3)} \times \boldsymbol{t}_2^{(2)} \times \boldsymbol{t}_3^{(1)}]_6$
		$[\boldsymbol{t}_1^{(3)} \times \boldsymbol{t}_1^{(3)} \times \boldsymbol{t}_4^{(2)}]_6,\ [\boldsymbol{t}_1^{(5)} \times \boldsymbol{t}_2^{(2)} \times \boldsymbol{t}_3^{(1)}]_6,\ [\boldsymbol{t}_2^{(4)} \times \boldsymbol{t}_2^{(2)} \times \boldsymbol{t}_2^{(2)}]_6,\ [\boldsymbol{t}_1^{(3)} \times \boldsymbol{t}_1^{(5)} \times \boldsymbol{t}_4^{(2)}]_6$
		$[\boldsymbol{t}_1^{(7)} \times \boldsymbol{t}_2^{(2)} \times \boldsymbol{t}_3^{(1)}]_6,\ [\boldsymbol{t}_1^{(3)} \times \boldsymbol{t}_1^{(3)} \times \boldsymbol{t}_1^{(3)} \times \boldsymbol{t}_3^{(1)}]_6,\ [\boldsymbol{t}_2^{(4)} \times \boldsymbol{t}_1^{(7)} \times \boldsymbol{t}_3^{(1)}]_6,\ [\boldsymbol{t}_1^{(3)} \times \boldsymbol{t}_1^{(7)} \times \boldsymbol{t}_4^{(2)}]_6$

Appendix B. Symmetries of the Coordinate Frame

B.1. Cubic holohedral group

The cubic holohedral $\mathrm{O_h}$ group is a natural symmetry group of the three-dimensional coordinate system because the forty-eight group elements are: the eight reverses of the axis arrows for each out of six permutations of axes. The three Bohr rotations, R_1, R_2, R_3, and the inversion P can serve as generators of this group (see Ref. 18 and Sect. 4.4 in Ref. 29). In general, the $\mathrm{O_h}$ group has ten irreducible representations,[32] namely

- four one-dimensional, denoted as $\mathrm{A}_1^{\pm}$, $\mathrm{A}_2^{\pm}$,
- two two-dimensional, denoted as $\mathrm{E}^{\pm}$,
- four three-dimensional, denoted as $\mathrm{F}_1^{\pm}$, $\mathrm{F}_1^{\pm}$.

The tensor representations D^{λ} of the O(3) orthogonal group can be decomposed into the following irreducible representations of $\mathrm{O_h}$, namely

- irreps E^{+}, F_2^{+} for $\lambda = 2$,
- irreps A_2^{-}, F_1^{-}, F_2^{-} for $\lambda = 3$.

B.2. Cubic coordinates

The decomposition of the real and imaginary parts, $a_{\lambda\mu}$, $b_{\lambda\mu}$ of the spherical components of tensors $\boldsymbol{\alpha}_\lambda$ into cubic coordinates is:

- $\lambda = 2$

$$\mathrm{E}^{+}\begin{cases} e_{20} = a_{20} \\ e_{22} = a_{22}, \end{cases} \qquad \mathrm{F}_2^{+}\begin{cases} g_{2x} = -b_{21}, \\ g_{2y} = -a_{21}, \\ g_{2z} = b_{22}, \end{cases} \tag{B.1}$$

- $\lambda = 3$

$$\begin{aligned} &\mathrm{A}_2^{-} : \; b_3 \equiv b = b_{32}, \\ &\mathrm{F}_1^{-}\begin{cases} f_{3x} \equiv f_x = \sqrt{\tfrac{3}{8}}a_{31} - \sqrt{\tfrac{5}{8}}a_{33}, \\ f_{3y} \equiv f_y = \sqrt{\tfrac{3}{8}}b_{31} + \sqrt{\tfrac{5}{8}}b_{33}, \\ f_{3z} \equiv f_z = a_{30}, \end{cases} \\ &\mathrm{F}_2^{-}\begin{cases} g_{3x} \equiv g_x = \sqrt{\tfrac{5}{8}}a_{31} + \sqrt{\tfrac{3}{8}}a_{33}, \\ g_{3y} \equiv g_y = -\sqrt{\tfrac{5}{8}}b_{31} + \sqrt{\tfrac{3}{8}}b_{33}, \\ g_{3z} \equiv g_z = a_{32}. \end{cases} \end{aligned} \tag{B.2}$$

Curly brackets match the cubic coordinates belonging to given irreducible representations of $\mathrm{O_h}$.

References

1. Lord Rayleigh, *Proc. R. Soc.* **29**, 71 (1879).
2. S. Flügge, *Ann. Physik, Lpz.* **39**, 373 (1941).
3. A. Bohr, *K. Danske Vidensk. Selsk., Mat.-Fys. Medd.* **26**, No. 14 (1952).
4. S. T. Belyaev, *Nucl. Phys.* **64**, 17 (1965).
5. K. Kumar and M. Baranger, *Nucl. Phys. A* **92**, 608 (1967).
6. P. Ring and P. Schuck, *The Nuclear Many-Body Problem* (Springer, New York, 1980).
7. B. Podolsky, *Phys. Rev.* **32**, 812 (1928).
8. W. Pauli, *Handbuch der Physik*, Vol. 24 (Springer, Berlin, 1933), p. 120.
9. H. Hofmann, *Z. Phys.* **250**, 14 (1972).
10. M. Bender and P.-H. Heenen, *Phys. Rev. C* **78**, 024309 (2008).
11. T. R. Rodriguez and J. L. Egido, *Phys. Rev. C* **81**, 064323 (2010).
12. S. G. Rohoziński, *J. Phys. G: Nucl. Part. Phys.* **39**, 095104 (2012); *Phys. Scr.* **2013**, No. T154, 014016 (2013).
13. F. Asaro, F. S. Stephens, Jr. and I. Perlman, *Phys. Rev.* **92**, 1495 (1953).
14. S. G. Rohoziński, *Rep. Progr. Phys.* **51**, 541 (1988).
15. P. A. Butler and W. Nazarewicz, *Rev. Mod. Phys.* **68**, 349 (1996).
16. L. P. Gaffney *et al.*, *Nature* **497**, 199 (2013).
17. B. Bucher *et al.*, *Phys. Rev. Lett.* **116**, 112503 (2016).
18. L. Próchniak and S. G. Rohoziński, *J. Phys. G: Nucl. Part. Phys.* **36**,123101 (2009).
19. S. G. Rohoziński, M. Gajda, W. Greiner, *J. Phys. G: Nucl. Phys.* **8**, 787 (1982).
20. S. Nishiyama and J. da Providência, *Nucl. Phys. A* **923**, 51 (2014).
21. S. G. Rohoziński, *J. Phys. G: Nucl. Phys.* **4**, 1075 (1978).
22. A. Bohr and B. R. Mottelson, *Nuclear Structure* (Benjamin, New York, Amsterdam, 1969), Vol. 1, App. 1A.
23. J. P. Davidson, *Rev. Mod. Phys* **37**, 105 (1965).
24. S. G. Rohoziński, *J. Phys. G: Nucl. Part. Phys.* **16**, L173 (1990); *Phys. Rev. C* **56**, 165 (1997).

25. S. G. Rohoziński and L. Próchniak, *Acta Phys. Polon. B, Proc. Suppl.* **10**, 191 (2017).
26. I. Hamamoto, Xi-zhen Zhang and Hong-xing Xie, *Phy. Lett. B* **257**, 1 (1991).
27. A. R. Edmonds, *Angular Momentum in Quantum Mechanics* (Princeton Press, New Jersey, 1957).
28. D. A. Varshalovich, A. N. Moskalev, V. K. Khersonsky, *Quantum Theory of Angular Momentum: Irreducible Tensors, Spherical Harmonics, Vector Coupling Coefficients, 3nj Symbols* (World Scientific, Singapore, 1988).
29. J. M. Eisenberg and W. Greiner, *Nuclear Models*, Third edn. (North-Holland, Amsterdam, 1987).
30. A. Bohr and B. R. Mottelson, *Nuclear Structure* (Benjamin, Reading, 1975), Vol. 2.
31. E. Merzbacher, *Quantum Mechanics* (John Wiley, New York, 1961).
32. M. Hamermesh, *Group Theory and its Application to Physical Problems* (Addison-Wesley, Reading, 1964), Chapt. 9, Sect. 4.
33. A. Dobrowolski, K. Mazurek and A. Góźdź, *Phys. Rev. C* **94**, 054322 (2016).

Modeling Hybrid Stars

S. Schramm

Frankfurt Institute for Advanced Studies,
Ruth-Moufang-Str. 1, D-60438 Frankfurt am Main, Germany
schramm@fias.uni-frankfurt.de

The core of neutron stars might reach densities up to ten times nuclear matter saturation density, depending on the equation of state of the stellar matter. At such conditions it is natural to assume that deconfined quarks appear in the stellar core, leading to a so-called hybrid star. This chapter contains a discussion of general properties and problems concerning hybrid stars, which includes an analysis of the properties of quark matter necessary for generating heavy stars. In addition, a particular approach of modeling hadronic and quark matter within one model is presented, which allows for a first-order as well as continuous phase transition from a hadronic to a quark dominated phase.

1. Introduction

The understanding of strong interaction physics in the non-perturbative regime is still quite limited. This is due to the lack of a clear expansion scheme in some small quantity as is the case for small-coupling perturbative approaches. In addition, introducing extreme conditions like high temperatures and high densities complicates the situation even further as long as one does not reach the asymptotic regime where perturbative methods are again applicable. Such systems are theoretically and experimentally challenging, which is one of the reasons why there is so much interest in this field. Here, a key feature is the transition from hadronic to quark degrees of freedom for large values of densities/chemical potentials and temperatures.

The investigation of such a transition, the order of the phase transition and possible observable consequences is a central topic for many physics activities. Prominent among these is a whole range of ultrarelativistic heavy-ion collision experiments at various beam energies like at RHIC in Brookhaven, the LHC heavy-ion program at CERN, as well as the upcoming facilities FAIR at GSI, Darmstadt, as well as NICA in Dubna. The idea is that the fireball created in the collision generates a system of deconfined quarks and gluons that will pass through the phase transition back to the hadronic world during the expansion of the hot system. The actual transition point in the plane of temperature and chemical potential depends on the beam energy, where the highest energies probe the system close to vanishing chemical potential.

Another very intriguing approach to study effects of the quark–hadron phase transition region is the investigation of neutron stars. Depending on the assumed equation of state these stars might reach central densities of about 4 to 10 times nuclear matter saturation density. Such a system would correspond to a closely packed ensemble of baryons, suggesting a high probability that the baryons actually dissolve into their quark constituents. This transition might already happen at distinctly lower densities, as long as quarks do not appear in ground state nuclear matter, which would generate problems with general nuclear structure phenomenology.

2. Consequences of a Quark Core

There are many calculations of hadronic compact stars, including nucleons only or, in extended studies, hyperons and other excited states like Δ baryons. Depending on the stiffness of the underlying equation of state these models deliver star masses that cover the heavy observed neutron stars with masses around 2 solar masses.[1,2] For a survey of a range of model results see Ref. 3. However, it has been observed that additional hadronic degrees of freedom (or meson condensates) tend to soften the equation of state considerably, lowering the maximum mass to values below the observed ones. In particular with respect to stars with hyperons, so-called hyper stars, various possibilities to remedy this problem have been proposed in recent years, including strong repulsive forces between hyperons mediated by the vector meson with hidden strangeness, the ϕ meson,[4] and possible non-linear hyperon–nucleon interactions that prevent hyperon states from being populated in stellar matter.[5] Another possibility, suppressing exotic hadronic components inside of the star, is the transition of hadronic to quark degrees of freedom at densities before the softening of the equation of state (EOS) occurs, which appears to be a natural idea given the high density of the system.

A simple, and still commonly used, method to incorporate a quark phase in the description of a compact star is to combine an EOS of a hadronic model with a corresponding EOS for quarks, switching from one to the other at the crossing point of the two pressures as function of baryon chemical potential. This procedure corresponds to a Maxwell construction of a first-order phase transition from hadronic to quark phase. The result of such an approach is seen in Fig. 1. In this case a standard hadronic model (G300[6]) and a basic MIT bag model for the quark phase was used. The figure illustrates the impact of changing the MIT bag pressure B on the equation of state. A large value of B leads to a correspondingly large value of latent energy in the transition from the hadronic to the quark phase. Such a big jump has drastic consequences for neutron star properties. This can clearly be observed when calculating masses and radii of the family of neutron (or hybrid) stars by solving the Tolman–Oppenheimer–Volkov (TOV) equations for static spherical stars[7,8] varying the central stellar pressure. The results are shown in Fig. 2. The

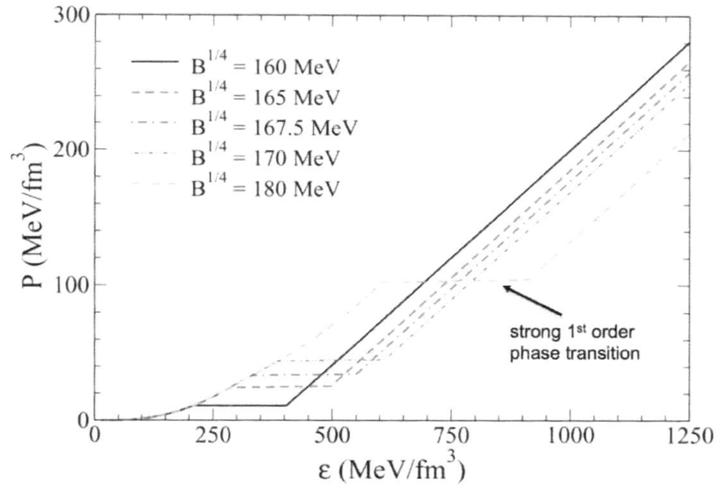

Fig. 1. Combined equation of state or a hadronic and quark phase showing the pressure as function of energy density. Large values of the bag pressure lead to a strong first-order phase transition.

kinks in the curves signal the appearance of a quark matter core. Increasing the value of the bag pressure leads to a reduced maximum star mass, in this case far below the observed 2 solar mass value. For the highest B value considered the curve shows a continuous drop of the stellar mass beyond the kink at lower radii, indicating unstable solutions. Thus, a strong phase transition to a quark phase leads to unstable hybrid star solutions and all stable stars are purely hadronic.

The conclusion from this rather general investigation is that assuming a standard quark phase in hybrid stars leads to reduced stellar masses in potential contradiction with observation. However, this study is based on a simple picture of the quark phase with non-interacting quarks. Taking into account repulsive forces between quarks leads to a stiffening of the equation of state. A simple approximation of such an effect is being achieved by using a quark equation of state of the form

$$P = \frac{3}{4\pi^2}(1-\alpha)\mu_q^4 - \frac{3}{4\pi^2}m_s^2\mu_q^2 - B \, , \tag{1}$$

where P denotes the pressure and μ_q is the quark chemical potential. m_s is the strange quark mass (up and down quarks are assumed to be massless) and α represents the perturbative correction from quark–quark interactions.[9] The α contribution reduces the chemical potential to an effective one, similar to a vector particle exchange shifting the corresponding chemical potential μ (for a more microscopic description of such an effect, see Ref. 10). The overall impact of such modification can be observed in the mass-radius diagrams depicted in Fig. 3. Depending on the values of α and B the star solutions after the appearance of quarks, i.e. to the left of the kink, do not exhibit the sharp drop in masses as before, but in this case reach values close to two-solar masses for one parameter set.

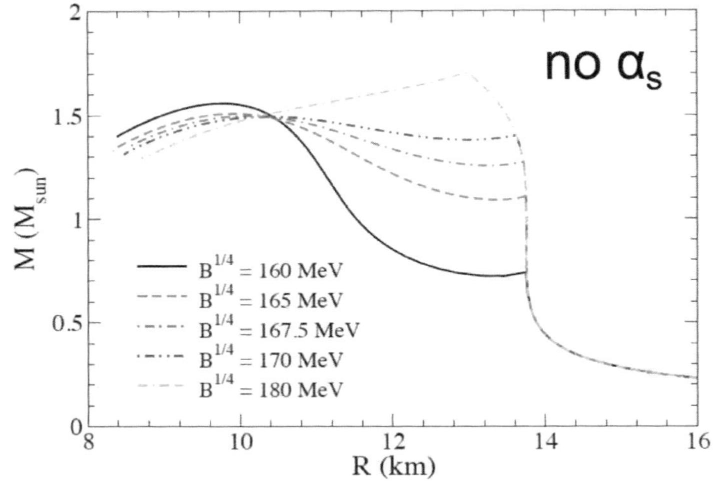

Fig. 2. Hybrid star mass–radius relation. Results for different bag pressures (see Fig. 1) of the quark phase are compared. Large values of B yield unstable hybrid star solutions.

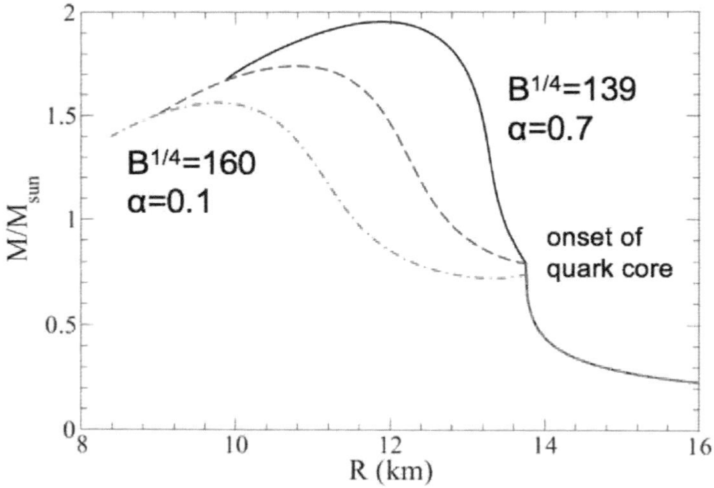

Fig. 3. Mass–radius relation for hybrid stars. Bag pressures as well as quark-quark interaction are varied leading to potentially large maximum star masses.

Such a behavior has been discussed some time ago in Ref. 11, where it was pointed out that by essentially modifying the quark equation of state to have a similar functional form as the hadronic one, the results for hybrid stars resemble the purely hadronic case, as only the equation of state enters the TOV equations without any impact of the type of constituents populating the star on the stellar masses or radii.

3. Unifying Quark and Hadron Descriptions

A problem of the previously discussed approaches of including quark degrees of freedom, results from the fact that in those cases hadronic and quark part are treated completely separately, only to be joined by matching the respective equations of state. One direct shortcoming of such a procedure derives from the fact that connecting the EOS inevitably leads to a jump in density and energy density, i.e. to a first-order phase transition. While at zero temperature the order of the hadron-quark transition is unknown, lattice QCD simulations have clearly established that at least at low chemical potential the system exhibits a smooth crossover during the transition. Therefore, a reasonable description of hadronic and quark matter has to be comprehensive, if the possibility of a crossover is to be included. One approach pursued at Frankfurt for several years is based on an effective chiral description of hadronic matter, termed chiral-mean-field (CMF) model. This model has been discussed in various publications.[4,12,13] The main ingredient is a relativistic mean field description of flavor-SU(3) multiplets for baryons and mesons with a linear coupling between the various baryonic fields and mesons like

$$L_{\mathrm{Int}} = - \sum_i \bar{\psi}_i [\gamma_0 (g_{i\omega}\omega + g_{i\phi}\phi + g_{i\rho}\tau_3\rho) + m_i^*]\psi_i, \qquad (2)$$

with effective baryon masses m_i^*:

$$m_i^* = g_{i\sigma}\sigma + g_{i\zeta}\zeta + g_{i\delta}\delta + \delta m_i \quad . \qquad (3)$$

The sum extends over the baryons from the SU(3) octet: protons, neutrons,$\Sigma^{0\pm}$, and Ξ^{0-}. The relevant meson fields include the scalar isoscalar fields σ, ζ (corresponding to a field with hidden strangeness), as well as the isovector δ, and the corresponding vector mesons ω, ϕ, and ρ.

As in σ models the self-interaction of the scalar fields leads to non-vanishing vacuum expectation values of the scalar mesons, here in the SU(3) approach σ_0 and ζ_0, where the former corresponds to the scalar non-strange quark condensate, whereas the latter can be viewed as the hadronic analogue of a strange quark condensate. Through the coupling term Eq. (2) baryonic masses are generated dynamically. In the dense medium the changing scalar condensates affect a reduction of the baryon masses and thus produce scalar attraction, whereas the baryon vector densities generate nonzero values for the 0-component of the vector fields producing repulsive interactions. Thus, the interactions are generated by the changing scalar and vector condensates represented by the corresponding mesonic mean fields. In this approach quarks are treated in the same way as additional degrees of freedom, also coupling to the mean fields with their own coupling constants but using again the same type of interaction Lagrangian, Eq. (2). While such an approach combines hadrons and quarks in a unified Lagrangian, it does not automatically ensure the correct degrees of freedom of a purely hadronic phase at low temperature T and chemical potential μ and a quarkic system at high T and/or μ. In order to suppress quarks at low densities and temperatures, phenomenologically approximating a deconfinement transition, in this approach quarks couple to an effective Polyakov-loop field Φ analogously to the so-called PNJL models for quarks.[14,15] This leads to a quark contribution to the grand canonical potential given by

$$\Omega_q = -T\frac{6}{(2\pi)^3}\sum_q \int d^3k \ln\left[1 + \Phi \exp\frac{E_i^* - \mu_i^*}{T}\right], \tag{4}$$

where the effective quark energies E_i^* and chemical potentials μ_i^* are modified due to the interactions of the quark with the mean fields in the same general way as for the case of the baryons in Eqs. (2) and (3). The dynamics of the Φ field is governed by a temperature-dependent potential using the standard PNJL parameterization in Ref. 16. The potential leads to a vanishing Polyakov loop field in the low-temperature phase, effectively suppressing quark degrees of freedom. In order to ensure that in the high temperature/density phase hadrons disappear from the system, an excluded volume v for hadrons is included in a thermodynamically consistent manner such that hadrons are suppressed in dense systems.[17–19] This has the advantage to other coupled quark–hadron approaches that at very large temperatures or densities the hadrons do not reappear, an usually artificial model side effect that is termed "reconfinement". With that procedure one is able to get a smooth (or first-order, depending on parameters) transition from hadrons to quarks. Hadronic models with excluded volume effects have the chronic problem that at high densities or temperatures the equation of state becomes non-causal, i.e. the speed of sound exceeds the speed of light. This happens because excluding more and more volume makes the equation of state extremely stiff. However, in the approach discussed here this problem does not occur as the system transits to a state of quarks, and gluons (represented by the Polyakov-loop field), that do not have an excluded volume term. Another practical approach of removing the "wrong" particles can be achieved by coupling the Polyakov loop to the masses of quarks and hadrons as it was done in Ref. 20.

The resulting stars for such an approach are depicted in Fig. 4. Here, a similar effect can be observed as has been discussed in the previous section. The large latent energy in the phase transition to quark matter leads to unstable stars with quark cores. This is somewhat smoothed out if one considers a Gibbs transition with a mixed phase as shown in the inset of the figure. In that case a star with a 2 km core of a mixed state of baryons and quarks is still stable.

One interesting point, that can be observed in this figure, is more clearly presented in Fig. 5. The shown curve represents the dependence of the star mass on the central baryon density. The results are obtained for the purely hadronic model without quark phase. The maximum mass is reached at about six times nuclear matter saturation density ρ_0. As one can observe, the curve is very flat around this value. That is, assuming that a transition to a soft equation of state consisting of exotic matter occurs at somewhat lower density, e.g. $4 \cdot \rho_0$, and due to the softness all stars with densities beyond this value become unstable (as is the case shown

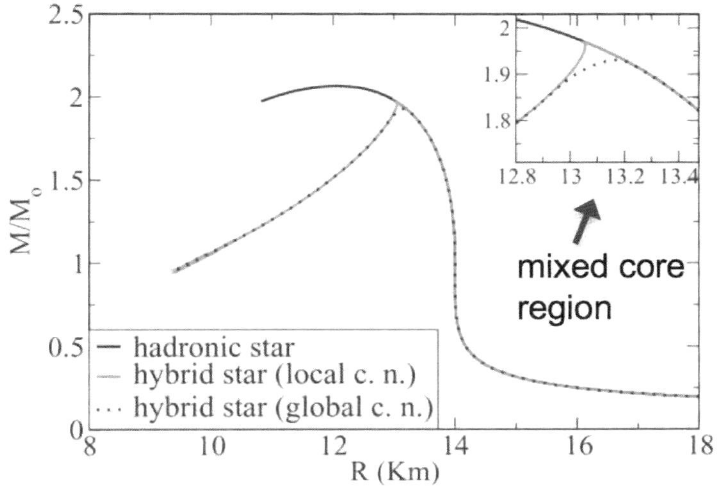

Fig. 4. Mass-radius relation for the CMF hadronic model including quarks as in Ref. 20. The onset of the quark phase cuts off the family of stable stars. The inset shows the impact of a Gibbs construction of the mixed phase on the stellar stability.

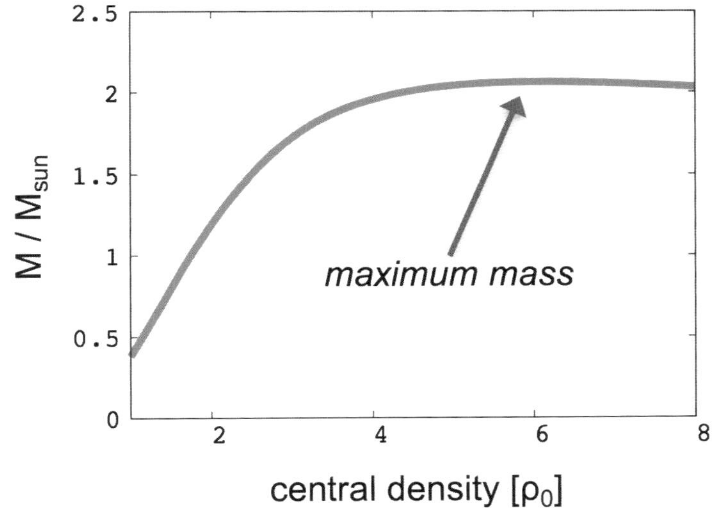

Fig. 5. Hadronic CMF star mass as function of central density in units of nuclear saturation density. The flatness of the curve around the maximum mass indicates the insensitivity of the maximum mass value on the onset of unstable solutions due to the appearance of exotic particles like deconfined quarks.

in Fig. 4), the maximum mass does not decrease very much. Only if the central density decreases by a factor of 2 or so, a substantial reduction of attainable star masses occurs.

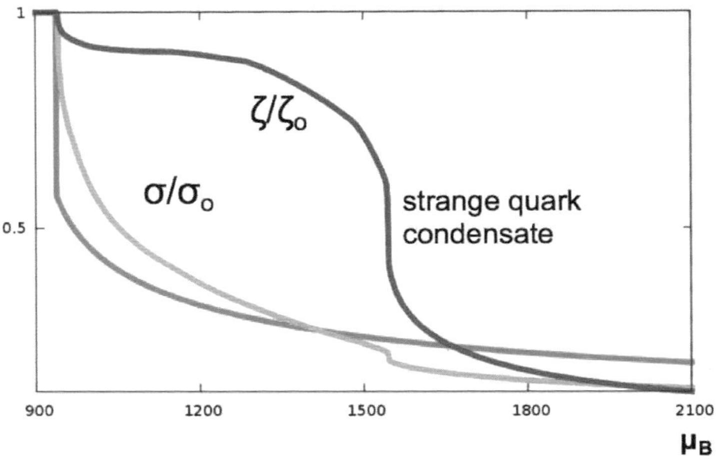

Fig. 6. Scalar fields corresponding to non-strange (σ) and strange (ζ) quark condensate as function of baryon chemical potential for stellar matter. A first-order phase transition is reflected by a sharp drop of the ζ field.

Whereas the previous calculation did not include a repulsive quark-quark vector interaction, a related study with a different parameter set including such an interaction can lead to interesting behaviour of the stellar matter, which is shown in Fig. 6.[21] Here one observes a smooth transition to a mixed phase with quarks (not shown in the figure) and then a first order transition, where the strange quark condensate, here represented by the field ζ, drops, leading to a strong increase of strangeness in the system.[21]

The corresponding particle densities as function of the total baryonic density can be seen in Fig. 7. The bar symbolizes the jump in density at the phase transition. Using the equation of state of such matter, the resulting star masses and radii are depicted in Fig. 8. Depending on the value of the quark interaction a second maximum mass at small radius can be observed. Such a configuration is usually termed twin star. Interestingly, there are two stable stars with a mass of about 1.6 solar masses and both stars are hybrid stars, i.e. they consist of baryons and quarks. The twin solution is characterized by a much larger strangeness content.

4. Conclusion and Outlook

In the previous section some basic concepts of modeling hybrid stars have been discussed. For a simple gas of non-interacting quarks, for instance in the form of the MIT bag model, where essentially all interaction effects are subsumed in the bag pressure term, the hybrid star solutions exhibit significantly reduced star masses due to the relative softness of the quark equation of state. Adding a repulsive quark interaction stiffens the equation of state, which leads to larger masses in agreement with observed values. A weakness of the approach, connecting hadronic and quark

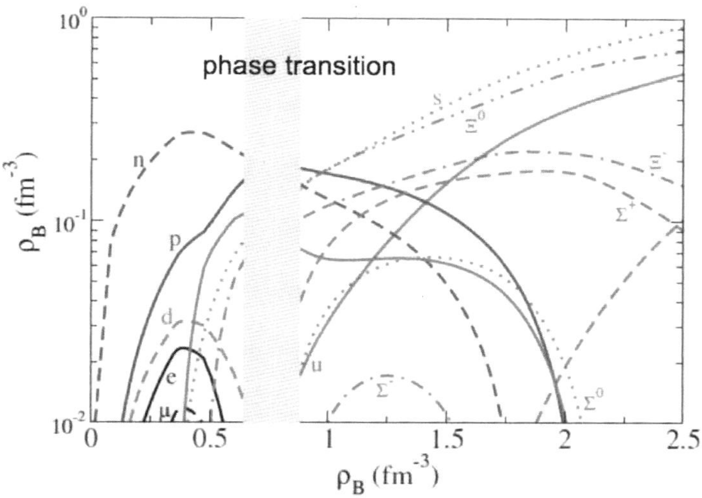

Fig. 7. Densities of particle species as function of total baryon density including hadronic and quark degrees of freedom.

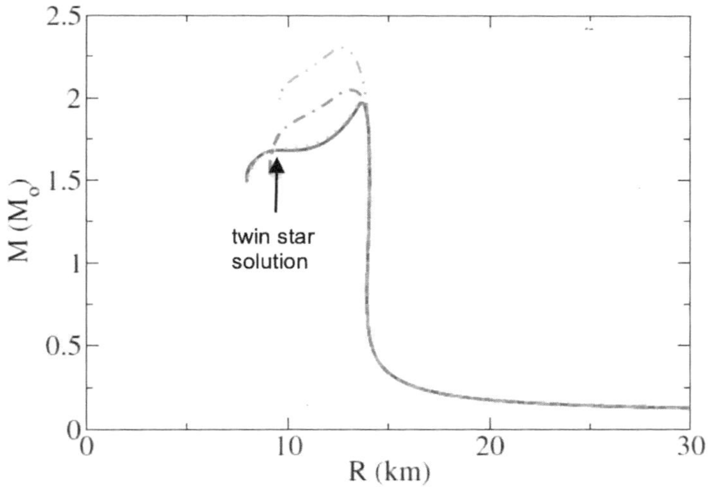

Fig. 8. Mass-radius diagram exhibiting twin-star solutions at small radii for specific strange-quark interaction strengths.

equations of state, is that these models are difficult to extend to a wider range of densities and temperatures. In particular, such models feature first-order phase transitions for all conditions, in clear disagreement with lattice QCD results at low chemical potential. As a remedy of this short-coming a unified description of hadrons and quarks within one model Lagrangian was discussed that allows for phase transitions of any kind. Depending on parameters one result shows twinned hybrid stars with equal mass but distinctly different radii, where the difference is triggered by the onset of a large increase of strangeness-carrying particles in the star.

One problem with all these approaches is that the high hybrid star masses rely on the introduction of strong repulsive quark-quark forces. However, as it has been shown in Refs. 22 and 23, such interactions lead to strong disagreement with lattice data at small chemical potentials. However, recently large hybrid star masses without the repulsive terms have been calculated. In this case, the star consists of a large mixed phase of baryons and quarks, where the repulsive forces of the baryons ensure the stability of the high-mass solutions.[24,25] More studies along this line has to be done in future work.

Such types of projects, pushing the conditions of matter to the extreme and trying to uncover drastically new effects, are just what Walter Greiner always enjoyed pursuing throughout his career, covering a wide variety of physics topics. His intuition and drive had a big impact on all of us who worked with him, and we will dearly miss this spirit of constantly seeking new challenges.

Acknowledgments

The author acknowledges support from the LOEWE center HIC for FAIR, as well as access to the computing facilities of the Center for Scientific Computing (CSC) at the Goethe University in Frankfurt am Main.

References

1. P. Demorest, T. Pennucci, S. Ransom, M. Roberts and J. Hessels, *Nature* **467**, 1081 (2010), doi:10.1038/nature09466.
2. J. Antoniadis *et al.*, *Science* **340**, 6131 (2013), doi:10.1126/science.1233232.
3. J. M. Lattimer, *Ann. Rev. Nucl. Part. Sci.* **62**, 485 (2012), doi:10.1146/annurev-nucl-102711-095018.
4. V. Dexheimer and S. Schramm, *Astrophys. J.* **683**, 943 (2008), doi:10.1086/589735.
5. D. Lonardoni, A. Lovato, S. Gandolfi and F. Pederiva, *Phys. Rev. Lett.* **114**, 092301 (2015), doi:10.1103/PhysRevLett.114.092301.
6. N. K. Glendenning, *Nucl. Phys. A* **493**, 521 (1989), doi:10.1016/0375-9474(89)90101-2.
7. R. C. Tolman, *Phys. Rev.* **55**, 364 (1939), doi:10.1103/PhysRev.55.364.
8. J. R. Oppenheimer and G. M. Volkoff, *Phys. Rev.* **55**, 374 (1939), doi:10.1103/PhysRev.55.374.
9. E. S. Fraga, R. D. Pisarski and J. Schaffner-Bielich, *Phys. Rev. D* **63**, 121702 (2001), doi:10.1103/PhysRevD.63.121702.
10. B. Franzon, R. O. Gomes and S. Schramm, *Mon. Not. Roy. Astron. Soc.* **463**, 571 (2016), doi:10.1093/mnras/stw1967.
11. M. Alford, M. Braby, M. W. Paris and S. Reddy, *Astrophys. J.* **629**, 969 (2005), doi:10.1086/430902.
12. P. Papazoglou, S. Schramm, J. Schaffner-Bielich, H. Stoecker and W. Greiner, *Phys. Rev. C* **57**, 2576 (1998), doi:10.1103/PhysRevC.57.2576.

13. S. Schramm, *Phys. Rev. C* **66**, 064310 (2002), doi:10.1103/PhysRevC.66.064310.
14. K. Fukushima, *Phys. Lett. B* **591**, 277 (2004), doi:10.1016/j.physletb.2004.04.027.
15. C. Ratti, M. A. Thaler and W. Weise, *Phys. Rev. D* **73**, 014019 (2006), doi:10.1103/PhysRevD.73.014019.
16. C. Ratti, S. Roessner, M. A. Thaler and W. Weise, *Eur. Phys. J. C* **49**, 213 (2007), doi:10.1140/epjc/s10052-006-0065-x.
17. J. Steinheimer, S. Schramm and H. Stocker, *Phys. Rev. C* **84**, 045208 (2011), doi:10.1103/PhysRevC.84.045208.
18. D. H. Rischke, M. I. Gorenstein, H. Stoecker and W. Greiner, *Z. Phys. C* **51**, 485 (1991), doi:10.1007/BF01548574.
19. J. Cleymans, M. I. Gorenstein, J. Stalnacke and E. Suhonen, *Phys. Scripta* **48**, 277 (1993), doi:10.1088/0031-8949/48/3/004.
20. V. A. Dexheimer and S. Schramm, *Phys. Rev. C* **81**, 045201 (2010), doi:10.1103/PhysRevC.81.045201.
21. V. Dexheimer, R. Negreiros and S. Schramm, *Phys. Rev. C* **91**, 055808 (2015).
22. J. Steinheimer and S. Schramm, *Phys. Lett. B* **736**, 241 (2014), doi:10.1016/j.physletb.2014.07.018.
23. J. Steinheimer and S. Schramm, *Phys. Lett. B* **696**, 257 (2011), doi:10.1016/j.physletb.2010.12.046.
24. A. Mukherjee, S. Schramm, J. Steinheimer and V. Dexheimer (2017), `arXiv:1706.09191 [nucl-th]`.
25. A. Mukherjee, J. Steinheimer and S. Schramm (2016), `arXiv:1611.10144 [nucl-th]`.

Elliptic Flow and the Nuclear Equation of State

W. Trautmann[1] and H. H. Wolter[2]

[1] *GSI Helmholtzzentrum für Schwerionenforschung GmbH, Planckstr. 1, D-64291 Darmstadt, Germany*

[2] *Fakultät für Physik, Universität München, Am Coulombwall 1, D-85748 Garching, Germany*

[1] *w.trautmann@gsi.de*, [2] *hermann.wolter@lmu.de*

New constraints for the nuclear equation of state at suprasaturation densities have been obtained by measuring collective particle flows in heavy-ion reactions at relativistic energies. Ratios and differences of neutron and hydrogen flows in ^{197}Au + ^{197}Au collisions at 400 MeV/nucleon were used in studies of the asymmetric-matter equation of state. The comparison with predictions of transport models favors a moderately soft to linear density dependence, consistent with ab-initio nuclear matter theories. Model predictions suggest that comprehensive data sets collected at higher bombarding energies will provide information on the asymmetric-matter equation of state in the density range up to two or three times the saturation value.

1. Introduction

Collective nuclear motion has always been a topic followed and advanced by Walter Greiner. His seminal papers on nuclear collective excitation from the 60s of last century have guided a generation of nuclear structure physicists.[1,2] The investigation and description of density oscillations of small amplitude have not only enhanced our knowledge about the collective degrees of freedom of atomic nuclei but have also revealed fundamental properties of nuclear matter. The nuclear compressibility is of importance for many phenomena in nuclear structure and nuclear reactions as well as for astrophysics.

To go beyond the small density interval probed with giant resonances requires nuclear reactions. In their famous shock-wave paper of 1974, Scheid, Müller and Greiner have studied new phenomena produced by the collective pressure in the shock-compressed overlap zone.[3] They have shown that heavy-ion reactions at sufficiently high energy provide us with the possibility to compress nuclear matter up to several times the saturation density ρ_0 encountered in the interior of heavy nuclei. The properties of nuclear matter at suprasaturation densities may thus be studied in laboratory experiments.[4] From the extensive search for observables suitable for probing the brief high-density phase of the collision, collective flows and sub-threshold production of strange mesons have appeared as most useful. A consensus has been reached that a soft equation of state (EoS), corresponding to a compressibility $K \approx 200$ MeV and including momentum dependent interactions, best describes the high-density behavior of symmetric nuclear matter. It is based on studies of flow and kaon production within the framework of transport theory.[4–6] This finding has very recently been confirmed in a new analysis of the high-precision flow data measured by the FOPI Collaboration at the GSI laboratory.[7,8] We may note here that the conclusion that K should be of the order of 200 MeV was presented by Greiner and coworkers 49 years ago, based on a theoretical analysis of the elastic-scattering excitation function for ^{16}O + ^{16}O measured by the Yale group.[9]

In recent years, motivated by the impressive progress made in observing properties of neutron stars and in understanding details of supernova explosion scenarios, the EoS of neutron-rich asymmetric matter has received increasing attention.[10–13] The symmetry energy, equal to the difference between the energies per nucleon of neutron matter and of symmetric matter, is seen as one of the biggest unknowns in this context. We have precise information on the symmetry energy near saturation density from the knowledge of nuclear masses. For densities below saturation, investigations of nuclear structure phenomena and of heavy ion reactions in the Fermi energy regime have constrained the symmetry energy considerably.[14–16] The importance of clustering at subsaturation densities has recently been demonstrated[17] and confirmed in comparison with experimental data from intermediate-energy reactions.[18–20] At suprasaturation density, however, the symmetry energy is still largely unknown for several reasons. Phenomenological forces, even though well constrained near saturation, yield largely diverging results if they are extrapolated to higher densities.[21,22] Many-body calculations with realistic potentials face the problem that three-body forces and short-range correlations are not sufficiently constrained at the higher densities at which their importance increases.[23–26] Even the magnitude of the kinetic contribution is possibly modified by a redistribution of nucleon momenta due to short-range correlations in high-density nuclear matter.[27–29]

The symmetry energy appears in nearly every aspect of nuclear structure and reactions and, as a consequence, a variety of constraints obtained with different methods have become available.[30] However, quantities as, e.g., the thickness of the neutron skin in heavy nuclei or the isospin transport in reactions of isospin-asymmetric nuclei in the Fermi-energy domain are predominantly sensitive to the strength of the symmetry term at densities near or below the saturation value.[11,14,15,31–34] The example shown by Brown[35] demonstrates that a very precise knowledge of the neutron-skin thickness of ^{208}Pb will be required if an extrapolation from density $\rho \approx 0.6\rho_0$, where the symmetry energy is well determined, to the saturation density

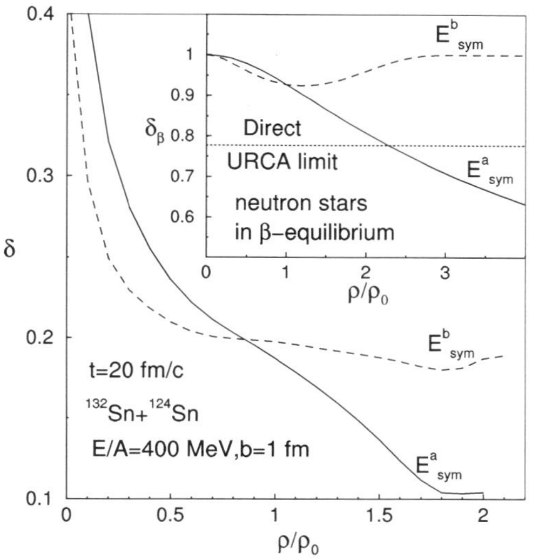

Fig. 1. Isospin asymmetry $\delta = (\rho_n - \rho_p)/\rho$ as a function of the normalized density ρ/ρ_0 at time $t = 20$ fm/c in the ^{132}Sn + ^{124}Sn reaction with a stiff ($E^a_{\rm sym}$) and with a soft ($E^b_{\rm sym}$) density dependence of the nuclear symmetry energy. The corresponding correlation for neutron stars in β-equilibrium is shown in the inset (reprinted with permission from Ref. 36; Copyright (2002) by the American Physical Society).

ρ_0 is attempted. This emphasizes the need for more direct high-density probes. As in the case of symmetric nuclear matter, collective flows and sub-threshold particle production are obvious candidates.

A strong motivation for exploring the information contained in isotopic flows was provided some time ago by Bao-An Li when he pointed to the parallels in the density-dependent isotopic compositions of neutron stars and of the transient systems formed in collisions of neutron-rich nuclei depending on the EoS input used in the calculations.[36] Figure 1 illustrates the remarkable fact that the same physical laws govern objects differing by 18 orders of magnitude in linear scale or 55 orders of magnitude in mass. Properties of exotic astrophysical objects may thus be inferred from data obtained in laboratory experiments, and vice versa. A main difficulty resides in the comparatively small asymmetry of atomic nuclei available for experiments. Symmetry effects are, therefore, always small relative to those of the dominating isoscalar forces. A partial cancellation of the latter may be expected in differences or ratios of observables between isotopic partners.

A further encouragement was provided by transport model calculations indicating that the elliptic flows of free neutrons and free protons respond differently to variations of the parametrization of the symmetry energy.[37] Elliptic flow refers to the second Fourier component of the azimuthal anisotropy of particle emissions. It has motivated a reanalysis of the FOPI-LAND data for ^{197}Au + ^{197}Au collisions at 400 MeV/nucleon, collected many years ago and used to demonstrate the existence of neutron squeeze-out in this energy regime.[38,39] Squeeze-out refers to a dominant out-of-plane emission of particles, relative to in-plane emission, and is considered as evidence for the pressure buildup in the collision zone. It should therefore be particularly sensitive to the high-density EoS. The analysis has favored a moderately soft to linear density dependence of the symmetry energy.[37,40]

This finding had a particular significance, in spite of a large statistical uncertainty. Rather different conclusions, ranging from a super-soft to a super-stiff behavior of the symmetry energy, had previously been reached in analyses of the π^-/π^+ production ratios, measured by the FOPI Collaboration[41] for the same ^{197}Au + ^{197}Au reaction, with different transport models.[42–45] In particular, the super-soft result, first presented by Xiao *et al.*,[43] has initiated a broad discussion of how it might be reconciled with other observations as, e.g., properties of neutron stars.[43,46,47] The FOPI-LAND elliptic-flow data were found to be inconsistent with this extreme assumption and, in fact, fairly independent of particular choices made for the model parameters used for the quantum-molecular-dynamics (QMD) transport-model calculations.[37,48]

The obvious need to improve the statistical accuracy beyond that of the existing data set has initiated a dedicated measurement by the ASY-EOS Collaboration of collective flows in collisions of ^{197}Au + ^{197}Au as well as of ^{96}Zr + ^{96}Zr and ^{96}Ru + ^{96}Ru. It was carried out in 2011 at the GSI laboratory with the LAND[49] detector coupled to a subset of the CHIMERA[50] detector array.[51] In the following sections, the present situation will be described in relation to the new results from the ASY-EOS experiment.

2. Present Knowledge of the Symmetry Energy

Our knowledge of the symmetry energy is originally based on nuclear masses whose dependence on the isotopic composition is reflected by the symmetry term in the Bethe–Weizsäcker mass formula. A density dependence is already indicated by the use of separate bulk and surface terms in refined mass formulae. In the Fermi-gas model, it is given by a proportionality to $(\rho/\rho_0)^\gamma$ with an exponent $\gamma = 2/3$, where $\rho_0 \approx 0.16$ nucleons/fm^3 is the saturation density. This kinetic contribution to the symmetry energy, however, amounts to only about 1/3 of the symmetry term of approximately 30 MeV for nuclear matter at saturation. The major contribution is given by the potential term reflecting properties of the nuclear forces.

Nuclear many-body theory has presented us with a variety of predictions for the nuclear equation of state.[4,11,22,52–54] The examples shown in Fig. 2 for the two cases of symmetric nuclear matter and of pure neutron matter demonstrate that, overall, the results are quite compatible among each other, except for densities exceeding saturation at which the predictions diverge. The symmetry energy $E_{\rm sym}$ can be defined as the coefficient of the quadratic term in an expansion of the energy per particle in the asymmetry $\delta = (\rho_n - \rho_p)/\rho$, where ρ_n, ρ_p and ρ represent the neutron, proton, and total densities, respectively,

$$E/A(\rho,\delta) = E/A(\rho,\delta=0) + E_{\rm sym}(\rho)\cdot\delta^2 + \mathcal{O}(\delta^4). \tag{1}$$

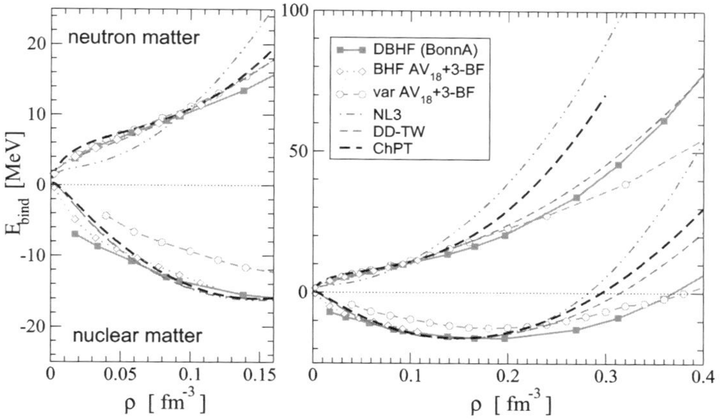

Fig. 2. EoS in nuclear matter and neutron matter as a function of density. BHF/DBHF and variational calculations with realistic forces are compared to phenomenological density functionals NL3 and DD-TW and to ChPT. The left panel zooms the low density range (from Ref. 22, reprinted with kind permission from Springer Science+Business Media).

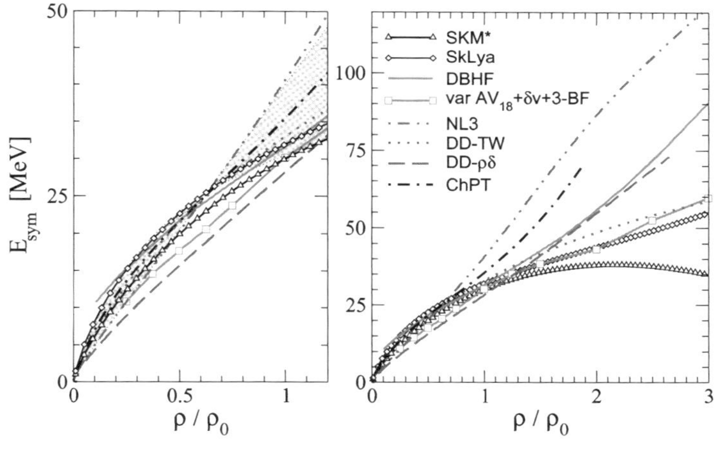

Fig. 3. Symmetry energy as a function of density as predicted by different models. The left panel zooms the low density range up to saturation. The full lines represent the DBHF and variational approaches using realistic forces (from Ref. 22, reprinted with kind permission from Springer Science+Business Media).

In the quadratic approximation, the symmetry energy is the difference between the energies of symmetric matter ($\delta = 0$) and neutron matter ($\delta = 1$). It is shown in Fig. 3 for a similar range of models as in Fig. 2. Also the symmetry energy diverges at high density, as expected from Fig. 2, while most empirical models coincide near or slightly below saturation, the density range at which constraints from finite nuclei are valid.

In calculations using realistic forces fitted to two- and three-nucleon data, the uncertainty is mainly related to the short-range behavior of the nucleon-nucleon force and, in particular, to the three-body and tensor forces.[23–26] The three-body force has been shown to make an essential but quantitatively small contribution to the masses of light nuclei.[55] The extrapolation of the partly phenomenological terms used there to higher densities is, however, highly uncertain.[25] The general effect of including three-body forces in the calculations is a stiffening of the symmetry energy with increasing density.[56,57] Short-range correlations become also increasingly important at higher densities. Results from very recent new experiments will, therefore, have a strong impact on predictions for high-density nuclear matter.[23,24,27–29] Data from neutron-star observations provide important constraints already now.[58] With the improvement of observational methods, they will become more stringent in the near future.[59]

At higher energies, the momentum dependence of the nuclear forces becomes important.[11,12,60,61] It is well known that nuclear mean fields are momentum dependent, as seen, e.g., in the energy dependence of the nuclear optical potential. The dominating effect is in the isoscalar sector but there is also an important isovector momentum dependence. It manifests itself as an energy dependence of the isospin-dependent part of the optical potential but can also be expressed in terms of a difference of the effective masses of protons and neutrons.[22] Even the ordering of these effective masses is still an open problem.[30,62,63] It has, moreover, been shown that the effective mass differences and the asymmetry dependence of the EoS are both influencing particle yields and flow observables, and that additional observables will be needed to resolve the resulting ambiguity.[12,60,61]

Transport theories needed for calculating the temporal evolution of nuclear reactions often use simplified descriptions of the composition-dependent part of the nuclear mean field. In the UrQMD of the group of Li and Bleicher,[64] the potential part of the symmetry energy is defined with two parameters, the value at saturation density, usually taken as 22 MeV in their calculations, and the power-law coefficient γ describing the dependence on density,

$$E_{\rm sym} = E_{\rm sym}^{\rm pot} + E_{\rm sym}^{\rm kin} = 22\ {\rm MeV} \cdot (\rho/\rho_0)^{\gamma} + 12\ {\rm MeV} \cdot (\rho/\rho_0)^{2/3}. \qquad (2)$$

In other codes the nuclear potential of Das *et al.* with explicit momentum dependence in the isovector sector is used.[65] There, as in the Boltzmann-Uehling-Uhlenbeck-type code IBUU04 developed by the groups of Li and Chen,[11,66] the density dependence of the symmetry energy is characterized by a parameter x appearing in the potential expressions. Examples of these parametrizations and of results obtained from the analysis of experimental reaction data are given in Fig. 4. The stiff ($E^a_{\rm sym}$) and soft ($E^b_{\rm sym}$) density dependences of Fig. 1 correspond approximately to the cases $\gamma = 1$ and $x = 1$ shown here. Density functionals of the Skyrme type are also increasingly used in transport model calculations.[63,67]

Parametrizations of this kind have the consequence that, once the symmetry energy at the saturation point is fixed, a single value at a different density or, alternatively, the slope or curvature at any density will completely determine the

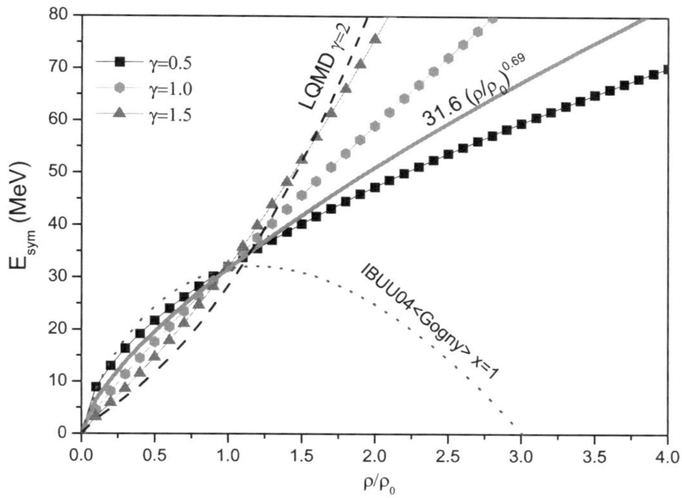

Fig. 4. Parametrizations of the nuclear symmetry energy as used in transport codes: three parametrizations of the potential term used in the UrQMD (Ref. 64) with power law coefficients $\gamma = 0.5, 1.0$, and 1.5 (lines with symbols as indicated), the result with $\gamma = 0.69$ obtained from analyzing isospin diffusion data with the IBUU04 (full line, Ref. 66), and the super-soft and stiff parametrizations obtained from analyzing the π^-/π^+ production ratios with the IBUU04 (dotted line, Ref. 43) and the LQMD (dashed line, Ref. 44) transport models (from Ref. 68, reprinted with kind permission from Springer Science+Business Media).

parametrization. Measurements of a variety of observables in nuclear structure and reactions have, in this manner, been used to obtain results for the density dependence of the symmetry energy. They are conventionally expressed in the form of the parameter L which is proportional to the slope at saturation,

$$L = 3\rho_0 \cdot dE_{\mathrm{sym}}/d\rho|\rho_0. \tag{3}$$

Most results with their errors fall into the interval 20 MeV $\leq L \leq$ 100 MeV and are compatible with a most probable value $L \approx 60$ MeV, roughly corresponding to a power-law coefficient $\gamma = 0.6$.[11,14,15,30,46,69] The full line in Fig. 4 represents, e.g., the result deduced by Li and Chen from the MSU isospin-diffusion data and the neutron-skin thickness in ^{208}Pb.[66] The corresponding slope parameter is $L =$ 65 MeV. Rather similar constraints have been deduced from recent investigations and observations of neutron-star properties.[26,57,58]

Considerable progress regarding the correlation of the symmetry energy with particular observables in different models has been made by the Florida and Barcelona groups.[70,71] In continuation of the work of Typel and Brown[72] and of Furnstahl,[73] a universal correlation between the thickness of the neutron-skin of ^{208}Pb and the L parameter has been found for empirical mean-field interactions.[71] The determination of the neutron-skin thickness by measuring the parity-violating contribution to electron scattering at high energy will thus offer a practically model-free access to the slope at saturation.[74,75]

3. Probing High Densities

Densities of two to three times the saturation density may be reached on time scales of $\approx$ 10–20 fm/c in the central zone of heavy-ion collisions at relativistic energies of up to $\approx$ 1 GeV/nucleon.[76,77] The resulting pressure produces a collective outward motion of the compressed material whose strength will be influenced by the symmetry energy in asymmetric systems.[4,78] At the same time, the excitation of Δ resonances in hard nucleon-nucleon scatterings leads to the production and subsequent emission of charged and neutral π and K mesons. Their property as messengers from the early reaction phase identifies them as potential probes for the high-density symmetry energy. Calculations show that the highest sensitivities may be expected at near-threshold energies.[6,42,79] In both cases, collective flows or meson production, isotopic differences or ratios of observables are useful to minimize the sensitivity to the isoscalar part in the EoS while maximizing that to the symmetry term.

The FOPI Collaboration has collected an extensive set of π^-/π^+ production ratios measured for the four reactions ^{40}Ca + ^{40}Ca, ^{96}Zr + ^{96}Zr, ^{96}Ru + ^{96}Ru, and ^{197}Au + ^{197}Au at several energies between 0.4 and 1.5 GeV/nucleon.[41] Theoretical analyses of these data, however, have come to conflicting conclusions, suggesting everything from a rather stiff to a super-soft behavior of the symmetry energy.[42–45] The reason may lie in the treatment of the pion in-medium and Δ dynamics[77,80,81] and in competing effects of the mean fields and the Δ thresholds[42,82] which will require further studies.[83] Cozma has demonstrated very recently that both types of observables, elliptic flow and pion production, can lead to compatible constraints for the stiffness of the symmetry energy, provided that care is taken to conserve the energy in inelastic collisions producing a Δ resonance.[80] It may be done at a local or global level by accounting for the potential energy of hadrons in the treatment of two-body collisions. The sensitivity inherent in the spectral dependence of pion production has been pointed out by several authors.[84–86] New information is expected to come from measurements performed by the SπRIT Collaboration[86,87] at RIKEN with radioactive Sn beams.

4. Transport Theory

Common to all high-density probes is the need to relate the observations made for the asymptotic outcome of the reaction to the short interval during which the compressed matter exists and impacts the subsequent evolution. Heavy-ion collisions are non-equililibrium processes and non-equilibrium theory is, therefore, needed for their interpretation. Hydrodynamical or statistical descriptions may be useful for describing certain stages of the reaction process. A unified description requires transport theory to follow the evolution from the initial state of the collision system to the final stage when the strong interactions cease to act between its components. Such approaches have first been developed in the group of Walter Greiner and at Michigan State University in the 80s,[88–90] and have since evolved into a very valuable tool for extracting physics information from heavy-ion experiments.

Two families of transport approaches with different philosophies have been developed in nuclear physics. Those of the Boltzmann–Vlasov type (often named Boltzmann–Uehling–Uhlenbeck (BUU) approaches) decribe the evolution in time of the one-body phase space density $f(\vec{r},\vec{p};t)$ under the influence of a mean field $U[f]$ and of two-body collisions

$$\frac{\partial f_1}{\partial t}+\frac{\vec{p}}{m}\nabla_{\vec{r}}f_1-\nabla_{\vec{r}}U\nabla_{\vec{p}}f_1= \tag{4}$$

$$\left(\frac{2\pi}{m}\right)^3\int d\vec{p}_2d\vec{p}_3d\vec{p}_4|\vec{v}_1-\vec{v}_2|\sigma_{\rm NN}(\Omega_{12})\delta(\vec{p}_1+\vec{p}_2-\vec{p}_3-\vec{p}_3)(f_3f_4\bar{f}_1\bar{f}_2-f_1f_2\bar{f}_3\bar{f}_4).$$

Here $f_i=f(\vec{r},\vec{p_i};t)$, $\bar{f}_i=(1-f_i)$, v_i are velocities, and $\sigma_{\rm NN}(\Omega)$ is the in-medium nucleon–nucleon (NN) cross-section. The potential $U[f]$ and the cross-section are either derived from an energy density functional, or are parametrized in order to test them relative to the data. If particle production, e.g. of pions and Δ resonances, is to be considered, additional physics input is needed: inelastic cross-sections, potentials of the new particles, their cross-sections for collisions with other particles and, possibly, the finite mass distributions of instable particles.

The solution of this equation is achieved with simulations using the test-particle (TP) method where the distribution function is represented in terms of finite elements as

$$f(\vec{r},\vec{p};t)=\frac{1}{N_{\rm TP}}\sum_{i=1}^{AN_{\rm TP}}g(\vec{r}-\vec{r}_i(t))\,\tilde{g}(\vec{p}-\vec{p}_i(t)); \tag{5}$$

here $N_{\rm TP}$ is the number of TPs per nucleon, $\vec{r}_i$ and $\vec{p}_i$ are the time-dependent coordinates and momenta of the TPs, and g and $\tilde{g}$ are the shape functions in coordinate and momentum space (e.g. δ-functions or Gaussians), respectively. Upon inserting this ansatz into the l.h.s. of Eq. (4), Hamiltonian equations of motion for the test particles are obtained, $\frac{{\rm d}\vec{r}_i}{{\rm d}t}=\vec{p}$ and $\frac{{\rm d}\vec{p}_i}{{\rm d}t}=-\vec{\nabla}_{r_i}U$. The collision term is simulated stochastically, by performing TP collisions with a probability depending on the cross section and obeying the Pauli principle for the final state according to blocking factors $\bar{f}_i=(1-f_i)$.

In the second family, the quantum molecular dynamics (QMD) model, the evolution of the collision is formulated in terms of the evolution of the coordinates $R_i(t)$ and momenta $P_i(t)$ of individual nucleons, similarly as in classical molecular dynamics, but with particles of finite width representing minimum nucleon wave packets. They move under the influence of NN forces. The method can also be viewed as being derived from the Time-Dependent Hartree (TDH) method with a product trial wave-function of single-particle states in Gaussian form. One obtains equations of motion of the same form as in BUU for the coordinates of the wave packets. Also a stochastic two-body collision term is introduced and treated in very much the same way as in BUU, but now for nucleons and the full NN cross-section. There are also relativistic formulations for both approaches using relativistic density functionals. Of the codes used in the analysis of the flow data discussed here UrQMD and TuQMD are relativistic codes, IQMD is non-relativistic. A review of the BUU method is given in Ref. 91 while the QMD method is reviewed in Ref. 92.

The main difference between the two approaches lies in the amount of fluctuations and correlations in the representation of the phase space distribution. In the BUU approach, the phase-space distribution function is seen as a smooth function of coordinates and momenta and can be increasingly better approximated by increasing the number of TPs. In the limit of $N_{\mathrm{TP}} \to \infty$, the BUU equation is solved exactly but the solution is deterministic and contains neither fluctuations nor correlations. If they are considered to be important, as is the case when looking at cluster and fragment production, they have to be introduced explicitely through the Boltzmann-Langevin equation with a fluctuation term on the r.h.s. of Eq. (4). In the QMD, fluctuations are present because of the finite number of wave packets in the representation. In addition, classical correlations are present if explicit two-body interactions are used. The fluctuations in QMD-type codes are regulated and smoothed by choosing appropriate width parameters of the wave packets. QMD can be seen as an event generator solving the time evolution of different events independently. Event-by-event fluctuations are not even suppressed in the limit of infinitely many events.

The results of simulations with the two methods are thus expected to be similar, though not necessarily identical, as far as one-body observables are concerned as, e.g., the flow observables discussed in this article. Larger differences are expected for observables depending on fluctuations and correlations, such as the production of clusters and intermediate-mass fragments. Generally, the description of observables going beyond the mean field level is a question under active discussion in transport theory. In the experiment, copious numbers of light clusters and fragments are observed in heavy-ion collisions, particularly at lower energies.

In addition to these more fundamental differences between existing transport approaches, there are also differences that are caused by different implementations of the highly complex transport theories. Analyses of experimental data with seemingly similar physics input have lead to rather different conclusions. The analyses of the FOPI π^-/π^+ ratios represent an example. In order to reach a better understanding of possible reasons, a code-comparison project has been started. In a first publication, results for a standardized heavy-ion collision with identical physics input were compared.[93] Eighteen commonly used transport codes, nine of BUU and nine of QMD type, were included. Quantitatively, the differences were found to depend on the incident energy and amounted to approximately 30% at 100 and 15% at 400 MeV/nucleon, respectively. The comparison is presently continued with calculations for infinite nuclear matter. There the different ingredients of the transport codes can be tested separately and compared to exact limits.

5. Elliptic Flow

A "flow of nuclear matter out of the regions of compressed densities" was already predicted by Scheid, Ligensa and Greiner 49 years ago when they investigated central collisions of ^{16}O nuclei.[9] The transverse emission of nuclear matter from the compressed interaction zone, the squeeze-out as it has been termed, has first been observed in experiments at the Bevalac.[94] In a sphericity analysis, the event shape in three dimensions was characterized by a kinetic-energy flow tensor whose main orientation with respect to the beam direction represents the collective sidewards or directed flow. A difference in the two minor axes indicates the existence of elliptic flow. At the bombarding energies of up to several GeV/nucleon investigated in these studies, a preferential emission of charged particles perpendicular to the reaction plane has been observed. The shadowing by the spectator remnants reduces the in-plane flow, so that the strength of the off-plane emission, as quantified by the azimuthal anisotropy, reflects the internal pressure.

It has become customary to express both, directed and elliptic flows, and also possible higher flow components by means of a Fourier decomposition of the azimuthal distributions measured with respect to the orientation of the reaction plane ϕ_R,[95,96]

$$\frac{dN}{d(\phi - \phi_R)} = \frac{N_0}{2\pi}\left(1 + 2\sum_{n\geq 1} v_n \cos n(\phi - \phi_R)\right), \tag{6}$$

where N_0 is the azimuthally integrated yield. The coefficients $v_n \equiv \langle \cos n(\phi - \phi_R)\rangle$ are functions of particle type, impact parameter, rapidity y, and the transverse momentum p_t; v_1 and v_2 represent the directed and elliptic flows, respectively.

Elliptic flow has become an important observable at other energy regimes as well. At ultrarelativistic energies, the observation of the constituent-quark scaling of elliptic flow is one of the prime arguments for deconfinement during the early collision phase, and the behavior as an almost perfect liquid of the formed strongly interacting quark–gluon plasma are deduced from the observed magnitude of collective motions.[97–100] It implies that elliptic flow develops very early in the collision which is valid also in the present range of relativistic energies, as confirmed by calculations.[101] Isotopic flow differences appear thus very suitable for studying mean-field effects at high density.

An excitation function of the elliptic flow of $Z = 1$ particles in ^{197}Au+^{197}Au collisions, compiled from various experiments,[102] is shown in Fig. 5. Squeeze-out perpendicular to the reaction plane, i.e. $v_2 < 0$, as a result of shadowing by the spectator remnants is observed at incident energies between about 150 MeV/nucleon and 4 GeV/nucleon with a maximum near 400 MeV/nucleon. At lower energies, the collective angular momentum in the mean-field dominated dynamics causes the observed in-plane enhancement of emitted reaction products. The figure also shows that elliptic flow can be measured quite precisely, as demonstrated by the good

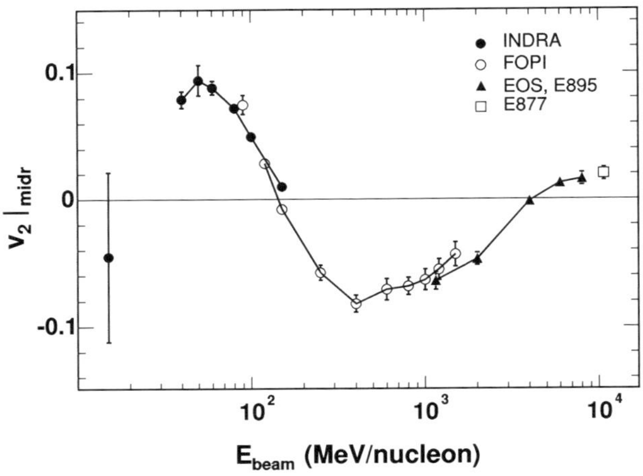

Fig. 5. Elliptic flow parameter v_2 at mid-rapidity for ^{197}Au+^{197}Au collisions at intermediate impact parameters ($\approx$ 5.5–7.5 fm) as a function of incident energy. The filled and open circles represent the INDRA and FOPI data[103,104] for $Z = 1$ particles, the triangles represent the EOS and E895 data[105] for protons, and the square represents the E877 data[106] for all charged particles (from Ref. 102, reprinted with kind permission from Springer Science+Business Media).

agreement of data sets from different experiments in the overlap regions of the studied intervals in collision energy.[7,102,103]

The precision in interpreting the measured particle flows has been demonstrated by the FOPI Collaboration in their extensive report by Reisdorf *et al.*[7] Calculations with the IQMD model were shown to account not only for the overall strength of the flow but also for the detailed dependence on rapidity and the change of sign of the v_2 parameter at rapidities $|y_0| \approx 0.7$; here y_0 is the rapidity normalized to the projectile rapidity in the c.m. system (Fig. 6). The approximately quadratic dependence of $v_2(y_0)$ has been fitted with two parameters, $v_2(y_0) = v_{20} + v_{22} \cdot y_0^2$, and a new quantity $v_{2n} = |v_{20}| + |v_{22}|$ has been introduced. It combines the information contained in the amplitude and the rapidity dependence of v_2. The dependence of v_{2n} on the incident energy in the interval 0.4 to 1.5 GeV/nucleon covered by FOPI is fairly flat and its discriminating power between the soft and stiff parametrizations of the symmetric-matter EoS appears rather convincing.[7,8]

6. Results from FOPI-LAND

The squeeze-out of neutrons has first been observed by the FOPI-LAND Collaboration who studied the reaction ^{197}Au + ^{197}Au at 400 MeV/nucleon.[38] The squeeze-out of charged particles reaches its maximum at this energy (Fig. 5), and similarly large anisotropies were observed for neutrons.[39] The neutrons had been detected with the Large-Area Neutron Detector, LAND,[49] while the FOPI Forward Wall, covering the forward hemisphere of laboratory angles $\theta_{\rm lab} \leq 30°$ with more than 700 plastic scintillator elements, was used to determine the modulus and azimuthal

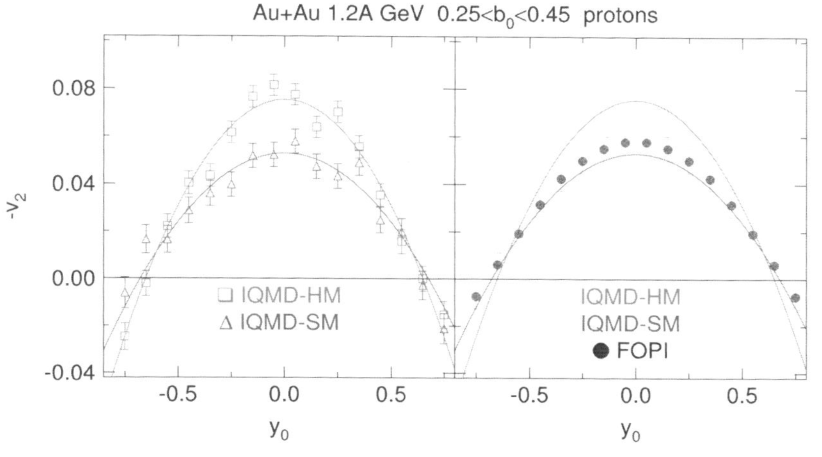

Fig. 6. Elliptic flow parameter $-v2$ (note the change of sign) of protons as a function of the normalized rapidity y_0 for ^{197}Au + ^{197}Au collisions at 1.2 GeV/nucleon and the indicated near-central interval of normalized impact parameters b_0. Left panel: IQMD calculations (symbols) for a hard (HM) and a soft (SM) EoS, both with momentum dependent forces, and the fit results (lines) assuming a quadratic dependence on y_o. Right panel: The obtained fit results (lines) in comparison with the experimental data (symbols) measured with the FOPI detector at the GSI laboratory (reprinted from Ref. 8, Copyright (2016), with permission from Elsevier).

orientation of the impact parameter.

The motivation for returning to the existing data set has been provided by studies performed with the UrQMD transport code for this fairly neutron-rich system ($N/Z = 1.49$). They indicated a significant sensitivity of the elliptic-flow parameters to the assumptions made for the density dependence of the symmetry energy.[37] In calculations with power-law coefficients $\gamma = 0.5$ and 1.5 (cf. Eq. (2) and Fig. 4), the relative strengths of neutron and proton elliptic flows were found to vary on the level of 15%.

The reanalysis of the data consisted mainly in choosing equal acceptances for neutrons and hydrogen isotopes with regard to particle energy, rapidity and transverse momentum (energy and momentum per nucleon for deuterons and tritons). The theoretical predictions have been obtained simulating the LAND acceptance and the experimental analysis conditions and were found to follow qualitatively the experimental data.

For the quantitative evaluation, the ratio of the flow parameters of neutrons versus protons or versus $Z = 1$ particles has been proposed as a useful observable.[37] Systematic effects influencing the collective flows of neutrons and charged particles in similar ways should thereby be minimized, on the experimental as well as on the theoretical side. In consideration of the systematic and experimental errors, a value $\gamma = 0.9 \pm 0.4$ has been adopted by the authors as best representing the power-law exponent of the potential term resulting from the elliptic-flow analysis.[37] It falls

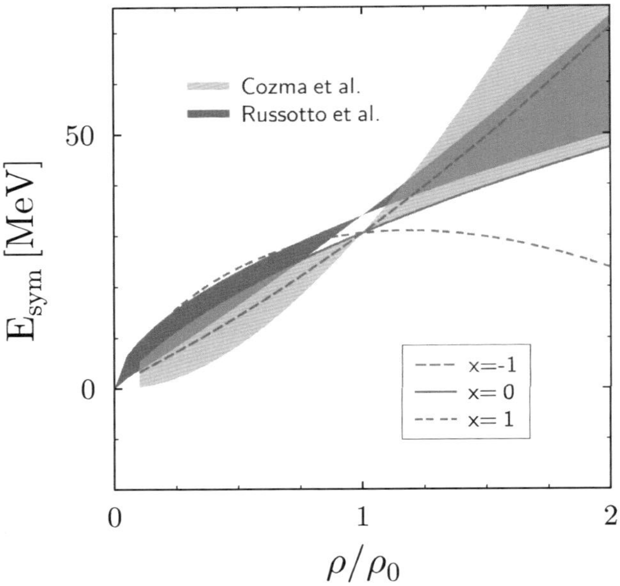

Fig. 7. Constraints on the density dependence of the symmetry energy obtained by Cozma *et al.* from comparing predictions of the Tübingen QMD for the neutron-proton elliptic-flow difference and ratio to FOPI-LAND experimental data (Ref. 107). The result of Russotto *et al.* (Ref. 37) is also shown, together with the Gogny-inspired parametrization of the symmetry energy for three values of the stiffness parameter: $x = -1$ (stiff), $x = 0$, and $x = 1$ (soft) (from Ref. 107).

slightly below the linear $\gamma = 1.0$ line shown in Fig. 4. The corresponding slope parameter is $L = 83 \pm 26$ MeV. Comparing with the many-body theories shown in Fig. 3, the elliptic-flow result is in good qualitative agreement with the range spanned by the DBHF and variational calculations based on realistic nuclear forces.

In an independent analysis, Cozma has used data from the same experiment and investigated the influence of several parameters on the difference and ratio of the elliptic flows of protons and neutrons using the Tübingen version of the QMD transport model.[107,108] They included the parametrization of the isoscalar EoS, the choice of various forms of free or in-medium nucleon–nucleon cross-sections, and model parameters as, e.g., the widths of the wave packets representing nucleons. The interaction developed by Das *et al.* with an explicit momentum dependence of the symmetry energy part was used.[65,66] In Fig. 7, the obtained constraint on the density dependence of the symmetry energy is presented. It is overall in agreement with the result reported by Russotto *et al.*[37] but has a larger uncertainty because the effects of exploring the full intervals of the above-mentioned theoretical parameters were taken into account in the error determination. A stiffness closely corresponding to the $x = -1$ scenario is clearly favored while a super-soft solution with $x = 1$ is ruled out.

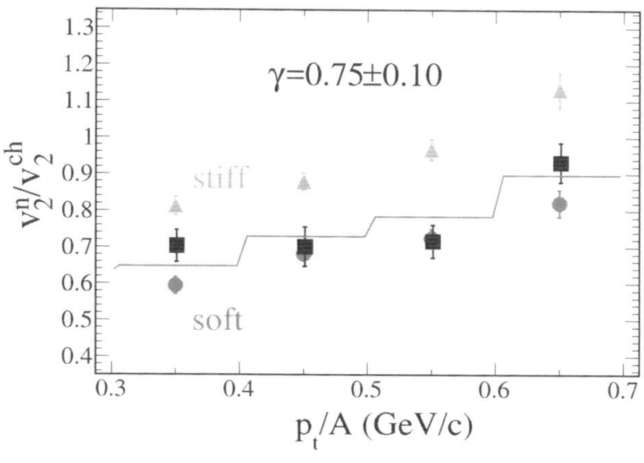

Fig. 8. Elliptic flow ratio of neutrons over all charged particles for central ($b < 7.5$ fm) collisions of ^{197}Au+^{197}Au at 400 MeV/nucleon as a function of the transverse momentum per nucleon p_t/A. The black squares represent the experimental data; the green triangles and purple circles represent the UrQMD predictions for stiff ($\gamma = 1.5$) and soft ($\gamma = 0.5$) power-law exponents of the potential term, respectively. The solid line is the result of a linear interpolation between the predictions, weighted according to the experimental errors of the included four bins in p_t/A and leading to the indicated $\gamma = 0.75 \pm 0.10$ (from Ref. 51).

7. Results from ASY-EOS

The experimental setup of the ASY-EOS experiment at the GSI laboratory followed the scheme developed for FOPI-LAND by using the Large Area Neutron Detector (LAND[49]) as the main instrument for neutron and charged particle detection. For the event characterization and for measuring the orientation of the reaction plane, three detection systems had been installed. The ALADIN Time-of-Flight (AToF) Wall[109] was used to detect charged particles and fragments in forward direction at polar angles up to $\theta_{\mathrm{lab}} \leq 7°$. Its capability of identifying large fragments and of characterizing events with a measurement of Z_{bound}[109] permitted the sorting of events according to impact parameter. Four double rings of the CHIMERA[110] multidetector carrying together 352 CsI(Tl) scintillators in forward direction and four rings with 50 thin CsI(Tl) elements of the Washington University Microball[111] array surrounding the target provided sufficient coverage and granularity for determining the orientation of the reaction plane from the measured azimuthal particle distributions. A detailed description of the experiment is available in Ref. 51.

The ratio v_2^n/v_2^{ch} obtained for the class of central ($b < 7.5$ fm) collisions as a function of the transverse momentum per nucleon p_t/A is shown in Fig. 8. The best description of the neutron-vs-charged-particle elliptic flow is obtained with a power-law coefficient $\gamma = 0.75 \pm 0.10$, where $\Delta\gamma = 0.10$ is the statistical uncertainty returned by the fit routine. It results from linearly interpolating between the predictions for the soft, $\gamma = 0.5$, and the stiff, $\gamma = 1.5$, predictions of the model within the range of transverse momentum $0.3 \leq p_t/A \leq 0.7$ GeV/c.

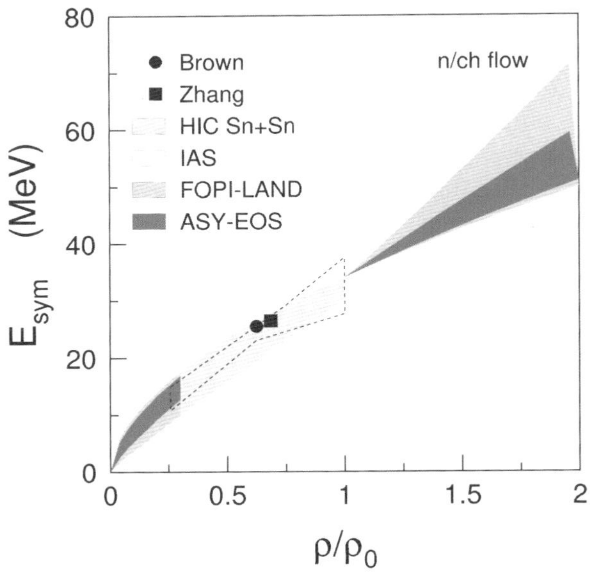

Fig. 9. Constraints deduced for the density dependence of the symmetry energy from the ASY-EOS data in comparison with the FOPI-LAND result of Ref. 37 as a function of the reduced density ρ/ρ_0. The low-density results of Refs. 14, 34, 35 and 112 as reported in Ref. 16 are given by the symbols, the gray area (HIC), and the dashed contour (IAS). For clarity, the FOPI-LAND and ASY-EOS results are not displayed in the interval $0.3 < \rho/\rho_0 < 1.0$ (from Ref. 51).

With all corrections and errors included, the acceptance-integrated elliptic-flow ratio leads to a power-law coefficient $\gamma = 0.72 \pm 0.19$. This is the result displayed in Fig. 9 as a function of the reduced density ρ/ρ_0. The new result confirms the former and has a considerably smaller uncertainty. It is also worth noting that the present parametrization is compatible with the low-density behavior of the symmetry energy from Refs. 14, 34, 35 and 112 that are included in the figure. The corresponding slope parameter describing the variation of the symmetry energy with density at saturation is $L = 72 \pm 13$ MeV.

The effective density probed with the elliptic flow measurement was explored in a study using again the Tübingen version[107] of the QMD model. The underlying idea consisted in performing transport calculations for the present reaction with two parametrizations of the symmetry energy that were chosen to be different for a selected range of density and identical elsewhere. The magnitude of the obtained difference between the two predictions for the elliptic flow ratio is considered a measure of the sensitivity to the selected density region.

It was found that the sensitivity achieved with the elliptic-flow ratio of neutrons over charged particles, the case chosen for the ASY-EOS analysis, reaches its maximum close to saturation density but extends beyond twice that value. Interestingly, the sensitivity of the neutron-vs-proton flow ratio has its maximum in the 1.4–1.5 ρ_0 region, i.e. at significantly higher densities than with light complex particles being included. This observation contains an important potential for future experiments. With efficient isotope separation, flow measurements may give access to the curvature of the symmetry energy at saturation, in addition to the slope.

8. Conclusion and Outlook

The "flow of nuclear matter out of the regions of compressed densities" studied by Walter Greiner and his group[9] many years ago has, in the meanwhile, evolved into a useful tool for the study of the nuclear equation of state (EoS) at high density. According to the predictions of transport models, the relative strengths of neutron and proton elliptic flows represent an observable sensitive to the symmetry energy at densities near and above saturation. By forming ratios or differences of neutron versus charged-particle flows, the influence of isoscalar-type parameters of the model descriptions can be minimized. As the interpretation of the FOPI pion ratios[41] is not yet conclusive, this observable is presently unique as a terrestrial source of information for the EoS of asymmetric matter at high densities.

The value $\gamma = 0.72 \pm 0.19$ obtained in the ASY-EOS experiment for the power-law coefficient of the potential part in the UrQMD parametrization of the symmetry energy and the slope parameter $L = 72 \pm 13$ MeV are equivalent to a symmetry pressure $p_0 = \rho_0 L/3 = 3.8 \pm 0.7$ MeVfm^{-3}. The latter may be used to estimate the pressure in neutron-star matter at saturation density. For an assumed asymmetry $\delta = (\rho_n - \rho_p)/\rho = 0.9$ in that part of the star, it amounts to 3.4 MeVfm^{-3} (Ref. 51), a value that compares well with the pressure obtained by Steiner *et al.*[113] from neutron-star observations.

The study of the parameter dependence has shown that the contributions of the isoscalar sector are still significant. It will, therefore, be important to improve the description of the nuclear interaction in transport models, to reduce the parameter ranges also in the isoscalar sector, to improve the algorithms used for clusterization, as well as going beyond the mean-field picture, including short-range correlations.[29] Moreover, it will be quite important to compare the experimental data with the predictions of several transport models, of both Boltzmann–Vlasov and molecular-dynamics type, in order to pursue the work towards a model-independent constraint of the high-density symmetry energy initiated in Refs. 107 and 108.

The presented experimental results and the theoretical study of the density range that has been probed, provide a strong encouragement for continuing the flow measurements with improved detection systems. Model studies indicate that the sensitivity to the stiffness of the symmetry energy is still significant at incident energies as high as 800 MeV or 1 GeV/nucleon. The density study suggests that the range of densities that can be probed in the laboratory may reach up to twice or three times the saturation value if higher precision and isotopic resolution for light charged particles can be achieved. Future experiments will, therefore, benefit from the improved capabilities of the NeuLAND detector[114] presently constructed as part of the R^3B setup for experiments at FAIR.

Acknowledgments

Stimulating and fruitful discussions with M. Di Toro, W. Reisdorf, Yongjia Wang, and with the authors of Ref. 108, M. D. Cozma, A. Le Fèvre, Y. Leifels, Qingfeng Li, J. Łukasik, and P. Russotto are gratefully acknowledged. HHW is partially supported by the Universe cluster of excellence of the DFG Germany.

References

1. M. Danos and W. Greiner, *Phys. Rev.* **134**, B284 (1964).
2. R. Ligensa, W. Greiner and M. Danos, *Phys. Rev. Lett.* **16**, 364 (1966).
3. W. Scheid, H. Müller and W. Greiner, *Phys. Rev. Lett.* **32**, 741 (1974).
4. P. Danielewicz, R. Lacey and W. G. Lynch, *Science* **298**, 1592 (2002).
5. C. Sturm *et al.*, *Phys. Rev. Lett.* **86**, 39 (2001).
6. C. Fuchs, A. Faessler, E. Zabrodin and Yu-Ming Zheng, *Phys. Rev. Lett.* **86**, 1974 (2001).
7. W. Reisdorf *et al.*, *Nucl. Phys. A* **876**, 1 (2012).
8. A. Le Fèvre, Y. Leifels, W. Reisdorf, J. Aichelin and Ch. Hartnack, *Nucl. Phys. A* **945**, 112 (2016).
9. W. Scheid, R. Ligensa and W. Greiner, *Phys. Rev. Lett.* **21**, 1479 (1968).
10. J. M. Lattimer and M. Prakash, *Phys. Rep.* **442**, 109 (2007).
11. Bao-An Li, Lie-Wen Chen and Che Ming Ko, *Phys. Rep.* **464**, 113 (2008).
12. M. Di Toro, V. Baran, M. Colonna and V. Greco, *J. Phys. G* **37**, 083101 (2010).
13. Topical Issue on Nuclear Symmetry Energy, *Eur. Phys. Jour. A* **50**, number 2 (2014).
14. M. B. Tsang, Yingxun Zhang, P. Danielewicz, M. Famiano, Z. Li, W. G. Lynch and A. W. Steiner, *Phys. Rev. Lett.* **102**, 122701 (2009).
15. M. B. Tsang *et al.*, *Phys. Rev. C* **86**, 015803 (2012).
16. C. J. Horowitz, E. F. Brown, Y. Kim, W. G. Lynch, R. Michaels, A. Ono, J. Piekarewicz, M. B. Tsang and H. H. Wolter, *J. Phys. G* **41**, 093001 (2014).
17. C. J. Horowitz and A. Schwenk, *Nucl. Phys. A* **776**, 55 (2006).
18. J. B. Natowitz *et al.*, *Phys. Rev. Lett.* **104**, 202501 (2010).
19. S. Typel, G. Röpke, T. Klähn, D. Blaschke and H. H. Wolter, *Phys. Rev. C* **81**, 015803 (2010).
20. L. Qin *et al.*, *Phys. Rev. Lett.* **108**, 172701 (2012).
21. B. A. Brown, *Phys. Rev. Lett.* **85**, 5296 (2000).
22. C. Fuchs and H. H. Wolter, *Eur. Phys. J. A* **30**, 5 (2006).
23. R. Subedi *et al.*, *Science* **320**, 1476 (2008).
24. O. Hen *et al.*, *Science* **346**, 614 (2014).
25. Chang Xu and Bao-An Li, *Phys. Rev. C* **81**, 064612 (2010).
26. A. W. Steiner and S. Gandolfi, *Phys. Rev. Lett.* **108**, 081102 (2012).
27. A. Carbone, A. Polls and A. Rios, *Europhys. Lett.* **97**, 22001 (2012).
28. Or Hen, Bao-An Li, Wen-Jun Guo, L. B. Weinstein and E. Piasetzky, *Phys. Rev. C* **91** 025803 (2015).
29. Gao-Chan Yong, *Phys. Rev. C* **93**, 044610 (2016).
30. Bao-An Li and Xiao Han, *Phys. Lett. B* **727**, 276 (2013).
31. V. Baran, M. Colonna, V. Greco and M. Di Toro, *Phys. Rep.* **410**, 335 (2005).
32. A. Klimkiewicz *et al.*, *Phys. Rev. C* **76**, 051603(R) (2007).
33. A. Tamii *et al.*, *Phys. Rev. Lett.* **107**, 062502 (2011).
34. Zhen Zhang and Lie-Wen Chen, *Phys. Lett. B* **726**, 234 (2013).

35. B. A. Brown, *Phys. Rev. Lett.* **111**, 232502 (2013).
36. Bao-An Li, *Phys. Rev. Lett.* **88**, 192701 (2002).
37. P. Russotto *et al.*, *Phys. Lett. B* **697**, 471 (2011).
38. Y. Leifels *et al.*, *Phys. Rev. Lett.* **71**, 963 (1993).
39. D. Lambrecht *et al.*, *Z. Phys. A* **350**, 115 (1994).
40. W. Trautmann and H. H. Wolter, *Int. J. Mod. Phys. E* **21**, 1230003 (2012).
41. W. Reisdorf *et al.*, *Nucl. Phys. A* **781**, 459 (2007).
42. G. Ferini *et al.*, *Nucl. Phys. A* **762**, 147 (2005); *Phys. Rev. Lett.* **97**, 202301 (2006).
43. Zhigang Xiao *et al.*, *Phys. Rev. Lett.* **102**, 062502 (2009).
44. Zhao-Qing Feng and Gen-Ming Jin, *Phys. Lett. B* **683**, 140 (2010).
45. Wen-Jie Xie, Jun Su, Long Zhua and Feng-Shou Zhang, *Phys. Lett. B* **718**, 1510 (2013).
46. Bao-An Li *et al.*, *J. Phys. Conf. Ser.* **312**, 042006 (2011).
47. De-Hua Wen, Bao-An Li and Lie-Wen Chen, *Phys. Rev. Lett.* **103**, 211102 (2009).
48. M. D. Cozma, *Phys. Lett. B* **700**, 139 (2011).
49. Th. Blaich *et al.*, *Nucl. Instrum. Methods Phys. Res. A* **314**, 136 (1992).
50. A. Pagano *et al.*, *Nucl. Phys. A* **734**, 504 (2004).
51. P. Russotto *et al.*, *Phys. Rev. C* **94**, 034608 (2016).
52. M. Baldo, C. Maieron, P. Schuck and X. Viñas, *Nucl. Phys. A* **736**, 241 (2004).
53. K. Fukukawa, M. Baldo, G. F. Burgio, L. Lo Monaco and H.-J. Schulze, *Phys. Rev. C* **92**, 065802 (2015).
54. C. Drischler, A. Carbone, K. Hebeler and A. Schwenk. *Phys. Rev. C* **94**, 054307 (2016).
55. R. B. Wiringa and S. C. Pieper, *Phys. Rev. Lett.* **89**, 182501 (2002).
56. F. Burgio, *J. Phys. G* **35**, 014048 (2008).
57. K. Hebeler, J. M. Lattimer, C. J. Pethick and A. Schwenk, *Phys. Rev. Lett.* **105**, 161102 (2010).
58. A. W. Steiner, J. M. Lattimer and E. F. Brown, *ApJ.* **722**, 33 (2010).
59. See, e.g., the NASA website for the Neutron star Interior Composition ExploreR Mission (NICER), https://heasarc.gsfc.nasa.gov/docs/nicer/
60. V. Giordano, M. Colonna, M. Di Toro, V. Greco and J. Rizzo, *Phys. Rev. C* **81**, 044611 (2010).
61. Zhao-Qing Feng, *Phys. Lett. B* **707**, 83 (2012).
62. Xiao-Hua Li, Wen-Jun Guo, Bao-An Li, Lie-Wen Chen, F. J. Fattoyev and W. G. Newton, *Phys. Lett. B* **743**, 408 (2015).
63. D. D. S. Coupland *et al.*, *Phys. Rev. C* **94**, 011601(R) (2016).
64. Qingfeng Li, Zhuxia Li, S. Soff, M. Bleicher and H. Stöcker, *J. Phys. G* **32**, 151 (2006); *ibid.* **32**, 407 (2006).
65. C. B. Das, S. Das Gupta, C. Gale and Bao-An Li, *Phys. Rev. C* **67**, 034611 (2003).
66. Bao-An Li and Lie-Wen Chen, *Phys. Rev. C* **72**, 064611 (2005).
67. Yongjia Wang, Chenchen Guo, Qingfeng Li, Hongfei Zhang, Y. Leifels and W. Trautmann, *Phys. Rev. C* **89**, 044603 (2014).
68. ChenChen Guo, YongJia Wang, QingFeng Li, W. Trautmann, Ling Liu and LiJuan Wu, *Science China Physics, Mechanics & Astronomy* **55**, 252 (2012).
69. J. M. Lattimer and Y. Lim, *Astrophys. J.* **771**, 51 (2013).
70. B. G. Todd-Rutel and J. Piekarewicz, *Phys. Rev. Lett.* **95**, 122501 (2005).
71. X. Roca-Maza, M. Centelles, X. Viñas and M. Warda, *Phys. Rev. Lett.* **106**, 252501 (2011).
72. S. Typel and B. A. Brown, *Phys. Rev. C* **64**, 027302 (2001).
73. R. J. Furnstahl, *Nucl. Phys. A* **706**, 85 (2002).

74. C. J. Horowitz *et al.*, *Phys. Rev. C* **85**, 032501(R) (2012).
75. S. Abrahamyan *et al.*, *Phys. Rev. Lett.* **108**, 112502 (2012).
76. Bao-An Li, *Nucl. Phys. A* **708**, 365 (2002).
77. Jun Xu, Lie-Wen Chen, Che Ming Ko Bao-An Li and Yu Gang Ma, *Phys. Rev. C* **87**, 067601 (2013).
78. V. Greco, V. Baran, M. Colonna, M. Di Toro, T. Gaitanos and H. H. Wolter, *Phys. Lett. B* **562**, 215 (2003).
79. X. Lopez *et al.*, *Phys. Rev. C* **75**, 011901(R) (2007).
80. M. D. Cozma, *Phys. Lett. B* **753**, 166 (2016).
81. N. Ikeno, A. Ono, Y. Nara and A. Ohnishi, *Phys. Rev. C* **93**, 044612 (2016).
82. Taesoo Song and Che Ming Ko, *Phys. Rev. C* **91**, 014901 (2015).
83. Bao-An Li, *Phys. Rev. C* **92**, 034603 (2015).
84. Jun Hong and P. Danielewicz, *Phys. Rev. C* **90**, 024605 (2014).
85. M. D. Cozma, *Phys. Rev. C* **95**, 014601 (2017).
86. M. B. Tsang *et al.*, *Phys. Rev. C* **95**, 044614 (2017).
87. R. Shane *et al.*, *Nucl. Instrum. Methods Phys. Res. A* **784**, 513 (2015).
88. G. F. Bertsch, H. Kruse and S. Das Gupta, *Phys. Rev. C* **29**, 673 (1984).
89. H. Kruse, B. V. Jacak, J. J. Molitoris, G. D. Westfall and H. Stöcker, *Phys. Rev. C* **31**, 1770 (1985).
90. J. Aichelin and H. Stöcker, *Phys. Lett. B* **176**, 14 (1986).
91. G. F. Bertsch and S. Das Gupta, *Phys. Rep.* **160**, 189 (1988).
92. J. Aichelin, *Phys. Rep.* **202**, 233 (1991).
93. Jun Xu *et al.*, *Phys. Rev. C* **93**, 044609 (2016).
94. H. H. Gutbrod *et al.*, *Phys. Rev. C* **42**, 640 (1990).
95. J.-Y. Ollitrault, preprint arXiv:nucl-ex/9711003 (1997).
96. A. M. Poskanzer and S. A. Voloshin, *Phys. Rev. C* **58**, 1671 (1998).
97. B. I. Abelev *et al.*, *Phys. Rev. Lett.* **99**, 112301 (2007).
98. R. Fries, V. Greco and P. Sorensen, *Annu. Rev. Nucl. Part. Sci.* **58**, 177 (2008).
99. A. Adare *et al.*, *Phys. Rev. C* **85**, 064914 (2012).
100. U. W. Heinz and R. Snellings, *Annu. Rev. Nucl. Part. Sci.* **63**, 123 (2013).
101. P. Danielewicz, *Nucl. Phys. A* **673**, 375 (2000) 375.
102. A. Andronic, J. Łukasik, W. Reisdorf and W. Trautmann, *Eur. Phys. J. A* **30**, 31 (2006).
103. J. Łukasik *et al.*, *Phys. Lett. B* **608**, 223 (2005).
104. A. Andronic *et al.*, *Phys. Lett. B* **612**, 173 (2005).
105. C. Pinkenburg *et al.*, *Phys. Rev. Lett.* **83**, 1295 (1999).
106. P. Braun-Munzinger and J. Stachel, *Nucl. Phys. A* **638**, 3c (1998).
107. M. D. Cozma, Y. Leifels, W. Trautmann, Q. Li and P. Russotto, *Phys. Rev. C* **88**, 044912 (2013).
108. P. Russotto, M. D. Cozma, A. Le Fèvre, Y. Leifels, R. Lemmon, J. Łukasik, Q. Li and W. Trautmann, *Eur. Phys. J. A* **50**, 38 (2014).
109. A. Schüttauf *et al.*, *Nucl. Phys. A* **607**, 457 (1996).
110. A. Pagano *et al.*, *Nucl. Phys. A* **734**, 504 (2004).
111. D. G. Sarantites *et al.*, *Nucl. Instr. and Meth. A* **381**, 418 (1996).
112. P. Danielewicz and J. Lee, *Nucl. Phys. A* **922**, 1 (2014).
113. A. W. Steiner, J. M. Lattimer and E. F. Brown, *Astrophys. J. Lett.* **765**, L5 (2013).
114. NeuLAND Technical Design Report, submitted to FAIR (2011).

Black Holes and High Energy Physics: From Astrophysics to Large Extra Dimensions

Michael Florian Wondrak,[1] Marcus Bleicher[2] and Piero Nicolini[3]

Frankfurt Institute for Advanced Studies (FIAS),
Ruth-Moufang-Str. 1, 60438 Frankfurt am Main, Germany
Institut für Theoretische Physik, Johann Wolfgang Goethe-Universität Frankfurt,
Max-von-Laue-Str. 1, 60438 Frankfurt am Main, Germany
[1]*wondrak@fias.uni-frankfurt.de,* [2]*bleicher@fias.uni-frankfurt.de,*
[3]*nicolini@fias.uni-frankfurt.de*

Up to now, Einstein's general theory of relativity has passed all experimental tests. But yet we know that it is not a fundamental theory and that it is incompatible with quantum theory. While several extended and improved gravitational theories on classical and quantum level are nowadays available, it is a great challenge to find experimental setups to check their signatures.

We discuss recent developments in direct observation of black holes comprising gravitational waves from black hole mergers, radio interferometry images of black hole shadows, and Hawking radiation of black holes in particle accelerators. These investigations cover the full black hole mass range from microscopic to stellar and supermassive black holes. We comment on the associated strong-field tests of Einstein's general theory of relativity and implications for quantum gravity.

Special emphasis lies upon the physics of large extra dimensions and black hole evaporation, the existence of a minimal black hole mass, and the cross sections of higher-dimensional black holes. We complete this short review with the latest experimental constraints at the Large Hadron Collider.

> "Theoretical physicists read in the book of nature
> — but experimentalists turn its pages."
>
> *Walter Greiner (1935–2016)*

1. Introduction

In these days, gravitational physics and astrophysics enter a new era: Black holes, which belong to the most astounding objects resulting from Einstein's general theory of relativity, come into reach of direct observation. Previous experimental studies of black holes have only been possible indirectly, e.g., by observing the matter-donating companion star in an X-ray binary system.[1,2] Using gravitational wave experiments and very-long-baseline interferometry, black holes can now be studied directly.[3,4] Although these astronomically observable black holes have masses from few to billions of solar masses ($M_\odot$), there is the speculative possibility of much smaller black holes with masses in the TeV range, i.e., potentially being produced in high-energy particle collisions.[5–9] Since improved experimental techniques have been developed, one can hope to distinguish in the near future between black holes resulting from Einstein's theory and those from alternative theories, including the case of completely different massive objects.[10]

In Sec. 2, we describe the possibility of testing the concept of general relativistic black holes and alternative ideas for new astronomical observations. We go on to review the idea of large extra dimensions with special emphasis on the production of black holes in accelerators and report on the experimental status in Sec. 3. In Sec. 4, we draw our conclusions and give an outlook.

2. Astrophysical Black Holes

The concept of black holes goes back to the 18th century. Based on Descartes' and Newton's corpuscular model, Mitchell suggested, in 1784, that even light could not escape from a massive object if its gravitational force is strong enough. The first black hole solution to Einstein's field equations has been derived by Schwarzschild in 1916, but it has been recognized as such by Finkelstein only in 1958. In the late 1960s, Hawking and Penrose showed the role of black holes as generic end-stage solutions for gravitational collapses within the Einstein's theory of relativity.[11]

On the observational side, however, the situations is less clear. The closest available hints are indirect. Observations suggest the existence of compact massive objects whose spatial extensions are too small to be composed of several ordinary astrophysical objects or neutron stars. For instance, studies of the trajectories of the closest stars to the galactic center showed that there is a total mass of $4.3 \times 10^6\,M_\odot$ confined in a region of 4.4×10^{10} m radius at the center of our Milky Way[12] (cf. Refs. 13 and 14). Nowadays, however, experimental techniques, such as gravitational wave detectors and radio interferometers, have drastically improved and may allow for a direct study of black hole properties.

The first gravitational wave signal has been detected in September 2015 by the LIGO and Virgo collaborations.[3] In the meantime, two further signals and one candidate signal with decreased signal-to-noise ratio have been recorded.[15–17] The analyses of these events strongly support black hole mergers as sources while they exclude the case of neutron star mergers. According to the simulations, the mergers took place in a distance between 1.3×10^9 ly and 2.9×10^9 ly and the participating black hole masses were in the range between $8\,M_\odot$ and $36\,M_\odot$.

The detected signals perfectly agree with predictions based on black holes in Einstein's general relativity up to 5σ.[16] Especially the attenuation of the signal in the ring-down phase (the phase after the actual merger when the final black hole settles) indicates a photon ring which is only present if the object is compact enough.[18] However, there is room for other objects to generate such a gravitational wave signal. For instance, according to Refs. 19 and 20, the precision is yet not sufficient to distinguish between general relativistic black holes and those of alternative theories of gravity.

In contrast to gravitational wave findings of stellar-mass black holes, the very-long-baseline interferometry (VLBI) concentrates on imaging of the supermassive species. The latter possess the largest spatial extensions and distinctive accretion disks so that they subtend the largest angles for an observer on Earth. The synchronized phase-sensitive recordings of many radio telescopes on several continents and subsequent data correlation resemble a radio telescope whose dish size lies in the order of Earth's diameter. The Event Horizon Telescope collaboration (EHT) reaches a resolution of 35 µas which corresponds to the apparent size of a tennis ball on the Moon as seen from Earth. The aim is to record the black hole shadow, i.e., the silhouette of the black hole in front of its background including gravitational lensing effects, and to resolve the origin of jets. The EHT had a promising measurement campaign in April 2017. The data analysis is expected to release the first results in 2018.

The primary focus of the EHT lies upon Sagittarius A* (Sgr A*), the black hole in the center of our Milky Way. As a radio source, its luminosity on the longest wavelengths (namely frequencies up to 1 THz) amounts to $\sim 100\,L_\odot$, where $L_\odot$ is the Sun's luminosity over the whole spectrum. Even though small on the scale of supermassive black holes, this brightness resulted in the discovery of Sgr A* by radio interferometry in 1974. Furthermore, at a distance of 2.6×10^4 ly, it is the closest supermassive black hole having a mass of $4.3 \times 10^6\,M_\odot$. The corresponding angle, 53 µas is the largest among all supermassive black holes and allows for radio observation.

The second object of interest is the black hole in the center of the giant elliptical galaxy M87. With a mass of $6 \times 10^9\,M_\odot$, it is the most massive black hole known. M87 lies in the center of the Virgo galaxy cluster. Even though its distance from Earth is 5×10^7 ly, it represents the center of the Virgo supercluster which the local group, including the Milky Way, belongs to. Because of its size, the black hole's angular diameter, 22 µas, is of the same order as the EHT resolution. M87 is of special interest due to its strongly collimated jet of high-energy particles which extends up to 2.5×10^5 ly from the central black hole.

The investigation of the shadow serves as an experimental testbed for gravitational theories. For Einstein's gravity black holes, the shape depends on the mass and angular momentum only. However, some alternative theories of gravity also lead to modifications of the shape, e.g., Refs. 21 and 22. One such theory is pseudo-complex general relativity (pc-GR) proposed by Hess and Greiner in order to prevent the occurrence of singularities.[10,23–25] Their ansatz algebraically generalizes spacetime coordinates to pseudo-complex, or hypercomplex, numbers which leads to new metric degrees of freedom. In contrast to general relativity, a new energy-momentum contribution arises which is similar to dark energy. Due to this repulsive pressure, a gravitational collapse finally comes to rest and produces a compact object of finite size, a so-called gray star, instead of a black hole. The predicted shadow of a gray star in pc-GR differs from the one of black holes in ordinary general relativity.[26,27] Thus this theory could be tested in the future when the EHT gains further resolution.

There is some hope that new, unexpected effects occur in the strong-field observations which have a quantum origin, e.g., probable echoes in gravitational wave signals,[28,29] and that those hint at a more fundamental theory of physical interactions.[30] Several reasons indicate that both the Standard Model of particle physics as well as Einstein's general relativity are not fundamental. Most obviously, we do not know a consistent theory which unifies all interactions. The Standard Model neither explains the non-zero neutrino masses nor the background of the external parameters. General relativity at classical level is plagued by long standing problems such as curvature singularities, the unknown nature of dark matter, and the cosmological constant problems (e.g., Refs. 31–33). Moreover, it lacks a direct quantization due to its non-renormalizable character.

At a recent conference[a] more than 80% of the participants were convinced that within the next decade there will be observational evidence for quantum gravity. It could be possible to gain this experimental access by a different approach. On the other side of the black hole mass spectrum, microscopic black holes might be produced in particle accelerators, due to increased energy capabilities, e.g., at the Large Hadron Collider as we will see in the next section.[34–47]

3. Large Extra Dimensions

For the discussion of gravity at small scales, we follow the lines of Ref. 47. The difference between the Standard Model interactions and general relativity manifests itself impressively in their dissimilar coupling strengths. While the electromagnetic, weak and strong forces differ by just 6 orders of magnitude, the gravitational interaction falls apart by further 33 orders. The ratio between the Fermi constant, G_{F}, and the gravitational constant, G_{N}, amounts to

$$\frac{G_{\mathrm{F}}}{G_{\mathrm{N}}} \approx 1.7 \times 10^{33}. \tag{1}$$

The weakness of gravity compared to the other forces is called the hierarchy problem.

The gravitational constant is related to the fundamental Planck mass, M_{Pl}, by

$$G_{\mathrm{N}} = \frac{1}{M_{\mathrm{Pl}}^2}, \tag{2}$$

where $M_{\mathrm{Pl}} \approx 1.2 \times 10^{16}\,\mathrm{TeV} \approx 2.2 \times 10^{-8}\,\mathrm{kg}$. Since G_{N} couples to two masses, m_1 and m_2, the term $G_{\mathrm{N}}\, m_1\, m_2$ has to be of the order of 1 for reaching a strength

[a]3rd Karl Schwarzschild Meeting on Gravitational Physics and the Gauge/Gravity Correspondence, Frankfurt am Main, July 2017.

comparable to the other interactions. Thus, quantum gravity is supposed to set in at the Planck scale.

The Standard Model is an excellent description of particle physics corroborated by experimental evidence — up to the LHC energy scale of $\sqrt{s} = 14\,\text{TeV} = 2.2 \times 10^{-6}\,\text{J}$. The gap of 15 orders of magnitude up to the Planck scale cannot be accessed experimentally yet. Since we do not know the physics in these energy regimes, one could simply assume that no new particles or physical effects emerge, which is named the big dessert approximation. Another idea to tackle the hierarchy problem is to find a mechanism which reduces the fundamental gravitational energy scale. This is realized in the extra dimension ansatz proposed by Arkani-Hamed, Dimopoulos, and Dvali together with Antoniadis, the so-called ADD model.[48–50]

The concept of introducing a higher dimensional spacetime goes back to Kaluza and Klein who tried to unify gravity and classical electrodynamics by adding an additional spatial dimension to form a 5-dimensional spacetime.[51–53] The Maxwell equations directly emerge from the higher dimensional Einstein field equations. Furthermore, the quantization of electric charge automatically appears by compactifying the extra dimension. Apart from the ADD model, extra dimensions are also employed in (9+1)-dimensional super string theory and (10+1)-dimensional M-theory.

ADD proposed that our familiar (3+1)-dimensional world is just a slice, a so-called D3-brane, which is embedded in a higher-dimensional bulk spacetime having k additional spatial dimensions. While Standard Model fields are confined to the brane, the extra dimensions are only accessible for gravitational fields. In this way, non-gravitational physics is the same while gravitational physics on small scales becomes much stronger. The strength of gravity can be understood by the new profile of the extra-dimensional Newton's law. However, in order to comply with short scale precision tests of gravity, which confirm the inverse square law down to $56\,\mu\text{m}$,[54] the additional dimensions have to be compactified.

The theory introduces two new degrees of freedom: the number of spatial extra dimensions, k, and the new fundamental gravitational energy scale,

$$M_D \equiv \left(\frac{1}{8\pi} \frac{1}{G_D} \right)^{1/(2+k)}, \tag{3}$$

where G_D is the gravitational constant in $D = ((3+k)+1)$ dimensions. Relating G_D to the ordinary 4-dimensional gravitational constant G_N determines the compactification radius, R_c, of the extra dimensions,

$$R_\text{c} = \frac{1}{2\pi} \left(\frac{M_\text{Pl}^2}{8\pi\, M_D^{2+k}} \right)^{1/k} \sim 10^{\frac{32}{k} - 19}\ \text{m}, \tag{4}$$

where we assumed $M_D \sim 1\,\text{TeV}$ in the last step. In this case, consistency with experiments requires $k \geq 3$ and R_c follows to be in the range [fm, nm].

Analogously to the static, uncharged Schwarzschild black hole, we can find a black hole solution to Einstein's field equations in the higher-dimensional spacetime.[b,57,58] The line element is given by

$$ds^2 = -\left(1 - \left(\frac{r_\mathrm{h}}{r}\right)^{1+k}\right) dt^2 + \frac{1}{\left(1 - \left(\frac{r_\mathrm{h}}{r}\right)^{1+k}\right)} dr^2 + r^2 \, d\Omega_{2+k}^2 \tag{5}$$

where $d\Omega_{2+k}^2$ is the usual angular line element in $2+k$ dimensions. The event horizon is located at

$$r_\mathrm{h} = \left(\frac{\Gamma\left(\frac{3+k}{2}\right) M_\mathrm{BH}}{(2+k)\,\pi^{(3+k)/2}\, M_D^{2+k}} \right)^{1/(1+k)} . \tag{6}$$

For microscopic black holes with masses $M_D \lesssim M_\mathrm{BH} \ll M_\mathrm{Pl}$, the event horizon is much smaller than the compactification radius R_c. This ensures the insensitivity of the black hole to the compactification of dimensions.

A fundamental concept in quantum physics is the delocalization of particles. The typical length scale of the probability distribution is given by the reduced Compton wavelength,

$$\lambda\!\!\!^{-}_\mathrm{C} = \frac{1}{m} . \tag{7}$$

Starting from this principle we naturally derive the existence of a lower mass limit for black holes. The mass of a black hole defines its horizon radius (6). In order not to leak out, the black hole has to be self-contained within its reduced Compton wavelength. If both characteristic scales agree, a black hole has the minimal mass $M_\mathrm{BH}^\mathrm{min}$ which amounts to

$$M_\mathrm{BH}^\mathrm{min} \sim M_D \sim \left(\frac{l_\mathrm{Pl}}{R_\mathrm{c}}\right)^{k/(2+k)} M_\mathrm{Pl}, \tag{8}$$

where $l_\mathrm{Pl} \approx 1.62 \times 10^{-35}$ m is the ordinary 4-dimensional Planck length. Of course, this argumentation assumes that the concepts of the Compton wavelength and of the classical horizon hold up to the fundamental energy scale M_D. The actual minimal mass will be determined by a theory of quantum gravity.[59–75]

If M_D lies in the range of TeV, microscopic black holes could be produced in particle accelerators.[5–8] The LHC center-of-mass energy for pp collisions lies at $\sqrt{s} = 14\,\mathrm{TeV} = 2.2 \times 10^{-6}\,\mathrm{J}$ at a design luminosity of $L = 10^{10}\,\mathrm{b}^{-1}\,\mathrm{s}^{-1}$ which has already been exceeded in the last year.[76] In head-on collisions of $^{208}_{82}\mathrm{Pb}$, the mean center-of-mass energy per nucleon pair amounts to 2.76 TeV. If two partons with normalized initial longitudinal momentum x_1 and x_2, respectively, collide at a center-of-mass energy of $\sqrt{\hat{s}} = \sqrt{x_1\, x_2\, s} = M_\mathrm{BH}$, they might turn into a black hole of that mass.

A natural ansatz for the production cross section at parton level is the black disk approximation[77] which uses the geometric cross section of the event horizon,

$$\hat{\sigma}(M_\mathrm{BH}, k) = \pi\, r_\mathrm{h}^2(M_\mathrm{BH}) . \tag{9}$$

[b]For a discussion about higher dimensional gravity field equations cf. Ref. 55. For the repercussions of higher dimensional scenarios in gravitational waves cf. Ref. 56.

However, this ansatz allows the formation of arbitrarily small black holes if the impact parameter in the parton–parton collision is just small enough. To cure this deficit, an improved cross section has been proposed in Ref. 78,

$$\hat{\sigma}(M_{\rm BH},\, k) = \pi l^2\, \Gamma\!\left(-1;\, \frac{l^2}{r_{\rm h}^2(M_{\rm BH})}\right)\, \theta_l\big(M_{\rm BH} - M_{\rm BH}^{\rm min}\big)\,. \tag{10}$$

It takes into consideration that at the energy level of the fundamental scale M_D, a finite resolution of spacetime of size l emerges as suggested by several theories, e.g., Refs. 79–85. $\Gamma(\alpha;\, x)$ denotes the upper incomplete gamma function

$$\Gamma(\alpha;\, x) \equiv \int_x^\infty {\rm d}t\, t^{\alpha-1}\, e^{-t}, \tag{11}$$

and $\theta_l(x)$ the modified Heaviside step function

$$\theta_l(x) \equiv \frac{1}{(4\pi\, l^2)^{\frac{1}{2}}} \int_{-\infty}^{x} {\rm d}y\, {\rm e}^{-\frac{y^2}{4l^2}}\,. \tag{12}$$

The differential cross section of black hole production with respect to the black hole mass $M_{\rm BH}$ depends on the parton distribution functions $f_i\big(x_j,\, Q^2\big)$ of parton type i in hadron j. It turns out to be

$$\frac{{\rm d}\sigma}{{\rm d}M_{\rm BH}} = \sum_{a,\, b} \int_0^1 {\rm d}x_1\, \frac{2M_{\rm BH}}{x_1\, s}\, f_a\big(x_1,\, M_{\rm BH}^2\big)\, f_b\!\left(\frac{M_{\rm BH}^2}{x_1\, s},\, M_{\rm BH}^2\right)\, \hat{\sigma}(M_{\rm BH},\, k)\,. \tag{13}$$

For instance, assuming an integral cross section of 10 nb, we would expect 10^2 black holes per second to be produced in pp collisions at the LHC design luminosity.[86,87]

Once a microscopic black hole is produced, it suffers a loss of mass due to the emission of Hawking radiation. The evaporation closely follows a blackbody spectrum where the temperature is given by the Hawking temperature[c,88]

$$T_{\rm BH} = \frac{\hbar c\, \kappa}{2\pi\, k_{\rm B}}. \tag{14}$$

Here κ denotes the black hole's surface gravity which grows the faster, the smaller the black hole becomes. This implies that the evaporation rate becomes increasingly large during the evaporation, with the smallest black holes producing the largest radiation densities.

The evaporation process of microscopic black holes can be classified in several stages: the balding phase, the spin-down phase, and the Schwarzschild phase. In the first one the black hole radiates away its electric and other gauge field charges and stabilizes towards a fast rotating state by emission of gravitational waves. In the second phase, its angular velocity slows down due to Hawking and Unruh–Starobinskii radiation. When the hole configuration switches to a static Schwarzschild one, the radiation is emitted with spherical symmetry and one speaks of third phase.

[c]Due to the gravitational potential, part of the emitted radiation is reflected so that an observer at spatial infinity actually perceives a graybody modification depending on the particle spin, angular momentum, and frequency.

The stages above are valid for black holes with masses (far) above the fundamental mass scale, M_D, and based on semiclassical analysis. Close to M_D, the description of so-called quantum black holes critically depends on the full theory of quantum gravity which is not yet available. However, several effective models agree in predicting a SCRAM[d] phase: Instead of a final explosion at an infinite temperature, the black hole turns into a light-weight remnant with vanishing temperature.[41,61,64,65,70,74,75,82–85,89–104]

If the hole is hot enough, the evaporation produces all Standard Model particles on nearly equal footing since no gauge interactions are involved. Due to the flavor and color degrees of freedom and including antiparticles, there are 36 different quarks compared with 6 different charged leptons. Therefore, around 75% of the primary emission are quarks and gluons. A detector, however, will detect a different composition. Because of mutual interactions of the emitted particles, plasmas around the black holes can arise, a chromosphere due to QCD reactions, and a QED-related photosphere.[38] Furthermore, the decay of instable particles will enhance the fraction of photons, neutrinos, electrons, and positrons. In experiments, data analysis concentrates on events with multiple high energetic objects. For semiclassical black holes, a lack of transverse energy is expected due to the emission of gravitons and neutrinos. In the case of quantum black holes, only few high-energy particles are supposed to be produced. The evaporation is expected to take place within 10^{-27} s for semiclassical black holes and even shorter for quantum black holes.[9] Conversely it has been argued that the SCRAM phase would elongate the evaporation time ($\sim 10^{-16}$ s), due to the reduced emission in the final stage of the evaporation.[105,106]

At the LHC, the ATLAS and CMS collaborations look for black hole signatures. Up to now, no hint for black holes has been found and there is no significant deviation from the expected Standard Model background, cf. Fig. 1. Theoretical predictions for the black hole production cross sections are obtained from event generators: BlackMax[107] and Charybdis 2[108–110] for semiclassical black holes, and QBH[111,112] for a model of quantum black holes. Comparison of these theoretically predicted with experimental upper limits for the cross sections leads to lower bounds on the minimal black hole mass, $M_{\mathrm{BH}}^{\mathrm{min}}$. This is illustrated in Fig. 2 for a CMS analysis of LHC Run 1 with $\sqrt{s} = 7\,\mathrm{TeV}$. The latest CMS analysis from Run 2 at $\sqrt{s} = 13\,\mathrm{TeV}$[9] has increased the constraints on the minimal black hole mass. For semiclassical black holes and for quantum black holes, $M_{\mathrm{BH}}^{\mathrm{min}}$ has to be to larger than 7 TeV to 9 TeV, cf. Fig. 3. This is consistent with the constraints derived by ATLAS.[113]

[d]This term has been adopted in Ref. 41 from the emergency shutdown of nuclear reactors. It originates from the Manhattan Project at Chicago Pile-1 in 1942 and Enrico Fermi is reported to coin the backronym *Safety Control Rod Axe Man*.

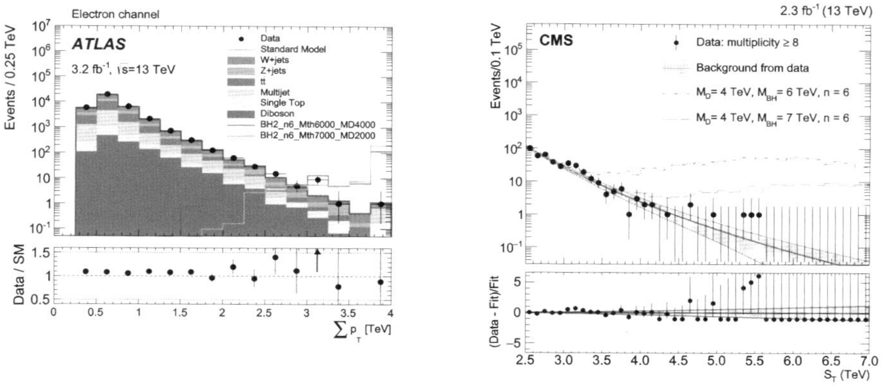

(a) (b)

Figure 1. LHC Run 2 data at $\sqrt{s} = 13\,\text{TeV}$ compared to Standard Model predictions. (a) The dots and their error bars show the experimental spectrum of events with high transverse momenta, p_{T}, objects. Only those events are considered which posses a high-p_{T} electron and at least 2 further objects with large p_{T}. The histogram indicates the theoretical contributions of different processes according to the Standard Model. The lower panel shows the ratio of the experimental counts to the Standard Model expectations. The blue and green curves display black hole evaporation spectra simulated by the Charybdis 2 generator for rotating semiclassical black holes in 6 extra dimensions. (b) An analogue graph based on the final particles' total transverse energy, S_{T}, for a combined jet, lepton, and photon multiplicity greater than 7. The lower panel shows the relative deviation of the data from the Standard Model predictions. The red and the green curve provide the simulated evaporation spectra of two semiclassical black holes for reference. Figures from Refs. 9 and 113.

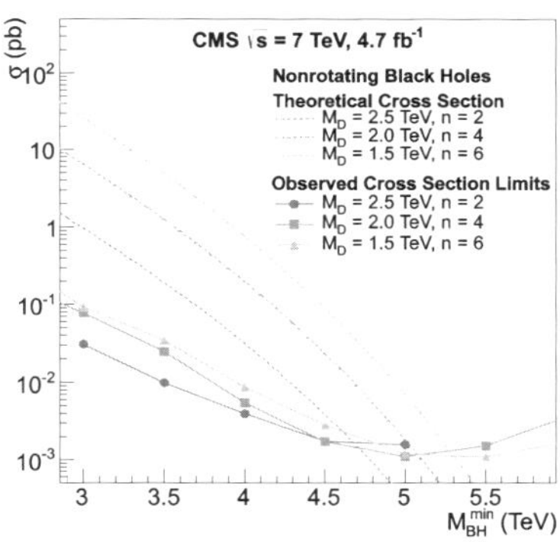

Figure 2. Experimental limits on the black hole production cross section as a function of the minimal black hole mass, $M_{\text{BH}}^{\text{min}}$, from LHC Run 1 at $\sqrt{s} = 7\,\text{TeV}$. The solid lines provide the experimentally derived upper bounds for a given fundamental energy scale, M_D, and number of extra dimensions, n. The dotted curves show the expected cross sections simulated by the BlackMax generator for complete evaporation. $M_{\text{BH}}^{\text{min}}$ is constrained to be larger than approx. $5\,\text{TeV}$. Figure from Ref. 114.

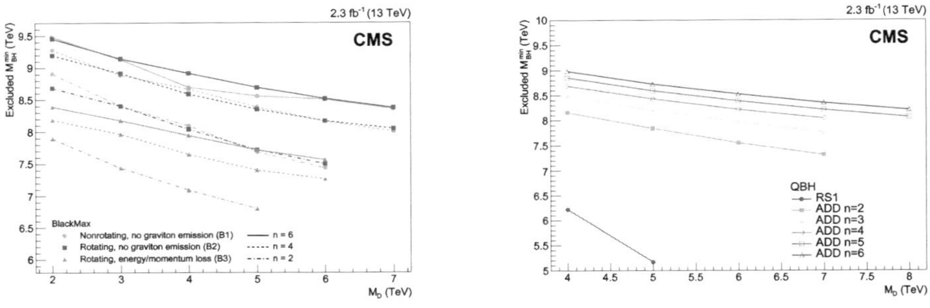

(a) (b)

Figure 3. Lower threshold for the minimal black hole mass, $M_{\mathrm{BH}}^{\mathrm{min}}$, from LHC Run 2 at $\sqrt{s} = 13\,\mathrm{TeV}$ as a function of the fundamental gravitational energy scale, M_D, for different numbers of extra dimensions, n. (a) Constraints for semiclassical black holes simulated with the BlackMax generator. (b) Constraints for quantum black holes simulated with the QBH generator. The limit for a Randall-Sundrum extra dimensional model (RS1) is indicated for comparison. Figures from Ref. 9.

4. Conclusions and Outlook

Recent experimental progress has started an exciting time for gravitational physics and astrophysics. The time of direct observations of black holes seems to become reality. Black holes in the microscopic, stellar, and supermassive regime can be targeted by dedicated improved investigation methods.

Theoretically, gravitational wave signals and black hole shadows can be predicted by most alternative gravitational theories. Experimentally, they can be recorded by corresponding observatories like LIGO and the EHT. By comparison, Einstein's general theory of relativity undergoes strong-field tests against alternative theories like pseudo-complex general relativity. Among possible classical deviations, maybe quantum gravity effects manifest themselves in some hidden details.

As a possible solution to the hierarchy problem, the ADD model proposes compactified large extra dimensions which are only accessible for gravitational interactions. Consequently, the fundamental gravitational energy scale, M_D, is drastically decreased; it could lie in the range of present particle accelerators. The LHC reaches energies which might allow the production of black holes. Their traces due to evaporation, namely additional events with a high multiplicity of large transverse momentum objects, have not yet been detected. Therefore the lower limit for the minimal black hole mass has been increased to 7 TeV to 9 TeV in LHC Run 2.

Further developments rely on improvements of the experiments. By decreasing the noise in gravitational wave observatories, neutron star mergers should come into reach. The gravitational wave signal is supposed to constrain the neutron star equation of state and thus the QCD one, too. A higher resolution of the EHT will bring up more details of accretion physics and jet launching. Accordingly, the tests on Einstein's general relativity will become tighter.

On the microscopic black hole side, black hole signatures might be uncovered in the next LHC Runs due to the expected increase of beam luminosity. We expect data to allow for conclusions about final stages of the black hole evolution, namely the quantum phase where semiclassical gravity breaks down. In particular, it would be a great step towards quantum gravity if there were evidence for a SCRAM phase.

Acknowledgments

Our interest in black hole physics was strongly motivated by our teacher and friend Walter Greiner. We dedicated this article to his memory. In addition to Walter's guidance and support further support was provided by the Frankfurter Förderverein für physikalische Grundlagenforschung, by the Stiftung Polytechnische Gesellschaft Frankfurt am Main, and DFG.

References

1. C. T. Bolton, *Nature* **235**, 271–273 (1972).
2. B. L. Webster and P. Murdin, *Nature* **235**, 37–38 (1972).
3. Virgo, LIGO Scientific Collaboration, *Phys. Rev. Lett.* **116**, 061102 (2016).
4. H. Falcke, F. Melia, and E. Agol, *Astrophys. J.* **528**, L13 (2000).
5. P. C. Argyres, S. Dimopoulos, and J. March-Russell, *Phys. Lett. B* **441**, 96–104 (1998), `arXiv:hep-th/9808138`.
6. T. Banks and W. Fischler, `arXiv:hep-th/9906038`.
7. S. Dimopoulos and G. L. Landsberg, *Phys. Rev. Lett.* **87**, 161602 (2001), `arXiv:hep-ph/0106295`.
8. S. B. Giddings and S. D. Thomas, *Phys. Rev. D* **65**, 056010 (2002), `arXiv:hep-ph/0106219`.
9. The CMS Collaboration, `arXiv:1705.01403 [hep-ex]`.
10. P. O. Hess, L. Maghlaoui, and W. Greiner, *Int. J. Mod. Phys. D* **19**, 1217–1232 (2010).
11. S. W. Hawking and R. Penrose, *Proc. Roy. Soc. Lond. A* **314**, 529–548 (1970).
12. S. Gillessen, F. Eisenhauer, S. Trippe, T. Alexander, R. Genzel, F. Martins, and T. Ott, *Astrophys. J.* **692**, 1075–1109 (2009).
13. A. M. Ghez *et al.*, *Astrophys. J.* **689**, 1044–1062 (2008).
14. R. Schödel, D. Merritt, and A. Eckart, *Astron. Astrophys.* **502**, 91 (2009).
15. Virgo, LIGO Scientific Collaboration, *Phys. Rev. Lett.* **116**, 241103 (2016).
16. Virgo, LIGO Scientific Collaboration, *Phys. Rev. X* **6**, 041015 (2016).
17. Virgo, LIGO Scientific Collaboration, *Phys. Rev. Lett.* **118**, 221101 (2017).
18. V. Cardoso, E. Franzin, and P. Pani, *Phys. Rev. Lett.* **116**, 171101 (2016). [Erratum: *Phys. Rev. Lett.* **117**, 089902 (2016)].
19. R. Konoplya and A. Zhidenko, *Phys. Lett. B* **756**, 350–353 (2016), `arXiv:1602.04738 [gr-qc]`.
20. M. D. Laurentis, O. Porth, L. Bovard, B. Ahmedov, and A. Abdujabbarov, *Phys. Rev. D* **94**, 124038 (2016).
21. J. W. Moffat, *Eur. Phys. J. C* **75**, 130 (2015).

22. A. Grenzebach, V. Perlick, and C. Lämmerzahl, *Int. J. Mod. Phys. D* **24**, 1542024 (2015).
23. P. O. Hess and W. Greiner, *Int. J. Mod. Phys. E* **18**, 51–77 (2009).
24. G. Caspar, T. Schönenbach, P. O. Hess, M. Schäfer, and W. Greiner, *Int. J. Mod. Phys. E* **21**, 1250015 (2012).
25. P. O. Hess, M. Schäfer, and W. Greiner, *Pseudo-Complex General Relativity*. FIAS Interdisc. Sci. Ser. Springer, Heidelberg, 2016.
26. T. Schönenbach, G. Caspar, P. O. Hess, T. Boller, A. Müller, M. Schäfer, and W. Greiner, *Mon. Not. Roy. Astron. Soc.* **430**, 2999–3009 (2013).
27. T. Schönenbach, G. Caspar, P. O. Hess, T. Boller, A. Müller, M. Schäfer, and W. Greiner, *Mon. Not. Roy. Astron. Soc.* **442**, 121–130 (2014).
28. J. Abedi, H. Dykaar, and N. Afshordi, `arXiv:1612.00266 [gr-qc]`.
29. J. Abedi, H. Dykaar, and N. Afshordi, `arXiv:1701.03485 [gr-qc]`.
30. G. 't Hooft, S. B. Giddings, C. Rovelli, P. Nicolini, J. Mureika, M. Kaminski, and M. Bleicher, in *2nd Karl Schwarzschild Meeting on Gravitational Physics (KSM 2015) Frankfurt am Main, Germany, July 20–24, 2015*. 2016. `arXiv:1609.01725 [hep-th]`.
31. N. Afshordi, D. J. H. Chung, M. Doran, and G. Geshnizjani, *Phys. Rev. D* **75**, 123509 (2007).
32. N. Afshordi, `arXiv:0807.2639 [astro-ph]`.
33. M. F. Wondrak, *The Cosmological Constant and its Problems: A Review of Gravitational Aether*, pp. 109–120. FIAS Interdisc. Sci. Ser. Springer International Publishing, Cham, 2017. `arXiv:1705.06294 [gr-qc]`.
34. G. L. Landsberg, in *Supersymmetry and unification of fundamental interactions. Proceedings, 10th International Conference, SUSY'02, Hamburg, Germany, June 17–23, 2002*, pp. 562–577. 2002. `arXiv:hep-ph/0211043`.
35. M. Cavaglia, *Int. J. Mod. Phys. A* **18**, 1843–1882 (2003), `arXiv:hep-ph/0210296`.
36. P. Kanti, *Int. J. Mod. Phys. A* **19**, 4899–4951 (2004), `arXiv:hep-ph/0402168`.
37. S. Hossenfelder, `arXiv:hep-ph/0412265`.
38. A. Casanova and E. Spallucci, *Class. Quantum Grav.* **23**, R45–R62 (2006), `arXiv:hep-ph/0512063`.
39. M. Bleicher, *Eur. J. Phys.* **28**, 509–516 (2007), `arXiv:physics/0703062`.
40. E. Winstanley, in *Conference on Black Holes and Naked Singularities Milan, Italy, May 10–12, 2007*. 2007. `arXiv:0708.2656 [hep-th]`.
41. P. Nicolini, *Int. J. Mod. Phys. A* **24**, 1229–1308 (2009), `arXiv:0807.1939 [hep-th]`.
42. M. Bleicher and P. Nicolini, *J. Phys. Conf. Ser.* **237**, 012008 (2010), `arXiv:1001.2211 [hep-ph]`.
43. M. Bleicher and P. Nicolini, *Astron. Nachr.* **335**, 605 (2014), `arXiv:1403.0944 [hep-th]`.
44. X. Calmet, *Mod. Phys. Lett. A* **25**, 1553–1579 (2010), `arXiv:1005.1805 [hep-ph]`.
45. S. C. Park, *Prog. Part. Nucl. Phys.* **67**, 617–650 (2012), `arXiv:1203.4683 [hep-ph]`.
46. P. Kanti and E. Winstanley, *Hawking Radiation from Higher-Dimensional Black Holes*, vol. 178 of *Fundam. Theor. Phys.*, pp. 229–265. Springer International Publishing, Cham, 2015. `arXiv:1402.3952 [hep-th]`.
47. M. F. Wondrak, P. Nicolini, and M. Bleicher, in *Frontier Research in Astrophysics - II Mondello, Palermo, Italy, May 23–28, 2016*. 2016. `arXiv:1612.08415 [hep-ph]`.
48. N. Arkani-Hamed, S. Dimopoulos, and G. Dvali, *Phys. Lett. B* **429**, 263–272 (1998).
49. I. Antoniadis, N. Arkani-Hamed, S. Dimopoulos, and G. Dvali, *Phys. Lett. B* **436**, 257–263 (1998).
50. N. Arkani-Hamed, S. Dimopoulos, and G. Dvali, *Phys. Rev. D* **59**, 086004 (1999).

51. T. Kaluza, *Sitzungsber. Preuß. Akad. Wiss.* 966–972 (1921).
52. O. Klein, *Nature* **118**, 516 (1926).
53. O. Klein, *Z. Phys.* **37**, 895–906 (1926).
54. E. G. Adelberger, J. H. Gundlach, B. R. Heckel, S. Hoedl, and S. Schlamminger, *Prog. Part. Nucl. Phys.* **62**, 102–134 (2009).
55. N. Dadhich, *Eur. Phys. J. C* **76**, 104 (2016), `arXiv:1506.08764 [gr-qc]`.
56. D. Andriot and G. Lucena Gómez, *J. Cosmol. Astropart. Phys.* **06**, 048 (2017), `arXiv:1704.07392 [hep-th]`.
57. F. R. Tangherlini, *Nuovo Cimento* **27**, 636–651 (1963).
58. R. C. Myers and M. J. Perry, *Ann. Phys. (N. Y.)* **172**, 304–347 (1986).
59. G. F. Giudice, R. Rattazzi, and J. D. Wells, *Nucl. Phys. B* **630**, 293–325 (2002), `arXiv:hep-ph/0112161`.
60. A. Aurilia and E. Spallucci, `arXiv:1309.7186 [gr-qc]`.
61. G. Dvali and C. Gomez, `arXiv:1005.3497 [hep-th]`.
62. G. Dvali, S. Folkerts, and C. Germani, *Phys. Rev. D* **84**, 024039 (2011), `arXiv:1006.0984 [hep-th]`.
63. G. Dvali, G. F. Giudice, C. Gomez, and A. Kehagias, *J. High Energy Phys.* **08**, 108 (2011), `arXiv:1010.1415 [hep-ph]`.
64. G. Dvali and C. Gomez, *J. Cosmol. Astropart. Phys.* **07**, 015 (2012), `arXiv:1205.2540 [hep-ph]`.
65. G. Dvali and C. Gomez, *Fortschr. Phys.* **61**, 742–767 (2013), `arXiv:1112.3359 [hep-th]`.
66. G. Dvali and C. Gomez, *Phys. Lett. B* **719**, 419–423 (2013), `arXiv:1203.6575 [hep-th]`.
67. G. Dvali and C. Gomez, *Eur. Phys. J. C* **74**, 2752 (2014), `arXiv:1207.4059 [hep-th]`.
68. G. Dvali, C. Gomez, R. S. Isermann, D. Luest, and S. Stieberger, *Nucl. Phys. B* **893**, 187–235 (2015), `arXiv:1409.7405 [hep-th]`.
69. R. Casadio and F. Scardigli, *Eur. Phys. J. C* **74**, no. 1, 2685 (2014), `arXiv:1306.5298 [gr-qc]`.
70. R. Casadio, O. Micu, and F. Scardigli, *Phys. Lett. B* **732**, 105–109 (2014), `arXiv:1311.5698 [hep-th]`.
71. R. Casadio, O. Micu, and P. Nicolini, *Minimum length effects in black hole physics*, vol. 178 of *Fundam. Theor. Phys.*, pp. 293–322. Springer International Publishing, Cham, 2015. `arXiv:1405.1692 [hep-th]`.
72. R. Casadio, O. Micu, and A. Orlandi, *Eur. Phys. J. C* **72**, 2146 (2012), `arXiv:1205.6303 [hep-th]`.
73. J. Mureika and P. Nicolini, *Eur. Phys. J. Plus* **128**, 78 (2013), `arXiv:1206.4696 [hep-th]`.
74. M. Isi, J. Mureika, and P. Nicolini, *J. High Energy Phys.* **11**, 139 (2013), `arXiv:1310.8153 [hep-th]`.
75. A. M. Frassino, S. Koppel, and P. Nicolini, *Entropy* **18**, 181 (2016), `arXiv:1604.03263 [gr-qc]`.
76. S. Pandolfi (LHC Team). `http://cds.cern.ch/record/2229040`. [Accessed 10 July 2017].
77. K. S. Thorne, "Nonspherical Gravitational Collapse - A Short Review," in *Magic Without Magic: John Archibald Wheeler*, J. Klauder, ed., pp. 231–258. W. H. Freeman & Co., San Francisco, 1972.
78. J. Mureika, P. Nicolini, and E. Spallucci, *Phys. Rev. D* **85**, 106007 (2012).
79. S. Hossenfelder, *Living Rev. Rel.* **16**, 2 (2013), `arXiv:1203.6191 [gr-qc]`.

80. M. Sprenger, P. Nicolini, and M. Bleicher, *Eur. J. Phys.* **33**, 853–862 (2012), arXiv:1202.1500 [physics.ed-ph].
81. P. Nicolini, A. Smailagic, and E. Spallucci, *Phys. Lett. B* **632**, 547–551 (2006), arXiv:gr-qc/0510112.
82. A. Bonanno and M. Reuter, *Phys. Rev. D* **62**, 043008 (2000), arXiv:hep-th/0002196.
83. L. Modesto, J. W. Moffat, and P. Nicolini, *Phys. Lett. B* **695**, 397–400 (2011), arXiv:1010.0680 [gr-qc].
84. L. Modesto, *Class. Quantum Grav.* **23**, 5587–5602 (2006), arXiv:gr-qc/0509078.
85. P. Nicolini, arXiv:1202.2102 [hep-th].
86. M. Bleicher, S. Hofmann, S. Hossenfelder, and H. Stöcker, *Phys. Lett. B* **548**, 73–76 (2002).
87. S. Hossenfelder, S. Hofmann, M. Bleicher, and H. Stöcker, *Phys. Rev. D* **66**, 101502 (2002).
88. S. Hawking, *Commun. Math. Phys.* **43**, 199–220 (1975).
89. G. Denardo and E. Spallucci, *Nuovo Cimento* **B44**, 381–393 (1978).
90. R. Balbinot and A. Barletta, *Class. Quantum Grav.* **5**, L11–L13 (1988).
91. P. Nicolini, *J. Phys. A* **38**, L631–L638 (2005), arXiv:hep-th/0507266.
92. P. Nicolini, A. Smailagic, and E. Spallucci, *ESA Spec. Publ.* **637**, 11.1 (2006), arXiv:hep-th/0507226.
93. P. Nicolini and E. Spallucci, *Class. Quantum Grav.* **27**, 015010 (2010), arXiv:0902.4654 [gr-qc].
94. A. Bonanno and M. Reuter, *Phys. Rev. D* **73**, 083005 (2006), arXiv:hep-th/0602159 [hep-th].
95. R. J. Adler, P. Chen, and D. I. Santiago, *Gen. Rel. Gravit.* **33**, 2101–2108 (2001), arXiv:gr-qc/0106080.
96. B. J. Carr, J. Mureika, and P. Nicolini, *J. High Energy Phys.* **07**, 052 (2015), arXiv:1504.07637 [gr-qc].
97. L. Modesto, *Phys. Rev. D* **70**, 124009 (2004), arXiv:gr-qc/0407097.
98. P. Nicolini and E. Spallucci, *Adv. High Energy Phys.* **2014**, 805684 (2014), arXiv:1210.0015 [hep-th].
99. P. Chen, Y. C. Ong, and D.-H. Yeom, *Phys. Rep.* **603**, 1–45 (2015), arXiv:1412.8366 [gr-qc].
100. P. Nicolini and D. Singleton, *Phys. Lett. B* **738**, 213–217 (2014), arXiv:1409.5069 [gr-qc].
101. E. Spallucci and A. Smailagic, arXiv:1605.05911 [hep-th].
102. E. Spallucci and A. Smailagic, arXiv:1601.06004 [hep-th].
103. E. Spallucci and A. Smailagic, *Phys. Lett. B* **743**, 472–477 (2015), arXiv:1503.01681 [hep-th].
104. X. Calmet, *Mod. Phys. Lett. A* **29**, 1450204 (2014), arXiv:1410.2807 [hep-th].
105. R. Casadio and P. Nicolini, *J. High Energy Phys.* **11**, 072 (2008), arXiv:0809.2471 [hep-th].
106. P. Nicolini and E. Winstanley, *J. High Energy Phys.* **11**, 075 (2011), arXiv:1108.4419 [hep-ph].
107. D.-C. Dai, G. Starkman, D. Stojkovic, C. Issever, E. Rizvi, and J. Tseng, *Phys. Rev. D* **77**, 076007 (2008).
108. C. M. Harris, P. Richardson, and B. R. Webber, *J. High Energy Phys.* **08**, 033 (2003).
109. C. M. Harris and P. Kanti, *J. High Energy Phys.* **10**, 014 (2003).
110. J. A. Frost, J. R. Gaunt, M. O. P. Sampaio, M. Casals, S. R. Dolan, M. A. Parker, and B. R. Webber, *J. High Energy Phys.* **10**, 014 (2009).
111. D. M. Gingrich, *J. Phys. G* **37**, 105008 (2010).

112. D. M. Gingrich, *Comput. Phys. Commun.* **181**, 1917–1924 (2010).
113. The ATLAS Collaboration, *Phys. Lett. B* **760**, 520–537 (2016), `arXiv:1606.02265v1 [hep-ex]`.
114. The CMS Collaboration, *J. High Energy Phys.* **04**, 061 (2012), `arXiv:1202.6396v2 [hep-ex]`.